Terrestrial Conservation Lagerstätten

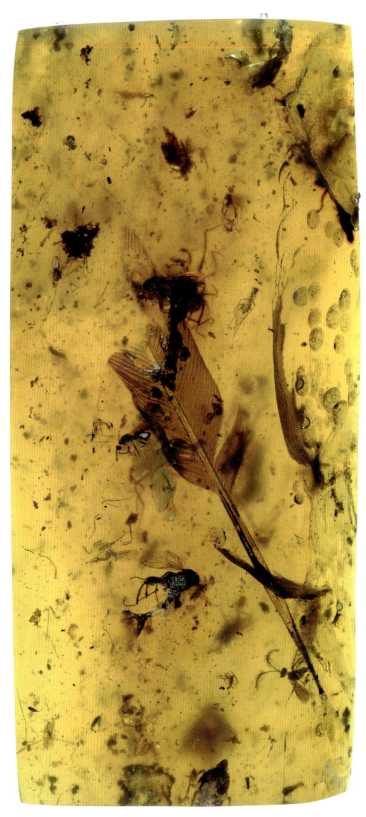

Cretaceous Burmese amber containing a feather and insects, including flies (Diptera) and wasps (Hymenoptera). The width of the feather is 7 mm.

Terrestrial Conservation Lagerstätten

Windows into the Evolution of Life on Land

Edited by

Nicholas C. Fraser
National Museums Scotland, Edinburgh

and

Hans-Dieter Sues
Smithsonian Institution, Washington

Published by
Dunedin Academic Press Ltd
Hudson House, 8 Albany Street
Edinburgh EH1 3QB, Scotland

London Office: 352 Cromwell Tower, Barbican
London EC2Y 8NB

For further details of other Dunedin
Earth and Environmental Sciences titles see
www.dunedinacademicpress.co.uk

ISBN 9781780460147 (Hardback)

© Dunedin Academic Press 2017

The right of Nicholas C. Fraser, Hans-Dieter Sues and the contributors to be identified as the authors of this work has been asserted by them in accordance with sections 77 and 78 of the Copyright, Designs and Patents Act 1988

All rights reserved.
No part of this publication may be reproduced or transmitted in any form or by any means or stored in any retrieval system of any nature without prior written permission, except for fair dealing under the Copyright, Designs and Patents Act 1988 or in accordance with the terms of a licence issued by the Copyright Licensing Society in respect of photocopying or reprographic reproduction. Full acknowledgment as to author, publisher and source must be given. Application for permission for any other use of copyright material should be made in writing to the publisher.

British Library Cataloguing in Publication data
A catalogue record for this book is available from the British Library

Typeset by Makar Publishing Production, Edinburgh
Printed in Poland by Hussar Books

CONTENTS

Introduction	/ vii	
Chapter 1	The Rhynie and Windyfield cherts, Early Devonian, Rhynie, Scotland / 1	
	Nigel H. Trewin and Hans Kerp	
Chapter 2	The East Kirkton Lagerstätte: a window onto Early Carboniferous land ecosystems / 39	
	Jennifer A. Clack	
Chapter 3	Triassic life in an inland lake basin of the warm-temperate biome: the Madygen Lagerstätte (southwest Kyrgyzstan, Central Asia) / 65	
	Sebastian Voigt, Michael Buchwitz, Jan Fischer, Ilja Kogan, Philippe Moisan, Jörg W. Schneider, Frederik Spindler, Andreas Brosig, Marvin Preusse, Frank Scholze and Ulf Linnemann	
Chapter 4	The Solite Quarry – a window into life by a late Triassic lake margin / 105	
	Nicholas C. Fraser, David A. Grimaldi, Brian J. Axsmith, Andrew B. Heckert, Cynthia Liutkus-Pierce, Dena Smith, Alton C. Dooley Jr.	
Chapter 5	The Yanliao Biota: a trove of exceptionally preserved Middle-Late Jurassic terrestrial life forms / 131	
	Xing Xu, Zhonghe Zhou, Corwin Sullivan, and Yuan Wang	
Chapter 6	The Jehol Biota: an exceptional window into Early Cretaceous terrestrial ecosystems / 169	
	Zhonghe Zhou, Yuan Wang, Xing Xu, and Dong Ren	
Chapter 7	The Santana Formation / 215	
	David M. Martill and Paulo M. Brito	
Chapter 8	The Messel Pit Fossil Site / 257	
	Stephan F. K. Schaal	
Chapter 9	Extraordinary Lagerstätten in Amber, with particular reference to the Cretaceous of Burma / 287	
	David A. Grimaldi and Andrew J. Ross	
Index	/ 343	

INTRODUCTION

Ever since Darwin first commented on this issue in his 1859 *magnum opus* evolutionary biologists have been concerned about the inherent incompleteness of the fossil record. Although our knowledge of the diversity of life in 'deep time' has dramatically improved since the mid-nineteenth century, many lineages of extant animals and plants still have only sparse, if any, fossil documentation. Even groups with hard-parts that would render them suitable for fossilization often only have a fairly limited record. Thus, although the fossil record has traditionally been viewed as critical to the reconstruction of the evolutionary history of life, many biologists have questioned its utility for this and other purposes.

Fortunately, from time to time, discoveries of occurrences of exceptionally preserved fossils, known as conservation *Lagerstätten* (*Konservat-Lagerstätten*; Seilacher, 1970), shed much new light on the past diversity of life. (The German word *Lagerstätte* – lode place or, more freely translated, motherlode – originally denotes a deposit of a commercially valuable substance such as ore or coal.) Due to unusual depositional conditions, these rare deposits preserve animals and plants that are otherwise poorly documented or altogether unknown in the fossil record, as well as often non-mineralized parts of organisms in exquisite detail. Famous examples of such conservation *Lagerstätten* include the Middle Cambrian Burgess Shale of British Columbia (Canada) with its diverse communities of soft-bodied metazoans, the Upper Jurassic Solnhofen Formation of Bavaria (Germany), renowned for exquisitely preserved specimens of the oldest known bird *Archaeopteryx*, pterosaurs, and insects, and the Lower Cretaceous Yixian Formation of Liaoning (China), which has yielded feathered non-avian dinosaurs, birds, mammals, and the earliest undisputed angiosperms.

Rapid burial by sediment is one of the key initial processes in the preservation of a living organism as a fossil. Therefore, not surprisingly, marine fossils in general are much more widespread and better known than terrestrial ones, since water is the prime agent of sediment transport. Moreover, some of the most diverse places on earth are coastal waters. Thus it is only to be expected that marine *Lagerstätten* are also more frequently encountered than continental representatives.

The rarity of terrestrial *Lagerstätten* bestows an extra special celebration of their occurrence that we present in this volume in the form of selected conservation *Lagerstätten* for terrestrial animals and plants from throughout the Phanerozoic worldwide.

There has been enormous progress in recent years in, firstly, documenting the biodiversity of such extraordinary fossil deposits and, secondly, elucidating the geological conditions for, and biogeochemical processes behind, the formation of conservation *Lagerstätten*. Interestingly, in many instances heat appears to have played a critical role in 'fixing' many of the original organisms in some way or another so that they have not been scavenged nor decayed, and now preserve details of soft parts that can only be marvelled at: the microtrichiae on the antennae of Triassic insects from the Solite Quarry, the cell structure of some of the first land plants at Rhynie, the last meals of Jurassic salamanders in the Yanliao biota or the muscle fibres and gill filaments of a Cretaceous fish in the Santana Formation of Brazil.

The Early Devonian Rhynie and Windyfield cherts of Scotland (chapter 1) provide insights into one of the earliest terrestrial fossil sites known. The small town of Rhynie is world-renowned for its exquisitely preserved early land plants and arthropods. The unique depositional environment is associated with silica-bearing hot springs, which played a significant role in the remarkable preservation of the biota.

North of Rhynie, the famous Old Red Sandstone of Caithness is also known for exquisitely preserved fossils, but in this case, apart from a few plants and the occasional millipede, this is mostly an aquatic assemblage of fishes. Yet travel a couple of hundred miles south to the rolling Borders countryside and we encounter another of Scotland's palaeontological treasures at East Kirkton (chapter 2). These Early Carboniferous deposits preserve a wealth of early tetrapods. In addition, a variety of plants and arthropods are preserved in the non-marine limestone. They include scorpion remains that appear just like modern representatives of the group. Once again, heat has played a critical role in their preservation, for these limestones are associated with lava flows. The discovery of the East Kirkton was the result of painstaking field work by the remarkable fossil hunter, Stan Wood.

Moving east and climbing up the stratigraphic column takes us to the Middle or Late Triassic Madygen Formation of Kyrgyzstan (chapter 3). Somewhat perversely, it is perhaps best known for two unusual, but not particularly beautifully preserved, reptiles. *Sharovipteryx* and *Longisquama* are certainly very unusual fossils that have been the subject of much speculation regarding their evolutionary relationships. Yet the Madygen Formation contains extraordinarily rich assemblages of fossil insects (mostly isolated wings) and plants that are often somewhat forgotten. Renewed fieldwork by Voigt and colleagues is helping to redress the balance, and they are providing a much better understanding of the depositional environment.

Only a little younger than the Madygen, the cyclical black shales of the Solite Quarry of the American mid-Atlantic are famous for their insect remains. Indeed, it is the only site in the world yielding a wealth of completely preserved Late Triassic insects. The site was first discovered by Paul Olsen in the mid-1970s and it is now recognized as the oldest known truly aquatic insect assemblage, documenting the oldest records of a number of living insect orders and families. It also mirrors the Madygen in preserving two very distinct tetrapods, *Tanytrachelos* and *Mechistotrachelos*. However, neither has (so far) been considered instructive either in the origin of flight or the origin of birds, and so have much less notoriety than their Madygen cousins.

Of all the palaeontological discoveries in recent years, none have inspired the general public as have the feathered dinosaurs of China. Remarkably, since the discovery of *Sinosauropteryx* a procession of new forms have come to light, and also from a variety of localities and representing different ages. The oldest are found in the so-called Daohugou Biota of Late Jurassic age. The lacustrine mudstones of the Daohugou extend from Inner Mongolia into Liaoning Province. This biota comprises mammals, feathered dinosaurs, pterosaurs, lizards, and a diversity of early salamanders, as well as hundreds of beautifully preserved insects and plants.

The first feathered dinosaur was described from the Jehol Group, which is Early Cretaceous in age (chapter 6). Also known from northeastern China, the Jehol fauna is best known in the media for its remarkable feathered dinosaurs and birds. Unfortunately, these often overshadow the equally important and spectacular array of other forms including plants, insects, fishes, amphibians and mammals. Again, volcanic activity and suffocating ash layers have played a role in preserving the remains.

Across the vast stretches of the Pacific from China and into the southern hemisphere, the Crato and Santana formations of Brazil are also Early Cretaceous in age. In this instance the sediments were deposited under rather different conditions in a shallow inland sea, and the fossils are preserved in limestone nodules. Although the nodules are particularly rich in fish, they also contain diverse terrestrial faunal elements including pterosaurs and other reptiles, amphibians, and insects. Some of the fossils have remarkable three-dimensional preservation of soft tissues.

We return to Europe and the UNESCO World Heritage Site of the Messel pit in northern Germany, which is notable for its diversity of vertebrates and insects. The remarkable vertebrate assemblage continues to yield new information with the basal primate, *Darwinius masillae*, being the latest significant find. The fossils are preserved in lacustrine 'oil shale' that was deposited in a volcanic caldera. They often include preserved soft parts in vertebrates and the colour patterns of insect wings.

Finally, insects preserved in amber are perhaps the most iconic of all the terrestrial *Lagerstätten* with many barely altered from when they were living, perfectly entombed in their graves of hardened resin. Baltic (Eocene) and Dominican (Oligocene to Miocene) ambers have been known for many years, but the Miocene Mexican and the Cretaceous Burmese ambers have been exploited more recently. Grimaldi and Ross survey all the ambers worldwide but concentrate on the remarkable inclusions coming to light in the Burmese amber, including a previously unsuspected diversity of rare vertebrate remains.

Membership of the world list of terrestrial *Konservat-Lagerstätten* is an exclusive club, and essentially all have been covered in this volume by their respective experts. Only the Green River (Eocene, USA) Formation might be added. Perhaps best known for their vast numbers of beautifully preserved fish fossils, these fine-grained lake deposits are also known as a source of numerous insects and plants. In recent years this formation has become increasingly important as an occurrence of phylogenetically important terrestrial vertebrates, including early bats and a diversity of birds. However, we refer the reader to Lance Grande's recent publication (2013) for a comprehensive coverage of this assemblage, including the very latest finds. No doubt chance discoveries will continue to be made that will lead to one or two additions to this exclusive list. For instance, other sites in the Scottish Borders discovered by Stan Wood are beginning to yield well-preserved remains and are shedding much-needed light on the interval known as Romer's Gap, the time in the earliest Carboniferous when vertebrate life really began to gain ground (Clack, 2012). Moreover, with work continuing on all the localities discussed in the following pages, it is inevitable that many of their secrets are yet to be discovered. Until then we consider the assemblages described in this book to be the most remarkable and enlightening continental faunas and floras from the past 410 million years.

References

Clack, J.A. 2012. *Gaining Ground: the Origin and Evolution of Tetrapods*. Indiana University Press, Bloomington.

Grande, L. 2013. *The Lost World of Fossil Lake: Snapshots from Deep Time*. The University of Chicago Press, Chicago.

Seilacher, A. 1970. Begriff und Bedeutung der Fossil-Lagerstätten. Neues Jahrbuch für Geologie und Paläontologie, Monatshefte 1970:34–39.

Chapter 1

The Rhynie and Windyfield cherts, Early Devonian, Rhynie, Scotland

Nigel H. Trewin[1] and Hans Kerp[2]

1 Nigel Trewin is an Emeritus Professor of Aberdeen University and Research Associate of National Museums Scotland, UK.
2 Hans Kerp is the Professor leading the Palaeobotanical Research Group at the University of Münster, Germany.

Abstract

The Early Devonian (c.407 Ma) Rhynie chert of Aberdeenshire, Scotland contains a remarkable biota of early terrestrial and freshwater environments. The communities of embryophyte plants, algae, fungi, arthropods and other invertebrates were preserved *in situ* by the action of hydrothermal hot springs depositing silica. Superb 3D cellular preservation of plants has allowed full life-histories to be documented. Fungi played major roles with symbiotic, parasitic and saprophytic forms. The arthropods include the earliest-known representatives of several groups such as harvestman spiders, mites and springtails. Rare early wingless insects are also present. The process of silicification could be so rapid that soft tissues were preserved, as in germinating spores and the oldest nematode worm. This deposit preserves a snapshot of early continental micro-environments, and provides insight into the evolution of terrestrial ecosystems. Early Devonian terrestrial and freshwater ecosystems were complex, with many features of modern environments already well established.

Introduction

The Rhynie chert of Aberdeenshire is world-famous for the excellent preservation of an Early Devonian biota of terrestrial and freshwater plants and animals. There are no natural exposures of either the Rhynie or Windyfield cherts, and material has been obtained as float in the fields, or from field boundary walls. Trenching and cored drilling have been used to determine the succession and to obtain *in situ* chert. Discovered by William Mackie (Fig 1) in 1912, the fossils are preserved in chert formed by silicification of hot-spring sinters. In 1989 a new fossiliferous deposit was found about 700 m from the original site and is known as the Windyfield chert. The two localities are thought to be at different levels in the local Old Red Sandstone (ORS) stratigraphy; however, the biotas from the two sites are essentially the same, but there is a greater richness of arthropods at the Windyfield site.

Figure 1 Dr William Mackie, who discovered the Rhynie chert.

The preservation of the biota is highly variable, but a significant quantity of the chert contains plants and animals three-dimensionally preserved with superb cellular details in plants, such as micorrhizae, and delicate setae and book lungs in arthropods. The earliest known occurrences of several animal groups such as nematodes, harvestman spiders and springtails are preserved in the chert. The seven sporophyte plant genera represent the earliest terrestrial flora with three-dimensional cellular preservation. The gametophyte generations of some of the plants are also known; hence *Aglaophyton* and *Rhynia* are the earliest plants for which the full life-history from spore to gametophytes to mature sporophyte is documented.

The biota in many of the chert beds was preserved *in situ*, with some beds containing terrestrial plants preserved in growth position, displaying aerial axes arising from rhizomes that traverse plant litter and sandy substrates. Plants are also present in all states of decay and invasion by fungi. Arthropods are found clinging to plant stems and within empty sporangia. Cherts that preserved the contents of freshwater pools contain aquatic crustaceans, charophyte algae and washed-in plant debris coated by microbial mats. Hence the chert preserves an early terrestrial and freshwater ecosystem where interactions between plants, fungi and animals are preserved.

History of exploration

Dr William Mackie (1856–1932) is credited with the discovery of the Rhynie chert. He was a graduate of Aberdeen University, having taken a degree in Arts with Honours in Natural Science in 1878, and in Medicine in 1888. Whilst in medical practice in Elgin he maintained an interest in geology, publishing over 40 articles, including pioneering work on sandstones and heavy mineral analysis. His interest in the Rhynie area arose from curiosity with the igneous rocks described in the Geological Survey Memoir (Grant Wilson and Hinxman, 1890). It was whilst mapping the area that he noted cherty rocks in walls and scattered in fields close to the basin margin fault. Mackie first collected the plant-bearing chert in 1912 (Horne *et al.*, 1916), and made thin sections that revealed plants with superb cellular preservation, and also remains of arthropods.

Mackie was an astute observer and recognized that some plants had decayed before silicification took place, and noted that 'chalcedonic aggregates' had formed around plant material. Mackie (1913) figured two thin sections of the chert, one with the plant now known as *Rhynia*, and the other *Asteroxylon*. Mackie did not name either plant, but indicated that Robert Kidston (Fig 2A) would describe them. Mackie (1913, p. 225) states that he realized, with hindsight, that he had found a piece of plant-bearing chert 'more than 35 years ago'; thus the Rhynie chert story might have started much earlier. It is of interest to note that others also missed the chert. The Reverend Alexander Mackay (1815–1895) was the first Free Church minister at Rhynie, had a keen interest in geology, and collected fossils locally. One specimen shown to Hugh Miller was considered to preserve tracks of a crustacean (Miller, 1857, p. 435). Archibald Geikie (1879) had established the succession in the area, and none others than Murchison and Ramsay of the Geological Survey were guests of Mackay at the manse at 17, The Square, Rhynie, prior to the 1858 meeting of the British Association for the Advancement of Science in Aberdeen. They all missed the Rhynie chert.

Mackie had not found the plant-bearing chert *in situ*, and thus its stratigraphic position was uncertain. Kidston, renowned for his work on Carboniferous plants, had strong links with the Geological Survey and was their palaeobotany adviser (Lang, 1925; Edwards, 1984). He may well have instigated the trenching carried out in 1913 by Mr Tait, fossil collector to the Geological Survey. Tait found the chert *in situ*, showing that the cherts lay within the ORS as mapped by Mackie (Horne *et al.*, 1916). Further trenching was carried out, supported by the British Association and The Royal Society, and chert was found in three trenches, clearly interbedded with sandstones and shales of the Dryden Shales of the ORS. Full details of this trenching programme are summarized with a map in Trewin (2004). Kidston had access to material from all the beds exposed in Tait's number 1 trench (Kidston and Lang, 1917, p. 763), and also material collected in the fields.

Kidston planned to collaborate with D. T. Gwynne-Vaughan in describing the Rhynie plants, but the premature death of Gwynne-Vaughan in 1915 led to W. H. Lang (Fig 2B) joining Kidston, a collaboration that produced the classic five-part series of papers (Kidston and Lang, 1917, 1920a,b, 1921a,b) on the flora of the chert. Kidston and Lang (1917) described *Rhynia* in Part 1, but in Part 2 (1920a) they split this genus into

Figure 2 A. Robert Kidston; B. William Lang.

two species *R. gwynne-vaughanii* and *R. major*, the latter now being *Aglaophyton* (D. S. Edwards, 1986). They also described *Hornea* (now *Horneophyton*; see Barghoorn and Darrah, 1938). *Asteroxylon* was described in Part 3 (1920b), together with illustrations of *Nothia* that Kidston and Lang thought might be the fertile axes of *Asteroxylon*. Restorations of *Rhynia*, *Aglaophyton* (*R. major*), *Hornea* (*Horneophyton*) and *Asteroxylon* were presented in Part 4 (1921a). Part 5 (1921b) includes algae, fungi and nematophytes.

The arthropods were described by several authors, notably Hirst (1923) on the trigonotarbids and mites, Scourfield (1926, 1940a) on the crustacean *Lepidocaris*, and Hirst and Maulik (1926a) on the springtail *Rhyniella*. Many of the arthropods were discovered by the Reverend William Cran (1856–1933), a native of Rhynie. He was a microscopist and examined the chert by breaking it into small chips that were semi-transparent. This laborious process produced very small or partial arthropods, and the alga *Palaeonitella crani*. Cran's name is also celebrated in the species name for the mite *Protocaris crani* and the genus name for the euthycarcinoid *Heterocrania rhyniensis*.

By 1930 the fauna and flora as known at the time had been described, and it had been established that these were of terrestrial origin, but that freshwater pools must have been present to support the crustacean *Lepidocaris* and the alga *Palaeonitella*. Mackie (1913, pp. 233–4) had suggested that the environment included geysers and hot springs, active 'during the decadent phases of local volcanic action'. He postulated that hot springs were responsible for the cherts, and gave a good description of geyserite. His general interpretation has stood the test of time. The scientific sensation caused by the Rhynie chert flora and fauna rapidly passed into text books (Scott, 1920; Seward, 1931), but the stream of new discoveries dried up, and little more was published, so that Tasch (1957) relied on work published prior to 1930 in his evaluation of the palaeoecology of the cherts.

The re-awakening of research on the Rhynie chert was due to Geoffrey Lyon (1918–1999; Fig 3A). He lectured in botany at Cardiff University prior to his retirement in 1973. He published on germinating spores preserved in the chert (Lyon, 1957), the recognition of book lungs in *Palaeocharinus* (Claridge and Lyon, 1961), description of *Nematoplexus* (Lyon, 1962), and the recognition of the fertile structures of *Asteroxylon* (Lyon, 1964). Thus Geoffrey Lyon showed that there was more to be found in the chert. The International Botanical Congress was held in Edinburgh in 1964, and an excursion visited Rhynie, where two trenches had been dug to expose the chert. This period of trenching (Fig 4A) lasted from 1963 to 1971, and produced material for three PhD students (El-Saadawy, 1966; Bhutta, 1969; David S. Edwards, 1973). Lyon also found the zosterophyll *Trichopherophyton* at this time, later published in collaboration with Dianne Edwards (Lyon and Edwards, 1991). This trenching phase resulted in the collection of large quantities of chert, much still unstudied in museum storage. The problem is that skilled labour and supporting finance are required to cut and section the chert, and there is no guarantee of finding something new to describe. The chert does not easily yield its secrets.

When Lyon retired in 1973 he moved to Rhynie, where he purchased the old manse at 17, The Square, once occupied by Mackay and visited by Ramsay and Murchison. He also owned the Rhynie chert site, and later gifted it to Scottish Natural Heritage, stipulating that the land was only to be used for agriculture and scientific research. Geoffrey Lyon made the most significant contribution to the palaeontology of the Rhynie chert since Kidston and Lang.

The next Rhynie sensation was the description of gametophytes by Winfried Remy (Fig 3B) and co-workers at Münster. The Remy family visited Lyon at Rhynie in 1977, 1979 and 1980, and Lyon gave Remy the chert block in which he found the first gametophytes (Remy and Remy, 1980a,b; Remy and Hass, 1991a–c). Winfried Remy died in 1995, but Hagen Hass continued to work cutting and examining chert at Münster, and undoubtedly it is the quantity of chert examined and the years of experience that have resulted in so many new finds by the Münster team. Noteworthy contributions are the work with Tom Taylor, Michael Krings and Nora Dotzler on fungi (Hass *et al.*, 1994; Taylor *et al.*, 1992a,b, 1995, 1999, 2004; Krings *et al.*, 2007b; Dotzler *et al.* 2006, 2009) and cyanobacteria (Krings *et al.*, 2007a, 2009;

Figure 3 A. Geoffrey Lyon in the Rhynie chert field with Tap o'Noth in the background. **B.** Winfried and Renate Remy in their garden in Münster, photograph P. G. Gensel.

Dotzler *et al.*, 2007). Other notable finds have been the harvestman *Eophalangium* (Dunlop *et al.*, 2003, 2004), the crustacean *Ebullitocaris* (Anderson *et al.*, 2004), the gametophytes of *Rhynia* (Kerp *et al.*, 2004) and nematodes (*Palaeonema*) within a decayed axis of *Aglaophyton* (Poinar *et al.*, 2008).

Research at Aberdeen University into the Rhynie hot spring system started with the discovery that the silicified rocks in the region, and the Rhynie chert in particular, are enriched in antimony, arsenic and gold and are the surface expression of a Devonian precious-metal-bearing hydrothermal system (Rice and Trewin, 1988). This discovery resulted in commercial drilling in 1988 to assess the economic potential of the silicified zone near the basin-margin fault. Aberdeen University hired the drilling rig used for the mineral exploration and sank a 35 m borehole that produced the first cored section of the chert-bearing sequence and the interbedded lithologies (Trewin and Rice, 1992; Powell *et al.*, 2000a;

Figure 4 A. Trenching at Rhynie in 1966, photograph by A.G. Lyon. B. Drilling at Rhynie, 1997.

Trewin, 2004). The plant succession in the borehole was described in her dissertation by Powell (1994) and in Powell *et al.* (2000a). It became clear that the succession of plants in the cherts first described by Kidston and Lang, and interpreted by Tasch (1957), is not a constant feature of the cherts (Trewin and Wilson, 2004).

The Windyfield chert had been discovered as float blocks during regional exploration and a new zosterophyll plant had been recognized and subsequently described as *Ventarura* by Powell *et al.* (2000b). Trenching at Windyfield in 1997 revealed pods of Windyfield chert *in situ* in hydrothermally altered shales. It was this chert that produced new arthropods described by Anderson and Trewin (2003), Fayers (2003), Fayers *et al.* (2005) and Fayers and Trewin (2003, 2004, 2005). These finds increased the known arthropod fauna from the cherts by 40%.

A second drilling programme in 1997 (Fig 4B) funded by the Carnegie Trust, SNH, and Aberdeen University (Rice *et al.*, 2002; Trewin and Wilson, 2004) succeeded in drilling a 210 m cored hole through the Rhynie cherts to basement. This proved the low-angle basin-margin fault and also showed that the fault acted as a conduit for hot fluids. A total of about 600 m of hole was drilled in 1997, but the faulted nature of the area and the fact that several holes terminated in fault gouge due to the drill sticking, meant that there were still, and still are, problems in interpretation of stratigraphy and structure. Further fieldwork on structure and stratigraphic interpretation following trenching and geophysical work resulted in the interpretation of the basin as a strike-slip controlled basin (Rice and Ashcroft, 2004).

Geoffrey Lyon had taken quiet interest in the drilling operations and new palaeontological finds, particularly the new plant *Ventarura*. However, his health was failing in 1997 and he moved to Aberdeen, where he died in a nursing home in 1999. He left a generous bequest to Aberdeen University to continue work on the geology and palaeontology of the Rhynie area. The Rhynie Research Project provided funding for five PhD students, of whom S. R. Fayers and R. Kelman contributed to the palaeontology. L. I. Anderson and S. R. Fayers also worked as postdoctoral researchers on the project. Modern hot-spring areas such as Yellowstone National Park and New Zealand have provided much valuable information to aid in the interpretation of the Rhynie environments and preservation of the biota (Trewin *et al.*, 2003; Trewin and Wilson, 2004). Work at Cardiff on experimental taphonomy (Channing and Edwards, 2004, 2009) has also greatly aided the interpretation of textures seen in the Rhynie chert.

Research still continues on the palaeontology and environments of the Rhynie and Windyfield cherts, and undoubtedly more finds will be made in the future.

However, it is worth noting as a cautionary message that the last trench to be dug at Rhynie in 2003, on the occasion of a Rhynie Conference reported in a volume edited by Trewin and Rice (2004), has not produced any new species. This is despite the fact that an excellent sequence of plant-bearing chert was sampled, with thin sections made from each bed. The details of the 2003 trench are reported in Trewin and Fayers (2016). On the other hand, a small block of chert picked up during a routine field survey contained a thin band with the new crustacean *Castracollis* (Fayers and Trewin, 2003). Thus chert from float can contain new material, maybe derived from cherts not sampled by the trenching and drilling programmes. The moral is certainly that the more chert is examined, the more organisms will be found.

Geological context of occurrence

In the Early Devonian the Rhynie Basin was situated within the Old Red Sandstone (ORS) continent 20–30° south of the Equator, and Scotland moved north such that it crossed the Equator in the Carboniferous. The area was far from any marine influence, within the ORS continental area. Later, in the Middle Devonian, there was a brief connection between the Orcadian Basin and marine areas in Estonia, but fully marine conditions were never established in the Orcadian Basin *sensu lato*. Thus the Rhynie Basin lay within an eroding continental area at the south of the wider Orcadian Basin.

Rhynie Basin structure

The Rhynie Basin contains a fill of conglomerates, sandstones, shales and minor extrusive andesitic lavas and tuffs of the Lower ORS of Early Devonian age. The basin can be considered as a small southern outlier of the larger Orcadian Basin. The ORS rests unconformably on metamorphic Dalradian basement of chlorite-grade psammites and pelites that have been intruded by Ordovician basic rocks (Fig 5).

The Rhynie area is poorly exposed, and interpretations of the geological setting of the Rhynie cherts have evolved from the early Geological Survey work of Horne *et al.* (1886) and Grant Wilson and Hinxman (1890), through the revisions of Horne *et al.* (1923) and Read (1923). The British Geological Survey (BGS) resurveyed the area late in the twentieth century (British Geological Survey (BGS), 1993; Gould, 1997). In the northern part of the basin where the Rhynie chert is situated these interpretations generally show an unconformable relationship between the ORS and basement on both sides of the basin, but a faulted western margin in the southern part of the basin.

More recently Rice *et al.* (2002), using trenching and borehole evidence, have shown that at the Rhynie chert site the western margin of the basin is bounded by a low angle fault which controlled ORS sedimentation. Thus the whole western margin of the Rhynie basin is fault-bounded. In the Rhynie area the basin-margin fault zone has undergone hydrothermal alteration over a distance of 2 km, with K-feldspar/illite alteration and the introduction of quartz/feldspar veins with minor calcite, pyrite and fluorite. Enrichments in gold, antimony and arsenic confirm the presence of a hydrothermal hot-spring system, and similar enrichments in Rhynie chert samples (Rice and Trewin, 1988; Rice *et al.*, 1995) prove the connection between the cherts and the hydrothermal system. The deep borehole at the Rhynie chert site cut the low-angle basin margin fault, which was heavily silicified; thus it appears that the hydrothermal fluids used the fault plane as a conduit to the surface. The hydrothermal system is illustrated in Figure 6. Despite the information gained from trenching and drilling in the region of the hydrothermal system, there were still uncertainties regarding the overall structure of the northern part of the basin. Thus Rice and Ashcroft remapped the northern part of the basin using traditional methods supported by trenching to bedrock in key areas and use of a ground magnetic survey. This investigation (Rice and Ashcroft, 2004) revealed that the area has three fault trends: NE–SW, NNE–SSW and N–S. Three sets of open folds are also present. The basin margins are generally fault-controlled, and the older interpretations of a simple half-graben structure cannot be supported. Rice and Ashcroft (2004) argued that the basin formed within a regional strike-slip system in the Early Devonian, and that sedimentation was strongly controlled by basin formation.

The Rhynie basin is linked by faults to the Turriff Basin of ORS to the NE, which contains Lower ORS conglomerates, sandstones and minor mudstones, overlain unconformably by Middle ORS with contrasting clast compositions to those in the Lower ORS. Thus there is evidence of continued structural activity affecting erosion and deposition in the area into the Middle Devonian.

Stratigraphy

The stratigraphic sequence in the basin (Fig 5) appears to vary from south to north, and, in the Rhynie locality, Rice and Ashcroft (2004) interpret a thickness of 700 m of the Dryden Flags Formation that includes the chert-bearing units, and consists mainly of grey-green laminated shales and thin sandstones. To the south of the Boghead Fault, the fluvial Quarry Hill Sandstone underlies the Dryden Flags, but it appears to be absent north of the fault, or only represented by thin sandstones and shales. On Quarry Hill 2 km to the south, the sandstone was quarried and remaining faces show massive and cross-bedded sandstone of fluvial origin,

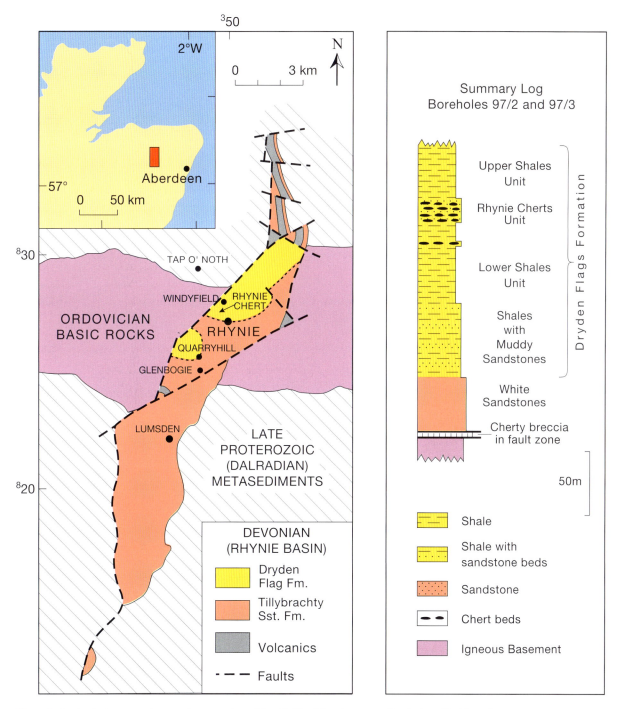

Figure 5 Geological map and general succession of the Old Red Sandstone at Rhynie (modified from Trewin and Wilson, 2004).

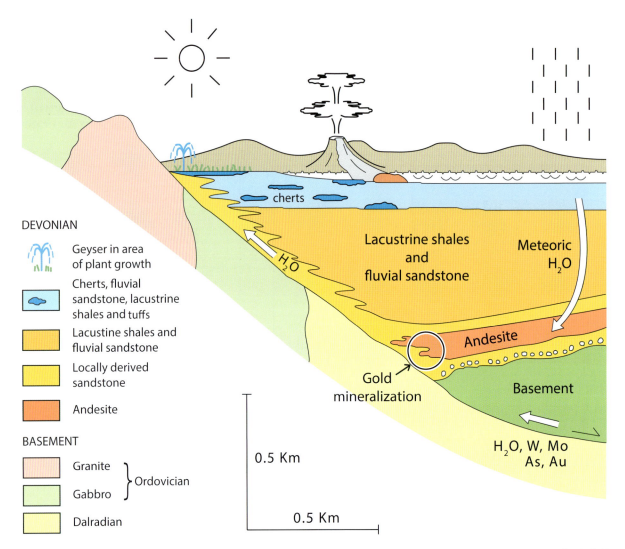

RECONSTRUCTION OF THE SUBSURFACE ENVIRONMENT DURING CHERT DEPOSITION

Figure 6 The Rhynie hydrothermal system (modified from Baron et al., 2004).

representing a series of channel deposits. Thin-bedded sandstones with mud partings have occasional large *Diplichnites* traces (Fig 7A), and large 'plant' fragments have also been found, including a sand-filled cast of a 'trunk' (Fig 7B) described by Newlands (1913). These are possibly casts of large nematophytes.

The lower part of the succession, the Tillibrachty Sandstone Formation, is also variable, including conglomerate units and some red shales with carbonate nodules. A general fluvial environment with mudstones in overbank areas is inferred, the nodules being indicative of evaporative conditions to produce caliche soil profiles. Andesitic lavas are also present within this formation, and are found in the north of the basin. In the zone of hydrothermal alteration at Rhynie, andesitic lavas, possibly extruded into wet sediment, and the Longcroft Tuffs are present, both in a hydrothermally altered condition. Thus it can be shown that extrusive igneous activity predated the hydrothermal alteration that was the source of the hot springs responsible for the Rhynie and Windyfield cherts.

The Rhynie Cherts Unit lies within the Dryden Flags Formation, and can be traced as a distinct unit in boreholes (Trewin and Wilson, 2004). The borehole section described by Powell et al. (2000a) revealed 53 chert beds in 35.4 m of core, with chert making up 4.2 m of the

Figure 7 A. Arthropod trackway *Diplichnites*, and **B.** Sandstone cast of probable nematophyte. Scale bar 10 cm. Both from the Quarry Hill Sandstone, Quarry Hill, Rhynie.

bed-to-bed correlation of cherts, and no correlation in the sequences of plants recorded. Drilling to basement at the Rhynie chert site did not reveal any earlier episodes of hot spring activity.

Rice and Ashcroft (2004) placed the Windyfield chert within the Milton Flags Member of the Dryden Flags Formation. Thus it is thought to lie about 550 m above the Rhynie cherts. Given the degree of faulting and lack of exposure in the area, this is an interpretation from mapping results. There is no biostratigraphical reason to think that the Rhynie and Windyfield cherts are significantly different in age, based on spores (Wellman, 2006).

Sedimentary Environment

The sedimentary environment at the time of deposition of the cherts is illustrated in Figure 8. The hot springs and geysers were probably concentrated at the basin-margin fault, and outwash from the springs flowed to the east away from the fault. To the west Dalradian basement was being eroded. In the basin an axial river flowed to the north, bounded by alluvial plains that were inundated when the river flooded. Thus there was a zone of interaction between fluvial and hot-spring processes, and an alternation of deposits (Fig 9). Waters from the hot spring cooled and deposited silica as they flowed away from the vents, probably forming a broad, irregular sinter apron. Silica buildup results in rapid changes in outwash from hot springs, thus floodplain pools and dry hollows can be invaded or abandoned by hot-spring outwash.

Observations in Yellowstone National Park (Trewin *et al.*, 2003) show that streams sourced from hot springs continue to deposit silica at temperatures of 20–28°C and a pH of 8.7. In these conditions plants are growing with roots in silica sinter, and charophytes and insect nymphs are abundant in the water. At Rhynie, similar conditions probably existed in the distal regions of outwash from the hot springs. The plumbing system of hot-spring areas changes rapidly, and thus hot water can invade previously cool pools. Hot-spring vents may become dormant, and are then colonized by plants and animals. At Rhynie, a high water table must have been maintained, and reducing conditions prevailed to preserve organic matter. Thus the best preservation probably took place in a marshy environment. There would have been a range in water composition and temperature in the environment from freshwater streams to hot,

succession. The thickest chert bed representing a single silicification event is 0.31 m thick, and the thickest composite chert at 0.76 m contained six beds. Trewin and Wilson (2004) compared the successions in three boreholes over a distance of 45 to 65 m. The Rhynie Cherts Unit can be recognized over this distance, but there is no

Figure 8 Reconstructed view of the Rhynie Basin margin showing hot springs erupting from the fault line and a geyser vent. Hot-spring water overflows to a marshy area where silicification of the biota took place.

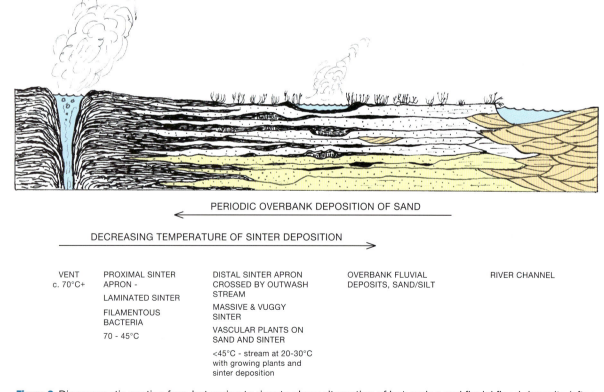

Figure 9 Diagrammatic section from hot spring to river to show alternation of hot-spring and fluvial flood deposits (after Trewin and Wilson, 2004).

silica-rich water from the hot springs. This variation is typical of hot-spring areas, and extremes can be found on a sub-metre scale (Trewin et al., 2003).

The Windyfield chert site appears to have been close to a hot-spring vent, and the shales of the Dryden Flags Formation have been contorted and hydrothermally altered. Drilling near the Windyfield site revealed sequences of shales and thin sandstones interpreted as overbank fluvial deposits and beds deposited by floods in shallow lakes (Fayers and Trewin, 2004). A few beds contain coarse debris including clasts of local Dalradian basement and also volcaniclastic grains of local origin. It appears that andesitic lavas and tuffs were being eroded during the period of hydrothermal activity, since silicified volcanic grains also occur in the cherts. A large specimen of a geyser vent rim with splash texture has also been found at the site (Trewin, 1994; Fayers and Trewin, 2004, Fig 3A). The Windyfield cherts occur as pods in soft white to blue altered shale, and show common brecciation and resealing features. There is also a greater content of detrital material in the Windyfield cherts when compared with the Rhynie cherts, and reworked chert is also present. The chert pods appear to have formed in hollows on an irregular surface. Some of these hollows became small ponds briefly colonized by plants and arthropods that were preserved when invaded by hot waters. The Windyfield cherts are highly localized, and were not found in any boreholes in the Windyfield area. Only a small volume of chert has been found at the Windyfield site, and the associated vent was probably short-lived and affected only a small area.

Biostratigraphic and radiometric ages

The biostratigraphic age of the Rhynie and Windyfield cherts based on palynology is late Pragian to earliest Emsian (Wellman, 2006). There is no detectable difference between the two sites, and indeed, all samples from the Dryden Flags Formation are indicative of the *polygonalis-emsiensis* Spore Assemblage Biozone (PE SAB) of the scheme of Richardson and McGregor (1986). On the zonal scheme of Streel et al. (1987) the assemblages lie within the *polygonalis-wetteldorfensis* Oppel Zone (PoW OZ). Wellman (2006) considers it possible that the assemblage belongs to the youngest of five interval zones within PoW OZ, placing it close to the Pragian/Emsian boundary, but urges caution in that the Rhynie assemblage is of low diversity and local origin, and difficult to compare with more diverse and distant lowland assemblages.

In terms of radiometric ages, Rice et al. (1995) reported a $^{40}Ar/^{39}Ar$ date of $395 \pm 12\,Ma$ based on a bulk chert sample. More recently, Mark et al. (2011) produced a $^{40}Ar/^{39}Ar$ age of $403.9 \pm 2.1\,Ma$ on feldspar from hydrothermal quartz-feldspar veins. Given that the hydrothermal system fed the Rhynie cherts, and the sample is from a vein less than 200 m from the nearest chert site, this is considered a reliable age for the Rhynie chert. Parry et al. (2011) produced a U/Pb date of $411 \pm 1.3\,Ma$ on zircons and titanite from an andesitic lava in the fault zone. This lava is probably from within the Tillibrachty Sandstone Formation, and although the lava cannot be directly related to the local chert-bearing succession it is unlikely to be much older than the cherts. When the Ar/Ar date of Mark et al. is converted to the U/Pb equivalent, ages of $407.1 \pm 2.2\,Ma$ and $407.6 \pm 2.2\,Ma$ result, depending on the method used (see Mark et al., 2013). An U/Pb date of around 407 Ma places the Rhynie chert close to the Pragian/Emsian boundary on the Devonian timescale of Becker et al. (2012), conforming well with the biostratigraphic age.

Preservation of Biota

The Rhynie biota is preserved in chert that formed as a silicification of sinters deposited from silica-rich waters that issued from hot springs and geysers near the basin-margin fault. The hot spring waters deposited silica as they cooled, forming a variety of types of sinter that have been converted to chert in early diagenesis. The chert takes several forms that are described as botryoidal, parallel-laminated, lenticular, nodular, massive and brecciated (Trewin, 1994, 1996). The common bed form comprises a central part where cellular preservation of plants is good, and plant axes may be preserved in growth position (Fig 10A). The upper and lower margins of the beds generally show a gradation to, or sharp stylolitized contact with, silicified sandstone in which plants have been compressed to carbonaceous sheets, and on which stylolitization has taken place during burial compaction (Fig 10B). Thus the cherts with three-dimensional preservation were formed prior to any compaction, virtually at the surface, and were sufficiently lithified to resist further compaction during burial. The brittle nature of the chert

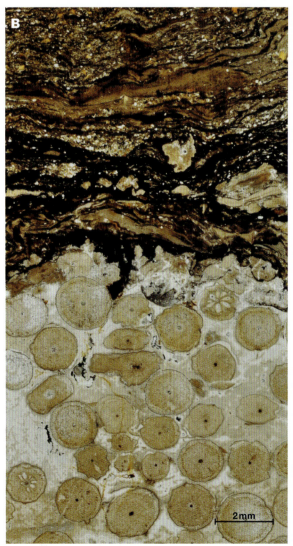

Figure 10 A. Rhynie chert bed with well-preserved plants in growth position, and gradation to cherty sandstone at margins. **B.** Chert bed margin showing sharp stylolitized division between chert with 3D preservation and cherty sandstone with carbonaceous plant compressions.

beds within softer shale and sandstone resulted in fracturing of the chert during burial and faulting; thus pieces of chert excavated from trenches seldom exceed 0.5 m in any dimension.

The starting point for consideration of the taphonomy of the plants is an environment with rapid variation in conditions on a micro-scale. The terrestrial sporophyte plants are preserved in every state from full cellular preservation, to plant litter fragments of dried and shriveled axes, curled cuticle fragments and xylem strands (Trewin and Fayers, 2016). Thus the full natural cycle of growth and decay in a dry terrestrial environment is recorded. Growth and decay also took place in waterlogged environments where different features are seen, such as plastic compaction of decaying axes and fungal infestation. These features are summarized in Figure 11.

The aforementioned features of growth and decay were modified by the silicification process. In the best preservation, the original cellular structure is retained, with the space between and within cells filled with silica (Fig 12A). The organic material of the cell walls may remain, may be silicified, or may be represented by a void, presumably after loss of organic matter. The remaining void may be filled with a later generation of silica, and may be stained with brown iron hydroxides, probably derived from the oxidation of pyrite in the chert. It is frequently the iron stain that highlights the details seen in thin section. Dehydration of cellular tissue, particularly the weaker phloem between the xylem strand and the outer cuticle, can result in extreme shrinkage (Fig 12B) of plant axes after the plant has been coated with silica, but prior to silicification of the plant tissue. The resulting voids may be filled with silica or remain empty.

Plant axes that were dry and shriveled prior to silicification (Fig 12C) tend to be preserved with star-shaped cross-sections, and cells have collapsed. Silica coats the structure and no further dehydration takes place during silicification. Compaction of plants occasionally took place prior to silicification in a waterlogged environment (Fig 12D), and extreme compaction is seen in the cherty sandstones at the margins of chert beds, where only partial silicification may occur (Fig 12E).

Details of experimental silicification of plants in hot springs at Yellowstone have been described by Channing and Edwards (2004), producing deposition of amorphous opal A on and within plant tissue within 30 days. After 330 days tissues were silicified

PLANT DEGRADATION PRIOR TO SILICIFICATION

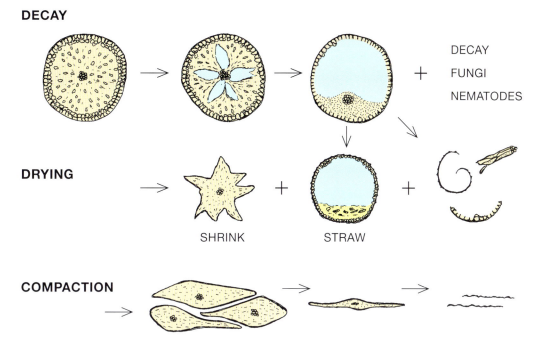

Figure 11 Diagram to illustrate some of the plant preservation features due to decay, drying and compaction prior to silicification.

sufficiently to replicate structures and form a strong matrix that prevented collapse of tissues. This is clearly a common route for silicification, but is relatively slow and does not account for all the features seen in the chert.

Petrographic features in the Rhynie chert suggest that very rapid processes of silicification were in operation. Features such as the preservation of soft tissue of nematodes, germinating spores, cell division, and sperm cells caught in the act of expulsion from a male gametophyte have to be preserved virtually instantaneously (Kerp *et al.*, 2004, Fig 9e). In modern nematodes the development from a fertilized egg cell to hatching from the egg shell takes about 800 minutes; several stages have been found in the chert. In modern plants arbuscules of mycorrhizae exist for only a week or less before deteriorating as a result of the reaction of the host plant. Also the numerous germinating spores suggest a rapid preservation, because the young gametophyte consists of a very thin membrane filled with cell plasma. In the case of preservation of sperm cells emerging from an antheridium, it was possibly the wetting agency of a silica gel that caused the cells to be released, and the high viscosity of the gel prevented the dispersion of the cells. Such preservation would have been impossible if the cells had been released into water.

In chert containing aerial plant axes in growth position, it is usual to find that silicification took place under water and that the axes were at least in part supported by microbial mats (Fig 13A). An area of terrestrial plant growth was flooded to a depth of a few centimetres and the plants continued to grow. Microbial mats formed in the shallow pond that had been created, supporting the plant axes prior to silicification. Similar situations occur today in Yellowstone, where outwash water from springs can be ponded, and plants partly immersed in water, and then surrounded by bacterial growth (e.g. outwash from Daisy Geyser; Trewin and Wilson, 2004, Fig. 7).

It is also clear that the silicification process was not always a single event. Complex internal silica fills in voids left by plant decay show several generations of fill (Fig 13B). Chert with well-preserved plants could have significant original void space, and the space can be partly filled with geopetal sediments, which contain angular broken fragments of the plants (Fig 13C). The brittle fractures of the plants indicate that silicification took place prior to the infiltration of internal sediment, which was itself later silicified. Thus silicification took place very close to the surface, within range of sediment infiltration (Trewin and Fayers, 2016).

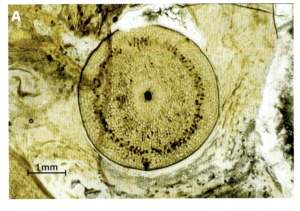
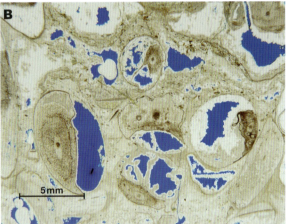
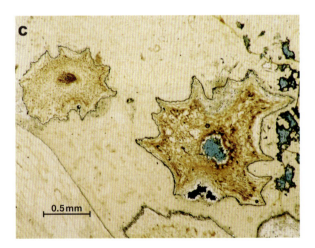
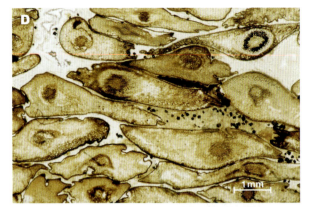

Figure 12 Plant preservation features in the Rhynie chert. **A.** Complete cellular preservation of a *Rhynia* axis. **B.** Dehydration of axes whilst surrounded with sinter; blue resin fills pore space left following shrinkage and decay of plant axes. **C.** Star-shaped plant axes dried and shriveled prior to silicification event. **D.** Plants partially compacted in waterlogged conditions prior to silicification event. **E.** Extreme compaction and partial silicification of plant axis in cherty sandstone.

It is likely that the best preservation occurred under water in reducing conditions, thus allowing preservation of organic matter. The presence of pyrite in the chert also suggests reducing conditions. The temperature of silicification of the plants was probably in the range of 20–30°C, and it appears that plants were not killed by the silica-rich waters. Experiments by Cady and Farmer (1996) showed that organic matter is better preserved when silicified below 35°C. However, silicification of microbial mats probably takes place at temperatures up to 59°C, and some cherts from Windyfield show filamentous microbial textures that were silicified at high temperatures, consistent with a position close to a vent (Fayers and Trewin, 2004).

The observations of Trewin *et al.* (2003) from Yellowstone NP showed that charophyte algae, plants and arthropods live in streams fed by hot springs in which silicification is actively taking place. Temperatures in these streams range from 26–28°C during the summer, but are cooler during the winter. Observations also show that where plants are engulfed and coated by sinter and later exposed in a subaerial setting, internal cellular preservation is rare, since the oxidizing environment results in rapid decay of the organic matter. This process leaves a hollow 'straw' of

Axes of *Ventarura* in pods of Windyfield chert (Fig 14A) are well preserved at the base of the pod, encrusted by microbial mats (Fig 14B) and associated with aquatic crustaceans, hence within a pool environment. When an axis is traced to the top of the chert

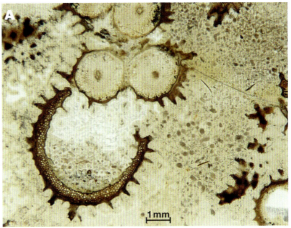

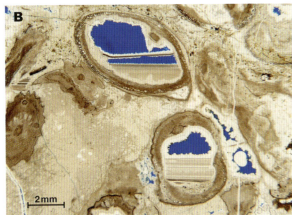

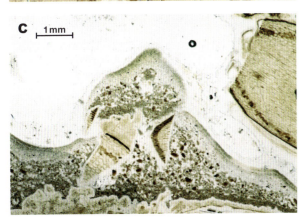

Figure 13 General preservation features. **A.** Support by bacterial sheets between *Rhynia* axes in growth position. **B.** Geopetal silica partly filling cavities left by plant decay; porosity impregnated with blue resin. **C.** Internal silica sediment in cavity in chert, including plant fragments showing brittle fracture that were silicified prior to the introduction of internal sediment to the highly porous sinter.

silica. Thus, good cellular preservation seems to require subaqueous reducing conditions, and lack of exposure to oxidizing conditions before the silicification process is complete, and the chert is impermeable.

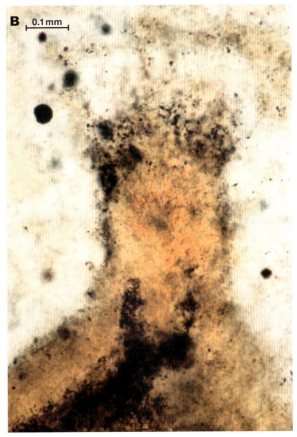

Figure 14 Microbial features. **A.** Microbial mat with micostromatolites grown on fragments of *Ventarura* axes and small coprolites (*Bacillafaex constipata*) preserved in clotted chert matrix. **B.** Detail of microbial micro-stromatolite showing filaments.

pod it may change to a hollow straw, showing that the organic material was destroyed above the water level. Geopetal features generally mark the change in preservation.

The arthropods found in the chert favoured either terrestrial (e.g. trigonotarbids, mites, opilionids), or freshwater (e.g. crustaceans *Lepidocaris*, *Castracollis*, and euthycarcinoids) environments. Preservation of the freshwater fauna depended on the invasion of a pool by hot spring water. The arthropods are then preserved in chert with a clotted texture containing numerous tiny coprolites (Fig 14A), part of the organic debris at the bottom of the pool (Trewin et al., 2003). The terrestrial arthropods occur as fragments washed into pools, and also in association with plants. Rarely, *Palaeocharinus* is found clinging to plant axes, and mites occur in empty sporangia. Many fragmentary arthropod specimens are moulted parts, but others are carcasses, and soft tissue such as book lungs is preserved.

It can be concluded that the excellent fossil preservation in the Rhynie cherts is due to the situation in a subsiding basin with a permanently high water table at or close to the surface. Plants could be silicified by silica coating and permeation in water, and also by immersion in a silica gel. This was combined with a reducing environment allowing preservation of organic matter. It is worth noting that in modern hot-spring marsh environments diatoms are abundant, and along with other modern vegetation such as grasses, efficiently fix silica from hot spring waters. It is possible that in the Early Devonian higher silica concentrations prevailed and led to more rapid silicification of the biota than is observed today.

Study Methods

Fractured surfaces of the chert reveal virtually nothing beyond the likely presence of plant axes appearing as darker cross-sections in lighter chert, generally with a bluish colour. Some chert is opaque white, and some specimens of Windyfield chert are brownish. Cran examined small, rough chips of chert to find microorganisms, or parts of them, but these are very difficult to study in detail. Chert that appears promising in hand specimen is cut on a diamond saw at the desired orientation, and the cut surfaces carefully examined under a binocular microscope. For general examination of chert, a series of thin sections is made to cover the full thickness of a bed. Depending on the chert texture, sections about twice normal petrographic thickness are usually best. It is advisable to examine slices to be used for thin sections at all stages in the process to prevent grinding away a 'surprise' find.

The plant *Ventarura* was reconstructed by slicing chert in 3 mm thick slabs and following the course of axes through the slices. Tracings of plant intersections on slab surfaces were transferred to transparent Perspex sheets that were assembled into a transparent block through which axes could be traced.

In the past, studies of plants were conducted by etching surfaces (with hydrofluoric acid, requiring stringent safety measures), taking an acetate peel of the surface and then regrinding and etching the surface. The peels then document closely spaced serial sections from which the three-dimensional structure could be reconstructed, as in the study of *Aglaophyton* by D. S. Edwards (1986). This method works only if organic matter is preserved in the chert, thus not all chert is suitable. Moreover, fine details such as thin-walled cortical cells are often not visible in peels, which only show the most durable parts such as tracheids and the cuticle. Other disadvantages of this method are that the specimen is destroyed in the process, and that arthropods are generally not recognized.

Much more time-consuming, but far more rewarding for detailed palaeontological work, is making high-quality thin sections (Hass and Rowe, 1999). Carefully selected small blocks of chert are mounted on the slide with thermoplastic cement, which starts to melt at $c.80°C$ and flows freely at $c.140°C$. The sectioned chert slices have a thickness of up to 3 mm, depending on the transparency of the chert. After a first microscopic examination the slide is ground down further to reduce the amount of chert overlying the fossil. Usually, there is also a rather thick layer of chert between the glass slide and the fossil, often with inclusions masking the fossil. The slide is heated until the cement has become fluid again and the chert slice can be taken off to be mounted upside down. After the cement has hardened, the other side of the chert is ground down. This procedure can be repeated several times until the thin section has the desired thickness, without too much chert above and below the object. In this way, obscuring structures in the chert can be minimized, and only in this way was it possible to examine whole organisms such as mites, and features of antheridia showing individual sperm cells and sperm cells just being released. Making a single slide may thus take up to two days, but it is really worth the effort! Slides are not covered with a cover slip, but are covered with oil for examination and photography. Thin sections are photographed in transmitted or incident light, depending on the thickness of the slide, and the clarity of the chert and the details. In many cases incident light with a piece of milk glass under the slide is to be preferred because this makes details visible that cannot be seen in transmitted light (Kerp and Bomfleur, 2011).

With semi-transparent chert it is now possible to take a series of images at different levels within the chert, and then combine these images into a stack to produce an image that is in focus through the whole thickness of the specimen, as demonstrated by Kamenz et al. (2008) in a study of the book lungs of *Palaeocharinus*. The technique has been extended to produce three-dimensional stereo images of Rhynie arthropods (Haug et al., 2009, 2012). There is a large quantity of Rhynie chert in museum collections and beneath the grass at Rhynie, but time-consuming effort is required to reveal the contents. A good supply of diamond saw blades is also needed, because the blades wear out rapidly when cutting chert.

Biodiversity
General features

A floral and faunal list for the cherts is provided in Tables 1–3 (see pp. 35-8), together with references to the original descriptions of species. This list expanded rapidly at the time of trenching and drilling, and following the discovery of the Windyfield chert. At Aberdeen University, the dissertation studies of Powell, Kelman and Fayers, plus the postdoctoral work by Anderson and Fayers, resulted in the discovery of new elements of the biota. At Münster, Hagen Hass was also examining material obtained by Remy, and from the Aberdeen field effort. The Windyfield chert has been a productive source of new material. Many of these new finds are represented by single specimens, or have only been found in a single block of chert. Some new finds have come from loose blocks of chert that might come from chert deposits not sampled in the trenches or boreholes. Thus the finding of new macro-elements of the biota has become increasingly difficult, particularly at the Rhynie site. However, there is considerable potential for further description of fungi and micro-organisms, as demonstrated by Krings, Dotzler, Taylor and their co-workers.

Microbes and flora
Microbes

Microbes, in a general sense, undoubtedly played an important part in the Rhynie ecosystem, as they do in modern hot-spring deposits where different types are able to survive at different temperatures, and give spectacular black, orange and green coloration to sinter deposited by hot springs. In fossil examples, 'streamer structure' (e.g. Jones et al., 2004, Fig 2) of filamentous bacteria is frequently preserved, but is not seen in the Rhynie chert because the chert does not split along bedding surfaces. Instead, microbial lamination is seen on cut surfaces. This comprises layers of fine filaments, each around a micrometre in diameter, visible in thin sections of chert. In some specimens, layers of flat-lying filaments may alternate with filaments growing vertically, presumably as a phototactic response. Microbial lamination is present draped between sporophyte plant axes that are in growth position, and it appears that the support given by microbial sheets was responsible for the preservation of axes in this position (Fig 13A). Presumably an area of living plants was flooded, and microbial sheets grew within and over the surface of the water, whilst the plants continued to live. Eventually the water became choked with microbial material and plants became silicified *in situ*.

Another type of microbial preservation occurs in the form of micro-stromatolites coating plant axes that were immersed in pools. This is seen in the Windyfield chert, where axes of *Ventarura* in various states of decay are coated. The coating clearly post-dated the plant decay (Fig 14). Microbial cells only 2 μm in size can be seen preserved within degrading plants, as illustrated by Kerp and Hass (2009), and such cells probably played an important role in plant degradation. It is generally not possible to identify the filamentous forms with filaments commonly only 1–2 μm in diameter as members of the Archaea or Cyanobacteria. Both include thermophilic forms, with thermophilic Archaea generally living at higher temperatures. At Windyfield a specimen with the characteristic globular texture of sinter deposited in the splash zone of a geyser reveals filamentous textures in thin section that may have been thermophilic Archaea, surviving in hot water around a vent.

Krings et al. (2009) have described fascinating material in which cyanobacterial filaments have grown into axes of *Aglaophyton* through the stomata and spread within the plant both between and within cells.

They suggested that this kind of association may have been the precursor of symbioses between land plants and cyanobacteria, but it is certainly the oldest evidence for cyanobacteria growing within land plants. The same axes of *Aglaophyton* are also invaded by hyphae of the mycorrhizal fungus *Glomites rhyniensis*, illustrating the close association of plants, cyanobacteria and fungi in the Early Devonian.

Fungi

The preservation of fungi is one of the remarkable features of the chert, and has revealed that the main fungal strategies had already evolved by Early Devonian times (Taylor *et al.*, 2004). Saprophytic fungi live off dead organic matter and assist in breaking down organic matter to be used by other organisms or added to soil. One example is the chytrid fungus *Palaeoblastocladia milleri*, which is associated with *Aglaophyton* (Fig 15A). Parasitic fungi invaded living plants and algae, frequently producing a response in the host, as in the case of swollen cells of *Palaeonitella* infected with *Krispiromyces discoides* (Fig 15B) and the vascular land plant *Nothia* in which fungal cells and hyphae are sheathed with an encasement of cell wall material, and the walls between the infected and the healthy cells are thickened in order to separate the fungi from the protoplast and the healthy cells (Krings *et al.*, 2007b). Ascomycetes are regularly seen in the marginal parts of *Asteroxylon* axes (Fig 15C).

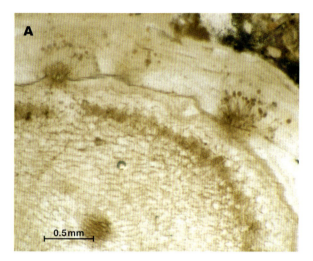
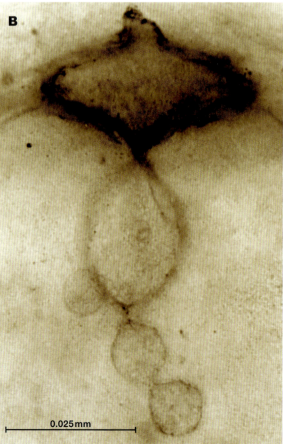
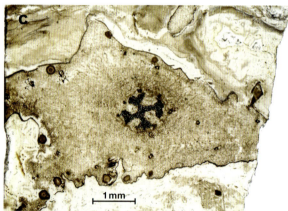
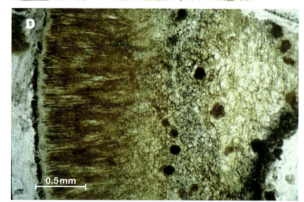

Figure 15 Fungi: A The saprophytic chytrid *Palaeoblastocladia milleri* on an axis of *Aglaophyton*. **B** The chytrid *Krispiromyces discoides* penetrating a cell wall of the alga *Palaeonitella*. **C** The ascomycete *Paleopyrenomycites devonicus* in an axis of *Asteroxylon*. **D** Section of *Prototaxites* showing the radial structure of the marginal zone.

Symbiotic fungi have a mutually beneficial association with plants, with fungal mycorrhizae living within the plant tissue. The fungus supplies water and nutrients to the plant, and the plant supplies carbohydrate from photosynthesis to the fungus. It is apparent that different fungi had specific plant hosts. In cross-sections of axes of *Aglaophyton* a dark band contains the arbuscules of *Glomites rhyniensis*, its mycorrhizal partner (Fig 15A, also seen in Fig 19A), as demonstrated by Taylor *et al.* (1995). The Rhynie chert is the oldest known deposit in which such relationships can be demonstrated. The fact that close relationships had formed between fungi and plants by the Early Devonian implies a long period of development prior to that time.

The problematic 'plant' known as *Prototaxites* (? = *Nematophyton, Nematoplexus*) (Fig 15D) consists of an inner meshwork of spirally coiled tubes, some of which show spiral thickenings. The tubes branch in dense 'knots' within the tissue. In some specimens tubes are oriented perpendicular to an external cuticle. In the Rhynie chert fragmentary specimens are found, originally described as *Nematophyton taiti* by Kidston and Lang (1921b), and *Nematoplexus rhyniensis* Lyon 1962. These structures, which are known elsewhere with a diameter of 1 m and height of 8 m, probably grew erect and have been interpreted as the giant fruiting bodies (saprophore) of a fungus (Hueber, 2001; Boyce *et al.*, 2007).

Lichens

Lichens represent a symbiosis between fungi and either green algae or cyanobacteria, and extant lichens are abundant on both rock and organic surfaces. It is thus surprising that only one example, *Winfrenatia*, has been found in the chert (Taylor *et al.*, 1997) (Fig 16).

Algae

Freshwater charopyhte algae are common in present-day freshwater streams, and even grow in the outwash of hot springs in Yellowstone National Park. *Palaeonitella* in the Rhynie chert (Fig 17) occupied a similar environment. This alga was originally described by Kidston and Lang (1921b), but discovery of reproductive organs, antheridia and gyrogonites allowed Kelman *et al.* (2004) to provide a detailed description and show that it had features relating it to both the extant Chareae and Nitellaea. They considered it as having been more closely related to Chareae but retaining features of the more basal Nitelleae. This alga is small when compared with present-day forms, and would have grown as small bush-like tufts a few millimetres high in fresh water. It was a delicate structure, so it is not commonly preserved. The filamentous algae *Mackiella* and *Rhynchertia* described by Edwards and Lyon (1983) occur in chert in association with *Palaeonitella* and were aquatic in habitat.

Figure 16 *Winfrenatia*, the only lichen described from the chert.

Figure 17 The alga *Palaeonitella*.

Vascular plants

The seven genera of embryophytic sporophyte plants known from the Rhynie and Windyfield cherts have most recently been reviewed by Edwards (2004). They are *Aglaophyton*, *Rhynia*, *Horneophyton*, *Nothia*, *Ventarura*, *Trichopherophyton*, and *Asteroxylon*. The plants are studied in thin section and on cut surfaces (see methods section). Thus it is difficult to make composite reconstructions, and ideas have evolved with time. The general forms of some of the plants are illustrated with models (Fig 18) that attempt to show the characteristic features of branching and the manner in which sporangia were carried. *Trichopherophyton* is the only plant for which a reconstruction is not available; it is a scarce plant and occurs in a fragmented state.

Figure 18 Models of (L to R) *Horneophyton*, *Aglaophyton* and *Rhynia* growing on a sinter surface with plant debris and cyanobacterial mats.

In the best material, extremely fine detail of the anatomy of the plants is seen. The upright axes of the plants comprise a xylem strand surrounded by phloem and a parenchymatous cortex. The epidermis is covered by a tough cuticle (Fig 19A). The anatomy and details of the cell structure characterize each plant. *Rhynia* and *Aglaophyton* have terminal fusiform single sporangia, whereas *Horneophyton* has terminal sporangia that are branched into two sometimes into four lobes. *Nothia*, *Asteroxylon*, *Trichopherophyton* and *Ventarura* have lateral sporangia consisting of two kidney-shaped valves. Stomata for gas exchange are well preserved (Fig 19C) on all plants. Detailed study of cells within the axes allows interpretation of factors such as water uptake and transport, as well as details of cell division in growing tips of axes. All plants, except for *Asteroxylon*, had naked branching aerial axes.

Horneophyton and *Nothia* were up to 20 cm high and both had underground rhizomes that were useful for storage. *Horneophyton* had bulbous rhizomes bearing rhizoids (Fig 19D), with each bulb supporting an aerial axis, whereas *Nothia* formed dense mats of underground axes from which aerial axes emerged (Kerp et al., 2001). Due to their growth habit they generally grew on soft substrates such as sandy soils. Also, in these plants vegetative reproduction played an important role. The female gametophyte of *Horneophyton* is complex, consisting of a disc-like structure with up to 30 finger-like outgrowths containing the deeply sunken archegonia. Of *Nothia* only the antheridia-bearing gametophyte is known, which is equally complex. Initially it was unclear to which species the gametophytes belonged, and they were described as new taxa. Since then the gametophytes have been assigned to sporophytes, mainly on the basis of the structure of the tracheids of the conducting tissue.

Asteroxylon, which looked superficially surprisingly similar to the modern club moss *Huperzia*, is the largest and most complex vascular plant from the Rhynie chert. It is a common element, but occurs only in relatively few horizons. Preservation is often fragmentary and less exquisite than that of other plants. *Asteroxylon* probably grew up to 45 cm high, with aerial axes up to 12 mm thick. Aerial axes are characterized by a central strand of conducting tissues, in which the xylem is typically cross- to star-shaped in transverse section (Fig 19B). The axes give off densely but irregularly spaced leaf-like appendages, which are not vascularized and thus technically cannot be termed leaves. Vascular bundles extend from the central stele toward the appendages but consistently terminate in the outer cortex, just before reaching the periphery of the axis. The horizontal underground axes possess a rather complex anatomy and branching pattern and give off repeatedly forking axes that often extend relatively deep into the substrate. Sporangia, consisting of two kidney-shaped valves, are interspersed among the leaf-like appendages and attached to the axis with a short stalk (Kerp et al., 2013). Fertile zones on axes with leaf-like appendages and sporangia alternate with sterile zones with only leaf-like appendages, indicating a periodicity of growth seen in the clonal growth of the other plants.

Aglaophyton and *Rhynia* were 20–25 cm high and did not have underground parts. Both have a central vascular strand, which is, however, not well developed and provided only limited support. The upright aerial axes continued to grow upward until the support of the vascular strand became insufficient and the axes bent

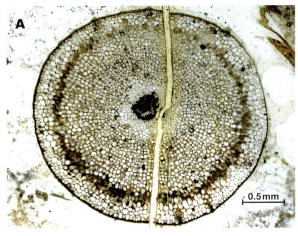

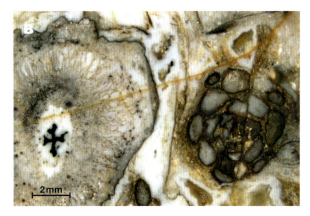

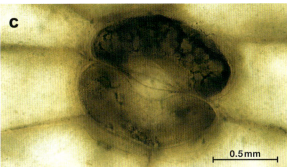

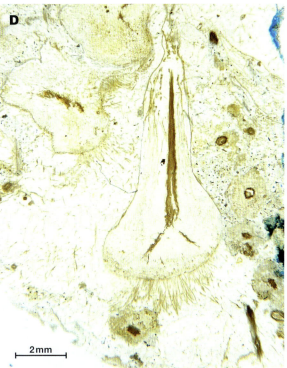

Figure 19 General sporophyte morphology. **A.** *Aglaophyton* axis cross section. **B.** Cross-sections of *Asteroxylon* to show cross-shaped xylem strand (left) and section of branch tip (right). **C.** Stomata of *Aglaophyton*. **D.** Rhizome and rhizoids of *Horneophyton*.

down. Where the axis touched the substrate, unicellular rhizoids developed, especially around the stomata, and then the axes bent up again, resulting in typical U-shaped axes. *Aglaophyton* and *Rhynia* both lived in symbiosis with vesicular arbuscular mycorrhizae, which were well developed in the upright axes. The mycorrhizae entered the axes and filled the intercellular spaces some four to five cells below the epidermis with hyphae, thus appearing as a dark ring (Fig 12A). The penetration of the living cells by hyphae, the formation of the so-called arbuscules as tree-like structures inside the cells, and the host reaction of the plant have been observed. Mycorrhizae doubtlessly played an important role in the uptake of water and nutrients of these plants, which could grow on various substrates, including barren sinter surfaces. All the better-known taxa had clonal growth, but little is known about the growth habits of *Trichopherophyton* and *Ventarura*. Sporangia of *Aglaophyton* are quite common, but much rarer in *Rhynia*, which formed dense stands and frequently has small bulbous outgrowths with rhizoids, apparently for vegetative reproduction. It may be that such isolated clonal plant patches became silicified, explaining the frequency of lenticular cherts dominated by one genus.

Due to the exquisite preservation in the chert, the full life cycles of *Aglaophyton* (Fig 20) and *Rhynia* can be documented. Features observed include germinating spores (Fig 20D), young developing gametophytes and mature gametophytes, being either male, bearing antheridia producing free-swimming sperm cells, or female with archegonia containing the egg cells. Amazingly, gametophytes (Fig 20E,G) have been preserved showing sperm cells emerging from an antheridium (Fig 20F), and also archegonia with the neck,

Figure 20 Sporophyte–gametophyte life-history. *Aglaophyton–Lyonophyton*. **A.** Diagram of life-history cycle *Aglaophyton*; **B.** Sporangium of *Aglaophyton*; **C.** Spore tetrads of *Aglaophyton*; **D.** Germinating spore; **E.** *Lyonophyton* male gametophyte; **F.** Sperm cell release from antheridium; **G.** *Lyonophyton* female gametophyte; **H.** Neck and egg canal in archegonium.

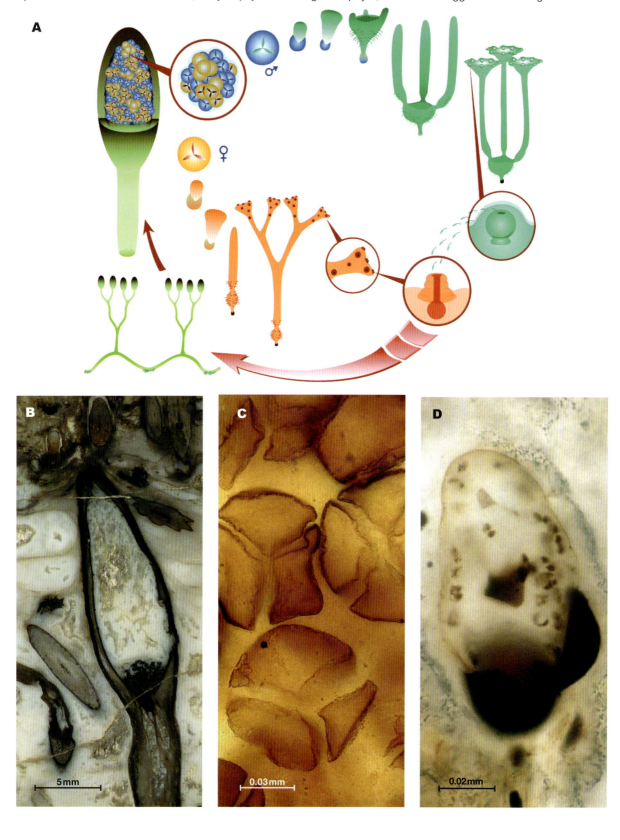

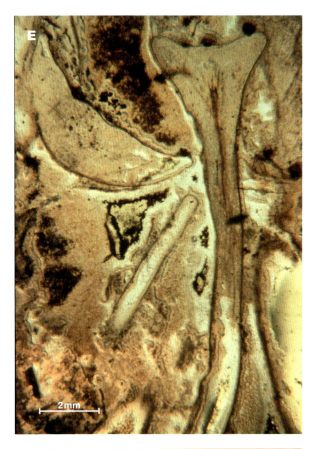

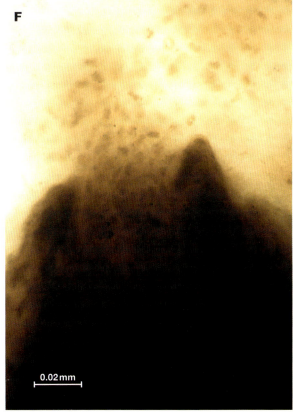

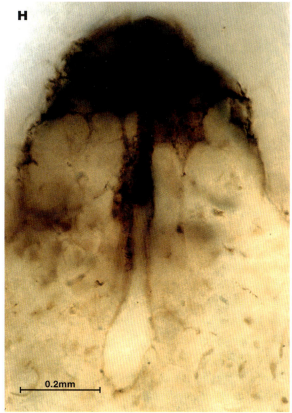

neck canal and egg chamber (Fig 20H). These are the earliest plants for which the alternation of sporophyte and gametophyte generations can be demonstrated. The gametophytes were very similar to the sporophytes in having a vascular strand, an epidermis with stomata and a cuticle, but they were much smaller, varying from less than one up to a few centimetres. The gametophytes of *Aglaophyton* (Fig 20E,G) and *Rhynia*, having apically slightly bowl-shaped to flattened axes, were relatively simple compared with those of *Horneophyton*.

Fauna

'Worms'

The preservation of soft-bodied 'worms' requires unusual circumstances, particularly in a terrestrial environment, but soft tissue is preserved in the cherts. In one specimen, nematodes (*Palaeonema*; Fig 21) were found within a decaying axis of *Aglaophyton* (Poinar et al., 2008). Numerous nematodes in various stages of development are present, and they were presumably living on the decay products of the plants and associated fungi. These are the oldest known nematodes, and they were performing the same ecological role as their modern descendants. Also present in the chert are rare structures that are possibly polychaete jaws.

Arthropods

The cherts contain a varied arthropod fauna that can be divided into terrestrial and aquatic forms (Fig 22A,B, Table 3). In general, arthropods are uncommon in the chert, but there are exceptions. Trigonotarbids are the most common arthropods, and can be found as complete individuals with preserved details of internal structure such as book lungs, and fragmentary material from moults. There are three currently recognized species of *Palaeocharinus* (Fig 23A, B), and a fourth awaits description. Most are less than 15 mm long. Juveniles have been found within empty plant sporangia that would have been a suitable place for egg-laying. They were predatory with large fangs, hunting other arthropods amongst the thickets of *Rhynia* and *Aglaophyton*, and occupied a similar niche to present-day predatory spiders. No true spiders are

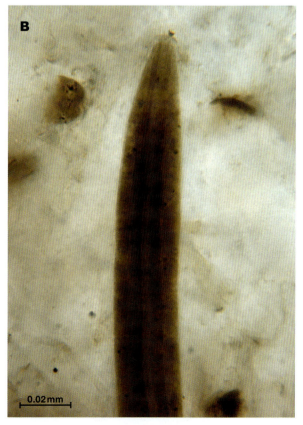

Figure 21 Nematode. **A.** The nematode *Palaeonema* within an axis of *Aglaophyton*. **B.** Detail of *Palaeonema*.

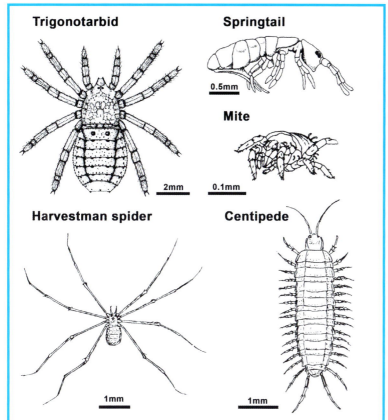
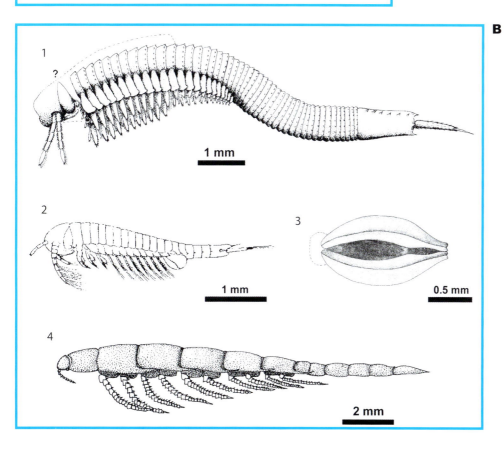

Figure 22 Arthropods from the Rhynie and Windyfield cherts. **A.** Terrestrial arthropods. **B.** Aquatic arthropods: 1 *Castracollis*; 2 *Lepidocaris*; 3 *Ebullitiocaris*; 4 *Heterocrania*.

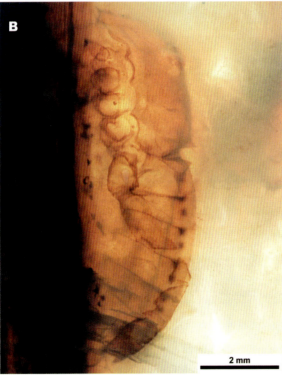
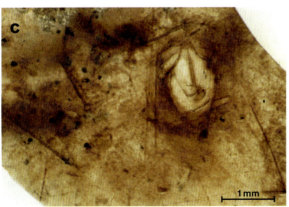

Figure 23 Arthropods from the cherts. **A.** Model of *Palaeocharinus tuberculatus*. Body of animal about 6 mm long. **B.** *Palaeocharinus* preserved clinging to a plant axis. **C.** The harvestman spider *Eophalangium*. **D.** Coprolite, *Lancifaex divisa*; product of a detritivore. **E.** Group of *Ebullitiocaris*. **F.** Model of the euthycarcinoid *Heterocrania*. Animal about 12 mm long. **G.** Eucrustacean nauplius larva. A, C, D from Rhynie chert; E, F, G from Windyfield chert.

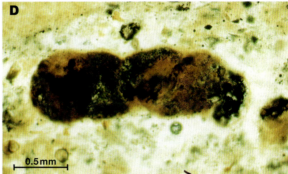

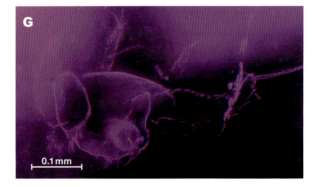

known from the chert, although *Saccogulus* is a possible candidate. *Palaeocharinus* was in competition with centipedes (e.g. *Crussolum*), which are represented by scarce fragmentary material. Several types were present, but material is too fragmentary for detailed description. An exciting find is the earliest known harvestman (opilionid), *Eophalangium* (Fig 23C). Both male and female specimens have been recognized, with internal organs present and a tracheal system for breathing (Dunlop *et al*., 2004). They are surprisingly similar to extant forms and would also have been predators.

Undoubtedly these predators considered smaller individuals of their own kind as prey, but there were also smaller arthropods present, notably the collembolan (springtail) *Rhyniella*, and several species of mites. A small diplopod *Eoarthropleura* was probably a detritivore. *Rhyniella* is often quoted in texts as the earliest 'insect', but modern taxonomists no longer consider collembolans 'insects'. However, *Rhyniognatha,* represented only by fragmentary mouthparts, has been assigned to the Pterygota, and *Leverhulmia* is also now regarded as a hexapod, possibly the earliest apterous insect (Fayers and Trewin, 2005). *Leverhulmia* is known only from a single specimen, and it has gut contents preserved, showing that it was a detritivore.

More evidence for terrestrial animal activity comes from coprolites found in the chert (Habgood, 2000; Habgood *et al*., 2004). Several types of coprolites are present, generally containing plant cuticle debris, spores and fungal material (*Lancifaex*) (Fig 23D, or consisting of amorphous rods (*Bacillafaex*) (Fig 14B). The range of coprolites described by Habgood *et al.* (2004) indicates the presence of detritivorous arthropods not yet found as body fossils. By comparison with other Early Devonian localities, millipedes would be expected to be present in the chert. It is also likely that larger arthropods were present, but they are unlikely to be recognized in thin sections. The large *Diplichnites* trackways from the Quarry Hill Sandstone (Fig 7A) show that large animals, possibly arthropleurids, were present.

From the aquatic realm three branchiopod crustaceans and a euthycarcinoid have been described (Fig 22B). Present-day branchiopods include 'fairy shrimps' and 'clam shrimps' that inhabit ephemeral pools, have a rapid life cycle and lay eggs that can withstand desiccation for many years. Thus they are adapted to regions where there is only intermittent water. It is possible that Devonian forms also had the same life style, and this would explain how they come to be found in small isolated pools near the Rhynie hot springs. *Lepidocaris* is the most common, and can be found in most chert where there is evidence that the environment was subaqueous. It is frequently associated with chert showing a clotted texture and containing micro-coprolites (Anderson and Trewin, 2003). *Castracollis* is known from over 30 individuals, but all came from a thin band in a single piece of chert picked up in a field some distance from Rhynie. The last is *Ebullitiocaris* (Fig. 23E), a strange creature largely encased in a shell less than a millimetre long. They occur in groups apparently attached to submerged plant material. Also characteristic of freshwater pools is the small euthycarcinoid *Heterocrania* (Fig 23F), an animal about 12 mm long with delicate legs that probably scavenged in the debris at the bottom of a pool. In the Windyfield chert, tiny larvae of arthropods are preserved (Fig 23G). They have features that suggest they are not the larvae of any of the known larger arthropods (Haug *et al.*, 2012).

There were certainly other arthropods present, such as *Rhyniemonstrum,* which is currently represented only by a few large segments with a distinctive ornament. With luck other specimens will be found so that it can be fully identified in the future.

Distribution of flora within Rhynie Cherts Unit

Kidston and Lang (1917, 1921b) recorded the distribution of plants within the original trenched sequence, and Tasch (1957) discussed in detail the succession of flora above the basal bed, which was the only bed from which Kidston and Lang recorded *Rhynia gwynne-vaughani*. Trenching in the 1963–1971 period showed that *Rhynia* was not confined to the basal bed and that there was not a regular floral succession through the cherts unit. The detailed study by Powell *et al.* (2000a) on a cored borehole through the cherts revealed 53 chert beds, 45 of which contained plants. A cycle of colonization could not be recognized, but *Rhynia* and *Horneophyton* were most common as initial colonizers. *Rhynia* occurred as the primary colonizer on sandy substrates and on plant litter in ten beds, and *Horneophyton* in nine beds, in four of which colonization was directly on a sinter surface. The diversity of plants in individual beds is low, only exceeding three in four of the beds.

Trewin and Wilson (2004) evaluated the plant successions in two other boreholes, confirming that no beds could be correlated over the distance of about 20 m between boreholes, and that plant genera have no preferred succession. The frequent occurrence of monotypic stands implies a short period of colonization prior to preservation. This is strengthened by the fact that spores were not found in 12 of the beds studied by Powell *et al.* (2000a), implying that the flora was preserved before the sporophytes had matured and released spores. Some beds also contain upright axes that were dead and shriveled in close proximity to fresh axes, both being silicified in a single event. This is also a possible indicator of seasonal growth, with the new growth sprouting amongst the dried axes of the previous year. Furthermore, the clonal growth of *Nothia* and *Horneophyton*, and the alternation of sterile and fertile zones on *Asteroxylon* axes, provide further indications of seasonal growth. Evidence from laminated Early Devonian lacustrine deposits (Trewin and Davidson, 1996) also supports a seasonal climate.

It must also be stressed that the plants were clonal forms, and one clonal patch may have excluded an adjacent patch of a different genus. Thus a borehole may intersect a single clonal colony in an area of greater diversity. Trenching has shown that beds can wedge out from 20 cm to zero over as many centimetres, and pods of chert can form around a single plant colony. Other beds continue laterally in trenches for over a metre with similar flora.

Powell *et al.* (2000a) also noted that eight of the 53 chert beds in the borehole (19C) do not contain plants. This may indicate that plant cover was incomplete, and the drill has intersected the bed at a point that had not been colonized by plants. Using the borehole data of Powell *et al.* (2000a), Trewin and Wilson (2004) estimated that 15% of the area was bare, 30% covered with plant litter and 55% had both litter and living plants. It seems probable that a short period between deposition of individual beds did not allow a climax plant community to develop. In present-day hot-spring areas, plants take time to colonize abandoned sinter surfaces, and isolated patches of colonizing species are a common feature on sinter surfaces and adjacent to hot-spring vents.

Plant ecology: Were the plants adapted to the hot-spring environments?

The diversity of plants in the Rhynie chert, living in an inland basin, is low when compared with other areas such as the lowland floodplain deposits of the ORS of South Wales (Wellman, 2006). The diversity of plants within the Rhynie chert is also low when compared with the spore diversity from over 100 productive samples within the Rhynie succession (Wellman, 2006). The distinctive spores of *Aglaophyton* and *Horneophyton* occur commonly in the basin succession, and these plants were not confined to the hot-spring environments (Wellman, 2004). They may have been pre-adapted to the environment, and able to survive the variations of water temperature and chemical composition, such as salt and metal content, that posed problems for other plants. The spores of other Rhynie chert plants are not so distinctive and cannot be confidently recognized in dispersed assemblages, but could have been similarly widespread. Thus the flora preserved in the chert may represent those elements of a larger flora that could survive in the area of the hot springs. The detailed anatomy of the Rhynie plants also lends support to adaptation to stressed conditions, and comparisons can be made with the physiological adaptations of modern plants growing under the influence of hot springs in Yellowstone National Park (Channing and Edwards, 2009).

It can also be argued that in a majority of chert beds, plant colonization took place on a sandy substrate of fluvial overbank deposits with plant litter present, and that there was no influence from the hot springs in the colonization of those surfaces. It was only the chance event of hot-spring outwash invading the area that caused silicification to occur. Thus the low diversity of plants may reflect the fact that the Rhynie cherts unit represents a small fraction of the Rhynie succession, and a very small geographic area. Hence a lower diversity would be expected when compared with the full succession and larger area. Both *Horneopyton* and *Rhynia* were capable of growing directly on sinter surfaces, and whilst it is difficult to ascertain the conditions at the time of growth, it is probable that conditions were stressed, and that the plants had some special adaptations to the conditions.

The Ecosystem

The Rhynie chert ecosystem can be documented in greater detail than most fossil ecosystems because the biota is preserved at, or very close to, the place where it lived, and micro-organisms such as fungi and bacteria are preserved in association with their hosts. Freshwater and terrestrial ecosystems form end-members of a continuum in which terrestrial material was washed into freshwater pools and terrestrial areas were frequently flooded. On land, contributors to litter and soil formation were plant tissues, fungi, bacteria, and the coprolites and tissues of arthropods. The main primary producers were the plants, bacteria and some fungi; these supported herbivorous arthropods such as collembolans and mites. Detritivores helped break down plant material, aided by fungi and nematode worms. The predators such as trigonotarbids and centipedes preyed on the smaller arthropods.

In the water algae and cyanobacteria were primary producers, and within ponds an organic-rich substrate formed from living bacterial mats and fungi, together with organic detrital debris and amorphous organic matter. Micro-herbivores extracted food by filter feeding (e.g. *Lepidocaris*, *Ebullitiocaris*), and *Castracollis* may have been predatory. Coprolites from these crustaceans accumulated along with other organic debris in low-energy pools. Cyanobacterial mats also formed at the water surface, and supported the upright axes of plants where terrestrial plant-rich areas were flooded by hot spring or floodwater.

A typical community of the terrestrial and aquatic biotas of the Windyfield chert is illustrated in Figure 24. Thus the cycle of growth, decay and recycling of nutrients is well preserved in the Rhynie chert, a unique feature of this deposit for pre-Carboniferous time.

Figure 24 Environmental reconstruction of a small pool surrounded by *Ventarura* with *Palaeocharinus*, harvestman spiders and centipedes. In the pool fallen *Ventarura* axes are coated with microbial mats. *Lepidocaris* and *Heterocrania* live in the pool, contributing coprolites to accumulating debris.

Time period of formation of the Rhynie Cherts Unit

Drilling has established that the Rhynie Cherts Unit represents a finite period of hot-spring activity. During that period about 50 chert beds were formed, representing individual events when hot-spring waters flooded a particular spot, as recorded in a borehole. The aggregate chert thickness in each borehole or trench varies between two and five metres. In modern hot-spring areas sinter can accumulate on outwash areas at rates in excess of 10 mm per year, thus the time spent in actual sinter deposition at Rhynie may have been as short as 50 years. To this period must be added the interbed fluvial deposits that record overbank flood events. Given the disrupted state of these beds it is not possible to count events, but the minimum must be about 50. Given that the climate was seasonal, it is possible that flood events were concentrated in a rainy season, and maybe hot-spring invasion took place in the dry season. However, this implies a very fast rate of deposition, which is quite possible.

Significance of deposit for understanding the history of Life

The Rhynie and Windyfield cherts provide us with a series of high-resolution 3D snapshots in time of an early terrestrial and freshwater environment. The superb preservation allows details of both plants and animals that are not usually preserved in the fossil record. In many ways the preservation is similar to that seen in amber, but rather than being engulfed in sticky plant resin, the plants and animals were trapped and silicified by hot-spring waters that invaded small areas of an ecosystem. It is the preservation of an *in situ* ecosystem that is the most remarkable feature of the cherts. This has enabled the study of the full life history, with sporophytes and gametophytes, of early terrestrial plants. Even the germination of spores and the ejection of sperm cells can be seen. The associations and interactions between fungi and plants can also be examined from saprophytic to parasitic and mutualistic (Taylor and Krings, 2005). Thus it has become clear that many of the plant/fungal interactions seen today were already developed in the Early Devonian. The arthropods also reveal that a complex community had developed in association with the early plants, and arthropod coprolites reveal information on diet. The arthropod fauna from the Windyfield chert, when combined with that of the Rhynie chert, gives the most diverse associated fossil arthropod fauna of terrestrial and freshwater origin from rocks of comparable age anywhere in the World.

Look in plant litter today and you will find harvestman spiders, springtails, centipedes and mites, and in the Early Devonian trigonotarbids fulfilled the ecological role now taken by some spiders. Fungi and nematodes also had a role in breaking down plant tissues and recycling nutrients. Many of the finds from the chert are the oldest known representatives of their group by many millions of years, a fact that confirms the paucity of the fossil record of small soft-bodied organisms. Thus it would be wrong to regard the Rhynie terrestrial community of plants and animals as 'primitive'. It was a remarkably complex ecosystem with many 'modern' features. Despite the fact that the plants are small and simple, the fungi and arthropods were well adapted to functions that continue to the present day, despite the enormous changes in flora that have taken place.

These observations force us to look back from the Early Devonian to imagine how long it had taken this terrestrial community to evolve. Spores from terrestrial liverworts have been reported from the early Middle Ordovician (*c*.472 Ma) (Rubenstein *et al.*, 2010) and one has to assume that small arthropods evolved features such as tracheae and book lungs to allow them to exploit the oldest terrestrial environments. Thus it can be concluded that the Rhynie chert biota was the result of over 60 million years of evolution in the terrestrial environment.

Acknowledgements

We thank all those who have given information and advice in the preparation of this chapter. NHT particularly thanks Lyall Anderson, Clare Powell, Steve Fayers and Ruth Kelman, the palaeontological members of the Rhynie Research Team at Aberdeen, for their contributions. We also wish to acknowledge the many discussions with Dianne Edwards and Alan Channing of Cardiff University, particularly in Yellowstone Park. HK wishes to thank all (former) members and collaborators of the Münster Rhynie Research team, especially Winfried Remy, Hagen Hass, Thomas N. Taylor, Michael Krings and Patricia Kearney. The German Science Foundation (DFG) is acknowledged for funding (grants KE584/13-1 and 13-2).

References

Anderson, L. I., and N. H. Trewin. 2003. An Early Devonian arthropod fauna from the Windyfield chert, Aberdeenshire, Scotland. Palaeontology 46:467–510.

Anderson, L. I., W. R. B. Crighton, and H. Hass. 2004. A new univalve crustacean from the Early Devonian Rhynie chert hot spring complex. Transactions of the Royal Society of Edinburgh: Earth Sciences 94:355–369.

Barghoorn, E. S., and W. C. Darrah. 1938. *Horneophyton*, a necessary change of name for *Hornea*. Harvard University Botanical Museum Leaflets 6:142–144.

Becker, R. T., F. M. Gradstein, and O. Hammer. 2012. The Devonian Period. Pp. 559–601. in F. M. Gradstein, J. G. Ogg, M. D. Schmitz, and G. M. Ogg (eds), A Geological Timescale 2012. Elsevier, Amsterdam.

Bhutta, A. A. 1969. Studies on the Flora of the Rhynie Chert. PhD dissertation. University College of South Wales, Cardiff.

Boyce, K. C., C. L. Hotton, M. L. Fogel, G. D. Cody, R. M. Hazen, A. H. Knoll, and F. M. Hueber. 2007. Devonian landscape heterogeneity recorded by a giant fungus. Geology 35:399–402.

British Geological Survey (BGS). 1993. Alford, Scotland. Sheet 76W. Solid geology 1: 50,000. British Geological Survey, Keyworth, Nottingham.

Cady, S. L., and J. D. Farmer. 1996. Fossilisation processes in siliceous thermal springs: trends in preservation along the thermal gradient. Pp. 150–173 in G. R. Brock and J. A. Goode (eds), Evolution of Hydrothermal Ecosystems on Earth (and Mars?). (Ciba Foundation Symposium 202.) Wiley, Chichester.

Channing, A., and D. Edwards. 2004. Experimental taphonomy: silicification of plants in Yellowstone hot-spring environments. Transactions of the Royal Society of Edinburgh: Earth Sciences 94:503–521.

Channing, A., and D. Edwards. 2009. Silicification of higher plants in geothermally influenced wetlands: Yellowstone as a Lower Devonian Rhynie analog. Palaios 24:505–521.

Claridge, M. F., and A. G. Lyon. 1961. Lung-books in the Devonian Palaeocharinidae (Arachnida). Nature 191:1190–1191.

Croft, W. N., and E. A. George. 1959. Blue-green algae from the Middle Devonian of Rhynie, Aberdeenshire. Bulletin of the British Museum (Natural History): Geology 3:341–353.

Dotzler, N., M. Krings, T. N. Taylor, and R. Agerer, R. 2006. *Scutellosporites devonicus* from the Rhynie chert. Mycological Progress 5:9–18.

Dotzler, N., T. N. Taylor, and M. Krings. 2007. A prasinophycean alga of the genus *Cymatiosphaera* in the Early Devonian Rhynie Chert. Review of Palaeobotany and Palynology 147:106–111.

Dotzler, N., C. Walker, M. Krings, H. Hass, H. Kerp, T. N. Taylor, and R. Agerer. 2009. Acaulosporoid glomeromycotan spores with a germination shield from the 400-million-year-old Rhynie chert. Mycological Progress 8:9–18.

Dubinin, V. B. 1962. [Class Acaromorpha: mites or gnathosomic chelicerate arthropods.] Pp. 447–473 in B. B. Rohdendorf (ed.), Osnovy Paleontologii Vol. 9. Nauka, Moscow. [Russian]

Dunlop, J. A. 1994. Palaeobiology of the Trigonotarbid Arachnids. PhD dissertation. University of Manchester, Manchester.

Dunlop, J. A. 1996. Systematics of the fossil arachnids. Revue Suisse de Zoologie, vol. hors série:173–184.

Dunlop, J. A., L. I. Anderson, H. Kerp, and H. Hass. 2003. Preserved organs of Devonian harvestmen. Nature 425:916.

Dunlop, J. A., L. I. Anderson, H. Kerp, and H. Hass. 2004. A harvestman (Arachnida: Opiliones) from the Early Devonian Rhynie cherts, Aberdeenshire, Scotland. Transactions of the Royal Society of Edinburgh: Earth Sciences 94:341–354.

Dunlop, J. A., S. R. Fayers, H. Hass, and H. Kerp. 2006. A new arthropod from the early Devonian Rhynie chert, Aberdeenshire (Scotland), with a remarkable filtering device in the mouthparts. Paläontologische Zeitschrift 80:296–306.

Edwards, D. 1984. Robert Kidston, this most professional palaeobotanist. A tribute on the 60th anniversary of his death. Forth Naturalist and Historian 8:65–93.

Edwards, D. 2004. Embryophytic sporophytes in the Rhynie and Windyfield cherts. Transactions of the Royal Society of Edinburgh: Earth Sciences 94:397–410.

Edwards, D. S. 1973. Studies on the Flora of the Rhynie Chert. PhD dissertation. University College of South Wales, Cardiff.

Edwards, D. S. 1986. *Aglaophyton major*, a non-vascular landplant from the Devonian Rhynie chert. Botanical Journal of the Linnean Society 93:173–204.

Edwards, D. S., and A. G. Lyon. 1983. Algae from the Rhynie chert. Botanical Journal of the Linnean Society 86:37–55.

El-Saadawy, W.. 1966. Studies in the Flora of the Rhynie Chert. PhD dissertation. University College of North Wales, Bangor.

El-Saadawy, W., and W. S. Lacey. 1979a. The sporangia of *Horneophyton lignieri* (Kidston and Lang) Barghoorn and Darrah. Review of Palaeobotany and Palynology 28:137–144.

El-Saadawy, W., and W. S. Lacey. 1979b. Observations on *Nothia aphylla* Lyon ex Høeg. Review of Palaeobotany and Palynology 27:119–147.

Engel, M. S. and Grimaldi, D. A. 2004. New light shed on the oldest insect. Nature 427: 627–630.

Fayers, S. R. 2003. The Palaeoenvironments and Biota of the Windyfield Chert, Early Devonian, Rhynie, Scotland. PhD dissertation. University of Aberdeen, Aberdeen.

Fayers, S. R., and N. H. Trewin. 2003. A new crustacean from the Early Devonian Rhynie chert, Aberdeenshire, Scotland. Transactions of the Royal Society of Edinburgh: Earth Sciences 93:355–382.

Fayers, S. R., and N. H. Trewin. 2004. A review of the palaeoenvironments and biota of the Windyfield chert. Transactions of the Royal Society of Edinburgh: Earth Sciences 94:325–339.

Fayers, S. R., and N. H. Trewin. 2005. A hexapod from the Early Devonian Windyfield chert, Rhynie, Scotland. Palaeontology 48:1117–1130.

Fayers, S. R., J. A. Dunlop, and N. H. Trewin. 2005. A new early Devonian trigonotarbid arachnid from the Windyfield chert, Rhynie, Scotland. Journal of Systematic Palaeontology 2:269–284.

Geikie, A. 1879. On the Old Red Sandstone of Western

Europe. Transactions of the Royal Society of Edinburgh 28: 345–452.

Gould, D. 1997. Geology of the Country around Inverurie and Alford. (Sheets 76E and 76W.) Memoir, British Geological Survey, Keyworth, Nottingham.

Grant Wilson, J. S., and L. W. Hinxman. 1890. Geology of the Area around Central Aberdeenshire Inverurie, Alford, Tarland. (Sheet 76.) Memoir, Geological Survey, Scotland.

Greenslade, P., and P. E. S. Whalley. 1986. The systematic position of *Rhyniella praecursor* Hirst & Maulik (Collembola), the earliest known hexapod. Pp. 319–323 in R. Dallai (ed), Second International Symposium on Apterygota. Siena.

Habgood, K. S. 2000. Integrated Approaches to the Cycling of Primary Produce in Early Terrestrial Ecosystems. PhD dissertation. Cardiff University, Cardiff.

Habgood, K. S., H. Hass, and H. Kerp. 2004. Evidence for an early terrestrial food web: coprolites from the Lower Devonian Rhynie chert. Transactions of the Royal Society of Edinburgh: Earth Sciences 94:371–389.

Harvey, R., A. G. Lyon, and P. N. Lewis. 1969. A fossil fungus from Rhynie chert. Transactions of the British Mycological Society 53:155–156.

Hass, H., and N. J. Rowe. 1999. Thin sections and wafering. Pp. 76–81 in T. P. Jones and N. J. Rowe (eds), Fossil Plants and Spores: Modern Techniques. Geological Society, London.

Hass, H., T. N. Taylor, and W. Remy. 1994. Fungi from the Lower Devonian Rhynie chert: mycoparasitism. American Journal of Botany 81:29–37.

Haug, J. T., C. Haug, A. Maas, S. R. Fayers, N. H. Trewin, and D. Waloszek. 2009. Simple 3D images from fossil and recent micromaterial using light microscopy. Journal of Microscopy 233:93–101.

Haug, C., J. T. Haug, S. R. Fayers, N. H. Trewin, C. Castellani, D. Waloszek, and A. Maas. 2012. Exceptionally preserved nauplius larvae from the Devonian Windyfield chert, Rhynie, Aberdeenshire, Scotland. Palaeontologia Electronica 15(2):24A.

Hawksworth, D. L. 2005. *Palaeopyrenomycites devonicus*, a 400 myr old pyrenomycete. Mycological Research 109:518.

Hirst, S. 1923. On some arachnid remains from the Old Red Sandstone (Rhynie chert Bed, Aberdeenshire). Annals and Magazine of Natural History, Series 9, 12:455–474.

Hirst, S., and S. Maulik. 1926a. On some arthropod remains from the Rhynie chert (Old Red Sandstone). Geological Magazine 63:69–71.

Hirst, S., and S. Maulik. 1926b. [Untitled contribution to 'Announcements and Inquiries' section.] Geological Magazine 63:288.

Honegger, R., D. Edwards, and L. Axe. 2013. The earliest records of internally stratified cyanobacterial and algal lichens from the Lower Devonian of the Welsh Borderland. New Phytologist 197:264–275.

Horne, J., H. M. Skae, D. R. Irvine, J. S. Grant Wilson, and L.W. Hinxman. 1886. Inverurie. Sheet 76. Geological Survey, Scotland.

Horne, J., J. S. Grant Wilson, and L. W. Hinxman. 1923. Huntly. Sheet 86. Geological Survey, Scotland.

Horne, J., W. Mackie, J. S. Flett, W. T. Gordon, G. Hickling, R. Kidston, B. N. Peach, and D. M. S. Watson. 1916. The plant-bearing cherts at Rhynie, Aberdeenshire. Pp. 206–216 in Report of the Eighty-Sixth Meeting of the British Association for the Advancement of Science. John Murray, London.

Jones, B., R. W. Renault, and M. R. Rosen. 2004. Taxonomic fidelity of silicified filamentous microbes from hot-spring systems in the Taupo Volcanic Zone, North Island, New Zealand. Transactions of the Royal Society of Edinburgh: Earth Sciences 94:475–483.

Kamenz, C., J. A. Dunlop, G. Scholtz, H. Kerp, and H. Hass. 2008. Microanatomy of Early Devonian book lungs. Biological Letters 4:212–215.

Karatygin, I.V., N. S. Snigirevskaya, and K. N. Demchenko. 2006. Species of the genus *Glomites* as plant mycobionts in Early Devonian ecosystems. Palaeontological Journal 40:572–579.

Kelman, R., M. Feist, N. H. Trewin, and H. Hass. 2004. Charophyte algae from the Rhynie chert. Transactions of the Royal Society of Edinburgh: Earth Sciences 94: 445–455.

Kerp, H., and B. Bomfleur. 2011. Photography of plant fossils – New techniques, old tricks. Review of Palaeobotany and Palynology 166: 117–151.

Kerp, H., and H. Hass. 2009. Ökologie und Reproduktion der frühen Landpflanzen. Berichte der Reinhold-Tüxen-Gesellschaft 21:111–127.

Kerp, H., H. Hass, and V. Mosbrugger, V. 2001. New data on *Nothia aphylla* Lyon, 1964 ex El Saadawy et Lacey, 1979: a poorly known plant from the Lower Devonian Rhynie chert. Pp. 52–82 in P. G. Gensel and D. Edwards (eds), Plants Invade the Land: Evolutionary and Environmental Perspectives. Columbia University Press, New York.

Kerp, H., N. H. Trewin, and H. Hass. 2004. Rhynie chert gametophytes. Transactions of the Royal Society of Edinburgh: Earth Sciences 94:411–428.

Kerp, H., C. H. Wellman, M. Krings, P. Kearney, and H. Hass. 2013. Reproductive organs and *in situ* spores of *Asteroxylon mackiei* Kidston & Lang, the most complex plant from the Lower Devonian Rhynie Chert. International Journal of Plant Sciences 174: 293–308.

Kidston, R., and W. H. Lang. 1917. On Old Red Sandstone plants showing structure, from the Rhynie chert bed, Aberdeenshire. Part I. *Rhynia gwynne-vaughani* Kidston & Lang. Transactions of the Royal Society of Edinburgh 51:761–784.

Kidston, R., and W. H. Lang. 1920a. On Old Red Sandstone plants showing structure, from the Rhynie chert bed, Aberdeenshire. Part II. Additional notes on *Rhynia gwynne-vaughani* Kidston and Lang; with descriptions of *Rhynia major*, n.sp., and *Hornia lignieri*, n.g., n.sp. Transactions of the Royal Society of Edinburgh 52:603–627.

Kidston, R., and W. H. Lang. 1920b. On Old Red Sandstone plants showing structure, from the Rhynie chert bed, Aberdeenshire. Part III. *Asteroxylon mackiei*, Kidston and Lang. Transactions of the Royal Society of Edinburgh 52:643–680.

Kidston, R., and W. H. Lang, 1921a. On Old Red Sandstone plants showing structure, from the Rhynie chert bed, Aberdeenshire. Part IV. Restorations of the vascular cryptogams, and discussion of their bearing on the general morphology of the pteridophyta and the origin of the organisation of land-plants. Transactions of the Royal Society of Edinburgh

52:831–854.

Kidston, R., and W. H. Lang. 1921b. On Old Red Sandstone plants showing structure, from the Rhynie chert bed, Aberdeenshire. Part V. The Thallophyta occurring in the peat-bed; the succession of the plants throughout a vertical section of the bed, and the conditions of accumulation and preservation of the deposit. Transactions of the Royal Society of Edinburgh 52:855–902.

Krings, M., and T. N. Taylor. 2013. *Zwergimyces vestitus* (Kidston et W. H. Lang) nov. comb., a fungal reproductive unit enveloped in a hyphal mantle from the Lower Devonian Rhynie chert. Review of Palaeobotany and Palynology 190:15–19.

Krings, M., H. Kerp, H. Hass, T. N. Taylor, and N. Dotzler. 2007. A filamentous cyanobacterium showing structured colonial growth from the Early Devonian Rhynie chert. Review of Palaeobotany and Palynology 146:265–276.

Krings, M., T. N. Taylor, H. Hass, H. Kerp, N. Dotzler, and E. J. Hermsen. 2007. Fungal endophytes in a 400-million-yr-old land plant: infection pathways, spatial distribution, and host responses. New Phytology 174:648–657.

Krings, M., N. Dotzler, and T. N. Taylor. 2009. *Globicultrix nugax* nov. gen. et nov. sp. (Chitridiomycota), an intrusive microfungus in fungal spores from the Rhynie chert. Zitteliana A, 48/49: 165–170.

Krings, M., H. Hass, H. Kerp, T. N. Taylor, R. Agerer, and N. Dotzler. 2009. Endophytic cyanobacteria in a 400-million-yr-old land plant: a scenario for the origin of a symbiosis? Review of Palaeobotany and Palynology 153:62–69.

Krings, M., N. Dotzler, J. E. Longcore, and T. N. Taylor. 2010. An unusual microfungus in a fungal spore from the Lower Devonian Rhynie Chert. Palaeontology 53:753–759.

Krings, M., T. N. Taylor, E. L. Taylor, H. Kerp, H. Hass, N. Dotzler, and C. J. Harper. 2012. Microfossils from the Lower Devonian Rhynie chert with suggested affinities to the Peronosporomycetes. Journal of Palaeontology 86:358–367.

Krings, M., T. N. Taylor, N. Dotzler, and C. J. Harper. 2013. *Frankbaronia velata* nov. sp., a putative peronosporomycete oogonium containing multiple oospores from the Lower Devonian Rhynie chert. Zitteliana A 53:23–30.

Lang, W. H. 1925. Robert Kidston 1852–1924. (Obituary notices of Fellows Deceased.) Proceedings of the Royal Society of London B 98:XIV–XXII.

Lyon, A. G. 1957. Germinating spores in the Rhynie chert. Nature 180:1219.

Lyon, A. G. 1962. On the fragmentary remains of an organism referable to the nematophytales, from the Rhynie chert, *Nematoplexus rhyniensis* gen. et sp. nov. Transactions of the Royal Society of Edinburgh 65:79–87.

Lyon, A. G. 1964. Probable fertile region of *Asteroxylon mackiei* K. and L. Nature 203: 1082–1083.

Lyon, A. G., and D. Edwards. 1991. The first zosterophyll from the Lower Devonian Rhynie chert, Aberdeenshire. Transactions of the Royal Society of Edinburgh: Earth Sciences 82:323–332.

Mackie, W. 1913. The rock series of Craigbeg and Ord Hill, Rhynie, Aberdeenshire. Transactions of the Edinburgh Geological Society 10:205–236.

Mark, D. F., C. M. Rice, A. E. Fallick, N. H. Trewin, M. R. Lee, A. Boyce, and J. K. W. Lee. 2011. ^{40}Ar/^{39}Ar dating of hydrothermal activity, biota and gold mineralisation in the Rhynie hot-spring system, Aberdeenshire, Scotland. Geochimica et Cosmochimica Acta 75:555–569.

Mark, D. F., C. M. Rice, and N. H. Trewin. 2013. Discussion on 'A high-precision U-Pb age constraint on the Rhynie Chert Konservat-Lagerstätte: time scale and other implications.' Journal of the Geological Society 168:863–872.

Miller, H. 1857. The Testimony of the Rocks: Or, Geology in Its Bearings on the Two Theologies. Shepherd and Elliot, Edinburgh, 502 pp.

Newlands, M. A. 1913. Note on a fossil plant from the Old Red Sandstone, Rhynie, Aberdeenshire. Transactions of the Edinburgh Geological Society 10 (Session 78, 1911–12):237.

Parry, S. F., S. R. Noble, Q. G. Crowley, and C. H. Wellman. 2011. A high precision U/Pb age constraint on the Rhynie Chert Konservat-Lagerstätte: time scale and other implications. Journal of the Geological Society, London 168:863–872.

Pia, J. 1927. Die Erhaltung der fossilen Pflanzen. Pp. 31–136 in M. Hirmer (ed), Handbuch der Paläobotanik. R. Oldenbourg Verlag, Munich.

Poinar, G. Jr., H. Kerp, and H. Hass. 2008. *Palaeonema phyticum* gen. n., sp. n. (Nematoda: Palaconematidae fam.n.), a Devonian nematode associated with early land plants. Nematology 10:9–14.

Powell, C. L. 1994. The Palaeoenvironments of the Rhynie Cherts. PhD dissertation. University of Aberdeen, Aberdeen.

Powell, C. L., D. Edwards, and N. H. Trewin. 2000a. Palaeoecology and plant succession in a borehole through the Rhynie cherts, Lower Old Red Sandstone, Scotland. Pp. 439–457 in P. F. Friend and B. P. J. Williams (eds), New Perspectives on the Old Red Sandstone. Geological Society of London Special Publication 180.

Powell, C. L., N. H. Trewin, and D. Edwards. 2000b. A new vascular plant from the Lower Devonian Windyfield chert, Rhynie, NE Scotland. Transactions of the Royal Society of Edinburgh: Earth Sciences 90:331–349.

Read, H. H. 1923. The Geology of the Country around Banff, Huntly and Turriff. (Sheets 86 and 96.) Memoir of the Geological Survey, Scotland.

Remy, W., and H. Hass. 1991a. Ergänzende Beobachtungen an *Lyonophyton rhyniensis*. Argumenta Palaeobotanica 8:1–27.

Remy, W., and H. Hass. 1991b. *Langiophyton mackiei* nov. gen., nov. spec., ein Gametophyt mit Archegoniophoren aus dem Chert von Rhynie (Unterdevon Schottland). Argumenta Palaeobotanica 8:69–117.

Remy, W., and H. Hass. 1991c. *Kidstonophyton discoides* nov. gen. nov. spec., ein Gametophyt aus dem Chert von Rhynie (Unterdevon, Schottland). Argumenta Palaeobotanica 8:29–45.

Remy, W., and R. Remy. 1980a. Devonian gametophytes with anatomically preserved gametangia. Science 208:295–296.

Remy, W., and R. Remy. 1980b. *Lyonophyton rhyniensis* n.gen. et nov. spec., ein Gametophyt aus dem Chert von Rhynie (Unterdevon, Schottland). Argumenta Palaeobotanica 6:37–72.

Remy, W., T. N. Taylor, and H. Hass. 1994. Early Devonian fungi: a blastocladalean fungus with sexual reproduction. American Journal of Botany 81:690–702.

Rice, C. M., and W. A. Ashcroft. 2004. Geology of the

northern half of the Rhynie Basin, Aberdeenshire, Scotland. Transactions of the Royal Society of Edinburgh: Earth Sciences 94:299–308.

Rice, C. M., and N. H. Trewin. 1988. A Lower Devonian gold-bearing hot-spring system, Rhynie, Scotland. Transactions of the Institute of Mineralogy and Metallurgy 97:141–144.

Rice, C. M., W. A. Ashcroft, D. J. Batten, A. J. Boyce, J. B. D. Caulfield, A. E. Fallick, M. J. Hole, E. Jones, M. J. Pearson, G. Rogers, J. M. Saxton, F. M., Stuart, N. H. Trewin, and G. Turner. 1995. A Devonian auriferous hot spring system, Rhynie, Scotland. Journal of the Geological Society, London 152:229–250.

Rice, C. M., N. H. Trewin, and L. I. Anderson. 2002. Geological setting of the Early Devonian Rhynie cherts, Aberdeenshire, Scotland: an early terrestrial hot spring system. Journal of the Geological Society, London 159:203–214.

Richardson, J. B., and D. C. McGregor. 1986. Silurian and Devonian spore zones of the Old Red Sandstone continent and adjacent regions. Geological Survey of Canada Bulletin 364:1–96.

Rubenstein, C. V., P. Gerrienne, G. S. de la Puente, R. A. Astini, and P. Steemans. 2010. Early Middle Ordovician evidence for land plants in Argentina (eastern Gondwana). New Phytologist 188:365–369.

Scott, D. H. 1920. Studies in Fossil Botany. Third Edition. Two volumes. A & C Black, London.

Scourfield, D. J. 1926. On a new type of crustacean from the Old Red Sandstone (Rhynie Chert Bed, Aberdeenshire) – *Lepidocaris rhyniensis*, gen. et sp. nov. Philosophical Transactions of the Royal Society of London B 214:153–187.

Scourfield, D. J. 1940a. Two new and nearly complete specimens of young stages of the Devonian fossil crustacean *Lepidocaris rhyniensis*. Proceedings of the Linnean Society 152:290–298.

Scourfield, D. J. 1940b. The oldest known fossil insect. Nature 145:799–801.

Scourfield, D. J. 1940c. The oldest known fossil insect (*Rhyniella praecursor* Hirst & Maulik) Further details from additional specimens. Proceedings of the Linnean Society 152:113–131.

Selden, P. A., W. A. Shear, and P. M. Bonamo. 1991. A spider and other arachnids from the Devonian of New York, and reinterpretations of Devonian Araneae. Palaeontology 34:241–281.

Seward, A. C. 1931. Plant Life Through the Ages. Cambridge University Press, Cambridge, 601 pp.

Shear, W. A., P. A. Selden, W. D. I. Rolfe, P. M. Bonamo, and J. D. Grierson. 1987. New terrestrial arachnids from the Devonian of Gilboa, New York (Arachnida, Trigonotarbida). American Museum Novitates 2901:1–74.

Shear, W. A., A. Jeram, and P. A. Selden. 1998. Centipede legs (Arthopoda, Chilopoda, Scutigeromorpha) from the Silurian and Devonian of Britain and the Devonian of North America. American Museum Novitates 3231:1–16.

Streel, M., K. Higgs, S. Loboziak, W. Riegel, and P. Steemans. 1987. Spore stratigraphy and correlation with faunas and floras in the type marine Devonian of the Ardenno-Rhenish regions. Review of Palaeobotany and Palynology 50:211–229.

Strullu-Derrien, C., P. Kenrick, P. Tafforeau, H. Cochard, J-L. Bonnemain, A. Le Hérissé, H. Lardeux, and E. Badel. 2014. The earliest wood and its hydraulic properties documented in c. 407-million-year-old fossils using synchrotron microtomography. Botanical Journal of the Linnean Society 175:423–437.

Tasch, P. 1957. Flora and fauna of the Rhynie chert: a paleo-ecological reevaluation of published evidence. University of Wichita Bulletin 36:1–24.

Taylor, T. N., and M. Krings. 2005. Fossil microorganisms and land plants: associations and interactions. Symbiosis 40:119–135.

Taylor, T. N., H. Hass, and W. Remy. 1992a. Devonian fungi: interactions with the green alga *Palaeonitella*. Mycologia 84:901–910.

Taylor, T. N., W. Remy, and H. Hass. 1992b. Parasitism in a 400 million-year-old green alga. Nature 357:493–494.

Taylor, T. N., W. Remy, H. Hass, and H. Kerp. 1995. Fossil arbuscular mycorrhizae from the Early Devonian. Mycologia 87:560–573.

Taylor, T. N., H. Hass, and H. Kerp. 1997. A cyanolichen from the Lower Devonian Rhynie chert. American Journal of Botany 84:992–1004.

Taylor, T. N., H. Hass, and H. Kerp. 1999. The oldest fossil ascomycetes. Nature 399:648.

Taylor, T. N., M. Krings, and H. Kerp. 2006. Hassiella monospora gen. et sp. nov. A microfungus from the 400 million year old Rhynie chert. Mycological Research 10:628–632.

Taylor, T. N., H. Hass, M. Krings, S. D. Klavins, and H. Kerp. 2004. Fungi in the Rhynie chert: a view from the dark side. Transactions of the Royal Society of Edinburgh: Earth Sciences 94:457–473.

Taylor, T. N., H. Hass, H. Kerp, M. Krings, and R. T. Hanlin. 2005. Perithecial ascomycetes from the 400 million-year-old Rhynie Chert: an example of ancestral polymorphism. Mycologia 97, 269–285.

Tillyard, R. J. 1928. Some remarks on the Devonian fossil insects from the Rhynie chert beds, Old Red Sandstone. Transactions of the Entomological Society of London 76:65–71.

Trewin, N. H. 1994. Depositional environment and preservation of biota in the Lower Devonian hot-springs of Rhynie, Aberdeenshire, Scotland. Transactions of the Royal Society of Edinburgh: Earth Sciences 84:433–442.

Trewin, N. H. 1996. The Rhynie cherts: an Early Devonian ecosystem preserved by hydrothermal activity. Pp. 131–149 in G. R. Brock and J. A. Goode (eds), Evolution of Hydrothermal Ecosystems on Earth (and Mars?). (Ciba Foundation Symposium 202.) Wiley, Chichester.

Trewin, N. H. 2004. History of research on the geology and palaeontology of the Rhynie area, Aberdeenshire, Scotland. Transactions of the Royal Society of Edinburgh: Earth Sciences 94:285–297.

Trewin, N. H., and C. M. Rice. 1992. Stratigraphy and sedimentology of the Devonian Rhynie chert locality. Scottish Journal of Geology 28:37–47.

Trewin, N. H., and R. G. Davidson. 1996. A Lower Devonian lake and its associated biota in the Midland Valley of Scotland. Transactions of the Royal Society of Edinburgh: Earth Sciences 86:233–246.

Trewin, N. H., and C. M. Rice (eds). 2004. The Rhynie hot-spring system: geology, biota and mineralisation. Transactions of the Royal Society of Edinburgh: Earth Sciences 86:283–521.

Trewin, N. H., and E. Wilson. 2004. Correlation of the Early Devonian Rhynie chert beds between three boreholes at Rhynie, Aberdeenshire, Scotland. Scottish Journal of Geology 40:73–81.

Trewin, N. H. and S. R. Fayers. 2016. Macro to micro aspects of the plant preservation in the Early Devonian Rhynie cherts, Aberdeenshire, Scotland. Earth and Environmental Sciences Transactions of the Royal Society of Edinburgh, 106: 67–80.

Trewin, N. H., S. R. Fayers, and R. Kelman. 2003. Subaqueous silicification of the contents of small ponds in an Early Devonian hot-spring complex, Rhynie, Scotland. Canadian Journal of Earth Sciences 40:1697–1712.

Wellman, C. H. 2004. Palaeoecology and palaeophytogeography of the Rhynie chert plants: evidence from integrated analysis of *in situ* and dispersed spores. Proceedings of the Royal Society of London B 271:985–992.

Wellman, C. H. 2006. Spore assemblages from the Lower Devonian 'Lower Old Red Sandstone' of the Rhynie outlier, Scotland. Transactions of the Royal Society of Edinburgh: Earth Sciences 97:167–211.

Whalley, P., and E. A. Jarzembowski. 1981. A new assessment of *Rhyniella*, the earliest known insect, from the Devonian of Rhynie, Scotland. Nature 291:317.

Tables

Table 1: Embryophytes from the Early Devonian (Pragian) Rhynie cherts.

Sporophyte	Gametophyte		Selected references
Rhynia gwynne-vaughanii	*Remyophyton delicatum*	♀ ♂	Kidston & Lang, 1917, 1920a; Edwards, 1986; Kerp *et al.*, 2004
Aglaophyton major	*Lyonophyton rhyniensis*	♀ ♂	Kidston & Lang, 1920a; Edwards, 1986; Remy & Hass, 1991a
Horneophyton lignieri	*Langiophyton mackiei*	♀ ♂	Kidston & Lang, 1920a; El-Saadawy & Lacey, 1979a; Remy & Hass, 1991b
Nothia aphylla	*Kidstonophyton discoides*	♂	Lyon, 1964; El-Saadawy & Lacey, 1979b; Remy & Hass, 1991c; Kerp *et al.*, 2001
Trichopherophyton teuchansii		?	Lyon & Edwards, 1991
Ventarura lyoni		?	Powell *et al.*, 2000b
Asteroxylon mackiei		?	Kidston & Lang, 1920b; Lyon, 1964; Kerp *et al.*, 2013

Note:
Plant spores specifically from the chert are not listed here. See Wellman (2004, 2006) for details of plant spores from the Rhynie succession.

Table 2: Fungi, bacteria and algae from the Early Devonian (Pragian) Rhynie cherts.

Flora		Selected references
Fungi		
Zygomycota	[1]*Palaeomyces agglomerata*	Kidston & Lang, 1921b
	[1]*Palaeomyces asteroxylii*	Kidston & Lang, 1921b
	[2]*Palaeomyces gordonii*	Kidston & Lang, 1921b
	[1]*Palaeomyces gordonii* var. *major*	Kidston & Lang, 1921b
	[1]*Palaeomyces horneae*	Kidston & Lang, 1921b
	[1]*Palaeomyces simpsonii*	Kidston & Lang, 1921b
Zygomycota or Glomeromycota	*Zwergimyces vestitus*	Kidston & Lang, 1921b (as *Palaeomyces vestita*); Krings & Taylor, 2013
Mucromycotina	*Palaeoendogone gwynne-vaughaniae*	Strullu-Derrien & Strullu, 2014 (in Strullu-Derrien et al., 2014)
Ascomycota	Unnamed perithecal ascomycete	Taylor et al., 1999
	Palaeopyrenomycites devonicus	Taylor et al., 2005
Chytridiomycota	*Krispiromyces discoides*	Taylor et al., 1992a
	Lyonomyces pyriformis	Taylor et al., 1992a
	Milleromyces rhyniensis	Taylor et al., 1992a
	Palaeoblastocladia milleri	Remy et al., 1994
	Globicultrix nugax	Krings, Dotzler et al., 2009
	Unnamed eucarpic and holocarpic forms	Taylor et al., 1992b; Hass et al. 1994
Glomeromycota	*Scutellosporites devonicus*	Dotzler et al., 2006
	Glomites rhyniensis	Taylor et al., 1995; Krings et al. 2007a,b
	Glomites sporocarpoides	Karatygin et al., 2006
	Palaeoglomus boullardii	Strullu-Derrien & Strullu, 2014 (in Strullu-Derrien et al., 2014)
	Unnamed acaulosporoid glomeromycotean spores	Dotzler et al., 2009
'Oomycota'	Unnamed mycelia	Harvey et al., 1969; Taylor et al., 2004
	Hassiella monospora	Taylor et al., 2006
	Frankbaronia polyspora	Krings et al., 2012
	Frankbaronia velata	Krings et al., 2013
Unclassified microfungus	*Kryphiomyces catenulatus*	Krings et al., 2010
Lichen	*Winfrenatia reticulata*	Taylor et al., 1997, also see Honegger et al., 2013
Cyanobacteria	*Archaeothrix contexta*	Kidston & Lang, 1921b
	Archaeothrix oscillatoriformis	Kidston & Lang, 1921b
	Kidstoniella fritschii	Croft & George, 1959
	Langiella scourfieldii	Croft & George, 1959
	Rhyniella vermiformis	Croft & George, 1959

		Rhyniococcus uniformis	Edwards & Lyon, 1983
		Croftalania venusta	Krings et al., 2007a
Bacteria (unicellular)		Schizophyta	Kidston & Lang, 1921b
Algae *s.l.*		Palaeonitella cranii	Kidston & Lang, 1921b; Pia, 1927; Edwards & Lyon, 1983; Kelman et al., 2004
		Mackiella rotunda	Edwards & Lyon, 1983
		Rhynchertia punctata	Edwards & Lyon, 1983
		Cymatiosphaera sp.	Dotzler et al., 2007
Nematophytes		Nematophyton taitii	Kidston & Lang, 1921b
		Nematoplexus rhyniensis	Lyon, 1962

Notes:
[1] The species of *Palaeomyces* described by Kidston & Lang (1921b) are based on hyphae and cannot be classified in the absence of reproductive structures, except for *Palaeomyces vestita* which has been reclassified as *Zwergimyces vestitus* (Krings and Taylor, 2013).
[2] *Palaeomyces gordonii* might be conspecific with *Scutellosporites devonicus* (Dotzler et al., 2006).

Table 3: Fauna and coprolites from the Early Devonian (Pragian) Rhynie cherts.

Fauna			**Selected references**
Arthropoda			
Arachnida			
	Opilionida	Eophalangium sheari	Dunlop et al., 2003, 2004
	Trigonotarbida	Palaeocharinus rhyniensis	
		(Palaeocharinus scourfieldi)	
		(Palaeocharinus calmani)	Hirst, 1923, Shear et al., 1987; Dunlop, 1994, 1996
		(Palaeocharinus kidstoni)	
		Palaeocharinus hornei	
		Palaeocharinus tuberculatus	Fayers et al., 2005
		Large, unnamed trigonotarbid	Fayers et al., 2005
	Araneae	[1](Palaeocteniza crassipes)	Hirst, 1923; Selden et al. 1991
	[2]Acari	Protacarus crani	
		Protospeleorchestes pseudoprotacarus	
		Pseudoprotacarus scoticus	Hirst, 1923; Dubinin, 1962
		Paraprotacarus hirsti	
		Palaeotydeus devonicus	
	Arachnida *incertae sedis*	Saccogulus seldeni	Dunlop et al., 2006
Myriapoda			
Diplopoda			
	Eoarthropleurida	Eoarthropleura sp.	Fayers & Trewin, 2004

Chilopoda		
Scutigeromorpha	*Crussolum* sp.	Shear *et al.*, 1998; Anderson & Trewin, 2003
	Unnamed scutigeromorph	Fayers & Trewin, 2004
Chilopoda *incertae sedis*	Unnamed centipede	Fayers & Trewin, 2004
Hexapoda		
Collembolla	*Rhyniella praecursor*	Hirst & Maulik, 1926a; Tillyard 1928; Scourfield, 1940b, c; Whalley & Jarzembowski, 1981; Greenslade & Whalley, 1986
Pterygota	*Rhyniognatha hirsti*	Hirst & Maulik, 1926a; Tillyard 1928; Engel & Grimaldi, 2004
Hexapoda *incertae sedis*	*Leverhulmia mariae*	Anderson & Trewin, 2003; Fayers & Trewin, 2004
Branchiopoda		
Lipostraca	*Lepidocaris rhyniensis*	Scourfield, 1926, 1940a
Calmanostraca *incertae sedis*	*Castracollis wilsonae*	Fayers & Trewin, 2003
Diplostraca	*Ebullitiocaris oviformis*	Anderson *et al.*, 2004
Eucrustacea	Nauplius larvae	Haug *et al.*, 2012
Arthropoda *incertae sedis*		
Euthycarcinida	*Heterocrania rhyniensis*	Hirst & Maulik 1926a, b; Anderson & Trewin 2003
Arthropoda *incertae sedis*	*Rhynimonstrum dunlopi*	Anderson & Trewin 2003
'Worms'		
Nematoda	*Palaeonema phyticum*	Poinar *et al.*, 2008
Polychaete?	Elements of jaw apparatus (undescribed)	
Coprolites		
	Lancifaex simplex	Habgood *et al.*, 2004
	Lanifaex divisa	Habgood *et al.*, 2004
	Lancifaex moniliforma	Habgood *et al.*, 2004
	Rotundafaex aggregata	Habgood *et al.*, 2004
	Bacillafaex constipata	Habgood *et al.*, 2004
	Bacillafaex mina	Habgood *et al.*, 2004

Notes.
[1] This animal is now considered a juvenile trigonotarbid (Selden *et al.*, 1991).
[2] The taxa of mites erected by Dubinin (1962) are in need of revision.
  Names in brackets are no longer considered valid species.

Chapter 2

The East Kirkton Lagerstätte: a window onto early Carboniferous land ecosystems

Jennifer A. Clack
University Museum of Zoology, Cambridge, UK.

Abstract

The late Viséan (early Brigantian) locality of East Kirkton near Bathgate in Scotland is noted for its wealth of fossils, where the unusual environment of the East Kirkton Limestone facilitated unusual preservation of animals and plants. These include the earliest known terrestrial fauna, comprising six new genera and seven new tetrapods species: *Balanerpeton*, *Eldeceeon*, *Eucritta*, *Kirktonecta*, *Silvanerpeton*, *Westlothiana*, and *Ophiderpeton kirktonense*. Among them are stem members of the crown groups of tetrapods; amphibians and amniotes. Actinopterygians, and a few acanthodians and chondrichthyans are restricted to the topmost horizons of the sequence. Seven groups of invertebrates are represented, including a new species of a terrestrial scorpion *Pulmonoscorpius kirktonensis*, the second oldest known opilionid *Brigantobunum*, eurypterids, and myriapods. A rich flora was present among which was a new genus *Stanwoodia*, named for the discoverer of the quarry. Set in a landscape of nearby volcanic activity, the environment represents a shallow lake into which warm or occasionally hot springs fed, but which was usually too mineral-rich to support an aquatic fauna. Volcaniclastic sediments such as tuffs preserve some of the fossils as individual bones, whereas other strata consist of black organic-rich mudstones, which preserve small but almost complete articulated tetrapods. Among the varied and complex lithologies are laminites that are often deformed and folded, and botryoidal or stromatolitic limestones that suggest bacterial or algal growths.

Introduction

When the well-known collector Stan Wood first rediscovered the site of the East Kirkton limestone quarry with its complement of tetrapod specimens, it aroused both astonishment and interest among vertebrate palaeontologists and geologists. The fossils represent the earliest known assemblage of terrestrial vertebrates anywhere in the world (Wood *et al.*, 1985; Milner *et al.*, 1986).

The first specimens to be described from this locality were 'Scouler's heids', later identified as the head plates of the large eurypterid *Hibbertopterus scouleri*. The quarry had been worked in the early nineteenth century, with its rocks used for local walls and farm buildings, but was eventually closed around 1844. It became overgrown with trees and a local dump for rubbish from the nearby housing estate that had grown up close by it. Although the existence of the quarry was well known to geologists (Muir and Walton, 1957), its palaeontological potential and significance remained unrealized.

Stan Wood's tenacity, perspicacity and gift for 'seeing' fossils were responsible for its rediscovery. Combining information from the matrix type, geological map data, and recorded localities of 'Scouler's heids' among the collections of the National Museum of Scotland, he first identified the likely rock-type – a highly distinctive one – forming local farm walls. At first he considered fish the most likely vertebrate finds from the East Kirkton Limestone, but one of the first-fruits turned out to be a tetrapod jaw. The discovery of a tetrapod skull was announced informally in 1984, followed by finds of several partially articulated skeletons in the walls, sometimes even as coping stones. Eventually, Stan tracked down the quarry itself, and began to explore the spoil heaps left after the quarry closed, assisted by Dr Tim Smithson. Not only tetrapods were recovered, but scorpions, a harvestman, and diverse fossil plants, as well as more eurypterids (Wood et al., 1985). Arrangements were later made to clear the quarry and excavate it systematically bed by bed, in collaboration with West Lothian District Council, the National Museum of Scotland (NMS) and the Nature Conservancy Council.

The NMS organized a research meeting in 1992, with an international team of more than 40 workers (Fig 1), including a field trip to the locality. The results were published as a volume of the Transactions of the Royal Society of Edinburgh entitled *Volcanism and early terrestrial biotas* in 1994 (Rolfe et al., 1994a). The volume included descriptions of the tetrapods, scorpions, eurypterids, myriapods, fishes and plants and as well as analyses of the sedimentology, geochemistry, palaeoenviroment, and comparisons with other sites of similar age or depositional circumstances (see also Milner et al., 1986; Clack, 2012). A detailed sedimentary log of the sequence showing the distribution of the flora and fauna was included in that volume, as well as a history of collecting there (Rolfe et al., 1994b). The reader will find additional aspects of the site in the 1994 volume that are not referred to here.

Since then, further material has been collected by Stan Wood and the NMS, and other, more detailed descriptions of tetrapods have been published. The geology and fossils from East Kirkton have formed the basis of several doctoral theses and many papers.

This chapter figures many tetrapod specimens for the first time, and introduces not only the seven previously named and described taxa but also some of those specimens that hint at a (literally) bigger picture. Many of the isolated elements clearly belong to substantially larger animals in contrast to the small size of the articulated skeletons. Also included here are specimen photographs introducing some of the fish, invertebrate and plant components of the biota.

Figure 1 The group of researchers who took part in the conference on East Kirkton in 1992. Stan Wood is standing at the extreme right of the picture. The picture was taken by M. J. Taylor in the rooms of the Royal Society of Edinburgh.

East Kirkton as a unique locality

Several factors make East Kirkton a particularly interesting and important site, two of which are closely interrelated. The preservation of the flora and fauna is highly unusual, resulting from volcanic activity close to a spring (or springs) that fed a lake. The waters of the lake were contaminated with high levels of trace elements inimical to most forms of life – strontium, magnesium, iron, manganese, and sulphur – as a consequence of the volcanism. They were also at different times warm or hot, and the tetrapods at least appear not to have lived in the water. This has meant, first, that the animals living around the lake, the terrestrial forms, were more likely than aquatic forms to be preserved when ash flows streamed down the volcanoes following floods. The ash deposits were washed down the slopes and swept up the tetrapod remains as isolated elements, or whole animals died and were buried in an instant. So, secondly, there have been a disproportionate number of almost complete articulated skeletons – many of them less than 250 mm in snout–vent length – found in the East Kirkton rocks, as compared with the more typical aquatic localities for Carboniferous tetrapods that tend to preserve large forms.

Probably as a consequence of how they were preserved, some of the bone material appears to have been 'cooked', so that although the articulated skeletons are informative and often visually attractive, details of the bones are often very difficult to discern. One suggestion for the preservation of these articulated skeletons derives from the predominance of algal remains in many of the horizons. It has been suggested that these algae formed thick mats over the water surface near the margins of the lake, and that unwary tetrapods might have strayed onto them, falling through into the warm or hot water as the mats thinned out further into the lake. Another scenario is that animals were driven into the lake by frequent forest fires. It is also possible that out-gassing of carbon dioxide asphyxiated animals where they lived, in a situation paralleled today, for example, in the Virunga National Park in the Ruwenzori Mountains of the Congo.

Another unusual facet of the site is that, as a result of the systematic excavation through the sequence, changes in the fauna, flora and environment can be broadly tracked through time. These factors will be treated in more detail below.

Geology of the East Kirkton Limestone

The systematic excavation undertaken by the NMS team cut a trench through the entire sequence in the quarry, from which a detailed lithostratigraphic table was drawn up (Fig 2). Rather than being designated as 'beds', the individual horizons – often only

Figure 2 The cleaned section through the East Kirkton Limestone with some of the units numbered as in the NMS log. The standing figure is the author's husband, Rob Clack.

a few millimetres thick – were numbered as 'units', each unit defined as a layer that separates easily from those above and below. Eighty-eight units were recognized (Rolfe et al., 1994b) numbered 1–88 from top to bottom. The entire exposure was not more than about 225 m thick, although not all of it was readily accessible. A series of borehole cores was also taken in the surrounding area. Together with previous studies the cores allowed the following descriptions and conclusions.

The East Kirkton Limestone forms the major part of a sequence of rocks exposed in a small quarry northeast of the town of Bathgate, West Lothian, Scotland (Fig 2). As far as can be judged from borehole cores, it is of limited extent and more or less confined to the quarry. Some 11 m of this laminated limestone are overlain by 2 m of blue-grey shale known as the Little Cliff Shale and 4 m of bedded tuffs, the Geikie Tuff. The sequence is capped by a basaltic lava flow (Clarkson et al., 1994; Smith et al., 1994; Goodacre, 1998). These rocks were laid down during the Brigantian stage (Cameron et al., 1988) of the Viséan as the lowest limestone of the Bathgate Hills Volcanic Formation, and represent the only non-marine sedimentary succession within this formation (Rolfe et al., 1994b; Goodacre, 1998). At that time there was extensive volcanic activity in what is now the Midland Valley of Scotland, lasting through most of the early Carboniferous. The area shows a very early example of intraplate alkaline basaltic magmatism (Clarkson et al., 1994).

One of the striking things about the East Kirkton Limestone is the variety of sediments comprising it, and the unusual lithologies that are found there (Fig 3). Laminites, botryoidal limestones, stromatolitic limestones, shales and tuffs are all present.

The laminites usually consist of carbonate and organic-rich couplets that may have been laid down seasonally. They show smooth to crenulated bedding often disrupted by slumping and often show brittle or plastic deformation (Goodacre, 1998). There are abundant radial fibrous spherules in some of the limestones, and silica laminae (cherts) are also present throughout the section. The dark organic-rich laminae include particles of cyanophyte and chlorophyte algae, as well as bacteria and possibly faecal pellets. Stromatolitic or botryoidal accretions often occur around organic material, such as plant stems, bone, ostracods, filamentous algae, unicellular remains, and other organic debris, although some may not have had such a particulate origin (Goodacre, 1998). Calcites seem to have been precipitated as a result of cyanobacterial photosynthesis. When the calcite came out of suspension, it seems to have formed blankets that suffocated the bacteria and buried the mats they formed. Manganese, strontium, magnesium, sulphur and iron found at high levels in the calcites are not typical of freshwater deposits, but suggest anaerobic conditions in the water (Walkden et al., 1994).

The siliceous cherty layers may contain spherules, and are often discontinuous. They may form thick or thin laminae and can show faulting and folding at small scales, but, because of their small grain size, can preserve organic remains in very fine detail. Siliceous laminae, although found throughout the sequence, are most common towards the base, especially in units 77–80.

Black mudstones, such as Unit 82, suggest that, at least at an early stage in its history, the lake was deep enough to maintain an anoxic bottom layer. It is here, and in other layers near the base of the sequence, that most of the articulated tetrapods have been found. Again, these black shales occur more commonly in the lower parts of the sequence.

Tuff deposits in the lake were water-lain rather than deposited by air-fall as was once imagined. They resulted from secondary pyroclastic flows that transported volcanogenic material into the lake (Durant, 1994). Volcanism is often associated with increased rainfall, which washed the tuff material into the lake. As it was washed down from the nearby volcanoes, it swept up wood fragments, dead leaves, eurypterid fragments and tetrapod bones, especially in the lower parts of the sequence. Here they often became coated with stromatolitic carbonate. The enriched mineral content of the waters of the lake was at one time considered the direct result of volcanic deposition. However, it appears that instead the minerals were leached out of the surrounding tuff deposits. The distribution of the fauna suggests that, at least for part of its history, the lake water was unsuitable for most organisms, except for ostracods, algae and bacteria.

Since the deposits were first found, there have been debates about whether or not there was hot spring activity during the history of the lake. Walkden et al. (1994) found that the calcite laminae had CO_2

Figure 3 Some of the unusual lithologies found at East Kirkton. **A)** A stromatolitic concretion around a plant stem. **B)** One of the tuff layers, marked by the pen, between two laminated layers. **C)** Polished specimen showing folded and disrupted layers and a nodular layer. **D)** A polished section through Unit 82. Photograph A courtesy of NMS; photographs B–D by JAC.

and O_2 isotopic signatures that indicate they were precipitated on the floor of a freshwater lake, but that they were not associated with hot spring activity. By contrast, McGill *et al.* (1994) found that the cherts showed an O_2 isotopic signature that corroborated the suggestion of thermal activity, with a temperature of about 60 °C. Work by Goodacre (1998) has confirmed the presence of two limestone mounds that represent hot spring vents, close to which the water could have reached temperatures of between 45 °C and 80 °C. The stromatolitic limestones and spherules around these mounds show an isotopic signature indicating higher temperatures, but by contrast, the laminae, presumably laid down further from the source of heat, suggest a lower temperature, averaging about 35 °C. In conclusion, it appears that the temperature of the water varied, the higher temperatures resulting from intermittent thermal activity. Further away from the springs, and during lulls in their activity, the water temperature would have been lower.

The volcanoes were perhaps no more than a few hundreds of metres high, forming nearby cinder cones that were only very occasionally active during the life of the lake. Nonetheless, the basaltic effluvia that they produced decayed into rich soil conducive to the growth of dense tropical forests, as it does today. The lake itself may have formed in an eroded tuff ring, or possibly a valley dammed by a lava flow. Although Goodacre (1998) found that the stromatolitic formations were comparable to those of Lake Tasek Dayang Bunting in Malaysia, taken as a whole, there seems no close modern analogue to the depositional environment of the East Kirkton Limestone (Clarkson et al., 1994).

The sequence of rocks exposed in the quarry was laid down in this small shallow lake, which was probably no more than a few metres deep and lasted less than a few tens of thousands of years. At the base of the sequence, black shales indicate a deeper lake than later in its life, with an anoxic hypolimnion. This is where most of the articulated tetrapod material is found. Later, the waters may have been shallower, but towards the top of the sequence, it appears that the lake became linked to a larger one known as Lake Cadell (Clarkson et al., 1994), allowing fishes to invade the area.

Tetrapods

Seven taxa of tetrapod have so far been described and named from East Kirkton: *Balanerpeton woodi* Milner and Sequeira, 1994; *Eldeceeon rolfei* Smithson, 1994; *Eucritta melanolimnetes* Clack, 1998; *Kirktonecta milnerae* Clack, 2011; *Ophiderpeton kirktonense* Milner, 1994; *Silvanerpeton miripedes* Clack, 1994; and *Westlothiana lizziae* Smithson and Rolfe, 1990, of which all except *Ophiderpeton* represent genera currently known only from this locality. Numerous isolated skeletal elements, often large, have also been recovered from the site. It is not often possible to associate these with the named taxa, but they will be considered below.

Balanerpeton woodi

By far the majority of tetrapod specimens represent *Balanerpeton*, a member of the temnospondyl amphibians widely considered to have given rise to modern amphibians (e.g. Ruta et al., 2003; Sigurdsen and Bolt, 2009, 2010). There are about 40 specimens referable to this genus. An almost complete skeleton was one of the first tetrapod specimens discovered by Stan Wood, found in a farm wall (Wood et al., 1985). Most of the specimens are found in black shale, spherulitic limestone or tuffs, and, based on the systematic excavation, most of the articulated material appears to come from near the base of the sequence in the black shale unit 82 (Rolfe et al., 1994b). The majority have skull lengths of about 40–50 mm, although one specimen represents a juvenile, with a skull length of only 25 mm (University Museum of Zoology, Cambridge (UMZC) T.1313) (Fig 4B). The skull of *Balanerpeton* shows typical temnospondyl features of enlarged palatal vacuities, a rounded otic notch, large orbits and a skull table pattern with contact between the supratemporal and postparietal bones (Fig 4). Some specimens show evidence of a temnospondyl-like rod-shaped stapes, suggesting the presence of a tympanic ear (Milner and Sequeira, 1994). The parasphenoid plate is short and broad with a small denticle-bearing patch, and the cultriform process is very narrow. The postcranial skeleton is quite robust with well-ossified limbs and a vertebral column with about 22 presacral vertebrae, each with a ventral intercentrum and paired dorsal pleurocentra. The belly region is covered with oval gastralia characterized by concentric growth rings. In the juvenile, the vertebral elements and gastralia are unossified. Therefore it is likely that the majority of specimens represent adults and that after reaching a certain size, they may have grown more slowly.

Silvanerpeton miripedes

In contrast to the large number of specimens of *Balanerpeton*, there are only about eight specimens referred to *Silvanerpeton*, the next commonest genus (Fig 5). Of these, half are known to derive from the base of the rock sequence, the black shale unit 82 (Rolfe et al., 1994b). Others may have been wall or spoil-heap specimens. Described as an anthracosaur by Clack (1994), a recent analysis of this species (Ruta and Clack, 2006) placed it just basal to a clade of embolomeres plus the Viséan genus *Eoherpeton*, usually considered an anthracosaur (henceforth,

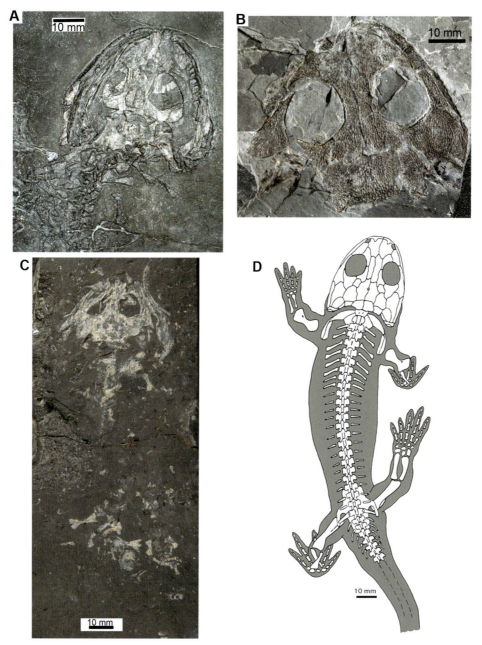

Figure 4 *Balanerpeton woodi*. **A)** Holotype showing palatal view and partial postcranium. Hunterian Museum specimen GLAHM V2051. **B)** UMZC T.1313, the smallest known example. **C)** UMZC T.1312, an isolated skull roof. **D)** Reconstructed skeleton of *Balanerpeton*, based on Milner and Sequeira (1994).

this combination of embolomeres, *Eoherpeton* and *Silvanerpeton* will be referred to by the informal term 'anthracosauroid'). *Silvanerpeton* attained a skull length of up to about 50 mm, with a snout – vent length of at least 215 mm. Features distinguishing this genus from *Balanerpeton* include a shallow 'otic' notch, a tabular horn, tabular–parietal contact, a closed palate and a lozenge-shaped interclavicle.

There are about 32 presacral vertebrae, consisting of a horseshoe-shaped pleurocentrum and a small wedge-shaped intercentrum. The distinctive hind foot shows an elongated fifth digit with five phalanges, a feature shared with other anthracosauroids, and each toe ends in a small triangular expansion. Gastralia are elongate with one end spatulate and the other more pointed, with a curved strengthening

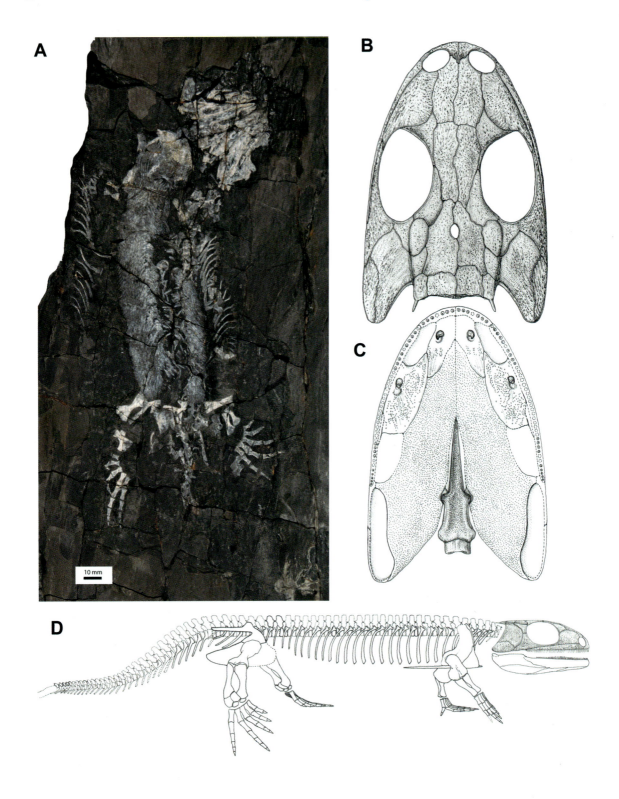

Figure 5 *Silvanerpeton miripedes*. **A)** Holotype UMZC T.1317. **B)** Skull roof reconstruction. **C)** Palatal reconstruction. **D)** Reconstructed skeleton. B–C from Ruta and Clack (2006).

ridge on the internal surface. Ruta and Clack (2006) suggested that *Silvanerpeton* was a predominantly terrestrial form whose relatively large feet, suggesting a hind-limb-dominated mode of locomotion, and whose tail was not obviously adapted for swimming.

Eldeceeon rolfei

Eldeceeon is another genus initially considered anthracosaur-like. However, it has yet to be included in a cladistic analysis, largely on account of the poor preservation of the cranial specimens available at the time. Only two specimens were known in 1992, the holotype from a wall and a second from Unit 76. These were the basis of the original description, although at least two more have been identified and are currently undergoing further study (Ruta and Clack, in progress) (Fig 6). Although similar in size to *Silvanerpeton*, *Eldeceeon* can be distinguished from that taxon on a number of characters (Smithson, 1994; Clack, 1994). It has a much shorter vertebral column of 24–26 presacral elements that are much more fully ossified than in *Silvanerpeton*, a rib cage in which the lumbar ribs are very short and poorly ossified or absent, an interclavicle with a long, narrow stem, a longer femur relative to head length and the absence of a tabular horn. Gastralia are more elongate and parallel-sided than in *Silvanerpeton*. For its body size, it has a very large and robust hind limb and foot with much more strongly ossified tarsal bones than in *Silvanerpeton*.

Eucritta melanolimnetes

Eucritta was the first new tetrapod taxon to be discovered, named, and described in the years following the 1994 volume (Clack, 1998, 2001). There are five known specimens, although the fifth was not included in the descriptions because it was acquired only subsequently from Stan Wood (Fig 7). Four of the specimens are from Unit 82, but the smallest, and the first to be recognized, is from a spoil heap.

The skull roof is remarkably similar to that of *Balanerpeton*, and isolated examples of the two can be hard to tell apart. There are differences in the crescentic shape of the postorbital, the anteriorly expanded frontals and the lack of a premaxillary process entering the almost square nasals of *Eucritta*. However, the palate is closed, unlike that of *Balanerpeton*, and the parasphenoid is narrowly wedge-shaped and denticulated along its length. Those palatal features are essentially primitive for tetrapods as a whole and are also found in *Silvanerpeton* and *Eldeceeon*. Features of the postcranial skeleton that also resemble those in the latter taxa include an ilium with both a dorsal process and an iliac blade, and it has a lozenge-shaped interclavicle like that of *Silvanerpeton* and embolomeres. The orbit has a distinct anteroventral extension absent in *Balanerpeton* and which appears to be unique to the genus. This mixture of primitive and derived features was hard to interpret, but cladistic analyses have usually resolved *Eucritta* as a basal baphetid (e.g. Clack, 1998, 2001; Ruta *et al.*, 2003; Milner *et al.*, 2009). The orbital extension could be seen as ancestral to the

Figure 6 *Eldeceeon rolfei*. **A)** Holotype NMS. G. 1986.39.1. **B)** NMS G. 1990.7 **C)** UMZC 2013.3.

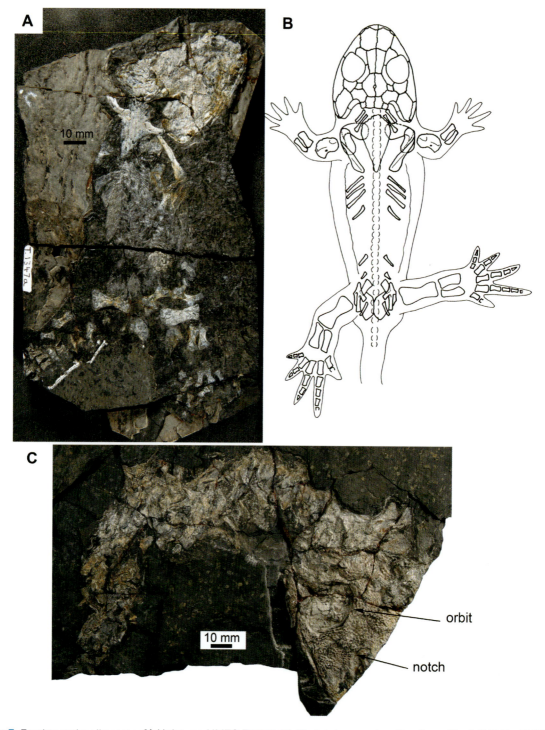

Figure 7 *Eucritta melanolimnetes*. **A)** Holotype UMZC T.1347. **B)** Skeletal reconstruction from Clack (1998). **C)** UMZC specimen 2000.1. Note that the head is orientated such that the head has been rotated by about 180 degrees to point inwards.

'key-hole' shaped orbit in baphetids. Baphetids have been associated in the past both with temnospondyls and with anthracosaurs, but they are essentially basal tetrapods whose phylogenetic position is equivocal. *Eucritta* has therefore been placed near the base of the crown group of tetrapods (Clack, 1998, 2001).

There is a range of sizes preserved among the five known specimens, which have a nasal–quadrate length (the premaxillae are not adequately preserved, so this is not the complete skull length) ranging from 31 to 84 mm (Clack, 2001). This distribution of sizes among so few specimens provides an interesting contrast with *Balanerpeton*. Furthermore, at a comparable size, the holotype is less well ossified than *Balanerpeton* in its vertebral column, limb bones and tarsus. It may be that *Eucritta* grew rapidly at this locality, and its growth may have continued to a larger size.

Westlothiana lizziae

Probably the most controversial taxon from East Kirkton is *Westlothiana* (Fig 8). When first described, it was hailed as 'the earliest known reptile' (Smithson, 1989), on the evidence of a single, essentially complete, skeleton of about 120 mm snout-vent length. This claim was based on the lack of a temporal or otic notch, meaning that the back of the skull was smooth and straight, the shortness of the skull relative to the body length with the post-temporal series of bones much reduced, and the large orbit placed more anteriorly in the skull than in more basal tetrapods. The vertebral column consisted of holospondylous vertebrae, with stubby, swollen neural arches sutured or fused to the centra in an amniote-like pattern. The single specimen appeared to show only two tarsal bones, interpreted as an astragalus and calcaneum, diagnostic of amniotes. Unfortunately, the skull was not sufficiently well preserved to allow interpretation of key regions such as the occiput and palate.

A second specimen later came to light. Like the first, it was found in Unit 82. It showed a better-preserved ankle. It was prepared as a mould by S. M. Andrews and a silicone peel was made from that. The skull of the holotype was then meticulously prepared to show areas of the palate and occiput. Combined, these specimens showed that interpretation was not so straightforward. The palate lacked a key amniote feature, the transverse flange of the pterygoid, but the skull possessed another one, a supraoccipital ossification (Smithson *et al.*, 1994). The ankle of the second specimen has three rather than two component bones. Therefore the two larger bones were probably not homologous to the

Figure 8 *Westlothiana lizziae*. **A)** UMZC specimen 2004.30. **B)** Skeletal reconstruction based on Smithson *et al.* (1994).

amniote pair. Its most unusual feature, and one that gave cause for caution over its interpretation, is the elongation of its vertebral column, with 36 presacral elements, and further emphasized by its very short limbs. Combined with what is clearly a long tail, at least 27 vertebrae but almost certainly many more, it has a rather unexpected appearance for an early amniote, or indeed such an early tetrapod. Only the limbless aïstopods (see below) and the adelogyrinids (Andrews and Carroll, 1991) are comparable.

Since the 1994 description several cladistic analyses have included *Westlothiana*, in which it has been placed as the sister-taxon to amniotes (Vallin and Laurin, 2004). Others have placed it with seymouriamorphs (Laurin and Reisz, 1999; Laurin, 2004) or as a basal lepospondyl (Ruta et al., 2003, Clack and Finney, 2005). Thus, in all of them *Westlothiana* appears somewhere on the amniote stem but in highly variable positions. Recently another specimen has been discovered that may help resolve its position, although preliminary study suggests that, as in the other two, the skull is not well preserved (Fig 8).

Ophiderpeton kirktonense

The position of *Ophiderpeton kirktonense* is anything but controversial, and it is the only tetrapod taxon from East Kirkton that can be placed in a previously known genus. That said, the position of aïstopods, the group to which it belongs, is a continuing mystery (Anderson, 2007; Ruta et al., 2003). Aïstopods are first known from a little earlier in the Viséan, from the Scottish locality of Wardie, and are notable for their snake-like body form. They lack any trace of limbs and girdles, and have up to 200 vertebrae, with a highly specialized and often bizarre skull structure. Four of the five described specimens of *O. kirktonense* consist of strings of articulated vertebrae, and only the holotype has a very poorly preserved skull, of which few details can be made out (Milner, 1994). The dentition is characteristic, being a series of between seven and ten widely spaced and recurved teeth in the maxilla. The vertebrae are elongate and spool-shaped with very low neural spines, and the ribs are simple and more or less straight. About 36 vertebrae are preserved on the holotype, but the total length of the column remains unknown. Milner suggested that this taxon was terrestrial, showing no trace of lateral lines on the skull bones, and being found in what is otherwise a predominantly terrestrial community.

All the described specimens were found either in one of the walls or in spoil heaps, but the holotype specimen is likely to have been from Unit 82.

A specimen of an aïstopod has been prepared by the collector Roger Jones, who has developed a refinement of the technique using acid to remove matrix. The unusual limey matrix in which this specimen is preserved makes it amenable to such treatment. Mr Jones has kindly allowed us to include a photograph of the specimen here (Fig 9A,B). It is to be hoped that at some stage this specimen will be available for scientific study, since it would yield much more detailed information than the currently known material.

Kirktonecta milnerae

A single specimen from Unit 82 represents this taxon, the most recent to be described from East Kirkton (Clack, 2011). This specimen is remarkable not only because of its small size – about 50 mm snout–vent length – combined with very well ossified hind limbs, but also because it shows evidence of soft tissues including an extremely well-preserved, dorsoventrally deep tail fin (Fig 9D). The small, posteriorly rounded skull with its short, anteriorly placed tooth row, the holospondylous vertebrae, and the striated gastralia suggest that it is a microsaur. As such, it is, to date, the earliest known member of the group, and the only one recorded from the United Kingdom. The specialized tail fin, unlike that of nectrideans, is not supported entirely by bony neural and haemal arches, although the haemal arches are broadened and elongated at the proximal end of the tail. The soft tissue evidence shows that the dorsal part of the fin was composed mainly of a fleshy web. The relatively large feet may have assisted in swimming, but also suggest that the animal was terrestrial for at least part of the time. Perhaps the fin web expanded dorsally on a seasonal basis, in a manner resembling modern urodeles.

Tetrapod remains of uncertain affinity

Among the many specimens of isolated tetrapod bones are some that hint at a wider diversity of animals present in the locality than is evident from the smaller, articulated skeletons. They include tooth-bearing elements, other skull bones, several vertebral elements such as centra and neural arches, isolated phalanges, and limb bones: many of these are of considerable size.

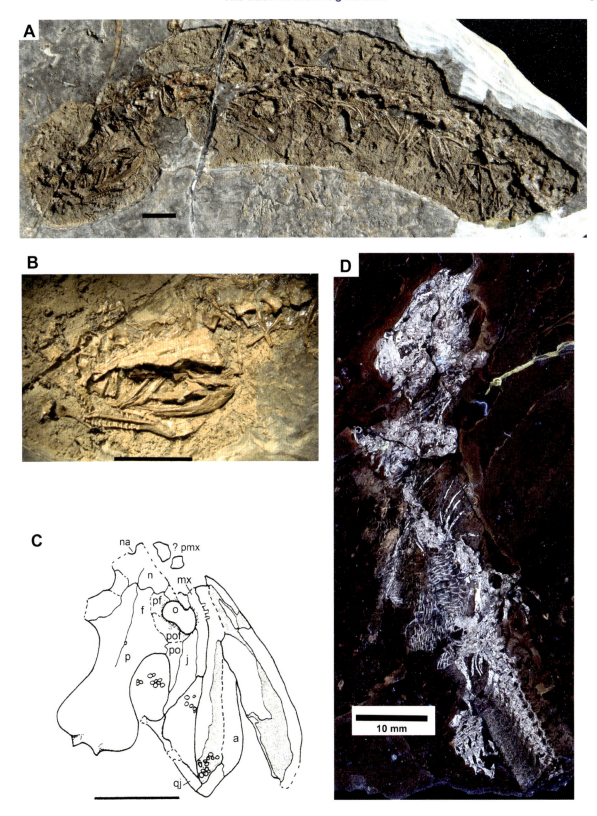

Figure 9 A–C *Ophiderpeton kirktonense*. **A)** Specimen in the collection of Mr Roger Jones, prepared using acid to remove matrix. **B)** Close up of the head of the same specimen. **C)** Diagram of the skull of NMS. 1988.3.1 from Milner (1994). **D)** *Kirktonecta milnerae*. Photograph in UV light by P. Crabb (NHM), from Clack (2011).

Among the more remarkable is a single, complete articulated digit, more than 120 mm long (NMS G.2006.15.1) from Unit 82 (Fig 10A). It consists of a metapodial bone and five phalanges including an ungual. It is not possible to be sure whether it was part of a manus or a pes. It could be a fourth pedal digit from one of the many known taxa or a fourth or fifth digit from one of the anthracosaur taxa such as *Silvanerpeton*, characterized by five phalanges on both pedal digits four and five. If so, they evidently reached a much larger size than are recorded in the articulated skeletons. The evidence suggests that *Balanerpeton* did not reach such a large size, but the proportions of the phalanges resemble those of a temnospondyl manual digit such as *Eryops*. The fourth manual digit of *Eryops* (from a skeletal model with a skull length of 400 mm) measures 110 mm and a temnospondyl of comparable size cannot be ruled out for this specimen. On the other hand, given the size range found among the few specimens of *Eucritta*, that taxon might have grown into a much larger adult. The digit is unlikely to have belonged to *Silvanerpeton*, whose unguals are more elongate and terminate in triangular expansions at the tip (Ruta and Clack, 2006). A very large femur (NMS G.1995.32.1) 130 mm long from Unit 65 might also belong to the same taxon as the digit (Fig 10E), as could a large fibula (NMS G.1987.7.108). The femur is exposed in dorsal view, so that features such as an adductor blade can only be exposed by further preparation or micro-CT scanning. As a guide to the size of the animal, the femur of *Eryops* with a skull length of 400 mm is 150 mm, and femur of *Limnoscelis* with a skull length of 260 mm is 120 mm. A skull length between about 300 and 350 mm might be estimated for the large East Kirkton animal.

A specimen showing an extensive series of large elongate gastralia and the head or entepicondyle of a correspondingly large humerus may also be associated with this taxon (NMS G.1985.4.16 Fig 10B), although an almost equally large humerus (NMS G.1987.7.49 Fig 10C) looks more like that of the early Carboniferous whatcheeriid *Pederpes* (Clack and Finney, 2005). Whatcheeriids comprise a group of tetrapods unknown in 1994, so they may also have been present in the fauna but remained unrecognized. Elongate gastralia are found in *Eucritta*, *Silvanerpeton* and *Eldeceeon*. Probably the most spectacular of the large appendicular elements is a complete pelvis retaining some of its original three-dimensional structure, recovered from one of the tuff horizons (NMS G.1993.67.1 Fig 10G). Identity as a temnospondyl is ruled out by its possession of both a post-iliac process and an iliac blade. It is unlikely to be *Eldeceeon*, whose pelvis has a particularly long post-iliac process even at small size, and that of NMS G.1993.67.1 appears complete. Identity as an anthracosauroid, baphetid or whatcheeriid remain possibilities.

Some of the large centra show a gastrocentrous condition that is similar in degree of ossification to that seen in the small individuals of *Eldeceeon*, but the possibility cannot be ruled out that either *Silvanerpeton* or *Eucritta* developed more fully ossified centra at larger sizes (e.g. NMS G.1987.7.100). It is clear that these elements do not belong to a temnospondyl. However, two flanged ribs have been suggested (NMS G 1987.7.41 and 42; Milner and Sequeira, 1994, fig 19) as belonging to a large temnospondyl. Recent work on whatcheeriids (Lombard and Bolt, 1995; Clack and Finney, 2005) has shown that flanged ribs are also found in the whatcheeriids; thus, either group could be the source of this rib. A third flanged rib (NMS G.1987.7.109) may be another example.

Several large isolated maxillae suggest the possibility of at least two taxa that reached large size in the vicinity of the lake. A maxilla with 19 recurved teeth alternating with spaces for a full count of at least 30 tooth positions is over 100 mm long. It has rather low-profile ornament and is associated with an appropriately proportioned palatal bone, probably a palatine (NMS G.2006.15.2; Fig 10F). A second example of what is probably the same taxon appears on a block with a disrupted skull of a small form, probably *Balanerpeton*. This specimen nicely shows the diversity in size of East Kirkton tetrapods (NMS G.1993.67.2 Fig 10D). A smaller maxilla, about 50 mm long, has 23 teeth and spaces but in a less regular pattern (NMS G.1987.7.46), and may represent a different taxon from that represented by the other large maxilla.

An 80 mm long dentigerous bone is possibly a large temnospondyl premaxilla. It has about 14 tooth positions, in which 8 teeth are present (NMS G.1995.32.2 (Fig 10H). *Balanerpeton* has 12 teeth in the premaxilla, and the primitive temnospondyl *Dendrerpeton* has 14 (Godfrey *et al.*, 1987). Anthracosauroids and whatcheeriids have only four or five teeth in the premaxilla, baphetids have 8–10, but the premaxilla of *Eucritta* is unknown. However, if this element is a temnospondyl premaxilla, it does not appear to show the

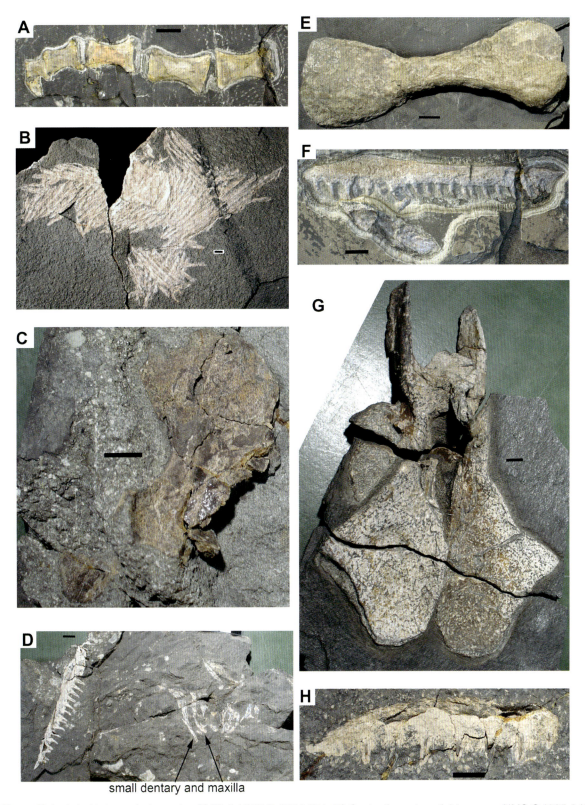

Figure 10 Isolated tetrapod elements. **A)** Digit NMS G.2006.15.1. **B)** Gastralia and partial humerus NMS G.1985.4.16. **C)** Humerus NMS. G. 1987.7.49. **D)** Maxilla and disrupted small skull, possibly *Balanerpeton* NMS G.1993.67.2. **E)** Femur NMS 1995.32.1. **F)** Maxilla and ?palatine NMS G.2006.15.2. **G)** Pelvis NMS G1993.67.1. **H)** ?Premaxilla NMS G.1995.32.2. Scale bars: 10 mm. Photographs by JAC.

characteristic nasal process that would be expected. The tooth distribution, shape and number, and the general proportions of the bone (assuming it to be complete) are not easily attributable to any of the known taxa, so that its identity must remain uncertain.

One specimen preserves a large lozenge-shaped interclavicle, both clavicles, and gastrocentrous centra (UMZC T.1314). A series of rib fragments on the slab may be associated. The generally lozenge-shaped interclavicle has an unusual feature in which the anterior margin is excavated into a rectangular notch in the midline. Based on its shape, the interclavicle cannot be attributed to *Eldeceeon*, which has a long and narrow posterior process, but it cannot be ruled out as belonging to a large *Silvanerpeton* or *Eucritta* in which the central part of the anterior margin has failed to ossify.

Perhaps the most interesting, but most frustrating specimen is UMZC 2000.3 (Fig 11), which shows a partial lower jaw, complete pterygoid/epipterygoid complex and a quadratojugal. It represents the largest animal to be recovered from Unit 82. The lower jaw has lost its dentary and most of the splenial bones, but shows a well-preserved articular, and a modest crest on the surangular. The mesial margin of the adductor fossa is marked by a strong ridge formed by the prearticular similar to that in the embolomere *Pholiderpeton* (Clack, 1987), and there appears to be a relatively large Meckelian fenestra in the posterior portion of the prearticular. Fragments of a denticulated coronoid are present.

The shape of the pterygoid/epipterygoid most closely resembles that of *Pholiderpeton* (Clack, 1987), and the presence of a Meckelian fenestra would accord with that. Future cladistic analysis should help resolve its phylogenetic position with respect to, for example, the anthracosauroid *Silvanerpeton*. The presence of a large embolomere is not reflected among the centra from East Kirkton, since disc-like centra are absent. However, the more primitive embolomere *Proterogyrinus* (Holmes, 1984) has horseshoe-shaped pleurocentra and wedge-shaped intercentra similar to those known from the site.

It is possible to estimate the size of the skull of which these bones formed part from the length of the pterygoid, and allowing for the presence of a quadrate, vomers and premaxillae. A length of 200 mm would not be unreasonable.

Overview of tetrapod distribution

It is doubtful whether we will ever be able to determine attribution of the larger individual elements in the fauna and whether they belong to East Kirkton taxa or others from further afield. Some general statements are nonetheless possible. Among the isolated larger material, almost none can be attributed to temnospondyls. The only possibilities so far noted are a couple of ribs, which might in fact be whatcheeriid, and a possible premaxilla. The larger elements are, then, almost all anthracosauroid, which many regard, along with lepospondyls and *Westlothiana*, as belonging on the amniote stem. This distribution may therefore reflect the idea that stem amniotes such as these were the predominant terrestrial members of the fauna. As isolated or scattered specimens, they may have been transported from more distant localities and not confined to the lake margins. Temnospondyls, on the other hand, may have been more dependent upon damp or wet conditions close to the lake, and this would explain their relative abundance compared with that of stem-amniote taxa. At the same time, the majority of tetrapod specimens are found in the lower half of the East Kirkton Limestone, where plant distribution and sediment types would suggest a drier environment than the upper half (see below under Plants), albeit a deeper lake. A detailed sedimentary analysis documenting which tetrapod taxa are found in which type of sediment might help to resolve this conundrum, but could prove difficult.

Other vertebrate taxa

Several groups of fish are represented in the East Kirkton sequence. Coates (1994) reported on acanthodian and actinopterygian specimens. They are restricted to the topmost horizons in the East Kirkton Limestone, the overlying Little Cliff Shale, and the lower part of the Geikie Tuff (Rolfe *et al.*, 1994b). The majority of vertebrate specimens have been found in a transition zone between the limestone and the shale.

The acanthodians, known from only three specimens, were restricted to the two latter horizons. They belong to two recognized families based on scale morphology. A very small, partially articulated individual was identified as an acanthodidid, and shows the form of fin spine and tooth structure associated with the genus *Acanthodopsis*. Another scale type has been identified as probably from a climatiid, but this would represent a very late record for the group. The

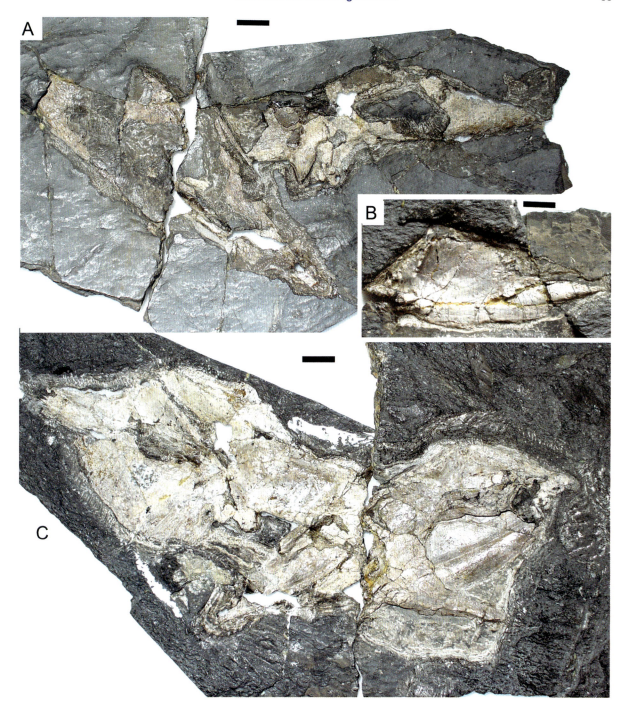

Figure 11 Skull elements of large tetrapod UMZC 2003.3. **A)** Pterygoid in buccal view and partial lower jaw in external view. **B)** Quadratojugal in internal view. **C)** Pterygoid and partial lower in internal view. Scale bars: 10 mm. Photographs by JAC.

specimen is poorly preserved, so that identification is uncertain.

At least six species of actinopterygians have been described from East Kirkton. Although they can be distinguished from each other, most of them are too poorly preserved to be referred to known genera, or indeed to be named as new genera or species. Most of these are in the form of isolated scales, and although most of these are similar to known genera, they are not identical. The one exception to this is one of five specimens referable to the genus *Eurynotus*, a deep-bodied form similar to fish living today among reefs and rocky shores (Fig 12A). This comes from the top of the East Kirkton Limestone, whereas others are from the Little Cliff Shale. *Eurynotus* is common in other Scottish localities of early Carboniferous age. Two other specimens represent juvenile animals that cannot be referred to any of the other species from the locality.

A few specimens represent chondrichthyans (Paton, 1994). Also found in the topmost units 32–37 of the sequence, they appear to be referable to known forms. There are fin spines probably referable to the hybodont *Tristychius* and teeth that may belong to the xenacanth *Diplodoselache*. Both these forms are known from the early Carboniferous of Scotland, from freshwater horizons, so their presence at East Kirkton is not surprising. They were probably large, versatile predators. Other chondrichthyan material is represented by so far unidentifiable prismatic cartilage.

One other kind of fish is present at East Kirkton, belonging to rhizodonts, a group of sarcopterygians. Large isolated scales and a possible tooth have been found from Unit 36, although they have not been identified or studied (Fig 12B & C). Rhizodonts were probably the dominant carnivores in the later stages of the life of the lake.

Coates (1994) noted that the greatest diversity of fishes was to be found distributed in horizons at the top of the East Kirkton Limestone and its transition into the Little Cliff Shale, in units 36 and 35 in the NMS sedimentary log. These horizons have also yielded some tetrapod elements. While some of the isolated fish scales were probably parts of organic debris washed in from elsewhere, and some were possibly from gut contents, the more complete specimens such as the example of *Eurynotus* probably came from *in situ* deposits.

Invertebrates

Seven different groups of invertebrates, six of which are arthropods, have been found at East Kirkton: crustaceans (ostracods); chelicerates (eurypterids, scorpions, opilionids, arachnids); mandibulates (myriapods - millipedes); and molluscs (bivalves).

Crustaceans

Ostracods are abundant elements of the fauna, but unfortunately most are not well enough preserved to show the fine detail of the carapace, which is needed

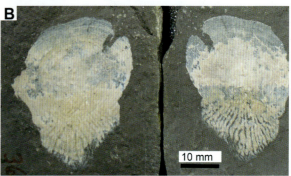

Figure 12 Fish remains. **A)** *Eurynotus* NMS G.1993.6.30. **B)** and **C)** Rhizodont scale in part and counterpart, NMS unregistered specimen. Photographs by JAC.

for identification. The animals are found mainly from the lower units 82–70 and from the upper units 37–27. These have been identified as paraparchitids and cyprids including the genus *Carbonita* (Rolfe et al., 1994b). Their stratigraphic distribution corresponds well with those of fish and tetrapod remains, although tetrapods are also found more commonly in the lower units. There could be an ecological signal here that indicates first, that the ostracods could feed on microorganisms in the lake water, and thus that the water quality was adequate to support them, and second, that being at the base of the invertebrate food chain, they could support the existence of larger invertebrates such as the probable sweep-feeder *Hibbertopterus* and vertebrates higher up the chain. There could also be a preservational bias to all the patterns.

Eurypterids

Three genera of eurypterid have been recorded from East Kirkton. *Hibbertopterus scouleri* (Hibbert, 1836) was the first to be discovered and named, in the nineteenth century, but subsequently two more, namely *Cyrtoctenus* Størmer and Waterston, 1968 and probably *Dunsopterus* Waterston, 1968, were recognized (Jeram and Selden, 1994). These latter could not be placed within known species because the remains were too incomplete. Indeed, *Cyrtoctenus* is known from a couple of leg combs, and the probable specimen of *Dunsopterus* consists only of a femur. Jeram and Selden (1994) suggested that these may all belong to large individuals of *Hibbertopterus*. The smaller specimens of *Hibbertopterus* could be exuviae, whereas the larger specimens attributed to *Cyrtoctenus* could be bodies of adult *Hibbertopterus* (Lamsdell *et al.*, 2010).

Hibbertopterus at the East Kirkton locality has a large domed, shield-like prosoma with a pair of centrally positioned eyes (Fig 13C). The segmented metasoma tapers sharply, ending in a spike-like telson. The largest individual from the site measures about 650 mm across its carapace, whereas the smallest are approximately 200 mm in total length. However, no very small individuals have been found. Although this could result from preservational bias against small and fragile juveniles, it could also suggest that the animals did not breed in the lake itself. Rather, they could have been feeding in pools around the lake, and moved there from more distant nursery pools (Jeram and Selden, 1994). Limb morphology of *Hibbertopterus* shows no evidence for specialized swimming ability, but whereas it does show some adaptations for terrestrial locomotion, it may have been too large to have lived entirely out of water. It could have been a paddler in the shallows (Clarkson *et al.*, 1994). *Cyrtoctenus*, with its dense leg-combs, was probably a filter feeder, and if those combs belonged instead to *Hibbertopterus*, it could be that the mode of feeding of these animals changed as they grew (Jeram and Selden, 1994).

Eurypterid cuticle fragments have been found in the lower part of the Little Cliff Shale and throughout the East Kirkton Limestone, although they are commoner in the more basal parts of the sequence.

Hibbertopterids are known from both Laurussia and Gondwana, but most remains are fragmentary and their relationships to other eurypterids are unclear. Material from the early Carboniferous is rare, and those from East Kirkton are among the better-preserved examples. Recent phylogenetic analysis has placed the hibbertopterids within a larger group, Hibbertopteroidea, a clade of Late Devonian to Permian genera. The range of *Hibberopterus* itself spans the entire Carboniferous, and most of the Permian (Lamsdell *et al.*, 2010).

Scorpions

East Kirkton has produced more articulated specimens of early Carboniferous scorpions than anywhere else in the world. Material of this age is very rare and usually poorly preserved, so the finds from East Kirkton are particularly important. The specimens include 'exquisitely well-preserved cuticles' (Jeram, 1994, p. 283), which range from tiny individuals to large adults with a length of about 700 mm (Fig 13A). Scorpion cuticle is found, like eurypterid cuticle, throughout the East Kirkton Limestone and most of the Little Cliff Shale, but is especially abundant in the tuff layers that also contain comminuted plant debris. Most of the material belongs to one species, *Pulmonoscorpius kirktonensis*, although other, highly fragmentary, specimens may indicate the presence of two other species of *Pulmonoscorpius* (Jeram, 1994).

Two specimens possess book-lungs, an adaptation that unequivocally shows that the scorpions were air-breathing land dwellers. They show large laterally placed compound eyes, implying that vision played a large part in prey search and capture, in contrast to modern forms that use mainly mechanoreception. *Pulmonoscorpius* was therefore likely to have been diurnal rather then nocturnal as are modern forms. Because of the rich record of scorpion cuticle representing different ontogenetic stages at East Kirkton, changes in morphology in an individual with age could be documented for the first time. It seems possible that among the many named genera of Carboniferous scorpions, at least some may turn out to be growth stages of others (Jeram, 1994).

Scorpions of the early Carboniferous reached much larger sizes than any extant terrestrial arthropod. Large size among arthropods of the Carboniferous has sometimes been used as one of the pieces of evidence suggesting a higher level of atmospheric oxygen than exists today. The arthopods usually cited for this are giant early dragonfly-relatives like *Meganeura*, from the late Carboniferous (Dudley, 1998). However, winged insects like these have not been discovered in Viséan rocks (see below), and, according to the latest estimates, oxygen levels at that time were quite similar to those of today (Berner, 2006). If that was the case, other explanations need to be sought for the large size of scorpions at East Kirkton.

Figure 13 Arthropods. **A)** Scorpion *Pulmonoscorpius kirktonensis*. NMS G. 1990.78.1. Shed cuticle of a juvenile. Specimen about 65 mm long. Photograph courtesy of Nigel Trewin. **B)** Opilionid *Brigantobunum listoni*. Hunterian Museum specimen GLAHM A2854. **C)** Eurypterid *Hibbertopterus scouleri*, part and counterpart specimen in an ash bed. UMZC 2013.4.1. Photographs B–C by JAC.

Opilionid

A single specimen of an opilionid (harvestman) has been found at East Kirkton (Wood et al., 1985) and is the second-oldest known fossil of this group. It comes from Unit 44, although it was found in one of the farm walls (S.P. Wood, pers. comm). Very little of its anatomy can be made out, but there was sufficient for Dunlop and Anderson (2005) to name a new genus and species, *Brigantobunum listoni*, and to describe its main features. Four long, slender legs are preserved, and a small globular body (Fig 13B). Dunlop and Anderson (2005) tentatively identified an ovipositor, similar to that found in a modern group of opilionids, the eupnoids. Basic leg anatomy is also a precise match for this living group in terms of the proportions of the segments and the bulbous swellings at the 'knee' joint. Harvestmen as a group seem to have evolved very early into recognizable crown groups and then changed little over hundreds of millions of years.

Millipedes

This group is represented by six specimens, none of which is complete, and most are not well preserved. One of them (Shear, 1994; Shear and Edgecombe, 2010) does not fit into any currently recognized order, and may represent a new one placed within the radiation of Helminthomorpha. It shows evidence of an ozopore, the opening of a gland that produces a stink gland such as is found in modern helminthomorphs. A previous suggestion for the appearance of stink gland in this group was that it coincided with the appearance of ants, which find it repugnant. However, ants do not appear in the fossil record until the Cretaceous, and thus Shear (1994) suggested an alternative. It is known that the product of these glands also repels amphibians, a feature that might have been advantageous to a myriapod in the East Kirkton environment. Two other interesting features are found on this specimen: spiracular openings and small backwardly pointing spines on each segment. Spiracles in myriapods are, like those in insects, used for air-intake during breathing. This specimen represents the earliest occurrence in the fossil record of such a feature in myriapods and indicates that the animal was fully terrestrial. The function of the backwardly pointing spines is less easy to interpret. Protection, strengthening, or use in burrowing are all possibilities.

The remaining specimens do not provide much detail, but there is enough to show that they do not fit readily into known groups. The record of early Carboniferous myriapods is very poor: only one other record is described from the United Kingdom, *Anthracodesmus* from the Tournaisian of Scotland (Wilson and Anderson, 2004; see also Smithson et al., 2012). Further finds would offer great potential for understanding myriapod diversity.

Bivalve molluscs

A very few records of bivalves, identified as the genus *Curvirimula*, have been found in the Little Cliff Shale and the Geikie Tuff. One record also comes from the topmost East Kirkton Limestone. The genus occurs elsewhere in non-marine facies (Rolfe et al., 1994b).

Plants

The flora of East Kirkton is represented by an abundance of fossils, although many are highly fragmentary. They show a variety of styles of preservation: coalified compressions, fusain (a kind of fossil charcoal) and permineralized, three-dimensionally preserved stems and branches. Studied in thin section, the latter give a valuable insight into the evolution of internal structures of stems and branches, showing the acquisition of the fluid-transporting systems of the plants.

The distribution of many of these fossils through the section is well documented as a result of the NMS excavation, and gives hints about changing conditions throughout the life of the lake, although caution in interpretation must be exercised here. Many of the fragments may have been transported from elsewhere. Some of the best material, found in the spoil heaps of the quarry or in farm walls, cannot be easily localized in the section (Scott et al., 1994). Generally speaking, the flora is similar to those from other early Carboniferous localities in Scotland.

Because the permineralized stems cannot easily be linked to the coalified impressions of leaves or other plant parts, the taxonomy of the plants from East Kirkton, and indeed generally for early Carboniferous floras, is to a degree unresolved. However, the groups identified include pteridosperms, progymnosperms and probable true gymnosperms, lycopsids, sphenopsids and ferns.

The East Kirkton Limestone includes plant material preserved in all the ways described above, with permineralized stems found predominantly in the tuffs, and large fronds and coalified axes found in the limestones and shales. Pteridosperms and gymnosperms dominate. An array of leaf types preserved as coalified compressions is attributable to several genera such as *Sphenopteridium*, *Spathulopteris* (Fig 14) and *Adianites*, usually considered pteridosperms. Gymnosperm-like plants are represented by the genera *Eristophyton*, *Bilignea*, *Lyginorachis*, and *Stanwoodia*, preserved as permineralized stems and branches. *Eristophyton* is recorded from some nearby early Carboniferous localities with branches up to 60 mm in diameter. *Pitus*, known from trunks up to 15 m high from other localities, and *Eristophyton*, were probably present around the lake as substantial trees, although it is not known whether they would have formed dense cover, or were more sparsely distributed (Galtier and Scott, 1994).

The genus *Stanwoodia* Galtier and Scott 1991 (Hunterian Museum specimens Pb4860 and CLAHM T522197) is unique to East Kirkton. It has been found at the base of the section as well as higher up, and is known from excellent permineralized branches. It shows a unique combination of primitive and derived anatomical features such as a thickened peridermal layer externally, which is the earliest evidence of a rhytidome (bark) in a fossil plant. There is considerable debate, however, as to whether *Stanwoodia* is a true gymnosperm such as a member of the Cordaitales, a progymnosperm, or a pteridosperm. It is possible that some of the leaf types found at East Kirkton may belong to *Stanwoodia* – or indeed the other gymnosperm-like taxa (Galtier and Scott, 1991).

Stanwoodia, and some of the other tree-forming taxa, show incomplete growth rings. These have been suggested as responses to possible stress conditions that, rather than being seasonal, are mediated by volcanism. Volcanism has, directly or indirectly, influenced the flora of East Kirkton by the production of burnt material preserved as fusain. Studies of this material

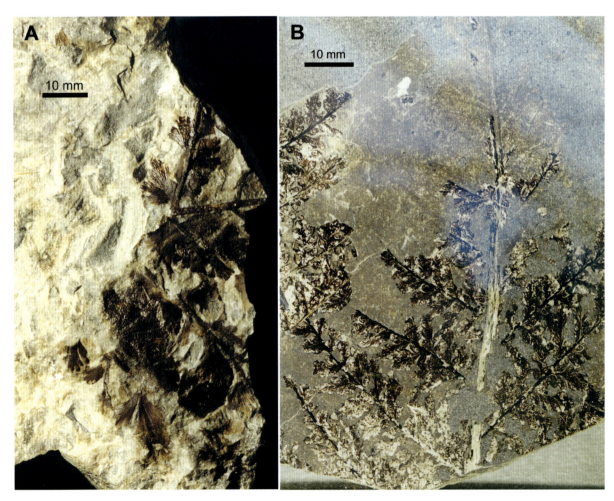

Figure 14 Plants. **A)** *Sphenopteridium crassum*. NMS G. 1993.14.33. **B)** *Spathulopteris obovata*. NMS specimen. Photographs copyright NMS.

have indicated that the temperature at which it burnt was up to 600 °C. It is not certain whether the volcanic activity was contemporaneous with the deposition of this fusain, or whether the fusain derives from reworked volcanic fallout from elsewhere (Brown *et al.*, 1994).

Whereas through the East Kirkton Limestones, all types of plant preservation have been found, only coalified compressions are found in Little Cliff Shale and Geikie Tuffs. In these two stratigraphic units, mainly lycopsid remains have been found. By comparison with the earlier parts of the sequence, this suggests that the environment became wetter, as lycopsids could probably live with their stigmarian roots in standing water. Early in the sequence, the abundance of fusain suggests frequent fires, with drier conditions generally. This distribution also fits with that of the fishes, which occur only in the higher parts of the sequence (Scott *et al.*, 1994).

Conclusions

The extremely complex history and environment of the lake and consequently, its highly varied lithology, has caused much confusion and debate, for example concerning the temperature of the lake water. The laterally variable extent and thickness of the sedimentary units, with their complex lithologies and conflicting signals, have also made it difficult or impossible to localize to precise horizons specimens collected from farm walls or spoil heaps. In some cases they can be attributed in very general terms to the basal, intermediate, or transitional (into the Little Cliff Shale) zones of the sequence, and this has contributed to a broad picture of the life and times of the lake.

A picture of the life of the lake

The picture that emerges is of a small shallow lake, probably quite isolated in location, filled for the most part with meteoric water, but fed occasionally by hot springs. A high mineral content resulted from periodic rainstorms that swept tuffs previously deposited on land nearby into the lake, or leached the minerals from them. The elevated levels of calcite leached from the surrounding land, the temperature of the water, and the activity of micro-organisms, facilitated the formation of bacterial or algal mats, and seasonally mediated changes formed laminations within the limestone. Silica, possibly derived from hot spring activity, sometimes replaced the carbonate layers. For the most part, the water was too rich in minerals for tetrapods to have inhabited it, but once their bodies fell into the lake, at least for the earlier part of its life, they were preserved almost intact by the anoxic conditions at the bottom. The decay of volcanic sediments provided nutrients for a rich flora, which early in the lake's history, consisted mainly of progymnosperms, gymnosperms and pteridosperms. These indicate relatively dry climatic and soil conditions. Plant material from these forests was regularly swept into the lake, where it also preserved well, often becoming mineralized with silica. Numerous invertebrates lived in the undergrowth, scorpions acting as significant daylight predators, myriapods burrowing through leaf litter, and eurypterids venturing in to the shallows perhaps to feed on ostracods. Tetrapods may have fed on this array of arthropods. Periodic wildfires razed the land surface from time to time, forming charcoal from woody remains, and perhaps pushing terrestrial tetrapods lakeward.

As the lake gradually filled, the occurrence of the black shales diminished, suggesting that the water became shallow enough not to produce an anoxic layer at its base. A layer of gypsum suggests that the lake might even have dried out entirely at least once. Eventually, although the original lake may almost have disappeared, layers of shaly limestone indicate the return of a wetter landscape. The end-point of this was the incursion from the nearby Lake Cadell into the remains of the East Kirkton lake, and the consequent influx of fishes. Wetter conditions are also implied by the change to a lycopsid-dominated flora, whose roots were adapted for living in standing water. Initial isolation of the lake and its later confluence with Lake Cadell may explain the fact that whereas the fish taxa found in the quarry are almost certainly representatives of other local genera, the tetrapods, so far as can be judged at present, are with a single exception, endemic genera.

There are seven named tetrapod taxa, and the author is currently studying material representing two more, so that East Kirkton continues to yield new information. The tetrapod fauna is extremely diverse for an early Carboniferous locality. These, plus the indication of the presence of other, larger forms, indicate what may be explosive evolution of terrestrial tetrapods during the early stages of the Carboniferous to occupy a wide range of habitats and niches.

Preservation of specimens

The ubiquitous presence of cherty laminae, while preserving exquisite details of plant anatomy, at the same time makes preparation of bone extremely difficult. Bones, teeth, and chert are often similar in colour and texture, all often having an amorphous but semi-translucent appearance. In most cases also, they rule out the possibility of acid preparation techniques, with a few notable exceptions (see above for *Ophiderpeton*). Mechanical preparation is usually difficult since the matrix is often harder than the bone that it preserves: however, it can be done with care to yield fresh surfaces. Added to this is the difficulty of interpreting the 'cooked', crushed, and disrupted skulls of the smaller vertebrates, in which bones are often split into part and counterpart such that distinguishing intermingled dorsal and ventral views is a challenge. The disjunct distribution of small, articulated skeletons and large, isolated elements makes any conclusions about attributions of the former to the latter, as well as assessing their faunal associations, a matter of a great deal of inference.

Other aspects of the lithology, for example the organic-rich laminae, have made possible the discovery of exquisitely preserved scorpion and eurypterid cuticle, with the results giving clues to the growth patterns of these animals. The tuff layers have favoured the preservation of fine three-dimensional specimens, providing hints of the fauna living further afield from the lake margins.

One important faunal group is missing from the East Kirkton sequences. Despite their preservation potential, not a single winged insect has been found in them to date. The earliest known of these is the Serpukovian *Delitzschala bitterfeldensis*, which is 10 to 15 million years younger than the East Kirkton Limestone (Brauckmann and Schneider, 1996). Three possibilities exist for explaining their non-preservation at East Kirkton: winged insects may not yet have evolved by the late Viséan; or they were still extremely rare and evolving elsewhere; or circumstances of environment or preservation meant that they were not represented in the fauna.

Other arthropods, myriapods and opilionids are very rare and shrimps, wingless 'insects', spiders and trigonotarbids are absent, even though they might be expected there. Bivalve molluscs are also rare, and though East Kirkton preserves a largely terrestrial fauna, no land gastropods have been found. As with insects, the earliest known land snails are early late Carboniferous in age (Hebert and Calder, 2004).

Altogether, the site of East Kirkton represents a completely unique set of circumstances of preservation and indeed original environment, with nothing exactly comparable in the present world. It has provided a tantalizing glimpse into the world of the early Carboniferous with its range of terrestrial flora and fauna about which very little was previously known.

Acknowledgements

I would like to dedicate this chapter to the late Mr Stanley P. Wood, whose perspicacity and determination laid the foundations for the East Kirkton project and who also discovered a large proportion of its extraordinary fauna. I would also like to thank Drs Tim Smithson and Marcello Ruta for discussions of the tetrapods and the fauna as a whole, Mr Roger Jones for permission to use his photograph of *Ophiderpeton*, and the staff of the National Museum of Scotland – Drs Nick Fraser, Stig Walsh and Andy Ross – for help with access to and study of the collections and provision of photographs. Current and former staff of the University Museum of Zoology have facilitated curation of specimens, with many of the specimens having been purchased with the aid of the PRISM fund formerly administered by the Museums, Libraries and Archives Council (now by the Arts Council) as well as from Crotch Fund administered by the UMZC. I thank Russell Stebbings for photography of UMZC tetrapod specimens, Neil Clark of the Hunterian Museum for allowing me to photograph specimens in their care, and Nigel Trewin for contributing the photograph of *Pulmonoscorpius*.

References

Anderson, J. S. 2007. Incorporating ontogeny into the matrix: a phylogenetic evaluation of developmental evidence for the origin of modern amphibians. Pp. 182–227 in J. S. Anderson and H.-D. Sues (eds), Major Transitions in Vertebrate Evolution. Indiana University Press, Bloomington and Indianapolis.

Andrews, S. M., and R. L. Carroll. 1991. The order Adelospondyli. Transactions of the Royal Society of Edinburgh, Earth Sciences 82:239–275.

Berner, R. A. 2006. GEOCARBSULF: a combined model for Phanerozoic atmospheric O_2 and CO_2. Geochimica and Cosmochimica Acta 70:5653–5664.

Brauckmann, C., and J. W. Schneider. 1996. Ein unter-karbonisches Insekt aus dem Raum Bitterfeld/Delitzsch (Pterygota, Arnsbergium, Deutschland). Neues Jahrbuch für Geologie und Paläontologie, Monatshefte 1996:17–30.

Brown R. E., A. C. Scott, and T. P. Jones. 1994. Taphonomy of plant fossils from the Viséan of East Kirkton, West Lothian, Scotland. Transactions of the Royal Society of Edinburgh, Earth Sciences 84:267–274.

Cameron, I. B., A. M. Aitken, M. A. E. Browne and D. Stephenson. 1998. Geology of the Falkirk district. Memoir of the British Geological Survey, 1:50 000 Geological Sheet 31E (Scotland). HMSO, London.

Clack, J. A. 1987. *Pholiderpeton scutigerum* Huxley, an amphibian from the Yorkshire coal measures. Philosophical Transactions of the Royal Society of London B 318:1–107.

Clack, J. A. 1994.. *Silvanerpeton miripedes*, a new anthracosauroid from the Viséan of East Kirkton, West Lothian, Scotland. Transactions of the Royal Society of Edinburgh, Earth Sciences 84:369–376.

Clack, J. A. 1998. A new Lower Carboniferous tetrapod with a mélange of crown group characters. Nature 394:66–69.

Clack, J. A. 2001. *Eucritta melanolimnetes* from the early Carboniferous of Scotland, a stem tetrapod showing a mosaic of characteristics. Transactions of the Royal Society of Edinburgh, Earth Sciences 92:72–95.

Clack, J. A. 2011. A new microsaur from the early Carboniferous (Viséan) of East Kirkton, Scotland, showing soft tissue evidence. Special Papers in Palaeontology 86:45–56.

Clack, J. A. 2012. Gaining Ground: The Origin and Evolution of Tetrapods. Second Edition. Indiana University Press, Bloomington and Indianapolis.

Clack, J. A., and S. M. Finney. 2005. *Pederpes finneyae*, an articulated tetrapod from the Tournaisian of western Scotland. Journal of Systematic Palaeontology 2: 311–346.

Clarkson, E. N. K., A. R. Milner, and M. I. Coates. 1994. Palaeoecology of the Viséan of East Kirkton, West Lothian, Scotland. Transactions of the Royal Society of Edinburgh, Earth Sciences 84:417–426.

Coates, M. I. 1994. Actinopterygian and acanthodian fishes from the Viséan of East Kirkton, West Lothian, Scotland. Transactions of the Royal Society of Edinburgh, Earth Sciences 84:317–328.

Dudley, R. 1998. Atmospheric oxygen, giant Paleozoic insects and the evolution of aerial locomotor performance. Journal of Experimental Biology 210:1043–1050.

Dunlop, J. A., and L. I. Anderson. 2005. A fossil harvestman (Arachnida, Opiliones) from the Mississippian of East Kirkton, Scotland. Journal of Arachnology 33: 482–489.

Durant, G. P. 1994. Volcanogenic sediments of the East Kirkton Limestone (Viséan), West Lothian, Scotland. Transactions of the Royal Society of Edinburgh, Earth Sciences 84:203–208.

Galtier, J., and A. C. Scott. 1991. *Stanwoodia*: a new genus of probably early gymnosperm from the Dinantian of East Kirkton. Transactions of the Royal Society of Edinburgh, Earth Sciences 82:113–123.

Galtier, J., and A. C. Scott. 1994. Arborescent gymnosperms from the Viséan of East Kirkton, West Lothian, Scotland. Transactions of the Royal Society of Edinburgh, Earth Sciences 84:261–266.

Godfrey, S. J., A. R. Fiorillo, and R. L. Carroll. 1987. A newly discovered skull of the temnospondyl amphibian *Dendrerpeton acadianum* Owen. Canadian Journal of Earth Sciences 24:796–805.

Goodacre, I. R. 1998. Microbial Carbonates in Lacustrine Settings: An Investigation into the Carboniferous East Kirkton Limestone. PhD thesis, University of Aberdeen.

Hebert, B. L., and J. H. Calder. 2004. On the discovery of a unique terrestrial faunal assemblage in the classic Pennsylvanian section at Joggins, Nova Scotia. Canadian Journal of Earth Sciences 41:247–254.

Hibbert, S. 1836. On the Freshwater Limestone of Burdiehouse in the neighbourhood of Edinburgh, belonging to the Carboniferous group of rocks. With supplementary notes on other freshwater limestones. Quarterly Journal of The Geological Society of London 33:169–282.

Holmes, R. 1984. The Carboniferous amphibian *Proterogyrinus scheeli* Romer, and the early evolution of tetrapods. Philosophical Transactions of the Royal Society of London B 306:431–524.

Jeram, A. J. 1994. Scorpions from the Viséan of East Kirkton, West Lothian, Scotland, with a revision of the infraorder Mesoscorpiona. Transactions of the Royal Society of Edinburgh, Earth Sciences 84:283–299.

Jeram, A. J., and P. A. Selden. 1994. Eurypterids from the Viséan of East Kirkton, West Lothian, Scotland. Transactions of the Royal Society of Edinburgh, Earth Sciences 84:301–308.

Lamsdell, J. C., S. J. Braddy, and O. E. Tetlie. 2010. The systematics and phylogeny of the Stylonurina (Arthropoda: Chelicerata: Eurypterida). Journal of Systematic Palaeontology 8:49–61.

Laurin, M. 2004. The evolution of body size, Cope's rule and the origin of amniotes. Systematic Biology 53:594–622.

Laurin, M., and R. R. Reisz. 1999. A new study of *Solendonsaurus janenschi*, and a reconsideration of amniote origins and stegocephalian evolution. Canadian Journal of Earth Sciences 36:1239–1255.

Lombard, R. E., and J. R. Bolt. 1995. A new primitive tetrapod, *Whatcheeria deltae*, from the Lower Carboniferous of Iowa. Palaeontology 38: 471–494.

McGill, R. A. R., A. J. Hall, A. E. Fallick, and A. J. Boyce. 1994. The palaeoenvironment of East Kirkton, West Lothian, Scotland: stable isotope evidence from silicates and sulphides. Transactions of the Royal Society of Edinburgh: Earth Sciences 84:223–238.

Milner, A. C. 1994. The aïstopod amphibian from the Viséan of East Kirkton, West Lothian, Scotland. Transactions of the Royal Society of Edinburgh: Earth Sciences 84:363–368.

Milner, A. C., A. R. Milner, and S. A. Walsh. 2009. A new specimen of *Baphetes* from Nýřany, Czech Republic and the intrinsic relationships of the Baphetidae. Acta Zoologica 90:318–334.

Milner, A. R., and S. E. K. Sequeira. 1994. The temnospondyl amphibians from the Viséan of East Kirkton, West Lothian, Scotland. Transactions of the Royal Society of Edinburgh: Earth Sciences 84:331–362.

Milner, A. R., T. R. Smithson, A. C. Milner, M. I. Coates, and W. D. I. Rolfe. 1986. The search for early tetrapods. Modern Geology 10:1–28.

Muir, R. O., and E. K. Walton. 1957. The East Kirkton Limestone. Transactions of the Geological Society of Glasgow. 12: 157–168.

Paton, R. L. 1994. Elasmobranch fishes from the Viséan of East Kirkton, West Lothian, Scotland. Transactions of the Royal Society of Edinburgh: Earth Sciences 84:329–330.

Rolfe, W. D. I., E. N. K. Clarkson, and A. L. Panchen (eds). 1994a. Volcanism and early terrestrial biotas. Transactions of the Royal Society of Edinburgh: Earth Sciences 84:1–464.

Rolfe, W. D. I., G. P. Durant, W. Baird, C. Chaplin, R. L. Paton, and R. J. Reekie. 1994b. The East Kirkton Limestone, Viséan, of West Lothian, Scotland: introduction and stratigraphy. Transactions of the Royal Society of Edinburgh: Earth Sciences 84:177–188.

Ruta, M., and J. A. Clack. 2006. A review of *Silvanerpeton miripedes*, a stem amniote from the Lower Carboniferous of East Kirkton, West Lothian, Scotland. Transactions of the Royal Society of Edinburgh: Earth Sciences 97:31–63.

Ruta, M., M. I. Coates, and D. Quicke. 2003. Early tetrapod relationships revisited. Biological Reviews 78:251–345.

Scott, A. C., R. Brown, J. Galtier, and B. Meyer-Berthaud. 1994. Fossil plants from the Viséan of East Kirkton, West Lothian, Scotland. Transactions of the Royal Society of Edinburgh: Earth Sciences 84:249–260.

Shear, W. A. 1994. Myriapodous arthropods from the Viséan of East Kirkton, West Lothian, Scotland. Transactions of the Royal Society of Edinburgh: Earth Sciences 84:309–316.

Shear, W. A., and G. D. Edgecombe. 2010. The geological record and phylogeny of the Myriapoda. Arthropod Structure and Development 39:174–190.

Sigurdsen, T., and J. R. Bolt. 2009. The lissamphibian humerus and elbow joint, and the origins of modern amphibians. Journal of Morphology 270:1443–1453.

Sigurdsen, T., and J. R. Bolt. 2010. The Lower Permian amphibamid *Doleserpeton* (Temnospondyli: Dissorophoidea), the interrelationships of amphibamids, and the origin of modern amphibians. Journal of Vertebrate Paleontology 30:1360–1377.

Smith, R. A., D. Stephenson, and S. K. Monro. 1994. The geological setting of the southern Bathgate Hills, West Lothian, Scotland. Transactions of the Royal Society of Edinburgh: Earth Sciences 84:18–196.

Smithson, T. R. 1989. The earliest known reptile. Nature 342:676–678.

Smithson, T. R. 1994. *Eldeceeon rolfei*, a new reptiliomorph from the Viséan of East Kirkton, West Lothian, Scotland. Transactions of the Royal Society of Edinburgh: Earth Sciences 84:377–382.

Smithson, T. R., and W. D. I. Rolfe. 1990. *Westlothiana* gen. nov.: naming the earliest known reptile. Scottish Journal of Geology 26:137–138.

Smithson, T. R., R. L. Carroll, A. L. Panchen, and S. M. Andrews. 1994. *Westlothiana lizziae* from the Viséan of East Kirkton, West Lothian, Scotland. Transactions of the Royal Society of Edinburgh: Earth Sciences 84:417–431.

Smithson, T. R., S. P. Wood, J. E. A. Marshall, and J. A. Clack. 2012. Earliest Carboniferous tetrapod and arthropod faunas from Scotland populate Romer's Gap. Proceedings of the National Academy of Sciences USA 109:4532–4537.

Størmer, L., and C. D. Waterston. 1968. *Cyrtoctenus* gen. nov., a large Palaeozoic arthropod with pectinate appendages. Transactions of the Royal Society of Edinburgh: Earth Sciences 68:63–104.

Vallin, G., and M. Laurin. 2004. Cranial morphology and affinities of *Microbrachis*, and a reappraisal of the phylogeny and lifestyle of the first amphibians. Journal of Vertebrate Paleontology 24:56–72.

Walkden, G. M., J. R. Irwin, and A. E. Fallick. 1994. Carbonate spherules and botryoids as lake floor cements in the East Kirkton Limestone of West Lothian, Scotland. Transactions of the Royal Society of Edinburgh: Earth Sciences 84:213–222.

Waterston, C. D. 1968. Further observations on the Scottish Carboniferous eurypterids. Transactions of the Royal Society of Edinburgh: Earth Sciences 68:1–20.

Wilson, H. M., and L. I. Anderson. 2004. Morphology and taxonomy of Paleozoic millipedes (Diplopoda: Chilognatha: Archipolypoda) from Scotland. Journal of Paleontology 78:169–184.

Wood, S. P., A. L. Panchen, and T. R. Smithson. 1985. A terrestrial fauna from the Scottish Lower Carboniferous. Nature 314:355–356.

Chapter 3

Triassic life in an inland lake basin of the warm-temperate biome – the Madygen Lagerstätte (southwest Kyrgyzstan, Central Asia)

Sebastian Voigt[1], Michael Buchwitz[2], Jan Fischer[1], Ilja Kogan[3,4], Philippe Moisan[5], Jörg W. Schneider[3,4], Frederik Spindler[6], Andreas Brosig[3], Marvin Preusse[7], Frank Scholze[3,4], and Ulf Linnemann[7]

1 Urweltmuseum GEOSKOP / Burg Lichtenberg (Pfalz), Germany.
2 Museum für Naturkunde Magdeburg, Germany.
3 TU Bergakademie Freiberg, Institut für Geologie, Germany.
4 Kazan Federal University, Institute of Geology and Petroleum Technologies, Russia.
5 Institut für Paläobotanik, Westfälische Wilhelms-Universität Münster, Germany.
6 Dinosaurier-Park Altmühltal, Denkendorf, Germany.
7 Senckenberg Naturhistorische Sammlungen Dresden, Museum für Mineralogie and Geologie, Germany.

Abstract

The Madygen Formation is a 560 m thick Middle to Late Triassic succession of lacustrine, fluvial, and alluvial deposits outcropping near the village of Madygen in southwest Kyrgyzstan, Central Asia. Following its discovery, stratigraphic discrimination, and geological exploration in the 1930s to 1950s, the Madygen Formation became the target of several palaeontological expeditions during the 1960s. Under the direction of Soviet palaeoentomologist Alexander G. Sharov numerous plant, insect, and fish fossils were unearthed – as well as the iconic reptiles *Longisquama* and *Sharovipteryx*. The geological fieldwork and study of the fossil record by research groups from Russia, Uzbekistan, and Germany has continued to this day and yielded a deeper insight into the life conditions of the ancient Madygen ecosystem, which can be characterized as follows: in the neighborhood of a large perennial lake whose favourable life conditions are indicated by a rich record of water plants, aquatic invertebrates, bony fishes, freshwater sharks and amphibians, as well as a diverse assemblage of invertebrate burrows. A densely vegetated floodplain sustained a highly diverse insect fauna, which in turn was preyed upon by small- to middle-sized tree- and ground-dwelling reptiles. The Madygen Formation's status as a Lagerstätte is based on the exceptional preservation of both aquatic and terrestrial biota, within siltstones deposited along the lake margins. Given their large body sizes, archosauriform reptiles and crocodylian-like reptiliomorphs, which functioned as apex predators in their respective sub-environments, are exempted from the taphonomic window of fine-grained lacustrine sediments – their remains occur instead within fluvial channel deposits. Apart from its role as the only better-known locality of continental Triassic biota in Central Asia, the high diversity and delicate preservation of land plants and insects makes Madygen a crucial study site for the evolution and ecological differentiation of these groups in the aftermath of the P/T extinction event. Preservation of skin impressions in some reptile fossils yields the rare possibility to infer patterns of skin evolution in the early Mesozoic.

Introduction

The stratotype area of the Madygen Formation in the northern foothills of the Turkestan Range in southwest Kyrgyzstan (Fig. 1A–B) is one of the world's richest assemblages of Triassic non-marine fossils, now commonly referred to as the Madygen Lagerstätte (Voigt *et al.*, 2006; Shcherbakov, 2008a; Kogan *et al.*, 2009; Sues and Fraser, 2010; Fischer *et al.*, 2011; Moisan *et al.*, 2011; Moisan, Voigt *et al.*, 2012; Moisan and Voigt, 2013). Formation and site are named after a small livestock herders' settlement near the centre of the remote outcrop area. Exploration of the Madygen Formation and its fossil content commenced in the 1930s (Brick, 1934, 1936; Kochnev, 1934), when the territories of the Soviet Union (to which Kyrgyzstan belonged) were systematically surveyed for natural resources. In 1945, after the end of World War II, the West Uzbek Geological Expedition and the Tashkent State University started systematic work in the Madygen area as part of the 1:200,000 geological mapping programme of Soviet Central Asia (Sixtel, 1949). Uzbek research in Kyrgyzstan lasted for more than two decades and resulted in numerous publications on the macroflora, stratigraphy, and depositional environment of the Madygen Formation (e.g., Sixtel, 1949, 1956, 1960a, b, 1961, 1962; Ikramov, 1967a, b, 1968, 1970, 1971, 1972).

The Palaeontological Institute of the Soviet (now Russian) Academy of Sciences in Moscow became involved in the Madygen research through the analysis of fossil insects (Becker-Migdisova, 1953). Between 1957 and 1971, field parties joined by Elena E. Becker-Migdisova, Inna A. Dobruskina, Alla V. and Maxim G. Minikh, Nestor I. Novojilov, Evgeny N. Kurochkin, Alexander G. Ponomarenko, Alexander P. Rasnitsyn, Alexander G. Sharov, Emilia I. Vorobyeva and many other well-known Russian palaeontologists gathered more than 20,000 fossils from the Madygen Formation. These finds included plants (Dobruskina, 1970a, b), crustaceans (Novozhilov, 1960), insects (Sharov, 1968; Ponomarenko, 1969; Rasnitsyn, 1969), fishes (Vorobyeva, 1967), and, most spectacular, unique small reptiles with skin impressions (Sharov, 1970, 1971a, b) that established the scientific importance of this Lagerstätte. In 1967, I. A. Dobruskina and M. G. Minikh prepared the first more detailed lithostratigraphic subdivision of the Madygen Formation and mapped the strata at a scale of 1:25,000 (Dobruskina, 1970a), in parallel to 1:50,000 mapping of the area by the South Kyrgyz Geological Expedition (Mamrenko, 1968). Fossil collecting at Madygen peaked in 1962–1967 during five field parties headed by A. G. Sharov. The untimely death of this outstanding entomologist in 1973 may have led to essential stagnation of palaeontological work in the area during the 1970s and 1980s. In autumn 1987, a Madygen field conference was organized by I. A. Dobruskina and M. G. Minikh in order to coordinate future research on the deposits. Implementation of projects, however, failed due to the break-up of the Soviet Union and the independence of Kyrgyzstan in 1991.

In 1993–1998, Madygen and adjoining areas were remapped at a scale of 1:50,000 by the South Kyrgyz Geological Expedition for mineral exploration purposes (Berezanskii, 1999). In spring 2005, almost 40 years after the main activities of A. G. Sharov and colleagues, a German expedition, initiated by Hartmut Haubold (Martin-Luther-Universität Halle-Wittenberg), rediscovered Madygen as a fossil site of global interest (Voigt *et al.*, 2006). Following up, the senior author built an integrative project on the environmental reconstruction of the ancient ecosystem at the Technische Universität Bergakademie Freiberg. During four field seasons in 2006–2009, the first detailed geological mapping at a scale of 1:5,000 of the Madygen area was undertaken, 1200 m of stratigraphic sections measured at the centimetre scale, five systematic fossil excavations conducted, and more than 3000 specimens of fossil plants, molluscs, crustaceans, insects, fishes, tetrapods, and trace fossils recovered from both previously known and many new sites (e.g., Fischer *et al.*, 2007; Kogan *et al.*, 2009; Voigt *et al.*, 2009; Béthoux *et al.*, 2010; Buchwitz and Voigt, 2010; Schoch *et al.*, 2010; Voigt and Hoppe, 2010; Alifanov and Kurochkin, 2011; Fischer *et al.*, 2011; Moisan *et al.*, 2011; Moisan, Voigt et al, 2012; Moisan, Labandeira *et al.* 2012; Moisan and Voigt, 2013). After a joint Russian–German expedition in the autumn of 2005 and a joint German–Russian expedition in 2006, Madygen fossil collecting has also been continued independently by the Palaeontological Institute of the Russian Academy of Sciences with field parties in 2007 and 2009.

The renaissance in research on the Madygen Lagerstätte since 2005 has yielded a remarkable amount of new data. Whereas previous work was

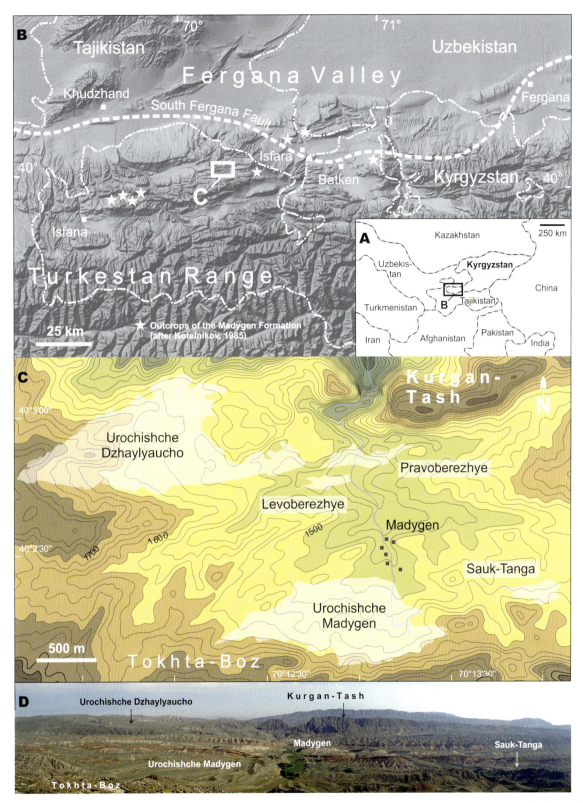

Figure 1 Geography of the Madygen Lagerstätte. **A**, position in Central Asia. **B**, distribution of Middle to Late Triassic rocks in Kyrgyzstan and adjacent Tajik and Uzbek territories. **C**, topographic map of the stratotype area of the Madygen Formation and isolated outcrops (light areas). **D**, panorama photographic mosaic of the stratotype area of the Madygen Formation (view from south).

mainly focused on taxonomic aspects of various organisms (e.g., Ivakhnenko, 1978; Minikh, 1981; Pritykina, 1981; Gorochov, 1987a, b, 2003, 2005; Storozhenko, 1992; Vishnyakova, 1998; Sytchevskaya, 1999; Novokshonov, 2001; Tatarinov, 2005), recent analyses have increasingly incorporated the sedimentological context and palaeoecological implications of the finds (Berner et al., 2009; Kogan et al., 2009; Buchwitz and Voigt, 2010; Voigt and Hoppe, 2010; Fischer et al., 2011; Moisan et al., 2011, Moisan, Voigt et al., 2012, Moisan, Labandeira et al., 2012; Moisan and Voigt, 2013). The aim of this work is to integrate these approaches in order to provide a more comprehensive picture of the Triassic Madygen ecosystem.

Geology
Location and regional geology

Sedimentary rocks referred to the Madygen Formation are rare in Central Asia, and crop out only in the northern foothills of the Turkestan Range (Fig. 1B). The Turkestan Range is an about 300 km long, east–west trending, Alpine-type mountain chain on the border of Kyrgyzstan and Tajikistan. It represents the westernmost extension of the southern Tien Shan and the physiographic boundary of the Fergana Valley to the south. The general geological map of the South Tien Shan (Kotelnikov, 1985) shows the type locality of the Madygen Formation and eight other small outcrop areas of Middle to Late Triassic rocks in the transitional zone between the Turkestan Range and the Fergana Valley. The majority of them are situated in the Batken Oblast, the southwestern-most political district of Kyrgyzstan. Two occurrences are situated near Isfara on Tadjik and Uzbek territory.

The stratotype area of the Madygen Formation corresponds to the built-up area of Madygen, a small livestock herders' settlement about 50 km west of Batken and 70 km southeast of Khudzhand (former Soviet Leninabad). Madygen is located at an altitude of 1460 to 1480 m above sea level in an upland valley framed by two prominent east–west trending Palaeozoic mountain chains, Tokhta-Boz (up to 2300 m high) to the south and Kurgan-Tash (up to 1800 m high) to the north (Fig. 1C–D). Prevailing vegetation of the area is an open mountain steppe shaped by a continental climate with hot summers and cold winters (cold semi-arid steppe climate; Kottek et al., 2006). Present-day farming is based on a short perennial streamlet, the Madygen creek, which arises at the foot of the Tokhta-Boz Mountains for running north where the water dissects the Kurgan-Tash Mountains in a steep gorge before it seeps into the ground less than 6 km from its source.

Sediments of the Madygen Formation crop out at five isolated places in the vicinity of Madygen village (Figs 1C–D, 2): (1) Urochishche (= landmark) Madygen, southwest of Madygen; (2) Urochishche Dzhaylyaucho, northwest of Madygen; (3) Sauk Tanga (= cold valley), east of Madygen; (4) Levoberezhye (= left side), north of Madygen at the left bank of the Madygen creek; and (5) Pravoberezhye (= right side), north of Madygen mainly at the right bank of the Madygen creek. In the three largest outcrop areas – Urochishche Madygen, Urochishche Dzhaylyaucho, and Sauk Tanga (Fig. 1D) – deposits of the Madygen Formation are tectonically bounded by heterogeneous Palaeozoic (Cambrian–Carboniferous) strata along the southern boundary and unconformably overlain with angular discordance by younger Mesozoic (Late Triassic–Early Jurassic or Cretaceous) sediments along the northern boundary (Fig. 2). This, and the polyphasic deformation of most beds of the Madygen Formation, provides evidence for a complex geological history of the area.

The Palaeozoic evolution of the southern Tien Shan is mainly the history of the Turkestan Ocean as an extended branch of the western Palaeopacific (Burtman, 1997, 2008). Throughout the early and middle Palaeozoic, the area was part of a shallow to deep marine environment with recurrent phases of sea-floor spreading. During the late Palaeozoic, the Turkestan Ocean was closed by subduction of oceanic crust and associated sediments beneath the northern Kazakh–Kyrgyz terrane. Crustal shortening continued until the collision with the southern Tarim–Alay terrane, whose northern margin then transformed into an assemblage of stacked thrust sheets. All Palaeozoic (Cambrian–Carboniferous) rocks in the vicinity of Madygen, including ocean floor basalt, deep marine siliceous shale, shelf carbonates, and flysch-type siliciclastics, are invariably fault-bounded units (Geyer et al., 2014). Deformation in an accretionary mélange is suggested by the proximity of the study area to the Turkestan Ocean suture (South Fergana Fault; Fig. 2). The late Palaeozoic continent–continent collision finally resulted in the formation of the ancestral

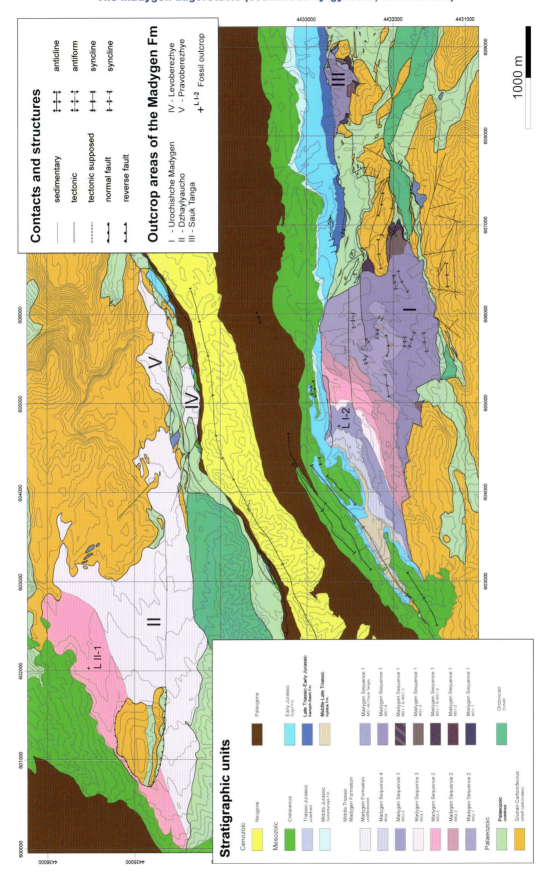

Figure 2 Geological map of the Madygen Formation stratotype area.

southern Tien Shan. Uplift and erosion of this Alpine-type mountain chain dominated the Permian Period in this part of Central Asia.

The orogenic belt of the ancestral southern Tien Shan was repeatedly reactivated during the Mesozoic due to far-field effects of collision-accretion events between peri-Gondwanan microplates, such as the Tibetan blocks, and the southern Eurasian margin (De Grave *et al.*, 2007, 2011). A basin-and-range-like physiography with precursors of the present-day Fergana Basin and Turkestan-Alay Range is therefore assumed for the region during Triassic–Jurassic times (Glumakov, 1988; De Grave *et al.*, 2011). Coal-bearing alluvio-fluvial and lacustrine deposits like those of the Madygen Formation reflect this period in the stratigraphic record. During most parts of the late Mesozoic and Palaeogene, the area was covered by a shallow epicontinental sea (Fergana Gulf; Hecker *et al.*, 1963). Related to the collision of India and Eurasia, the formation of the modern Tien Shan commenced in the Oligocene (Abdrakhmatov *et al.*, 1996; De Grave *et al.*, 2007). The ongoing uplift is accompanied by the accumulation of thick continental molasse deposits and deformation of all pre-Quaternary rocks.

From the regional geological history, it can be inferred that the deposition of the Triassic Madygen Formation took place in a continental environment. Palinspastic maps for the Triassic of Eurasia (Fedorenko and Miletenko, 2002) place the Madygen area at 34–38° northern latitude and at a minimum distance of about 600 km to the nearest marine shoreline. Recent analyses of oxygen and strontium isotopic data from tooth enameloid of juvenile hybodontid and actinopterygian fishes of the Madygen Formation yielded strong freshwater signals that support the interpretation as an inland depositional environment (Fig. 3; Fischer *et al.*, 2011). A freshwater system is also indicated by the low sulphur content of subbituminous coal preserved at the base of the Madygen Formation (Berner *et al.*, 2009). Palaeogeographic environmental maps for the Triassic of Eurasia (Fedorenko and Militenko, 2002) classify the area as a region of low mountains and plateaux suggesting altitudes between

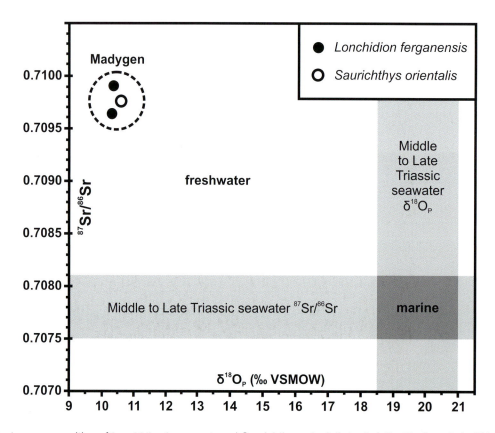

Figure 3 Isotope composition of *Lonchidion ferganensis* and *Saurichthys orientalis* teeth (after Fischer et al., 2011). Oxygen and strontium ratios of the Madygen samples do not match isotope signatures expected for the Middle Triassic marine realm but they are consistent with the interpretation of the Madygen lake system as a freshwater environment.

a few hundred and a little bit more than one thousand metres. Madygen was, then, likely situated in the core of an eroded high-mountain belt (ancestral South Tien Shan) with reactivation of Palaeozoic structures as the major relief-forming process.

Madygen Formation

Lithostratigraphy and depositional environment

The term 'Madygen Formation' was introduced for sediments with a supposedly Early Triassic flora intercalated between Devonian shelf carbonates and continental Jurassic strata (Kochnev, 1934). Further studies on fossil plants, palynomorphs, and bivalves suggested the presence of Permian deposits in the lower part of the Madygen Formation (Sixtel, 1949, 1960a, b). The idea of a series of Permian and Triassic strata was further developed by Ikramov (1967a, b, 1968) who divided the Madygen Formation into four lithostratigraphic subunits. However, this scheme did not gain acceptance because this author ignored the different tectonic history of each outcrop area. Dobruskina (1970a, 1995) proposed a Middle to Late Triassic (Ladinian–Carnian) age for the entire Madygen Formation based on macrofloral remains and introduced a subdivision of its succession into five members (T1–T5). Although her lithostratigraphic scheme is easier to follow than all previous concepts, it still requires further refinement because grain size, rock colour, and thickness were the only criteria used for the identification, correlation, and definition of individual beds, disregarding any information on the sedimentary facies. Here we introduce a new subdivision into four depositional sequences based on centimetre-scale sections measured for the entire 560 m thick succession of the Madygen Formation at Urochishche Madygen, the most completely preserved succession of all known outcrop areas.

The Madygen Formation at Urochishche Madygen (Fig. 1C–D) crops out in an area approximately 1.5 km by 3.0 km, where its folded, faulted, and tilted beds become generally younger from SE to NW (Fig. 2). The lithostratigraphic record can be subdivided into four genetic units, herein called Madygen Sequences 1–4 and abbreviated as MS1 to MS4 (Fig. 4). Boundary surfaces of these units indicate abrupt changes in the nature of deposition caused by tectonic impulses, climate changes, or both. MS1 to MS3 are further divided into genetic subunits (MS1–1 to MS1–4, MS2–1 to MS2–3, and MS3–1 to MS3–2), with each of them grouping sediments of similar sedimentary facies such as alluvial-fan, alluvial-plain, or lacustrine deposits (Fig. 5). MS4 represents a uniform series of deltaic-lacustrine sediments without need for further subdivision. Both the Madygen sequences and their potential subdivisions are mappable geological units. With respect to the small size of the study area, they are assumed to be largely isochronic.

At Urochishche Madygen the Madygen Formation starts with an up to 20 m-thick, clast-supported, oligomictic conglomerate (base of MS1–1; Fig. 4) that mainly consists of limonite-coated, reworked pebbles and cobbles of marine Palaeozoic limestone. The yellowish-gray, crudely stratified (0.3–0.5 m) sediment is interpreted as a flood flow deposit in an alluvial-fan environment (Fig. 5A–B). It is overlain by a 23 m-thick succession of interbedded matrix-supported polymictic conglomerates and sandy to gravelly mudstones (top of MS1–1). The wide range of grain size (clay to cobble), poor sorting, overall massive appearance, and rusty- to reddish-brown colour of these sediments indicates formation by debris and mud flows in an alluvial fan or proximal alluvial plain (Fig. 5A). A 7 m-thick horizon of interbedded conglomerates, sandstones, and mudstones above (MS1–2) is considered to be a distinct unit based on its bluish-grey color, lower average grain size, better sorting, and the common presence of coaly detritus and root traces. These in turn are indicative of channel and overbank deposits in a basin with fluvial drainage and scattered to dense vegetation (Fig. 5B). Toward the top the fluvial deposits pass into an 80 m-thick greyish-brown to dark grey, laminated, organic-rich mudstone (MS1–3) with notable quantities of plant fossils, aquatic invertebrates, and fish remains (Fischer *et al.*, 2011). Coal seams (1–130 cm thick) and abundant root traces occur in the basal 10 m of this unit, which is capped by an up to 30 cm-thick layer of cone-shaped fibrous calcite. Lithology and fossil content of this subunit indicate a low-energy lacustrine environment with wide boggy shorelines (Fig. 5D). The organic-rich mudstone is overlain by a 110 m-thick succession of tightly interbedded, greenish-grey to brownish-grey conglomerates, sandstones, and siltstones (MS1–4). Its laminated to wavy-laminated fine-grained sediments are particularly rich in plant remains, coaly detritus, and penetrative traces of aquatic invertebrates (Voigt and Hoppe, 2010; Moisan *et al.*, 2011), but also contain crustaceans, insect wings, fishes, and tetrapod remains (Voigt *et al.*, 2006; Kogan *et al.*, 2009; Fischer *et al.*,

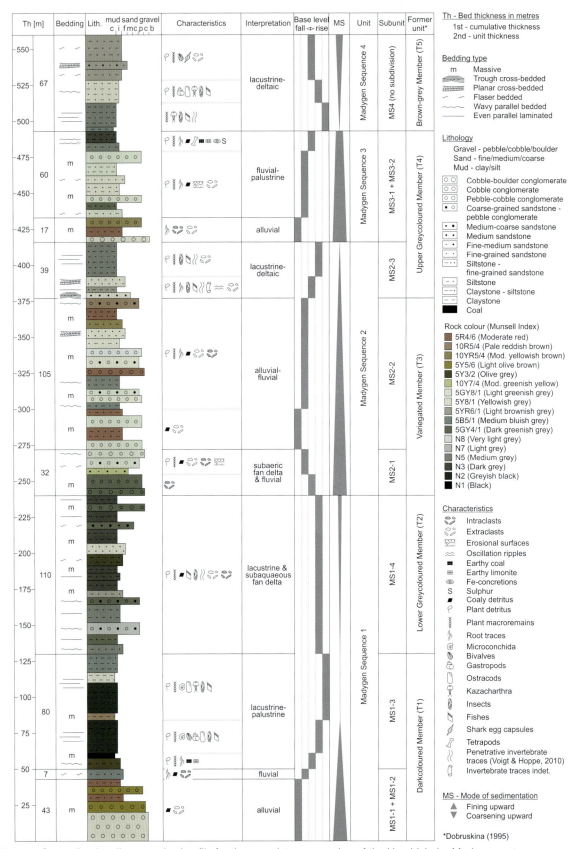

Figure 4 Generalized sedimentary log/profile for the complete succession of the Urochishche Madygen outcrop area.

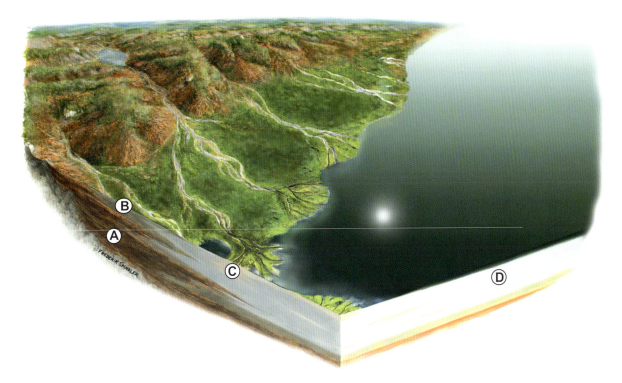

Figure 5 Block image reconstruction of the Madygen Formation palaeoenvironment. The sedimentary strata are characterized by distinct depositional systems: alluvial fan (**A**), flood plain (**B**), river delta (**C**) and lake bottom (**D**).

2011). All features of MS1–4 point to a (fan-)deltaic-lacustrine environment (Fig. 5C) with the recurrent input of coarse-grained material.

Grain-size significantly increases at the basis of MS2, which begins with a 32 m-thick succession of clast-supported, poorly sorted, pebbly to cobbly fanglomerates (basal 14 m) and massive to poorly stratified conglomerates with intercalations of flaser- and wavy-bedded sandstone (top 18 m). This subunit is assigned to MS2–1 and records the transition from a lacustrine fan-delta to a fluvial deposit (Fig. 5B–C). It is overlain by 105 m of alternating poorly lithified, matrix-supported conglomerates and sandy to gravelly, massive mudstones with minor intercalations of clast-supported conglomerates, better-sorted sandstones, and root-bearing mudstones (MS2–2). The bright red, green, brown, and bluish colors support the interpretation of alluvial to fluvial deposits (Fig. 5B). The following, 39 m-thick succession (MS2–3) commences with a series of root-bearing trough to tabular cross-bedded sandstone and fossiliferous mudstones. Abundant and diverse plant, mollusc, crustacean, insect, fish, and invertebrate trace fossils characterize the 9 m-thick horizon as a complex of delta-plain, delta-front, and prodelta deposits (Fig. 5C). The much thicker upper part of MS2–3 is dominated by laminated mudstone deposited under more open-lacustrine conditions (Fig. 5D) although macrofloral remains, terrestrial insects, and single-grain laminae and pockets of coarse sand and granules provide evidence for periods of enhanced recurrent terrigenous input.

A distinct erosional unconformity forms the base of MS3 with oligomictic conglomerate of reworked Palaeozoic limestone pebbles and cobbles that deeply incise the lacustrine mudstone of MS2–3. The conglomerate marks the beginning of a 17 m-thick succession (MS3–1) of variegated, matrix-supported conglomerates and sandy to gravelly mudstones with few intercalations of better-sorted sediments. These alluvial to fluvial deposits (Fig. 5A–B) grade upward into 60 m of tightly interbedded massive to stratified conglomerates, sandstones, mudstones, coaly mudstone, and coal (MS3–2) that are characterized by greyish-brown to dark-grey colours, abundant plant remains, coaly detritus, and extensively rooted fine clastics. Complexly interstratified gravelly to muddy deposits in the uppermost part of this subunit yield important macrofloral and tetrapod remains (Schoch *et al.*, 2010; Moisan *et al.*, 2011). The inferred depositional setting is a moderate to poorly drained alluvial-plain environment with small, meandering, frequently migrating channels, hydromorphic soils, and peat-accumulating backswamps (Brosig, 2012; Fig. 5B).

There is a sharp boundary between the rooted coaly mudstone of MS3–2 and MS4 (67 m in total thickness) The base of MS4 is marked by 3 m of laminated, pale bluish-grey claystone that yields an offshore lacustrine fauna. The characteristic claystone is in turn followed by a laminated, greyish-brown mudstone transitioning to a flaser- to cross-bedded sandstone. The latter were deposited in a lacustrine environment during deltaic progradation (Fig. 5C). Fine-grained deposits of MS4 locally contain significant amounts of fossil plants, molluscs, crustaceans, insects, fish remains, and invertebrate traces (Dobruskina, 1995; Shcherbakov, 2008a; Voigt and Hoppe, 2010; Fischer *et al.*, 2011; Moisan *et al.*, 2011; Moisan and Voigt, 2013).

Between three and 150 mm thick layers of whitish-grey to whitish-pink bentonitic claystone have been recorded in all lacustrine beds of Urochishche Madygen (10 layers in MS1–3, 12 in MS1–4, four in MS2–3, and one in MS4). Zircon and feldspar crystals suggest that these rocks originated by *in situ* diagenetic alteration of volcanic ash-fall tuff. They probably correspond to the montmorillonitic claystones mentioned by Sixtel (1960b) and Ikramov (1967a, 1971).

The Madygen Formation at Urochishche Madygen combines fining- and coarsening-upward cycles with marked reversals in the mode of deposition that suggest sedimentation in a dynamic basin with temporarily steep relief gradients (Fig. 5). Alluvial fan, alluvial plain and lake deposits of MS1–1 to MS1–3 indicate a continuous rise of the base level (= lake table) during the early stage of basin formation. Lacustrine sedimentation represented by MS1–4 was affected by an overall steady input of coarse-grained fluvial sediment through a nearby fan-delta. At the beginning of MS2 deposition, the (fan-)delta prograded and (within the limits of the outcrop area) filled the lake of MS1. Alluvial to fluvial deposition of MS2–2 was subsequently replaced by the deltaic-lacustrine sedimentation of MS2–3 reflecting another period of rising base level. Progradation of alluvial fans at the base of MS3 interrupted lacustrine development and introduced a third period of base level rise. At the boundary between MS3 and MS4, the depositional environment instantly changed from wet alluvial plain to profundal lake conditions, indicating an abrupt rise of the water table triggered by flooding of the basin or earthquake-related subsidence of the basin floor. Finally, MS4 records a continuous base-level fall due to delta progradation. The dynamics and relative instability of the depositional system are also demonstrated by the fact that the prevailing direction of clastic input changed from N–NE (present-day orientation) during the formation of MS1 to MS2–1 to S–SW during the formation of MS2–2 to MS4, as inferred from various current indicators and the lateral variation in bed thickness (Brosig, 2012; Preuße, 2012).

At Urochishche Dzhaylyaucho (Figs. 1C–D, 2) the Madygen Formation is difficult to assess because of limited outcrops. Only the uppermost part of the section is adequately exposed: an up to 100 m thick succession of deltaic-lacustrine deposits (Fig. 5C–D) that have been correlated with unit MS2–3 at Urochishche Madygen (= lowermost part of T4 according to Dobruskina, 1970a, 1995). The deltaic-lacustrine strata of Urochishche Dzhaylyaucho are conformably underlain by several hundred metres of coal-bearing alluvial–fluvial deposits that still await detailed sectioning and mapping.

At Sauk Tanga (Figs. 1C–D) the Madygen Formation comprises only beds of MS1 (Fig. 2). Although all strata of this outcrop area are heavily affected by Neogene to present-day compressional deformation, alluvial-fluvial beds (Fig. 5B) correlative with MS1–1 and MS1–2 as well as coal beds and organic-rich lacustrine mudstone (Fig. 5D) of MS1–3 are clearly discernible in places. The majority of the outcrop area, however, is made up of fine- to coarse-grained, sometimes remarkably fossiliferous lacustrine deposits of MS1–4, which are separated from overlying Late Triassic to Early Jurassic continental deposits by an angular unconformity. At the northeastern margin of the outcrop area this border forms a tectonic contact including wedges of Palaeozoic rocks. Dobruskina (1970a, 1995) calculated the maximum thickness of the Madygen Formation at Sauk Tanga to be approximately 200 m. A reliable overall thickness is difficult to determine because all units of MS1 at this place are invariably fault-bounded to each other and therefore incomplete.

Revisional stratigraphic work is pending for the remaining two outcrop areas, Levoberezhye and Pravoberezhye (Fig. 2). Ikramov (1967a, b, 1968) and Dobruskina (1970a, 1995) referred the strata of both places to the basal part of the Madygen Formation.

The small size of the outcrop areas, their complicated structure, strong deformation and apparently poor fossil content may be the main reason why relatively little attention has been paid to these beds to date.

There are no data that define the spatial relationship of the five outcrop areas during deposition of the Madygen Formation. N–S compression is likely considering the general direction of the prevailing tensional regime in the region. Based on correlative lacustrine beds at Urochishche Madygen and Sauk Tanga (MS1–3 and MS1–4) and at Urochishche Madygen and Urochishche Dzhaylyaucho (MS2–3), the minimum extent of the Madygen lake amounts to between three and five square kilometres. It is possible that Madygen just represents marginal remnants of an extended early Mesozoic lake that covered significant parts of the present-day Fergana Valley (Fischer et al., 2011).

Biostratigraphy and radiometric age
The age of the Madygen Formation has long been contentious, with estimates ranging from the Late Permian to the Early Jurassic. Most reliable age constraints are based on macrofloral and insect remains because they represent the most frequently collected and best-studied fossils of the occurrence. Based on pleuromeian lycophytes, Brick (1934, 1936) assigned the Madygen Formation to the Early Triassic. Then Sixtel (1949, 1960b, 1962) identified elements of the Palaeozoic Cathaysian flora and postulated that the formation consists of two series, a lower Late Permian and an upper Early Triassic one. All of the studied insect remains, however, showed Mesozoic affinities (Becker-Migdisova, 1953; Sharov, 1968; Ponomarenko, 1969; Rasnitsyn, 1969), and the abundant phyllopod crustaceans (Kazacharthra) in the upper part of the formation were even considered to indicate an Early Jurassic (Liassic) age (Novozhilov, 1970). Dobruskina (1970a) uncovered the misidentification of supposed Palaeozoic plants and demonstrated that the beds of all Madygen outcrop areas have an essentially uniform early Mesozoic flora. After revisionary studies of contemporaneous Eurasian floras (Dobruskina, 1970b, 1974, 1975, 1980), she hypothesized a Middle to Late Triassic (Ladinian–Carnian) age for the Madygen Formation (Dobruskina, 1982). This datum has been adopted by almost all subsequent studies on the occurrence (e.g., Shcherbakov, 1984; Zherikhin and Gratshev, 1993; Sytchevskaya, 1999; Gorochov, 2005; Schoch et al., 2010).

The Madygen insect fauna is a typical Anisian–Carnian assemblage based on its exceptional diversity, the abundance of post-Palaeozoic taxa and the presence of Triassic endemics (Shcherbakov, 2008a–c). A late Middle Triassic age is most likely considering that Madygen (1) represents the second-earliest dipteran fauna after the early Anisian Grès à Voltzia of the Vosges, France, (2) contains the earliest known Hymenoptera, (3) but in contrast to the Carnian assemblage of Ipswich, Australia, lacks truly aquatic Heteroptera. Moreover, the Madygen Blattodea, Orthoptera, and Hemiptera are represented by more primitive taxa than the otherwise most similar Carnian insect faunas of South Africa and Australia (Shcherbakov, 2008b). A Middle Triassic age is also suggested by the monospecific assemblage of *Almatium gusevi* (Chernyshev, 1940), inasmuch as this taxon seems to be at the base of the radiation of Kazacharthra, an extinct group of branchiopod crustaceans currently known only from the late Middle and Late Triassic of Central Asia (Chen et al., 1996). The presence of basal archosauromorph reptiles whose remains are the most abundant tetrapod fossils in fluvial deposits of MS3–2, in combination with a relict reptiliomorph (Schoch et al., 2010), makes an age of the Madygen Formation younger than Carnian unlikely.

Initial U-Pb zircon dating of volcanic ash layers has been undertaken and 18 grains were analysed. The sample was taken from a pyroclastic horizon within lacustrine sediments of MS1–3 (Urochishche Madygen) and analysis was carried out by LA-ICP MS at Senckenberg, Dresden. The preliminary concordia age of six concordant zircons was found to be 237+/-2 Ma. The U-Pb age corresponds to a Middle to Late Triassic (late Ladinian or early Carnian) age (Mundil et al., 2010; Cohen et al., 2013).

Fossil preservation and taphonomic peculiarities
The stratotype of the Madygen Formation has been repeatedly referred to as a Lagerstätte (Voigt et al., 2006; Shcherbakov, 2008a; Kogan et al., 2009; Sues and Fraser, 2010; Fischer et al., 2011; Moisan et al., 2011; Moisan, Voigt et al., 2012; Moisan and Voigt, 2013), i.e., a rock body with unusually rich palaeontological information (Seilacher et al., 1985). This status is justified because the Madygen Formation has yielded the richest Triassic insect fauna of the world including representatives of almost all contemporaneous groups (Shcherbakov, 2008a), one of the most diverse early

Figure 6 Plant macrofossils. **A**, mass occurrence of the thalloid liverwort *Ricciopsis ferganica*; **B**, non-elongated corm of *Isoetites madygensis* with leaves/sporophylls; **C**, serrated leaves of *Isoetites sixteliae* with a fine midvein; **D**, air channels in serrated leaves of *Isoetites sixteliae*; **E**, ligulate sporophylls of *Lepacyclotes zeilleri*; **F**, leaf of the lycopsid *Mesenteriophyllum kotschnevii* showing the midvein and undulated leaf margin; **G**, leaf of the cycadalean *Pseudoctenis lanei*; **H**, foliage of the peltasperm seed fern *Vittaephyllum*. Scale units equal 5 mm.

Mesozoic floras (Dobruskina, 1995), and tetrapod fossils with preservation of integumentary structures (Sharov, 1970, 1971a; Voigt et al., 2009; Alifanov and Kurochkin, 2011; Buchwitz, 2011).

Given the variety of depositional environments represented by the Madygen Formation, a differentiated view on the occurrence of fossil conservation within this succession is appropriate: Whereas alluvial fan deposits (MS1–1, MS2–2, MS3–1; Fig. 5A) are virtually devoid of fossils, fine-grained marginal lake deposits (parts of MS1–3, MS1–4, MS2–3, MS4; Fig. 5C) include horizons of particularly high fossil density and high-quality preservation. Profundal lake deposits (part of MS4; Fig. 5D) can be fossiliferous as well, and fluvial deposits (MS3–2; Fig. 5B) may yield layers of dense plant remains as well as small-scale concentrations of bone, but usually the fossil density and quality of preservation are not those of a Lagerstätte.

The most productive fossil site yet known within the Madygen Formation is the type locality of the reptilian taxa *Longisquama* and *Sharovipteryx* (locality L II/1, herein also called the Sharov Quarry, Fig. 2; corresponds to outcrop No. 14 in Dobruskina, 1995) on the northern slope of Urochishche Dzaylyaucho. This locality was discovered by Tatiana A. Sixtel, extensively exploited by Soviet palaeontologists during the 1960s, and restudied by the German and Russian expeditions between 2005 and 2009. LII/1 is marked by a succession of thinly alternating beds of brownish silty mudstone and greyish clayey mudstone. All features of the even to slightly uneven, parallel-laminated, partially carbonate-cemented rocks with abundant fish remains are in accordance with lacustrine deposits. Single-grain laminae and pockets of sand and granule, the predominance of terrestrial plants and insects (Dobruskina, 1995; Shcherbakov, 2008a), the presence of articulated skeletons of land-dwelling tetrapods, and drifted roots with still attached substrate, provide evidence of continuous fluvial input of terrigenous matter during formation of these beds. The fossil assemblage of LII/1 is at least partly allochthonous and may be interpreted to represent a prodeltaic depositional environment. Common invertebrate burrows (Voigt and Hoppe, 2010) and the lack of black shales suggest oxygenated conditions at the lake bottom. These observations contrast with an interpretation as oxbow-lake deposits proposed for the site by Shcherbakov (2008a).

Apart from the large quantity of preserved fossils, L II/1 is arguably the best example for the occurrence of Konservat-Lagerstätten conditions (*sensu* Seilacher, 1970; Seilacher et al., 1985; Briggs, 2003) in the Madygen Formation. Within its fine-grained sediments insect bodies as well as the skeletons of fishes and tetrapods may be preserved in full articulation, though some layers are rich in fragmentary specimens such as patches of fish scales and isolated insect wings. Insects are often preserved as thin organic films (Briggs, 2003; 'compression fossils' *sensu* Schweitzer, 2011). This type of preservation sometimes includes colouration patterns (Sharov, 1968; Shcherbakov, 2002, 2011), with the darker sections of the organic layer indicating former melanin-rich pigmentation (Schweitzer, 2011, Fig. 8A, G). Furthermore, the scleroticized cover wings of certain insect groups, such as cockroaches and beetles, may be preserved as a surface relief in addition to, or instead of, an organic layer (Fig. 8B–C, E–G). Wing impressions often preserve the wing curvature, sculpture, and venation pattern.

Tetrapod fossils unearthed at L II/1 usually display the structure of the skin preserved as a surface relief (Fig. 12). Unlike the scaly skin impressions of *Kyrgyzsaurus* (Alifanov and Kurochkin, 2011) and the folded 'wing membrane' imprints of *Sharovipteryx* (Sharov 1971a; Gans et al., 1987), the conspicuous skin appendages of *Longisquama* (Fig. 12A–C) are three-dimensionally preserved and marked by an upper side and an underside separated by a sedimentary core (Reisz and Sues, 2000; Voigt et al., 2009). A somewhat similar preservation state is found in fossil shark egg capsules (Fischer et al., 2011) whose imprints include sediment cores in varying degrees of flattening. These observations are indicative of a particular fossilization process (Buchwitz, 2011; Buchwitz and Voigt, 2012): after the interior of an appendage or capsule had been soaked with wet sediment, the conservation of the surface relief on one or both sides was facilitated by microbially enhanced precipitation of minerals within the pores of the sediment in contact with the surface of the decaying body ('death mask' preservation as discussed by Briggs, 2003). Organic remains of tetrapod skin have not yet been described.

It has to be noted that tetrapod fossils are not particularly abundant for L II/1: apart from appendage fragments of *Longisquama* (Voigt et al., 2009) only a single specimen of *Kyrgyzsaurus* (Alifanov and Kurochkin, 2011) has been discovered since the resumption of fieldwork in 2005. Unlike skeletal remains and isolated bones discovered within fluvial strata of Urochishche Madygen (Locality L I/2,

MS2–3; Figs. 2, 11), which lack skin preservation but display only minor deformation, specimens from L II/1 are usually strongly compressed and broken, and thus their classification often remains equivocal.

Some invertebrates such as bivalves, ostracods, and kazacharthrans (Fig. 7) are locally abundant in the Madygen Formation, but usually strongly deformed in response to the complex tectonic history of the area. Although lacustrine mudstones of MS4 at Urochishche Madygen contain thousands of kazacharthran head shields and abdominal sections, only very few specimens are more completely preserved and exhibit impressions of soft parts. This is in marked contrast to the detailed preservation of kazacharthrans from the Late Triassic of China (McKenzie et al., 1991; Chen et al., 1996). The rarity of soft-tissue preservation in the Madygen kazacharthrans might be due to efficient recycling by aquatic scavengers or microbial decay, supporting the assumption of an overall well-aerated lacustrine system.

Despite the fact that pollen, spores, and detached cuticular fragments were recorded in some palynomorph and organopetrographic samples (Voigt et al., 2006; Berner et al., 2009), organic material has never been observed on macrofloral remains from the Madygen Formation. Epidermal and cuticular features, however, may be well preserved as impressions in clay-rich lacustrine deposits of MS2–3 and MS4 (Moisan et al., 2011, Moisan, 2012; Moisan, Voigt et al., 2012; Moisan, Labandeira et al., 2012; and Moisan and Voigt, 2013; Fig. 6). Non-systematically varying thermal maturity of the sediments suggests that the preservational condition of organic matter in the Madygen Formation is site-specific (Berner et al., 2009).

Palaeontology

The total number of fossils recovered from the Madygen Formation may well exceed 30,000. Approximately 90% of the specimens are stored at the Borissiak Palaeontological Institute of the Russian Academy of Sciences in Moscow (abbreviated PIN). The second most important Madygen collection is housed at the Geological Institute of the Technische Universität Bergakademie Freiberg (abbreviated FG). Smaller amounts of material have been deposited at the Geological Institute of the Russian Academy of Sciences in Moscow, the Chernyshev Central Geological Survey and Research Museum in St. Petersburg, the Moscow State University of Geological Exploration, the Saratov State University, the Uzbek Geological Survey in Tashkent, and the Geological Institute of Martin-Luther-Universität Halle-Wittenberg (Dobruskina, 1995; Voigt et al., 2006).

Flora

Palynomorphs

Palynomorphs (spores and pollen) are poorly documented from the Madygen Formation. Ikramov (1967a) mentioned palynological studies of Yu. M. Kuzichkin that indicate a Permian age for the lower part of the succession. Berezanskii (1999) noted, without details, that spores and pollen are preferentially preserved in dark claystones of the Madygen Formation. According to Dobruskina (1995), samples for palynological analyses were taken during the 1987 field conference but, to our knowledge, results were never published. Lacustrine mudstone from Urochishche Madygen and Urochishche Dzhaylyaucho sampled during the first German Madygen expedition yielded translucent to opaque woody phytoclasts and some well-preserved cuticules but no palynomorphs (Voigt et al., 2006). A thin section of subbituminous coal from the base of MS1–3 at Urochishche Madygen yielded micro- and macrospores, pollen, and cutinite (Berner et al., 2009).

Non-vascular plants

The only reliable record of non-vascular plants from the Madygen Formation are bryophytes (Shcherbakov, 2008a; Moisan, Voigt et al., 2012). *Thallites insolites* Sixtel, 1962, once supposed to be a thallophyte (Dobruskina, 1995), is now referred to the endemic lycopsid *Mesenteriophyllum kotschnevii* Sixtel, 1961 (Moisan, Voigt et al., 2012). Whether the finds of Charales, mentioned by Sixtel (1960a), came from the Madygen Formation remains unclear. Bryophytes have been observed in lacustrine mudstones of MS2–3 and MS4 at Urochishche Madygen and Urochishche Dzhaylyaucho. The flora comprises three new fossil bryophyte taxa (Moisan, Voigt et al., 2012): the thalloid liverwort *Ricciopsis ferganica* Moisan, Voigt et al., 2012 (Fig. 6A), and two leafy moss gametophytes, *Muscites brickiae* (Moisan, Voigt et al., 2012) and *Muscites* sp. All localities with mass occurrences of *Ricciopsis ferganica* occur within marginal lacustrine deposits, supporting the idea that this fossil liverwort formed mats of tangled thalli floating in low water along the lake shoreline. The relative abundance of bryophytes at Madygen suggests that they played an important role in the ecosystem and probably fostered the diversity of aquatic Coleoptera and Heteroptera, amphibiotic insects (Ephemeroptera, Odonata, Plecoptera), and other invertebrates by providing well-oxygenated open-water microhabitats (Shcherbakov, 2008a).

Vascular plants

The Madygen Formation contains representatives of all major groups of Mesozoic vascular plants except angiosperms. Most abundant are pteridosperms and sphenopsids, while lycopsids, ferns, cycadophytes, ginkgoaleans and conifers together account for less than one-third of the plant material collected during Soviet times (Dobruskina, 1995). At the generic level, the Madygen fossil flora comprises about as many endemic as widespread plant taxa. Macrofloral remains are most common in lacustrine sediments (Dobruskina, 1970a, 1995; Voigt et al., 2006) and fine-grained fluvial overbank deposits (Moisan et al., 2011). Detailed palaeobotanical studies have been done for lycopsids, pteridosperms, and cycadophytes (Dobruskina, 1974, 1975; Moisan et al., 2011; Moisan, 2012; Moisan and Voigt, 2013).

Sphenopsids are dominated by the cosmopolitan genus *Neocalamites* Halle, 1908 and two kinds of strobili tentatively assigned to *Neocalamostachys* Kon'no, 1962 and *Echinostachys* Brongniart, 1828. Mass occurrences of *Neocalamites* in marginal lacustrine deposits of MS2–3 imply that this sphenopsid formed dense shore vegetation. Lycopsids turned out to be more common than stated by Dobruskina (1974, 1995), constituting a mixture of subarborescent and herbaceous forms with characteristics of primarily (semi-)aquatic habitats. *Isoetites madygensis* Moisan and Voigt, 2013 (Fig. 6B), *Isoetites sixteliae* Moisan and Voigt, 2013 (Fig. 6C–D) and *Lepacyclotes zeilleri* Retallack, 1997 (Fig. 6E) were herbaceous lycopsids with the first two taxa being almost identical to the extant quillwort *Isoëtes* Linnaeus, 1753. The endemic taxa *Ferganodendron sauktangensis* Dobruskina, 1974, *Pleuromeiopsis kryshtofovichii* Sixtel, 1962 and *Mesenteriophyllum kotschnevii* Sixtel, 1961 (Fig. 6F) resemble *Pleuromeia*-type lycopsids of the Early Triassic. Ferns are relatively rare in the Madygen Formation and are mainly represented by the widespread genera *Danaeopsis* Heer, 1865 and *Cladophlebis* Brongniart, 1849.

Pteridosperms (seedferns), as the most diverse group of plants of the Madygen Formation, include endemic genera (e.g., *Edyndella* Mogutcheva, 1973, *Madygenopteris* Sixtel, 1956, *Madygenia* Sixtel, 1956, *Uralophyllum* Kryshtofovich et Prynada, 1933, *Vittaephyllum* Dobruskina, 1975; Fig. 6H) and widespread taxa such as *Lepidopteris* Schimper, 1869, *Peltaspermum* Harris, 1937, *Ptilozamites* Nathorst, 1878, *Sagenopteris* Presl, 1838, and *Scytophyllum* Bornemann, 1856. Ginkgoales are numerous and represented by well-known genera: *Baiera* Braun, 1843, *Sphenobaiera* Florin, 1936, *Ginkgoites* Seward, 1919, and *Glossophyllum* Kräusel, 1943. All sites with ginkgoaleans are dominated by isolated leaves of *Glossophyllum*. Cycadophytes comprise the cycadalean *Pseudoctenis lanei* Thomas, 1913 (Fig. 6G) and the bennettitaleans *Pterophyllum pinnatifidum* Harris, 1932 and *P. firmifolium* Ye, 1980 (in Wu et al., 1980). The Madygen cycadophytes show cuticular features (trichomes and leaves densely covered with papillae) that are commonly considered to indicate xeromorphy. As the same features also occur in hygrophytic and halophytic modern plants, alternative functional interpretations have been proposed, e.g., self-cleaning of the leaf surface, providing mechanical stability to wind speed, and defence against phytophagous insects (Moisan et al., 2011). Conifers are not diverse and relatively uncommon except for *Podozamites* Braun, 1843 and some reproductive organs and seeds of unspecified gymnosperm affinity. *Mesenteriophyllum* (Fig. 6F), previously referred to conifers (Sixtel, 1961; Dobruskina, 1995), is now considered a form taxon for isolated leaves of subarborescent lycopsids (Moisan, 2012; Moisan and Voigt, 2013).

Fauna

Lower aquatic invertebrates

The record of lower aquatic invertebrates from the Madygen Formation comprises microconchids, bivalves, and gastropods (Fig. 7A–C). Microconchids are represented by casts of small (1–3 mm in diameter) planispiral tubes that form loose to dense aggregations on various plant remains (Fig. 7C). Although the alternative classification as spirorbiform tubeworms cannot be tested because calcareous tubes are not preserved in the Madygen specimens, we tentatively assign the finds to microconchids because spirorbiform polychaetes are neither known from pre-Jurassic beds nor from

Figure 7 Aquatic invertebrates. **A**, freshwater bivalves with concentric sculpture patterns are very common in back swamp lake deposits; **B**, tiny gastropods have been found in shallow lake deposits, either isolated or associated with plant remains; **C**, plant remains in close association with coiled shells of microconchids; **D**, conchostracan valve from prodeltaic-lacustrine mudstone; **E–F**, ostracods from shallow lake deposits; **G–H**, remains of triopsid-like Kazacharthra from shallow lake deposits; **I**, decapod fossil from prodeltaic-lacustrine mudstone. Scale units equal 5 mm (**A–C**, **G–I**) and 1 mm (**D–F**).

freshwater deposits (Taylor and Vinn, 2006; Vinn and Mutvei, 2009; Vinn, 2010). All microconchids of the Madygen Formation come from organic-rich mudstones of MS1–3 at Urochishche Madygen and Sauk Tanga. The presence of sessile encrusting organisms supports the idea that these beds formed in a lake with an extremely low sedimentation rate.

Fossil freshwater bivalves from Madygen have been repeatedly mentioned (Sixtel, 1960a, b; Ikramov, 1967a, b, 1968; Kolesnikov, 1980; Troitsky et al., 1987) but none of these records can be definitely referred to the Madygen Formation (Dobruskina, 1995). We recorded bivalves at four different levels in the Madygen Formation. Thousands of thin, strongly deformed shells occur in a 5 m-thick horizon of dark-grey mudstone near the base of MS1–3 at Urochishche Madygen. The oval, up to 25 mm long valves with concentric growth lines seem to form a monospecific assemblage (Fig. 7A).

Autochthony is likely given the fine-grained matrix and the presence of articulated shells. Fragments of similar bivalves were found in lacustrine and deltaic-lacustrine deposits of MS1–4 at Sauk Tanga and MS2–3 and MS4 at Urochishche Madygen.

The distribution of gastropods largely corresponds to that of bivalves, suggesting that both shared similar aquatic habitats. Numerically dominating are small (3–5 mm long, 2–3 mm wide) ovoid snails (Fig. 7B) resembling the living pulmonate freshwater gastropod *Radix balthica* (Linnaeus, 1758) in shape and size. Steinkerns are locally abundant in clayey and silty lacustrine mudstones of MS1–3 (Sauk Tanga, Urochishche Madygen) and MS4 (Urochishche Madygen) with individuals being either loosely scattered in the sediment or attached to plant remains. It is assumed that these small gastropods fed like many modern mud snails on organic detritus and algae. A fragment of a much larger and probably different kind of gastropod was found in shallow lacustrine deposits of MS2–3 at Urochishche Madygen.

All discussed remains of microconchids, bivalves, and gastropods are currently under study. For the sake of completeness it should be noted that Shcherbakov (2008a) mentions statoblasts of freshwater bryozoans (Plumatellidae, Phylactolaemata) from the Madygen Formation but without reference to material or localities.

Crustaceans

Crustaceans are represented by conchostracans, ostracods, kazacharthrans, and a malacostracan decapod (Fig. 7D–I), but none of these groups has ever been formally described from the Madygen Formation. The occurrence of conchostracans at Sauk Tanga (Novozhilov and Kapelka, 1960) actually refers to material from the overlying Late Triassic–Early Jurassic Kamysh-Bashi Formation (Dobruskina, 1995). Poorly preserved conchostracans were recently found in limonite-cemented sandy siltstone from the base of MS1–4 at Urochishche Madygen. The only other record of this group is the concave print of a single valve from prodeltaic-lacustrine mudstone of MS2–3 at Urochishche Dzhaylyaucho (Fig. 7D). With its large umbonal area and the low number of growth lines, this specimen resembles *Sedovia fecunda* Novozhilov, 1958, which Kozur and Weems (2010) considered a subjective junior synonym of *Vileginia tuberculata* (Novozhilov, 1946) within the late Anisian *Diaplexa tigjanensis* Zone. More and better-preserved material is needed for reliable taxonomic and stratigraphic conclusions. Conchostracans are rare in the Madygen Formation; this might be due to competition with other crustaceans (e.g., ostracods), predation pressure by kazacharthrans and fish, or unfavourable physico-chemical conditions.

Ostracods are known in large number from lacustrine deposits of MS1–3 (Urochishche Madygen), MS1–4 (Urochishche Madygen, Sauk Tanga), MS2–3 (Urochishche Dzhaylyaucho) and MS4 (Urochishche Madygen). In order of decreasing abundance they have been observed in clayey to silty mudstone, sandy siltstone, and limonite-cemented, probably reworked sandy siltstone. Most samples show only disarticulated valves in external view. If preserved, shells are thin and neither physically nor chemically extractable without destruction. Two basic types of ostracods can be discerned: (1) An up to 1.41 mm long form with pear- to club-shaped valves is only known by a few steinkerns, all preserved in sandy siltstone (Fig. 7E). (2) More common are up to 1.25 mm long, rounded-rectangular shells with reticulate ornaments forming monotypic assemblages in clayey to silty mudstones (Fig. 7F). Schudack (2008) proposed a close relationship to extinct Permian freshwater ostracods of eastern European Russia, specifically the darwinulacean *Suchonella anybensis* Kashevarova, 1986 (Late Permian, Volga-Ural region; Madygen type 1) and the cytheracean *Permiana* Sharapova, 1948 (Middle to Late Permian, south Timan and Volga-Ural region; Madygen type 2). Many modern freshwater ostracods live as detritivores on or in the substrate and are potential prey of fishes, dipteran larvae, and gastropods (Carbonel *et al.*, 1988).

Kazacharthra are an extinct order of early Mesozoic branchiopods similar to the extant Notostraca (tadpole shrimps) and represent the first noticed and most common crustaceans of the Madygen Formation (Mamrenko, 1968; Dobruskina, 1970a, 1995; Novozhilov, 1970; Voigt *et al.*, 2006; Shcherbakov, 2008a; Voigt and Hoppe, 2010). The projected number of remains can locally exceed 1000 specimens per square metre. Typically they are preserved as isolated head shields and abdominal sections of variable length (Fig. 7G–H). Specimens of kazacharthrans may represent remains of carcasses (true body fossil), moulted exoskeletons (exuvia), or casts of the one or the other (impression). All remains from Madygen are of consistent morphology and have been assigned to a single taxon, *Almatium gusevi* (Chernyshev, 1940) (Novozhilov, 1970; Dobruskina, 1970a, 1995). Substantial evidence for another species, *Jeanrogerium sornayi* Novozhilov, 1959, mentioned by Shcherbakov (2008a) according to an unpublished note of N. I. Novozhilov, is still missing. Remains are common in clayey to silty lacustrine mudstone of MS1–4 and MS4 at Urochishche Madygen. A single specimen each is known from clayey to silty lacustrine mudstone of MS2–3 at Urochishche Madygen and Urochishche Dzhaylyaucho (L II/1). The distribution of kazacharthrans in the Madygen Formation is not readily understood except for the fact they are restricted to lacustrine mudstones (Voigt et al., 2006; Voigt and Hoppe, 2010). Madygen *Almatium* probably avoided active shorelines with high input of terrigenous matter. Living notostracans are considered as predatory omnivores feeding on protozoans, bacteria, plants, and small invertebrates but also larval stages of their own species (Fox, 1949; Fryer, 1985; Chen et al., 1996). A similar lifestyle can be assumed for kazacharthrans.

The partial exuvia of an unidentified decapod from MS2–3 at L II/1 is the only evidence of malacostracan crustaceans in the Madygen Formation (Voigt et al., 2006). Preserved are part of head and thorax, and the first pereiopods with 1 cm long chelae (Fig. 7I). Considering the deltaic-lacustrine interpretation of the preserving bed, this exoskeleton may have belonged to a terrestrial or semi-terrestrial decapod.

Insects

Insects are the most abundantly collected fossils from the Madygen Formation. Including recent finds, the total number of specimens certainly exceeds 25,000. Although not fully studied, the material has already been assigned to more than 500 species with representatives of almost all major groups of Triassic insects (Rasnitsyn and Quicke, 2002; Shcherbakov, 2008a, 2011). Insect remains have been found in lacustrine deposits of all outcrop areas, but more than 80% of the archived material comes from a single locality (L II/1) at Urochishche Dzhaylyaucho. Most specimens are represented by isolated wings of flying adults; articulated remains come from marginal lacustrine deposits.

The Madygen insect fauna is numerically dominated by Coleoptera (Fig. 8A–B), Blattodea (Fig. 8E–G), and Hemiptera, which constitute at least four out of five finds at any insect locality within the Madygen Formation. Beetles have been assigned to 73 species in 34 genera and 11 families, 90% of which belong to the Archostemata, the most ancient group of coleopterans (Ponomarenko, 1969, 1977, 2002; Arnoldi et al., 1991; Shcherbakov, 2008a). The remarkable diversity of the Madygen beetles suggests adaptation to a variety of ecological niches. The cupedid Archostemata, whose 30 species represent the all-time diversity high of this important family, were mostly phytophages that lived in rotting wood feeding on fungi and decaying plant matter (Arnoldi et al., 1991). Most representatives of schizophorid Archostemata (17 species) are interpreted as amphibiotic-aquatic or riparian detritivores and carnivores including large forms up to 35 mm in length and specialized mollusc-eaters (Arnoldi et al., 1991; Shcherbakov, 2008a). The remaining coleopteran groups Archostemata, Adephaga and Polyphaga are rarer; they include algophagous, xylodetritophagous, carnivorous and spermophagous amphibiotic–aquatic and terrestrial forms. Obreniid polyphagans are presumed to have developed in gymnosperm strobili (Zherikhin and Gratshev, 1993). Cockroaches are known by 26 species in 11 genera and 6 or 7 families of Palaeozoic and Mesozoic affinity (Vishnyakova, 1998; Papier and Nel, 2001; Vršanský et al., 2002; Shcherbakhov, 2008a). Most of the exceptionally abundant blattodean remains belong to Phylloblattidae and Caloblattinidae. The most diverse group of Madygen insects are the Hemiptera, with representatives of 19 families of Auchenorrhyncha (cicada, hoppers) (Fig. 8D), Sternorrhyncha (plant lice), and Coleorrhyncha (true bugs) (Shcherbakov and Popov, 2002; Shcherbakov 2008a, 2011). The dominating Cicadomorpha include a number of forms that apparently had a cryptic appearance living hidden in dense vegetation. With bizarrely shaped tegmina (modified leathery forewings), dorsal projections on the thorax and tegmen, well-developed surface sculpture, and dull coloration, these insects may have mimicked thorns, bracts, seed-bearing organs, seeds, or buds of their host plants in order to reduce predation risk by small arboreal insectivorous reptiles like *Sharovipteryx*, *Longisquama*, or *Kyrgyzsaurus* (Shcherbakov, 2011).

Subdominant insect groups include Mecoptera (scorpionflies), Miomoptera, Neuroptera (lacewings), Orthoptera (grasshoppers), Phasmatodea (stick insects), Protorthoptera/Grylloblattodea, and Titanoptera (Fig. 8C) (Shcherbakov, 2008a).

Figure 8 Insects. Non-aquatic insects are the most common animal remains in the Madygen Formation, including representatives of the orders Coleoptera (**A**, **B**), Titanoptera (**C**), Auchenorrhyncha (**D**), and Blattodea (**E–G**). The formation is famous for articulated fossil insects (**F**, a cockroach showing the body, all four wings and the legs preserved) as well as for the preservation of colour patterns (**D**, wings of a cicada; **G**, pronotum of a cockroach). Scale units equal 5 mm.

Remarkably abundant and diverse are grasshoppers (>1400 specimens; 11 families, 74 genera, 109 species) and their close relatives, the Titanoptera (~250 specimens; 3 families, 8 genera, 21 species). Elytra from Madygen are among the earliest records of disruptive colour patterns and stridulatory structures in Orthoptera (Gorochov and Rasnitsyn, 2002). Similar modifications of the forewings are documented for the spectacular Titanoptera (Fig. 8C), whose largest members could attain a wingspan of at least 40 cm (Sharov, 1968; Grimaldi and Engel, 2005). Only a few specimens of Dermaptera (earwigs), Diptera (true flies), Glosselytrodea/Jurinida, Hymenoptera (wasps), and Psocoptera (booklice) have been found. Madygen represents the second earliest dipteran fauna and the earliest record of Hymenoptera (Rasnitsyn, 1969; Shcherbakov et al., 1995; Shcherbakov, 2008a). Entirely amphibiotic insects, Ephemeroptera (mayflies), Odonata (dragonflies), Plecoptera (stoneflies), and Trichoptera (caddisflies), are uncommon (altogether not more than 200 specimens) but relatively diverse (Sukatsheva, 1973; Sinitshenkova 2000, 2002; Pritykina, 1981; Shcherbakov, 2008a).

Fishes

The Madygen ichthyofauna (Figs. 9–10) is composed exclusively of endemic species of actinopterygians, sarcopterygians and chondrichthyans. The most abundant and diverse group of fishes are actinopterygians, of which six genera and species in four families have been described to date (Sytchevskaya, 1999): the palaeoniscids *Ferganiscus osteolepis* and *Sixtelia asiatica*, the 'perleidids' *Alvinia serrata* and *Megaperleidus lissolepis*, the evenkiid *Oshia ferganica*, and the saurichthyid *Saurichthys orientalis*. Numerically dominating are remains of the 10–20 cm long palaeoniscid *Ferganiscus osteolepis* Sytchevskaya and Yakovlev, 1999 (in Sytchevskaya 1999) (Fig. 9A). Judging from its fusiform body, this species was a pelagic form that possibly lived on insects or crustaceans. Slightly less abundant is the 6–12 cm long *Sixtelia asiatica* Sytchevskaya, 1999 (Fig. 9B). It resembles *Ferganiscus* in its cranial anatomy, but its deeper body, alternately positioned fins and short, ventrally placed mouth suggest a benthic habit. The small, rather deep-bodied *Alvinia serrata* Sytchevskaya, 1999 and its larger counterpart *Megaperleidus lissolepis* Sytchevskaya, 1999 have been compared to the Triassic genus *Perleidus* De Alessandri, 1910 but these affinities are questionable (see Lombardo, 2001). Growing up to 45 cm, *Oshia ferganica* Sytchevskaya, 1999 (Fig. 9D–E) was a predator characterized by an unusually long dorsal fin with more than 70 unforked fin rays. Together with related forms from the Triassic of Siberia, China, USA and Sweden, *Oshia* has been placed in the order Scanilepiformes by Sytchevskaya (1999), which is referred to as stem-group neopterygians by Xu and Gao (2011). At about 45 cm length, *Saurichthys orientalis* (Fig. 9G) is a rather small representative of the genus *Saurichthys* Agassiz, 1834—a taxon that is known from Triassic marine and freshwater deposits worldwide (Romano et al., 2012). *S. orientalis* exhibits six rows of thick ganoid scales (Kogan et al., 2009) and is most similar in squamation pattern and fin morphology to some Early Triassic saurichthyids as well as to younger Triassic *Saurichthys* species found in freshwater deposits (*S. gigas* Woodward, 1890, *S. gracilis* Woodward, 1890 and *S. parvidens* Wade, 1935, from the Anisian of Australia; *S. huanshenensis* Chou & Liu, 1957, from the ?Rhaetian of North China; see Romano et al., 2012). This might be indicative of separate saurichthyid evolution in freshwater habitats, favouring the retention of heavy body armour and fin ray segmentation. Given its long and slender body, far posteriorly placed symmetrical dorsal and anal fins, strengthened tail, and elongated jaws with powerful dentition, *Saurichthys* is regarded as an efficient piscivorous ambush predator.

Several additional actinopterygians, including articulated remains, are still under description (Fig. 9C, F). Furthermore, fish microremains gained from processing of bulk samples (Franeck et al., 2012, 2013) also may contain additional forms, and together they indicate a much greater diversity of the ichthyofauna.

Sarcopterygians are represented by the dipnoan *Asiatoceratodus sharovi* Vorobyeva, 1967, and scattered scales of an undetermined coelacanth. *A. sharovi* is known from up to 30 cm-long complete skeletons, disarticulated cranial material and isolated tooth plates (Fig. 10A–B). Like most lungfishes, the species was probably omnivorous, able to live on plants, soft- to hard-shelled invertebrates and small fishes. The presence of coelacanths is based on several scales exhibiting the characteristic structure of macroscopic collagen fibre bundles running parallel within concentric growth rings and enameloid ridges in the posterior part of the external surface (Fig.

Figure 9 Fishes I: Actinopterygii. **A–C**, Actinopterygian skeletons representing juveniles of *Ferganiscus osteolepis* (**A**), *Sixtelia asiatica* (**B**) and a still unidentified form (**C**). **D–E**, Remains of the large predator *Oshia ferganica*, scales with furrows and grooves (**D**) and short fin ray segments with longitudinal ridges (**E**); **F**, undescribed actinopterygian; **G**, *Saurichthys orientalis*, the only cosmopolitan genus of Madygen osteichthyans. Scale units equal 5 mm.

Figure 10 Fishes II: Sarcopterygii (**A–D**) and Chondrichthyes (**E–H**). In addition to isolated bones and tooth plates (**A**), the dipnoan *Asiatoceratodus sharovi* is represented by well-preserved complete specimens (**B**), a rarity for the post-Palaeozoic record. The operculum (**C**) might belong to a coelacanth, which is clearly present on the basis of a typical scale (**D**). The hybodont shark *Lonchidion ferganensis* is represented by numerous juvenile tooth crowns (**E**), while fragments of dermal denticles (**F**) might indicate the presence of a xenacanthid shark. Moreover, hybodontid egg capsules of the type *Palaeoxyris alterna* (**G**) and the putative xenacanthid capsule type *Fayolia sharovi* (**H**) prove the co-occurrence of at least two different taxa of sharks in the Triassic Madygen lake. Scale units equal 5 mm (**A–D, G–H**) 0.3 mm (**E–F**).

10C–D). Actinistia are common members of Triassic marine and freshwater fish faunas (Cloutier and Forey, 1991; Schultze, 2004), and thought to have pursued their prey over rough-bottomed waters (Lombardo and Tintori, 2005).

Two kinds of oviparous chondrichthyans have been documented. Firstly, the hybodontid shark *Lonchidion ferganensis* Fischer et al., 2011 with its related egg capsule type *Palaeoxyris alterna* Fischer et al., 2011 and, secondly, the egg capsule type *Fayolia sharovi* Fischer et al., 2011 which probably originated from a xenacanthid. *L. ferganensis* refers to small (approximately 1 mm long) juvenile teeth that can be divided into three distinct morphotypes. Together these morphotypes formed a heterodont crushing and grinding type of dentition (Fig. 10E). A single large adult tooth demonstrates, to a certain degree, ontogenetic heterodonty. The species is most likely durophagous, and this is supported by the fact that the majority of teeth have been found in mollusc-rich lacustrine deposits of MS1–3 at Urochishche Madygen. The up to 12 cm-long egg capsule, *Palaeoxyris alterna* Brongniart, 1828 (Fig. 10G), is composed of six clockwise-spiralled helicoidal bands and displays a three-fold division into a fusiform body, a short pointed beak, and a long, slender pedicle with parallel ribs that are typical of Mesozoic representatives of *Palaeoxyris* Brongniart, 1828. The Madygen species, named for its unique pattern of bands with alternating width, is most similar to *Palaeoxyris* egg capsules from the early Anisian Grès à Voltzia of the Vosges, France.

Assignment of *Palaeoxyris* to hybodontid sharks is based on the concordant stratigraphical range of relevant fossils (Crookall, 1932; Zidek, 1976; Fischer et al., 2010). The presence of the egg capsule with *Lonchidion* teeth in the Madygen Formation provides further evidence for this association. The second egg capsule type, *F. sharovi* (Fig. 10H), attained a length of 7 cm and is characterized by a fusiform body of two clockwise helicoidal spiralled bands accompanied by distinct scar-lines. *Fayolia* Renault et Zeiller, 1884 indicates the presence of a yet unknown xenacanthid shark in the Madygen fauna based on the widely accepted assignment for the producer of this egg type (Renault and Zeiller, 1888; Pruvost, 1930; Schneider and Reichel, 1989). The occurrence of xenacanthids is further supported by rare xenacanthiform-like dermal denticles (Franeck et al., 2013; Fig. 10F).

Fish remains occur in all lacustrine deposits of the Madygen Formation but have been preferentially collected from MS2–3 at locality L II/1. The fossil fish assemblage seems to be site-independently dominated by juveniles, in particular with regard to certain actinopterygians (*Ferganiscus*, *Sixtelia*, *Alvinia*, and other undescribed forms) and chondrichthyans (*Lonchidion*). Early ontogenetic stages are inferred from small body size, incomplete squamation, and the low degree of ossification and ornamentation in scales and dermal bones. If the rareness of adult remains is not a taphonomic effect, the Madygen lake, and especially its vegetated shorelines, might have been preferred breeding areas, providing sufficient shelter and food for juveniles. This idea is supported by the widespread occurrence of *Palaeoxyris*, both laterally and vertically, in the Madygen Formation without any appreciable record of adult remains. This is also highly indicative of the lake acting as a long-term shark nursery (Fischer et al., 2011).

Tetrapods

Six genera and species of land vertebrates not known from any other fossil site have been described from the Madygen Formation: the amphibian *Triassurus sixtelae* Ivakhnenko, 1978, the basal reptiliomorph *Madygenerpeton pustulatum* Schoch, Voigt and Buchwitz, 2010, the synapsid *Madysaurus sharovi* Tatarinov, 2005, and three diapsids, *Longisquama insignis* Sharov, 1970, *Sharovipteryx mirabilis* (Sharov, 1971a), and *Kyrgyzsaurus bukhanchenkoi* Alifanov and Kurochkin, 2011. With the exception of *Madygenerpeton*, the type material of all tetrapods comes from the prodeltaic mudstones of MS2–3 at L II/1. *Madygenerpeton* was found in fluvial channel deposits of MS3–2 at L I/2. This locality features several successive layers with bone concentrations, which have also yielded a cynodont lower jaw and teeth as well as partially articulated and disarticulated archosauriform remains that may belong to several taxa of different body size. A possible new specimen of *Triassurus* was recently discovered in profundal lacustrine mudstone at the base of MS4 at Urochishche Madygen. Further minor occurrences of tetrapod fossils have been documented within lacustrine sediments of MS1–3 and MS1–4 at Urochishche Madygen and Sauk Tanga.

Triassurus was described as an early urodelan (Ivakhnenko, 1978) but this classification was questioned (Estes, 1981). Milner (1993) listed the broad skull with large squamosal-pterygoid bar, low number

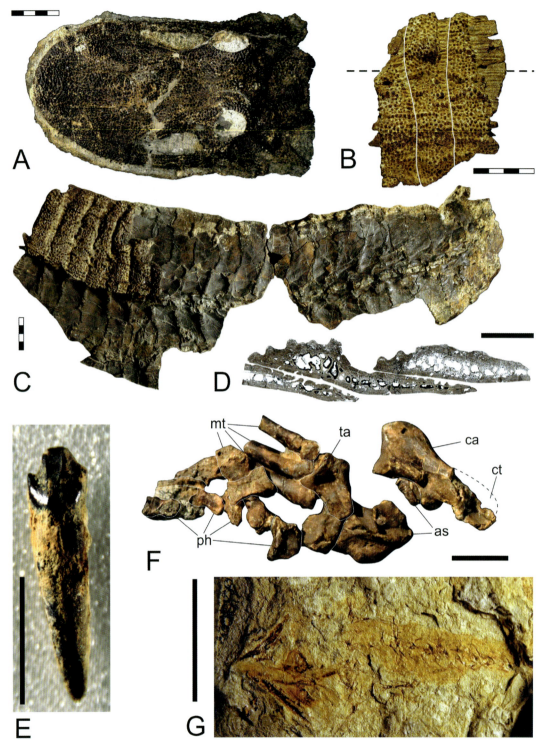

Figure 11 Tetrapod fossils from the fluvial deposits (tetrapod locality L I/2; MS 3–2) and the uppermost lacustrine deposits (MS 4) of Urochishche Madygen. **A–D**, chroniosuchian reptiliomorph *Madygenerpeton pustulatum*: **A**, Skull (specimen FG 596/V/4, holotype); **B**, carapace segment consisting of three individual osteoderms (specimen FG 596/V/8); **C**, anterior segments and internal cast of the carapace (specimen FG 596/V/7); **D**, thin-section of three neighbouring dorsal osteoderms (specimen FG 596/V/13); the position of the section is illustrated by a hatched line in B. **E**, Tooth associated with an undescribed cynodont lower jaw. **F**, tarsus and pes of an undescribed small archosauromorph. **G**, undescribed amphibian skeleton similar to *Triassurus sixtelae*. Abbreviations: as, astragalus, ca, calcaneum, mt, metatarsal, ph, phalanx, ta, distal tarsal. Scale units equal 5 mm.

of presacral vertebrae, short ribs, and slender limbs with long humeri and femora as similarities to Lissamphibia. The new incomplete specimen of an amphibian is of similar size to and has been tentatively assigned to *T. sixtelae*; it is currently under study (Fig. 11G).

Madygenerpeton pustulatum is known from a 12 cm-long skull roof with pustular ornamentation (Fig. 11A), carapace segments consisting of jointed osteoderms (Fig. 11B–D), isolated osteoderms, and a fragmentary interclavicula. The species is a late-surviving member of the Chroniosuchia, a clade of reptiliomorphs mainly reported from the latest Palaeozoic and Early to Middle Triassic of the Russian pre-Urals (Golubev, 2000; Shishkin *et al.*, 2006). *Madygenerpeton* is similar to Permian chroniosuchians in its possession of antorbital fenestrae and a broad, bony back shield (carapace). The latter consists of an anteroposterior row of individual bone plates (osteoderms), which were attached to the vertebrae and connected to each other by highly complex joints. Even though the comparison to crocodylian shields suggests that the chroniosuchian back shields supported the vertebral column and trunk-carrying system during raised walk on land (Clack and Klembara, 2009; Buchwitz and Voigt, 2010; Buchwitz, Witzmann *et al.*, 2012), there are some further indications for a rather waterbound lifestyle in the case of *Madygenerpeton*: (1) Despite its similarly high width, the carapace of *Madygenerpeton* was laterally more flexible than those of related Permian chroniosuchians due to its derived articulation facets on the carapace segments. Thus, *Madygenerpeton* was probably an undulating swimmer (Buchwitz and Voigt, 2010). (2) The osteoderms of *Madygenerpeton* have a greater compactness than those of some related Permian chroniosuchians and arguably also served as an anti-buoyancy device typical for shallow-water swimmers (Buchwitz, Witzmann *et al.*, 2012). (3) *Madygenerpeton* had a more crocodile-like skull with a massive akinetic snout region, which had a parabolic outline, tiny nares, and lacked a premaxillary fontanelle (Schoch *et al.*, 2010). These autapomorphies were probably linked to a change in the prey-catching capability (Buchwitz, Foth *et al.*, 2012).

Madysaurus sharovi has been classified as an aberrant member of the procynosuchian lineage (Tatarinov, 2005). As a basal cynodont it could have been closely related to Late Permian taxa and might, like *Madygenerpeton*, represent the Triassic relict of a group that invaded Central Asia during the latest Palaeozoic. The new therapsid material from LI/2 is a fragmentary 9 cm-long lower jaw and associated postcanine teeth bearing crowns with several cusps and long roots (Fig. 11F). The relationship of the new specimen to *Madysaurus sharovi* remains to be determined.

Longisquama insignis is based on a 15 cm-long anterior portion of a skeleton (Fig. 12A) with seven impressions of elongated dorsal appendages diverging from the midline of its back. Seven more specimens of isolated appendages without contact to bone have been found at L II/1 (e.g. Fig. 12B-C; reviewed in Buchwitz and Voigt, 2012). Sharov's (1970) classification of *Longisquama* as an archosaur was questioned in later studies, which proposed prolacertiform archosauromorphs (Peters, 2000), lepidosauromorphs (Unwin *et al.*, 2000), and the basal diapsid clade 'Avicephala' (Senter, 2004) as alternatives. Sharov (1970) regarded the dorsal appendages of *Longisquama* as elongated scales that might have been related to the origin of avian feathers. This hypothesis was adopted by Jones *et al.* (2000) who argued that a tree-climbing basal archosaur bearing feathers homologous to avian feathers could be a reasonable alternative to a theropod ancestor of birds. The structural complexity of *Longisquama* appendages, which is only topped by vaned feathers, suggests that they are the product of the same crucial evolutionary innovation in the archosauromorph skin that also led to the wide range of elongated skin appendages found in Mesozoic archosaurs, but does arguably not imply that *Longisquama* was closely related to birds (Voigt *et al.*, 2009; Buchwitz and Voigt, 2012). Whereas the interpretation of *Longisquama* as an insectivorous tree-dweller (e.g., Alifanov and Kurochkin, 2011; Shcherbakov, 2011) remains to be tested, the idea that its dorsal appendages functioned as a device for parachuting or gliding flight (Sharov, 1970; Haubold and Buffetaut, 1987; Martin, 2004) is unlikely for mechanical reasons and because there is no evidence for a wing-like arrangement of the appendages in the holotype (Unwin and Benton, 2001; Voigt *et al.*, 2009; Fig. 13).

Sharovipteryx mirabilis is characterized by relatively long hind limbs and a large uropatagium (wing membrane between legs and tail) but the shape and size of other wing membranes are not readily deducible from the only known specimen (Gans *et al.*, 1987; Dyke *et al.*, 2006; Fig. 12E). It probably represents the earliest

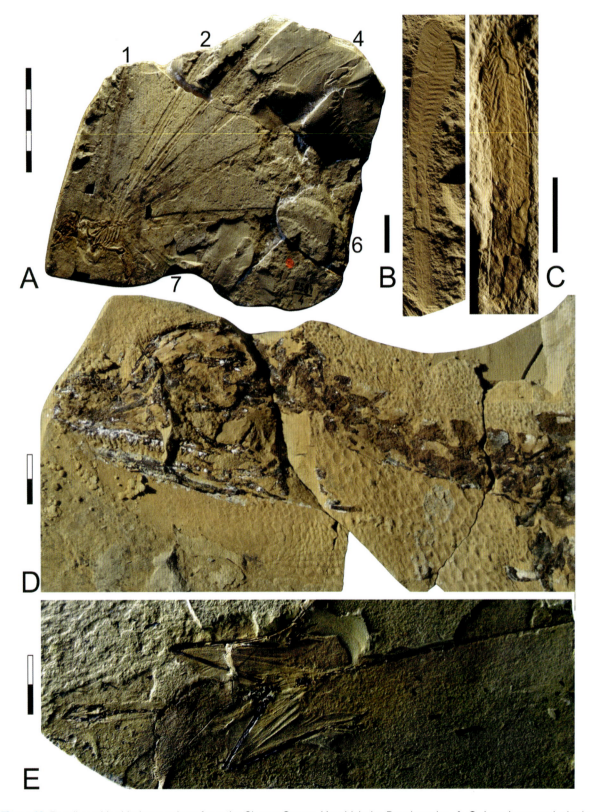

Figure 12 Reptiles with skin impressions from the Sharov Quarry, Urochishche Dzaylyaucho. A–C, *Longisquama insignis*, specimens PIN 2584/4 (A, holotype), 2584/7 (B), and FG 596/V/3 (C). D, *Kyrgyzsaurus bukhanchenkoi*, holotype PIN 2584/12. E, *Sharovipteryx mirabilis*, holotype PIN 2584/8. Scale units equal 10 mm.

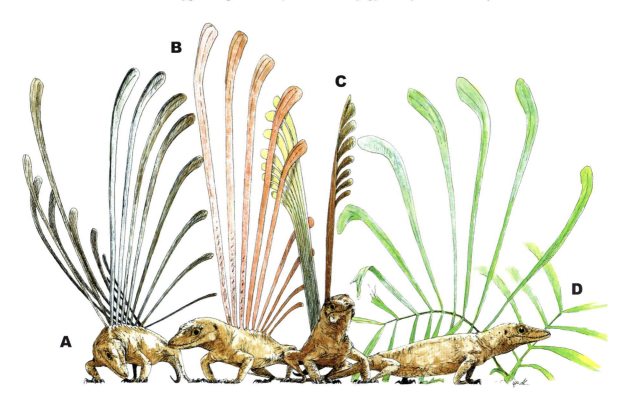

Figure 13 Various restorations of *Longisquama insignis*. **A**, double row of detached appendages; after a painting that Russian palaeoentomologist A. G. Ponomarenko (pers. comm) prepared for an exhibition of Madygen fossils during the 1970s. **B**, single row appendages as envisioned by Sharov (1970), represents the most realistic reconstruction; **C**, two rows of tightly set appendages forming two wings; this idea occurred for the first time in a popular science book by Halstead (1975); **D**, restored without any dorsal appendages but depicted as a *Longisquama* carcass washed together with foliage (e.g., Paul, 2001, p. 64).

example for a limb-supported gliding flier (Schaller, 1985). According to aerodynamic model calculations, a double delta-wing membrane shape would have been the most advantageous for *Sharovipteryx*, providing a low gliding angle and landing speed (Dyke *et al.*, 2006). A relationship to prolacertiform archosauromorphs has been suggested (Tatarinov, 1989; Peters, 2000; Unwin *et al.*, 2000); some authors also find indications for a pterosaur affinity of *Sharovipteryx* (Peters, 2000; Unwin, 2000).

The holotype of *Kyrgyzsaurus bukhanchenkoi* (Fig. 12D) comprises the anterior part of an articulated skeleton, surrounded by impressions of scaly skin. *Kyrgyzsaurus* has a short skull with relatively large orbitae. The dentition consists of dense, mostly straight teeth with broad tips, which is consistent with an insectivorous diet. *Kyrgyzsaurus* was probably a basal archosauromorph but its classification as a drepanosaurid (Alifanov and Kurochkin 2011), largely based on the observation of highly expanded spines on the first, fourth, and fifth thoracic neural arches, seems equivocal as the presumed fourth and fifth spines are more likely to represent long bones of a relatively small forelimb. The scales of the skin impression differ in shape and arrangement: longitudinal rows of larger oval scales within a meshwork of small, round to rectangular elements occur on the dorsal side, whereas enlarged hexagonal to rounded scales cover the ventral side and lateral neck. Dots spread across the skin imprint have been interpreted as small osteoderms (Alifanov and Kurochkin, 2011).

A small, partially articulated hind limb with a vaguely crurotarsan-like ankle (Fig. 11E) and an associated pelvis, both from fluvial deposits of L I/2, signify the presence of another small archosauromorph that is derived in some aspects, such as the presence of a calcaneal tuber and an expanded articular surface of the humeral head, whereas the pelvic structure with a short and plate-like ischium and pubis forming a thyroid fenestra, and several further features, indicates a position well outside the Archosauriformes (*contra* preliminary assignment of Buchwitz, 2011). A thin mudflow layer at the same locality yielded an assemblage of more than 80 scattered bones and bone fragments including over 20 cm-long limb bones and jaw fragments with serrated teeth, among them a massive 18 cm-long maxilla. Pending further results, these fossils are tentatively referred to large archosauriform diapsids. They constitute a modern element of the Madygen tetrapod fauna and increase the rather sparse Asian record of Triassic terrestrial archosauromorphs, which includes finds from the Early to Middle Triassic of China (e.g. Young, 1973; Zhang, 1975; Wu, 1981; Wu and Russell, 2001; Li *et al.*, 2006) and from the Late Triassic of Thailand and Turkey (Buffetaut and Ingavat, 1982; Buffetaut *et al.*, 1988, 2009).

Trace fossils

Trace fossils are common in the Madygen Formation, but have only been described in recent studies (Voigt *et al.*, 2006; Voigt and Hoppe, 2010). Fine-grained fluvial, deltaic and near-shore lacustrine deposits often show pervasive penetration by downward or horizontally branched tubes filled with carbonaceous sediment, which were interpreted as root traces (Fig. 14A). Size, depth, orientation, branching pattern, and surficial morphology of tubular casts allow separation of at least six types of rhizoliths within the Madygen Formation, which may be referred to a wide range of low-growing and arborescent plants (Brosig, 2012). Matching root traces to certain plant fossils might be possible at Madygen since there are a number of sites where rhizoliths and autochthonous to parautochthonous plant remains occur in the same strata (Moisan *et al.*, 2011).

Horizontal networks of tiny, multiple-branched tubular burrows (Fig. 14B) constitute the most common trace fossils in entirely lacustrine deposits of the Madygen Formation (Voigt and Hoppe, 2010). The shallow, penetrative traces have been interpreted as combined dwelling and deposit-feeding structures of wormlike invertebrates. Maximum intensity of bioturbation is recorded in mudstones that probably formed around the sublittoral–profundal boundary, which usually coincides with the thermocline–chemocline level in modern stratified lakes. In analogy to extant oligochaetes and aquatic insect larvae, it is supposed

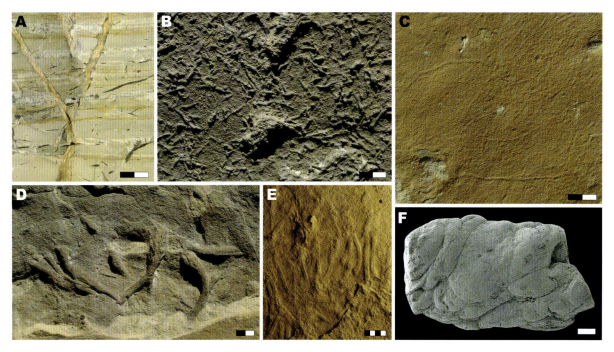

Figure 14 Trace fossils. **A**, Root traces in near-shore lacustrine deposits; **B**, Dwelling and deposit-feeding structures of wormlike invertebrates in lacustrine deposits; **C**, Traces of shallow subsurface miners in deltaic-lacustrine deposits; **D**, Unidentified sand-filled burrow systems from high-energy deltaic-lacustrine setting; **E**, Ribbon-like burrows from prodeltaic siltstone; **F**, Spiral-shaped fish (micro-)coprolite. Scale units equal 5 mm (**A–E**) and 0.1 mm (**F**).

that the producers of the Triassic traces were able to tolerate water with low dissolved oxygen – a trait that enabled them to exploit deeper parts of the lake while avoiding predation by carnivorous aquatic animals from well-aerated lake zones. This kind of trace is ubiquitous in lacustrine mudstone of MS1–4, MS2–3, and MS4 but it has not yet been observed in organic-rich lake deposits of MS1–3. Voigt and Hoppe (2010) stressed the overall similarity of the burrows with smooth-walled ophiomorphid trace fossils, particularly *Thalassinoides paradoxicus* (Woodward, 1830), while Knaust (2010) regarded it as synonymous with his new ichnogenus *Virgaichnus*.

Three other types of feeding, and maybe combined dwelling traces of wormlike aquatic invertebrates, are worth mentioning: (1) *Helminthoidichnites tenuis* Fitch, 1850 is represented by smooth curvilinear, unbranched furrows and ridges of up to 0.5 mm width in clayey to silty, rhizolithic mudstone of MS2–3 (Fig. 14C). All features point to traces of shallow subsurface miners in interdistributary bay deposits of the deltaic-lacustrine facies. (2) Unidentified sand-filled burrow systems of up to 5 mm width and unilaterally radiating branching tubes occur at the base of cross-bedded sandstones in MS2–2 and MS2–3 (Fig. 14D). These traces may be indicative of high-energy microhabitats within deltaic-lacustrine settings. (3) Enigmatic ribbon-like burrows with transverse segmentation have been observed in prodeltaic siltstone of MS4 (Fig. 14E). Voigt and Hoppe (2010) interpreted them as locomotion traces of branchiopod arthropods because these ichnia occur in strata with kazacharthran remains, and in some cases resemble bilobed arthropod trails. Preservation in concave and convex epirelief, however, suggests endostratal origin.

Small spiral-shaped coprolites of up to 2 mm in length and 1.4 mm in diameter were recorded in microfossil samples from the mollusc-rich mudstone of MS1–3 at Urochishche Madygen (Fig. 14F). The material comprises four incomplete specimens that share the heteropolar condition of unevenly spaced coils. Given the occurrence of excremental remains and juvenile teeth of *Lonchidion ferganensis* in the same bed, a relation to hybodontid shark is likely, but not established because this type of coprolite is known from lungfish and piscivorous actinopterygians as well (Jain, 1983; Northwood, 2005).

Traces of insect damage on fossil plants remain a largely unexploited field of ichnological information from the Madygen Formation. Arthropod feeding traces on fossil leaves of Madygen ferns, pteridosperms, ginkgoaleans, and gymnosperms are quite common (Zherikin, 2002; Shcherbakov, 2008a). Ellipsoidal egg scars on the leaves of *Isoetites madygensis* and *Isoetites sixteliae*, probably produced by dragonflies of the extinct suborder Archizygoptera, represent the first evidence of oviposition damage on fossil lycopsids (Moisan, Labandeira et al., 2012).

Implications of the Madygen palaeoecosystem
Biotic reconstruction

The Madygen Formation developed in a freshwater lake basin at mid-northern latitude and several hundred kilometres away from the nearest marine shoreline. Given its heterogeneous lithology, the Triassic Madygen lake was situated in a landscape of marked relief. Alluvial fans with temporarily subaqueous deposition of coarse-grained terrigenous material indicate that the lake was surrounded by nearby mountain ranges (Fig. 5). An open system at least several kilometres in extent has to be assumed, based on the distribution of lacustrine sediments and the presence of some widespread Triassic fishes such as *Saurichthys* or *Lonchidion* (Kogan et al., 2009; Fischer et al., 2011). According to palaeoclimatic reconstructions (Scotese, 2000), present-day Central Asia belonged to the warm-temperate climatic zone during the Middle to early Late Triassic. Apart from the rich and diverse biota of the Madygen Formation, it is the ubiquity of hydromorphic soils and the lack of mudcracks and other desiccation features like raindrop imprints, tetrapod footprints and evaporites, which suggest constantly moist or wet (humid) palaeoclimatic conditions (Brosig, 2012). Statoblasts of freshwater bryozoans mentioned by Shcherbakov (2008a) might be used as an argument for temperature seasonality, as resting devices of these invertebrates usually serve for hibernation (Wood, 1983; Oda, 1984). Although the annual average temperature was relatively high during the Triassic, air temperatures between winter and summer may have been significantly different in continental regions of middle and high latitude (Sellwood and Valdes, 2006).

Terrestrial habitats of the Triassic Madygen basin and the ecology of their biota are still poorly known. Coaly detritus and scarce root traces are the only fossils observed in alluvial fan deposits. They provide evidence that the proximal parts of the basin and the mountainous hinterland were at least covered by scattered vegetation (Fig. 5). More distal parts of the alluvial plain were characterized by shallow 20–60 m wide channels with thin lateral sand sheets that graded into extended (>1000 m) muddy floodplain fines. Mudflow deposits in overbank areas, numerous small distributary channels, and the absence of soil formation on the near-channel sand sheets indicate a rather unstable fluvial system with frequently migrating channels and common overflow (Brosig, 2012). Based on the distribution of root traces, vegetation was mainly restricted to overbank fines and shows little evidence for defined associations. The shoreline of backswamps was predominantly covered by sphenopsids, and silty mudstone sometimes reveals mass occurrences of a single root type of arborescent plants (Brosig, 2012). Sphenopsids, lycopsids, pteridosperms, ginkgoaleans, and cycadophytes were probably the dominating plants of the lowland areas that surrounded the Madygen lake (Moisan et al., 2011; Moisan, 2012). Living and dead plant matter supported a rich and diverse fauna of phytophagous (e.g., Coleoptera, Hemiptera, Mecoptera, Miomoptera, Orthoptera, Phasmatodea) and saprophagous insects (e.g., Blattodea, Coleoptera) (Arnoldi et al., 1991; Zherikhin and Gratshev, 1993; Shcherbakov, 2008a). Coleoptera, Diptera, Odonata, Protorthoptera/Grylloblattodea, and Titanoptera may have included many, or in some cases even exclusively, zoophagous species. Given the abundance of arthropods, it is likely that the system has supported a variety of small insectivorous tetrapods such as *Kyrgyzsaurus*, *Longisquama*, and *Sharovipteryx* (Shcherbakov, 2011). The feeding habits of recently recovered larger and supposedly terrestrial tetrapods (cynodont, archosauriform) have not yet been assessed (Buchwitz, 2011).

The Madygen lake, as the most important aquatic habitat of the system, included organisms representing at least five trophic levels (Fig. 15). Phytoplankton (of which there is no fossil evidence as yet) and macrophytes (e.g., *Ricciopsis*, *Neocalamites*, and some lycopsids) were presumably the main primary producers. An important external source of food must have been dead organic matter (plants, insects, tetrapods, etc.) transported from the land into the lake. Zooplankton (of which there is no fossil evidence as yet), microconchids, gastropods, bivalves, conchostracans, ostracods, kazacharthrans, certain insects (e.g., schizophorid

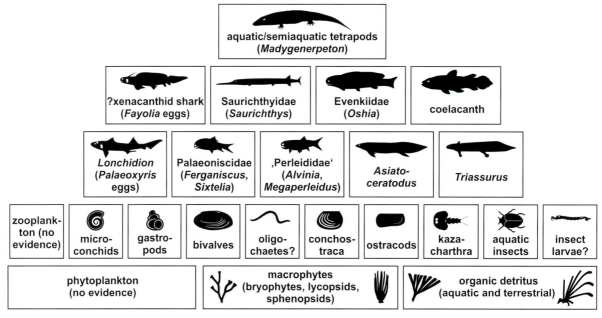

Figure 15 Ecosystem of the Madygen Formation: Trophic levels and their respective constituents as proposed for the Triassic Madygen lake environment.

beetles), and wormlike aquatic invertebrates such as oligochaetes or insect larvae (indirect evidence from trace fossils) are interpreted as primary consumers that in turn served as food for a variety of fishes including actinopterygians (Palaeoniscidae, 'Perleididae'), dipnoans (*Asiatoceratodus*), and durophagous sharks (*Lonchidion*). Four large carnivorous fishes, *Saurichthys*, *Oshia*, a so far unknown xenacanthid (suggested by *Fayolia*-type egg capsules), and the coelacanth can be considered tertiary consumers. The semi-aquatic reptiliomorph *Madygenerpton* may have been the apex predator of the Madygen lake.

On the basis of the lateral and vertical distribution of fish remains as well as the early ontogenetic stages of several members of the Madygen ichthyofauna, the following model is proposed for the Madygen lake environment (Fig. 16): *Saurichthys* (A) chased after juvenile actinopterygians (B) that lived and developed in densely vegetated shoreline habitats side by side with dipnoans (C). In shallow, well-aerated waters hybodontid and xenacanthid sharks attached their egg capsules, certainly with some degree of temporal partitioning, onto submerged driftwood or plants (D). Offspring thrived in the nutrient-rich littoral lake zone (E) until the juveniles had matured sufficiently to join the adult population. Coelacanths (F) probably dwelled in open waters, searching for food near the lake bottom, whereas larger actinopterygians such as *Oshia* (G) and adult chondrichthyans may have lived outside of the juvenile habitats, in deeper parts of the same lake, other freshwater systems or, in the case of the cosmopolitan *Saurichthys* and *Lonchidion*, perhaps even the Palaeo-Tethys (Kogan et al., 2009; Fischer et al., 2011).

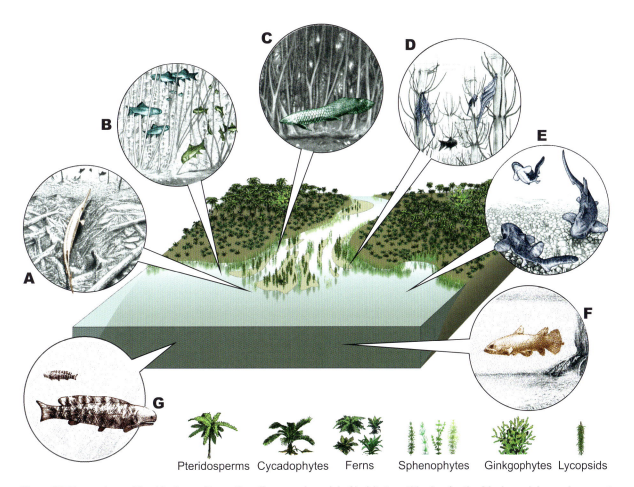

Figure 16 Ecosystem of the Madygen Formation. Proposed model of habitat partitioning for the Madygen lake environment illustrating *Saurichthys* (**A**), juvenile actinopterygians (**B**), dipnoans (**C**), *Palaeoxyris* egg capsules (**D**), hybodont shark offspring (**E**), coelacanths (**F**), and larger actinopterygians such as *Oshia* (**G**).

Evolutionary significance

With their delicate preservation of insect bodies and wing coloration patterns (Fig. 8), articulated dipnoans as well as tetrapod body surfaces and skin impressions, the lacustrine shales of the Madygen Formation, especially those of locality L II/1 (Urochishche Dzaylyaucho; Fig. 2), represent one of the few examples of Triassic Konservat-Lagerstätten (other examples: Fraser et al., 1996; Etter, 2002; Gall and Grauvogel-Stamm, 2005; Hu et al., 2011). Their importance is illustrated by the generally sparse Triassic record of unossified tetrapod skin structures: apart from the Madygen finds, some specimens of the basal archosauromorph *Tanytrachelos ahynis* (Olsen, 1979) from the marginal lacustrine deposits of Solite Quarry, Virginia (Casey et al., 2007), and several marine reptile fossils from the Monte San Giorgio, Switzerland (Sander, 1989; Renesto, 2005), skin preservation is only found in basal archosauromorph and archosaur footprints (e.g., Klein and Haubold, 2004; Bernardi et al., 2010) – unlike the later Mesozoic skin record, which is much richer and covers various reptilian groups, including plesiosaurs (Rusconi, 1948), ichthyosaurs (Lingham-Soliar, 1999), pterosaurs (e.g., Frey and Martill, 1999; Ji and Yuan, 2002; Kellner et al., 2010), various dinosaur clades (e.g., Czerkas, 1997; Christiansen and Tschopp, 2010; Xu et al., 2010 and references therein), and lepidosaurs (Caldwell and Dal Sasso, 2004; Evans and Wang, 2007, 2010).

Unlike the more completely ossified Palaeozoic dipnoans, fully articulated Triassic lungfishes are extremely rare. Only a few complete skeletons have been described from the Early Triassic of Madagascar (Lehman et al., 1959) and Angola (Antunes et al., 1990) and from the Anisian of Australia (Wade, 1935; Ritchie, 1981). Mesozoic and Cenozoic lungfishes are usually known by isolated tooth plates and, less frequently, by scattered skull bones or fragmentary skulls (Kemp, 1998; Schultze, 2004). Consequently, current systematic concepts are based either on tooth plates, skull morphology or both, which can lead to confusing interpretations (Cavin et al., 2007). In this context, the fragmentary and complete *Asiatoceratodus* specimens from the Madygen Formation provide one of the rare occasions to study whole-body morphology of Early Mesozoic dipnoans.

The insect diversity of Madygen is extraordinarily high compared to those of other Permian to Triassic localities, and marked by a high number of first and last occurrences (40% of all families), which leads to a particular late Middle Triassic peak in the family diversity curve, a typical Lagerstätten effect (Shcherbakov, 2008b).

Due to its Triassic geographic position as deposits of an inland basin along the northern rim of the Tethys, the Madygen Formation provides a link between East European land animal assemblages of the Early to Middle Triassic (cf. Ochev and Shishkin, 1989; Sennikov, 1996; Tverdokhlebov et al., 2002) and those of East Asia (cf. Lucas, 1993, 2001; Li et al., 2008) and Southeast Asia (Buffetaut et al., 2009). The Madygen Formation is of similar importance for plant palaeobiogeography, considering its position at the intersection of several Triassic floral provinces (Dobruskina, 1982, 1995).

Certain ecological and palaeobiological aspects of the Madygen lake and its inhabitants are otherwise not described from the early Mesozoic; the distribution pattern of egg capsules and shark teeth shows that habitat segregation and site fidelity of spawning grounds represents a behavioural pattern of sharks that came to be long before the rise of neoselachians (Fischer et al., 2011). Abundant trace fossils signify the settlement of less than well-aerated lake ground sediments by worms and larvae. Thus, the composition of trace fossil assemblages appears to reproduce a zonation of the lake bottom according to the degree of oxygen saturation (Voigt and Hoppe, 2010).

Conclusions

Since fieldwork recommenced in 2005, the following new aspects of the Madygen Formation, its ecosystem and palaeoenvironment have been revealed by the studies of the Moscow and Freiberg working groups:

1) Comprehensive geological remapping of the Urochishche Madygen and Sauk Tanga areas and the detailed description of their sedimentary successions have yielded a better understanding of the Madygen palaeoenvironment. The floodplain deposits in particular, with their rich plant fossil content, abundant hydromorphic soils and lack of desiccation features, are indicative of humid conditions without regular dry periods. Steep facial gradients, i.e. rapid vertical and lateral changes between lacustrine, fluvial and alluvial deposition systems, as well as numerous pyroclastic horizons, indicate that the Madygen

lake environment was located in a tectonically active region.

2) New plant taxa include several species of bryophyte water plants and cycadophytes with xeromorphic features. The enigmatic *Mesenteriophyllum* is now referred to lycopsids, which constituted a more common element of the Madygen flora than previously thought. The newly collected specimens yield a considerable record of plant–insect interactions.

3) Aquatic invertebrates are the least-studied organisms of the Madygen Formation. They are invaluable for palaeoecological and biostratigraphical analyses. There is a considerable diversity of newly recorded taxa including microconchids, freshwater bryozoans, conchostracans, and a malacostracan decapod.

4) Analogous to certain extant members of this insect group, several newly described species of cicadomorphs display wing surface structures that mimic different plant organs. Camouflage of this kind possibly functioned as a protection against predation by small tetrapods.

5) Newly discovered freshwater shark egg capsules, hybodont teeth, coelacanth scales, as well as further actinopterygian macro- and microfossils indicate a richer ichthyofauna than previously assumed (total of 13 to 16 taxa). Abundant finds of juvenile individuals and egg capsules in deposits of the lake margin suggest that it served as a spawning ground and habitat for young fish. Isotopic analysis of shark and *Saurichthys* tooth enamel has yielded a clearly non-marine signal.

6) Isolated and partially articulated skeletal finds from a newly discovered tetrapod locality within fluvial deposits include the semi-aquatic reptiliomorph *Madygenerpeton*, therapsid remains, and archosauromorph remains belonging to small- and middle-sized individuals (tooth length up to 4.5 cm). Recent fossils from lacustrine deposits include the partial skeleton and skin impression of the archosauromorph *Kyrgyzsaurus* and additional *Longisquama* skin appendages from the Sharov Quarry as well as remains of very small amphibians that probably belong to *Triassurus*.

7) Invertebrate trace fossils turned out to be a common feature within fine clastic lake sediments. Associations of varying composition reflect zonation of the lake bottom from shore to deep water.

8) A more complex picture of the Madygen ecosystem and its various environments has emerged: in the case of the floodplain sediments, differences in root system structure, plant fossil content and soil type reflect variable vegetation covers. Apart from allochthonous assemblages of animal and plant fossils from the Madygen forest, the deposits of the Madygen lake include a rich record of marginal and deep lacustrine benthic communities that reflect varying degrees of oxygen saturation, as well as variation in the input of organic and inorganic debris. Newly reported finds of aquatic producers (bryophytes), low-level consumers (invertebrate traces and body fossils) and high-level consumers (freshwater sharks, coelacanths, semi-aquatic reptiliomorphs) complement our knowledge of the structure of the lake community.

References

Abdrakhmatov, K. Ye., S. A. Aldazhanov, B. H. Hager, M. W. Hamburger, T. A. Herring, K. B. Kalabaev, V. I. Makarv, P. Molnar, S. V. Panasyuk, M. T. Prilepin, R. E. Reilinger, I. S. Sadybakasov, B. J. Souter, Yu. A. Trapeznikov, V. Ye. Tsurkov, and A. V. Zubovich. 1996. Relatively recent construction of the Tien Shan inferred from GPS measurements of present-day crustal deformation rates. Nature 384:450–453.

Agassiz, L. 1834. Abgerissene Bemerkungen über fossile Fische. Neues Jahrbuch für Mineralogie, Geognosie, Geologie und Petrefaktenkunde 1834: 379–390.

Alifanov, V. R. and E. N. Kurochkin. 2011. *Kyrgyzsaurus bukhanchenkoi* gen. et sp. nov., a new reptile from the Triassic of southwestern Kyrgyzstan. Paleontological Journal 45:639–647.

Antunes, M.T., J.G. Maisey, M. Monteiro Marques, B. Schaeffer, and K.S. Thomson. 1990. Triassic fishes from the Cassange Depression (R. P. de Angola). Lisboa: Ciêncas da Terra Número Especial 1: 1–64.

Arnoldi, L. V., V. V. Zherikhin, L. M. Nikritin, and A. G. Ponomarenko. 1991. Mesozoic Coleoptera. Oxonian Press, New Delhi.

Becker-Migidsova, E. E. 1953. [Two new representatives of Hemiptera from Madygen.] Doklady Akademiya Nauk SSSR 90:461–464 [Russian]

Berezanskii, A. V. 1999. [Geological structure and natural resources of the Sulyukta ore district (Tokhta-Boz area). Report about the results of geological mapping, geological exploration and general prospection in the scale of 1:50,000, conducted by the Tokhta-Boz party in 1993–1998.] South Kyrgyz Geological Expedition, Osh. [Unpublished report; Russian]

Bernardi, M., F. M. Petti, and M. Avanzini. 2010. A webbed archosaur footprint from the Upper Triassic (Carnian) of the Italian Southern Alps. New Mexico Museum of Natural History and Science Bulletin 51:65–68.

Berner, U., G. Scheeder, J. Kus, S. Voigt, and J. W. Schneider. 2009. Organic geochemical characterization of terrestrial source rocks of the Triassic Madygen Formation (Southern Tien Shan, Kyrgyzstan). Oil and Gas 35:135–139.

Béthoux, O., S. Voigt, and J. W. Schneider. 2010. A Triassic palaeodictyopteran from Kyrgyzstan. Palaeodiversity 3:119–123.

Bornemann, J.G. 1856. Über organische Reste der Lettenkohlegruppe Thüringens. Leipzig: Verlag Wilhelm Engelmann.

Braun, F.R. 1843. Beiträge zur Urgeschichte der Pflanzen. I. Die Fundorte von fossilen Pflanzen in der Umgegend von Bayreuth und Geschichte ihres Auffindens. In Münster, G. Graf zu (ed) Beiträge zur Petrefacten-Kunde, Bayreuth: Buchner'sche Buchhandlung, pp. 1–46.

Brick, M. I. 1934. A find of Lower Triassic flora in Central Asia. Doklady Akademia Nauk SSSR 4:412–414.

Brick, M. I. 1936. The first find of Lower Triassic flora in Middle Asia. Trudy Geologicheskii Instituta Akademiya Nauk SSSR 5:161–714. [Russian]

Briggs, D. E. G. 2003. The role of decay and mineralization in the preservation of soft-bodied fossils. Annual Review of Earth and Planetary Sciences 31:275–301.

Brongniart, A. 1828. Essai d'une Flore du grès bigarré. Annales des Sciences Naturelles 15: 435–60.

Brongniart, A. 1849. Tableau des genres de végétaux fossiles considérés sous le point de vue de leur classification botanique et de leur distribution géologique. Paris: Dictionnaire Universel d'Histoire Naturelle.

Brosig, A. 2012. Sedimentology of Middle Triassic – Lower Jurassic Fluviolacustrine Deposits in the Madygen Area (SW Kyrgyzstan, Central Asia). Unpublished diploma thesis. TU Bergakademie Freiberg, Freiberg.

Buchwitz, M. 2011. Taxonomy, Phylogenesis, and Palaeobiology of the Madygen Tetrapod Fauna. Doctoral dissertation. TU Bergakademie Freiberg, Freiberg.

Buchwitz, M., and S. Voigt. 2010. Peculiar carapace structure of a Triassic chroniosuchian implies evolutionary shift in trunk flexibility. Journal of Vertebrate Paleontology 30: 1697–1708.

Buchwitz, M., and S. Voigt. 2012. The dorsal appendages of the Triassic reptile *Longisquama insignis* – reconsideration of a controversial integument type. Paläontologische Zeitschrift 86:313–331.

Buchwitz, M., C. Foth, I. Kogan, and S. Voigt. 2012. On the use of osteoderm features in a phylogenetic approach on the internal relationships of the Chroniosuchia (Tetrapoda: Reptiliomorpha). Palaeontology 55:623–640.

Buchwitz, M., F. Witzmann, S. Voigt, and V. Golubev. 2012. Osteoderm microstructure indicates the presence of a crocodylian-like trunk bracing system in a group of armoured basal tetrapods. Acta Zoologica 93:260–280.

Buffetaut, E., and R. Ingavat. 1982. Phytosaur remains (Reptilia, Thecodontia) from the Upper Triassic of north-eastern Thailand. Geobios 15:7–17.

Buffetaut, E., M. Martin, and O. Monod. 1988. Phytosaur remains from the Cenger Formation of the Lycian Taurus (western Turkey): stratigraphical implications. Geobios 21: 237–243.

Buffetaut, E., G. Cuny, J. Le Loeuff, and V. Suteethorn. 2009. Late Palaeozoic and Mesozoic continental ecosystems of SE Asia: an introduction. Geological Society of London, Special Publications 315:1–5.

Burtman, V. S. 1997. Kyrgyz Republic. Pp. 483–492 in E. M. Moores and R. W. Fairbridge (eds), Encyclopedia of European and Asian Regional Geology. Chapman & Hall, London.

Burtman, V. S. 2008. Nappes of the southern Tien Shan. Russian Journal of Earth Sciences 10:1–35.

Caldwell, M. W., and C. Dal Sasso. 2004. Soft-tissue preservation in a 95 million year old marine lizard: form, function, and aquatic adaptation. Journal of Vertebrate Paleontology 24:980–985.

Carbonel, P., J.-P. Colin, D. L. Danielopol, H. Löffler, and I. Neustrueva. 1988. Paleoecology of limnic ostracodes: a review of some major topics. Palaeogeography, Palaeoclimatology, Palaeoecology 62:413–461.

Casey, M. C., N. C. Fraser, and M. Kowalewski. 2007. Quantitative taphonomy of a Triassic reptile *Tanytrachelos ahynis* from the Cow Branch Formation, Dan River Basin, Solite Quarry, Virginia. Palaios 22:598–611.

Cavin, L., V. Suteethorn, E. Buffetaut, and H. Tong. 2007. A new Thai Mesozoic lungfish (Sarcopterygii, Dipnoi) with an insight into post-Palaeozoic dipnoan evolution. Zoological Journal of the Linnean Society 149: 141–77.

Chen, P.-J., K. G. McKenzie, and H.-Z. Zhou. 1996. A further research into Late Triassic Kazacharthra fauna from Xinjiang Uygur Autonomous Region, NW China. Acta Palaeontologica Sinica 35:272–308.

Chernyshev, B. I. 1940. [Mesozoic Branchiopoda from Turkestan and Transbaikal.] Geologichesky Zhurnal Akademiya Nauk Ukrainsky SSR 7(3):5–43. [Russian]

Chou, H.-H., and H.-T. I. Liu. 1957. Fossil fishes from Huanshan, Shensi. Acta Palaeontologica Sinica 5: 295–305.

Christiansen, N.A., and E. Tschopp. 2010. Exceptional stegosaur integument impressions from the Upper Jurassic Morrison Formation of Wyoming. Swiss Journal of Geosciences 103:163–171.

Clack, J. A., and J. Klembara. 2009. An articulated specimen of *Chroniosaurus dongusensis*, and the morphology and relationships of the chroniosuchids. Special Papers in Palaeontology 81:15–42.

Cloutier, R., and P. L. Forey. 1991. Diversity of extinct and living actinistian fishes (Sarcopterygii). Environmental Biology of Fishes 32:59–74.

Cohen, K. M., S. Finney, and P. L. Gibbard. 2013. International Chronostratigraphic Chart v2013/01. International Commission on Stratigraphy, available at http://www.stratigraphy.org/ICSchart/ChronostratChart2013-01.jpg.

Crookall, R. 1932. The nature and affinities of *Palaeoxyris*, etc. Summary of Progress of the Geological Survey of Great Britain and the Museum of Practical Geology for 1931, 2: 122–140.

Czerkas, S. A. 1997. Skin. Pp. 669–675 in P. J. Currie and K. Padian (eds), Encyclopedia of Dinosaurs. Academic Press, San Diego.

De Alessandri, G. 1910. Studii sui pesci Triasici della Lombardia. Memorie della Società italiana di Scienza naturale 7: 1–147.

De Grave, J., M. M. Buslov, and P. Van den haute. 2007. Distant effect of India–Eurasia convergence and Mesozoic intracontinental deformation in Central Asia: constraints from apatite fission-track thermochronology. Journal of Asian Earth Sciences 29:188–204.

De Grave, J., S. Glorie, A. Ryabinin, F. Zhimulev, M. M. Buslov, A. Izmer, M. Elburg, F. Vanhaecke, and P. Van den haute. 2011. Late Palaeozoic and Meso-Cenozoic tectonic evolution of the southern Kyrgyz Tien Shan: constraints from multi-method thermochronology in the Trans-Alai, Turkestan-Alai segment and the southeastern Ferghana Basin. Journal of Asian Earth Sciences 44:149–168.

Dobruskina, I. A. 1970a. [The age of the Madygen Formation in relation to Permian and Triassic of Middle Asia.] Sovietskaya Geologiya 3(12):16–28. [Russian]

Dobruskina, I. A. 1970b. [The Triassic floras.] Trudy Geologicheskii Instituta Akademiya Nauk SSSR 208:158–212. [Russian]

Dobruskina, I. A. 1974. [The Triassic lycopsids.] Paleontologicheskii Zhurnal 1974(3):111–124. [Russian]

Dobruskina, I. A. 1975. [The role of peltaspermous pteridosperms in the Late Permian and Triassic floras.] Paleontologicheskii Zhurnal 1975(4):120–132. [Russian]

Dobruskina, I. A. 1980. [The stratigraphical position of the plant-bearing of the Eurasian Triassic.] Trudy Geologicheskii Instituta Akademiya Nauk SSSR 346:1–160 [Russian]

Dobruskina, I. A. 1982. [Triassic floras of Eurasia.] Trudy Geologicheskii Instituta Akademiya Nauk SSSR 365:1–196. [Russian]

Dobruskina, I. A. (1995) Keuper (Triassic) Flora from Middle Asia (Madygen, Southern Fergana). New Mexico Museum of Natural History and Science Bulletin 5:1–49.

Dyke, G. J., R. L. Nudds, and J. M. V. Rayner. 2006. Flight of *Sharovipteryx mirabilis*: the world's first delta-winged glider. Journal of Evolutionary Biology 19:1040–1043.

Estes, R. 1981. Gymnophiona, Caudata. Pp. 1–115 in P. Wellnhofer (ed), Handbuch der Paläoherpetologie, Part 2. Gustav Fischer Verlag, Stuttgart.

Etter, W. 2002. Monte San Giorgio: remarkable Triassic marine vertebrates. Pp. 221–242 in D. J. Bottjer, W. Etter, J. W. Hagadorn, and C. M. Tang (eds), Exceptional Fossil Preservation. Columbia University Press, New York.

Evans, S. E., and Y. Wang. 2007. A juvenile lizard specimen with well-preserved skin impressions from the Upper Jurassic/Lower Cretaceous of Daohugou, Inner Mongolia, China. Naturwissenschaften 94:431–439.

Evans, S. E., and Y. Wang. 2010. A new lizard (Reptilia: Squamata) with exquisite preservation of soft tissue from the Lower Cretaceous of Inner Mongolia, China. Journal of Systematic Palaeontology 8:81–95.

Fedorenko, O. A., and N. V. Miletenko (eds). 2002. Atlas of the Lithology-Paleogeographical, Structural, Palinspastic and Geoenviromental Maps of Central Asia. Scientific Research Institute of Natural Resources YUGGEO, Almaty.

Fischer, J., B. J. Axsmith, and S. R. Ash. 2010. First unequivocal record of the hybodont shark egg capsule *Palaeoxyris* in the Mesozoic of North America. Neues Jahrbuch für Geologie und Paläontologie, Abhandlungen 255:327–344.

Fischer, J., S. Voigt, and M. Buchwitz. 2007. First elasmobranch egg capsules from freshwater lake deposits of the Madygen Formation (Middle to Late Triassic, Kyrgyzstan, Central Asia). Freiberger Forschungshefte C524:41–46.

Fischer, J., S. Voigt, J. W. Schneider, M. Buchwitz, and S. Voigt. 2011. A selachian freshwater fauna from the Triassic of Kyrgyzstan and its implication for Mesozoic shark nurseries. Journal of Vertebrate Paleontology 31:937–953.

Fitch, A. 1850. A historical, topographical and agricultural survey of the county of Washington. Transactions of New York State Agricultural Society 9:753–944.

Florin, R. 1936. Die fossilen Ginkgophyten von Franz-Joseph-Land nebst Erörterungen über vermeintliche Cordaitales mesozoischen Alters. I. Palaeontographica 81B: 71–173.

Fox, H. M. 1949. On *Apus*: its rediscovery in Britain, nomenclature and habits. Proceedings of the Zoological Society of London 119:693–702.

Franeck, F., J. Fischer, I. Kogan, S. Voigt, and J. W. Schneider. 2013. Microvertebrate analyses extend the ichthyodiversity of the continental Triassic Madygen Formation, Kyrgyzstan. – Palaeobiology and Geobiology of Fossil Lagerstätten through Earth History. A Joint Conference of the "Paläontologische Gesellschaft" and the "Palaeontological Society of China", Göttingen, Germany, September 23-27, 2013, Abstract Volume: 52–53.

Franeck, F., J. W. Schneider, J, Fischer, I. Kogan, M. Buchwitz, F. Spindler, and S. Voigt. 2012. Microvertebrate remains from the non-marine Triassic Madygen Formation of Kyrgyzstan. Terra Nostra 2012/3:57–58.

Fraser, N. C., D. A. Grimaldi, P. E. Olsen, and B. Axsmith. 1996. A Triassic Lagerstätte from eastern North America. Nature 380:615–619.

Frey, E., and D. M. Martill. 1999. Soft tissue preservation in a specimen of *Pterodactylus kochi* from the Upper Jurassic of Germany. Neues Jahrbuch für Geologie und Paläontologie, Abhandlungen 210:421–441.

Fryer, G. 1985. Structure and habits of living branchiopod crustaceans and their bearing on the interpretation of fossil forms. Transactions of the Royal Society of Edinburgh B 76:103–113.

Gall, J.-C., and L. Grauvogel-Stamm. 2005. The early Middle Triassic 'Grès à Voltzia' Formation of eastern France: a model of environmental refugium. Comptes Rendus Palevol 4:637–652.

Gatesy, S. M., N. H. Shubin, and F. A. Jenkins, Jr. 2005. Anaglyph stereo imaging of dinosaur track morphology and microtopography. Palaeontologia Electronica 8:10A.

Gans, C., I. Darevski, and L. P. Tatarinov. 1987. *Sharovipteryx*, a reptilian glider? Paleobiology 13:415–426.

Glumakov, P. V. 1988. [Fergana Depression.] Pp. 5–39 in Anonymous (ed), [Tectonics, Formation, and Oil and Gas, Intermountain Basins of Central Asia and Kazakhstan.] Nauka, Moscow. [Russian]

Golubev, V. K. 2000. [Permian and Triassic chroniosuchians and biostratigraphy of the Upper Tatarian series in Eastern Europe.] Trudy Paleontologicheskogo Instituta Rossiya Akademiya Nauk 276:1–175. [Russian.]

Gorochov, A. V. 1987a. [New fossil orthopterans of the families Adumbratomorphidae fam. n., Pruvostitidae and Proparagryllacridae (Orthoptera, Ensifera) from Permian and Triassic deposits of the USSR.] Vestnik Zoologii 1:18–24. [Russian]

Gorochov, A. V. 1987b. [New and little known Mesotitanidae and Paratitanidae (Titanoptera) from the Triassic of Kyrgyzstan.] Vestnik Zoologii 4:20–28 [Russian]

Gorochov, A. V. 2003. New and little known Mesotitanidae and Paratitanidae (Titanoptera) from the Triassic of Kyrgyzstan. Paleontological Journal 37:400–406.

Gorochov, A. V. 2005. Review of Triassic Orthoptera with

descriptions of new and little known taxa. Paleontological Journal 39:272–279.

Gorochov, A. V., and A. P. Rasnitsyn. 2002. Superorder Gryllidea Laicharting, 1781. Pp. 293–303 in A. P. Rasnitsyn and D. L. J. Quicke (eds), History of Insects. Kluwer Academic Publishers, Dordrecht.

Grimaldi, D., and M. S. Engel. 2005. Evolution of the Insects. Cambridge University Press, New York.

Halle, T.G. 1908. Zur Kenntnis der mesozoischen Equisetales Schwedens. Kungliga Svenska Vetenskapsakademiens Handlingar 43:1–36.

Halstead, L. B. 1975. The evolution and ecology of the dinosaurs. P. Lowe, London.

Harris, T.M. 1932. The fossil flora of Scoresby Sound, East Greenland. Part 3: Caytoniales and Bennettitales. Meddelelser om Grønland 85: 1–133.

Harris, T.M. 1937. The fossil flora of Scoresby Sound, East Greenland. Part 5: Stratigraphic relations of the plant beds. Meddelelser om Grønland 113: 1–112.

Haubold, H. 2004. Überlieferungsbedingte Variation bei Chirotherien und Hinweise zur Ichnotaxonomie nach Beispielen aus der Mittel- bis Obertrias (Anisium-Karnium) von Nordbayern. Hallesches Jahrbuch für Geowissenschaften B 26:1–15.

Haubold, H., and E. Buffetaut. 1987. Une nouvelle interprétation de *Longisquama insignis*, reptile énigmatique du Trias supérieur d'Asie centrale. Comptes Rendus de l'Académie des Sciences, Série II 305:65–70.

Hecker, R. T., A. I. Ossipova, and T. N. Belskaya. 1963. Fergana Gulf of Paleogene sea of Central Asia, its history, sediments, fauna and flora, their environment and evolution. Bulletin of the American Association of Petroleum Geologists 47:617–631.

Heer, O. 1865. Die Urwelt der Schweiz. Zürich: Friedrich Schultheß.

Hu, S.-x., Q.-y. Zhang, Z.-Q, Chen, C.-y. Zhou, T. Lü, T. Xie, W. Wen, J.-y. Huang, and M. J. Benton. 2011. The Luoping biota: exceptional preservation, and new evidence on the Triassic recovery from end-Permian mass extinction. Proceedings of the Royal Society B 278:2274–2282.

Ikramov, I. R. 1967a. [Correlation of sections of the lower part of the Madygen Formation.] Trudy Tashkentskogo Gosudarstvennogo Universiteta Imeni V. I. Lenina 306:3–5. [Russian]

Ikramov, I. R. 1967b. [Petrographic composition and depositional conditions of the Madygen Formation.] Trudy Tashkentskogo Gosudarstvennogo Universiteta Imeni V. I. Lenina 306:6–14. [Russian]

Ikramov, I. R. 1968. [On the characterization of the Permo-Triassic deposits from Madygen.] Trudy Tashkentskogo Gosudarstvennogo Universiteta Imeni V. I. Lenina 325:27–34. [Russian]

Ikramov, I. R. 1970. [Depositional conditions of Permotriassic sediments in South Fergana on the example of the Madygen Formation.] Doctoral dissertation. Tashkent State University, Tashkent. [Russian]

Ikramov, I. R. 1971. [Once more about the age of the Madygen Formation.] Trudy Tashkentskogo Gosudarstvennogo Universiteta Imeni V. I. Lenina 407:156–161. [Russian]

Ikramov, I. R. 1972. [Distribution patterns of allogenic minerals of the Madygen Formation.] Trudy Tashkentskogo Gosudarstvennogo Universiteta Imeni V. I. Lenina 408:66–69. [Russian]

Ivakhnenko, M. F. 1978. [Urodelans from the *Triassic* and *Jurassic* of Soviet *Central Asia*.] Paleontologicheskii Zhurnal 1978(3):84–89. [Russian]

Jain, S. L. 1983. Spirally coiled 'coprolites' from the Upper Triassic Maleri Formation, India. Palaeontology 26:813–829.

Ji, Q., and C. Yuan. 2002. Discovery of two kinds of proto-feathered pterosaurs in the Mesozoic Daohugou Biota in the Ningcheng Region and its stratigraphic and biologic significances. Geological Review 48:221–224.

Jones, T. D., J. A. Ruben, L. D. Martin, E. N. Kurochkin, A. Feduccia, P. F. A. Maderson, W. J. Hillenius, N. R. Geist, and V. Alifanov. 2000. Nonavian feathers in a Late Triassic archosaur. Science 288:2202–2205.

Kashevarova, N. P. 1986. [Nonmarine Ostracoda.] In N. P. Kashevarova and I. I. Molostovskaya (eds) [An Atlas of the Characteristic Assemblages of the Permian Flora and Fauna of the Urals and the Russian Platform.] Nedra, Leningrad. [Russian]

Kellner, A.W., X. Wang, H. Tischlinger, D. A. Campos, D. W. E. Hone, and X. Meng. 2010. The soft tissue of *Jeholopterus* (Pterosauria, Anurognathidae, Batrachognathinae) and the structure of the pterosaur wing membrane. Proceedings of the Royal Society B 277:321–329.

Kemp, A. 1998. Skull structure in post-Paleozoic lungfish. Journal of Vertebrate Palaeontology 18: 43–63.

Knaust, D. 2010. Meiobenthic trace fossils comprising a miniature ichnofabric from Late Permian carbonates of the Oman Mountains. Palaeogeography, Palaeoclimatology, Palaeoecology 286:81–87.

Kochnev, E. A. 1934. [On the exploration of Jurassic coal-bearing deposits of Fergana.] [Materials on the Geology of Coal Deposits of Middle Asia] 5–6:1–53. [Russian]

Kogan, I., K. Schönberger, J. Fischer, and S. Voigt. 2009. A nearly complete skeleton of *Saurichthys orientalis* (Pisces, Actinopterygii) from the Madygen Formation (Middle to Late Triassic, Kyrgyzstan, Central Asia) – preliminary results. Freiberger Forschungshefte C532:139–152.

Kolesnikov, Ch. M. 1980. [Systematic, stratigraphic distribution and zoogeography of Mesozoic limnic bivalves of the USSR.] Pp. 9–65 in G. G. Martinson (ed), [Limnic Biota of Eurasian Freshwater Basins.] Nauka, Leningrad. [Russian]

Kon'no, E. 1962. Some species of *Neocalamites* and *Equisetites* in Japan and Korea: Tohoku University Scientific Reports, 2nd Series. Geology 5: 21–48.

Kotelnikov, V. I. 1985. [Geological map of Turkestan, Alai and Fergana Mountains with surrounding areas, South Tien Shan, Scale 1:50,000.] Ministerstvo Geologii SSSR, Leningrad. [Russian]

Kottek, M., J. Grieser, C. Beck, B. Rudolf, and F. Rubel. 2006. World map of the Köppen-Geiger climate classification updated. Meteorologische Zeitschrift 15:259–263.

Kozur, H. W., and R. E. Weems. 2010. The biostratigraphic importance of conchostracans in the continental Triassic of the northern hemisphere. Geological Society of London, Special Publications 334:315–417.

Kräusel, R. 1943. Die Ginkgophyten der Trias von Lunz

in Nieder-Österreich und von Neue Welt bei Basel. Palaeontographica 87B: 59–93.

Kryshtofovich, A. N., and V. D. Prynada. 1933. Contribution to the Rhaeto-Liassic flora of the Cheliabinsk brown coal basin, Eastern Urals. Transactions of the United Geological and Prospecting Service of USSR 346:1–40.

Lehman, J.-P., C. Château, M. Laurain, and M. Nauche. 1959. Paléontologie de Madagascar 27. Les poissons de la Sakamena moyenne. Annales de Paléontologie 45: 175–219.

Li, C., X.-C. Wu, Y.-N. Cheng, T. Sato, and L. Wang. 2006. An unusual archosaurian from the marine Triassic of China. Naturwissenschaften 93:200–206.

Li, J.-L., X.-C. Wu, and F. C. Zhang (eds). 2008. The Chinese Fossil Reptiles and their Kin. Science Press, Beijing.

Lingham-Soliar, T. 1999. Rare soft-tissue preservation showing fibrous structures in an ichthyosaur from the Lower Lias (Jurassic) of England. Proceedings of the Royal Society B 266:2367–2373.

Linnaeus, C. 1753. Species plantarum, exhibentes plantas rite cognitas, ad genera relatas, cum differentiis specificis, nominibus trivialibus, synonymis selectis, locis natalibus, secundum systema sexuale digestas. Laurentii Salvii., Stockholm.

Linnaeus, C. 1758. Systema naturae per regna tria naturae, secundum classes, ordines, genera, species, cum characteribus, differentiis, synonymis, locis. Editio decima. Tomus I. Laurentius Salvius, Stockholm.

Lombardo, C. 2001. Actinopterygians from the Middle Triassic of Northern Italy and Canton Ticino (Switzerland): anatomical descriptions and nomenclatural problems. Rivista Italiana di Paleontologia e Stratigrafia, 107(3):345–369.

Lombardo, C. and A. Tintori. 2005. Feeding specializations in Late Triassic fishes. Annali dell'Università degli Studi di Ferrara. Museologia Scientifica e Naturalistica, volume speciale 2005: 25–32.

Lucas, S. G. 1993. Vertebrate biochronology of the Triassic of China. New Mexico Museum of Natural History and Science Bulletin 3:301–306.

Lucas, S. G. 2001. Chinese Fossil Vertebrates. Columbia University Press, New York.

Mamrenko, A. V. 1968. [Geology and hydrogeology of the Belesynyk Chain, sheet 5–159G, and partial sheets 5–159B, 5–159V, 5–160A, 5–160V.] South Kyrgyz Geological Expedition, Osh. [Unpublished report; Russian]

Martin, L. D. 2004. A basal archosaurian origin of birds. Acta Zoologica Sinica 50:978–990.

McKenzie, K. G., P.-J. Chen, and S. Majoran. 1991. *Almatium gusevi* (Chernyshev, 1940): redescription, shield shapes, and speculation on the reproductive mode (Branchiopoda, Kazacharthra). Paläontologische Zeitschrift 65:305–317.

Milner, A. R. 1993. Late Triassic and Jurassic amphibians: fossil record and phylogeny. Pp. 5–22 in N. C. Fraser and H.-D. Sues (eds), In the Shadow of the Dinosaurs: Early Mesozoic Tetrapods. Cambridge University Press, New York.

Minikh, A. V. 1981. [*Saurichthys* from the Triassic of the USSR.] Paleontologicheskii Zhurnal 1981(1):105–113. [Russian]

Mogutcheva, N.K. 1973. [Early Triassic flora of the Tunguska basin.] Trudy Sibirskogo Naučno-issledovatel'-skogo Instituta Geologii, Geofiziki i Mineral'nogo Syr'â 154: 1–160. [Russian]

Moisan, P. 2012. Systematic and Palaeoecological Study of the Middle–Late Triassic Flora from the Madygen Formation, SW Kyrgyzstan. Doctoral dissertation. Westfälische Wilhelms-Universität, Münster.

Moisan, P., and S. Voigt. 2013. Lycopsids from the Madygen Lagerstätte (Middle to Late Triassic, Kyrgyzstan, Central Asia). Review of Palaeobotany and Palynology 192: 42–64.

Moisan, P., S. Voigt, J. W. Schneider, and H. Kerp. 2012. New fossil bryophytes from the Triassic Madygen Lagerstätte (SW Kyrgyzstan). Review of Palaeobotany and Palynology 187:29–37.

Moisan, P., C. C. Labandeira, N. A. Matushkina, T. Wappler, S. Voigt, and H. Kerp. 2012. Lycopsid–arthropod associations and odonatopteran oviposition on Triassic herbaceous *Isoetes*. Palaeogeography, Palaeoclimatology, Palaeoecology 344–345: 6–15.

Moisan, P., S. Voigt, C. Pott, M. Buchwitz, J. W. Schneider, and H. Kerp. 2011. Cycadalean and bennettitalean foliage from the Triassic Madygen Lagerstätte (SW Kyrgyzstan, Central Asia). Review of Palaeobotany and Palynology 164:93–108.

Mundil, R., J. Pálfy, P. R. Renne, and P. Brack. 2010. The Triassic timescale: new constraints and a review of geochronological data. Pp 41–59 in S. G. Lucas (ed), The Triassic Timescale. Geological Society, London, Special Publications 334 ,.

Nathorst, A.G. 1878. Bidrag till Sveriges fossila flora. Floran vid Höganäs och Helsingborg. Kungliga Svenska Vetenskapsakademiens Handlingar 16: 1–53.

Northwood, C. 2005. Early Triassic coprolites from Australia and their palaeobiological significance. Palaeontology 48:49–68.

Novokshonov, V. G. 2001. New Triassic scorpionflies (Insecta, Mecoptera) from Kyrgyzstan. Paleontological Journal 36:281–288.

Novozhilov, N. I. 1946. [New Phyllopoda from Permian and Triassic deposits of the Nordvik-Chatangsk area.] Nedra Arktiki 1:172–202. [Russian]

Novozhilov, N. I. 1958. [Change of faunal complexes and systematics of bivalved freshwater crustaceans (Conchostraca).] 20th International Geological Congress, Mexico City, 7: 191–212. [Russian]

Novozhilov, N. I. 1959. Position systématique des Kazacharthra (arthropodes) d'après de nouveaux matériaux des monts Ketmen et Sajkan (Kazakhstan SE et NE). Bulletin de la Société Géologique de France 1:265–269.

Novozhilov, N. I. 1960. [Subclass Gnathostraca.] Pp. 216–253 in Yu. A. Orlov (ed), Osnovy Paleontologii, Vol. 8. Nauka, Moscow. [Russian]

Novozhilov, N. I. 1970. [Conchostraca—Limnadioidea.] Nauka, Moscow. [Russian]

Novozhilov, N. I., and V. Kapelka. 1960. Crustacés bivalves (Conchostraca) de la série Daido de l'Asie Orientale dans le Trias supérieur de Madygen (Kirghizie Occidentale). Société Géologique du Nord, Annales 80:177–187.

Ochev, V. G., and M. A. Shishkin. 1989. On the principles of global correlation of the continental Triassic on the tetrapods. Acta Palaeontologica Polonica 34:149–173.

Oda, S. 1984. Hibernation of freshwater bryozoans. Dobutsu to Shinen 14:17–23.

Paul, G. S. 2001. Dinosaurs of the Air: the Evolution and Loss of

Flight in Dinosaurs and Birds. Baltimore and London: John Hopkins University Press.

Papier, F., and A. Nel. 2001. Les Subioblattidae (Blattodea, Insecta) du Trias d'Asie Centrale. Paläontologische Zeitschrift 74:533–542.

Peters, D. 2000. A reexamination of four prolacertiforms with implications for pterosaur phylogenesis. Rivista Italiana di Paleontologia e Stratigrafia 106:293–336.

Ponomarenko, A. G. 1969. [Historical development of archostematan beetles.] Trudy Paleontologicheskii Instituta Akademiya Nauk SSSR 125:3–240. [Russian]

Ponomarenko, A. G. 1977. [Suborder Adephaga.] Trudy Paleontologicheskii Instituta Akademiya Nauk SSSR 161:17–96. [Russian]

Ponomarenko, A. G. 2002. Order Coleoptera Linne, 1758. Pp. 164–176 in A. P. Rasnitsyn and D. L. J. Quicke (eds), History of Insects. Kluwer Academic Publishers, Dordrecht.

Presl, G. B. 1838. Pp. 1–220 in C. von Sternberg (ed) (1820–1838), Versuch einer geognostisch-botanischen Darstellung der Flora der Vorwelt. Leipzig.

Preuße, M. 2012. GIS-basierte Datenanalyse zur faziellen und strukturellen Entwicklung der triassischen Fossillagerstätte Madygen (SW-Kirgisistan, Zentralasien). Unpublished diploma thesis. TU Bergakademie Freiberg.

Pritykina, L. N. 1981. [New Triassic Odonata from Middle Asia.] Trudy Paleontologicheskii Instituta Akademiya Nauk SSSR 183:5–42. [Russian]

Pruvost, P. 1930. La faune continentale du terrain houiller de la Belgique. Mémoires du Musée Royal d'Histoire Naturelle de Belgique 44:133–141.

Rasnitsyn, A. P. 1969. [Origin and evolution of lower Hymenoptera.] Trudy Paleontologicheskii Instituta Akademiya Nauk SSSR 123:1–196. [Russian]

Rasnitsyn, A. P., and D. L. J. Quicke (eds). 2002. History of Insects. Kluwer Academic Publishers, Dordrecht.

Reisz, R. R., and H.-D. Sues. 2000. The 'feathers' of *Longisquama*. Nature 408:428.

Renault, B., and R. Zeiller. 1884. Sur un nouveau genre de fossiles végétaux. Comptes Rendus Hebdomadaires des Séances de l'Académie des Sciences:1391–1394.

Renault, B., and R. Zeiller. 1888. Sur l'attribution des genres *Fayolia* et *Palaeoxyris*. Comptes Rendus Hebdomadaires des Séances de l'Academie des Sciences 107:1022–1025.

Renesto, S. 2005. A new specimen of *Tanystropheus* (Reptilia, Protorosauria) from the Middle Triassic of Switzerland and the ecology of the genus. Rivista Italiana di Paleontologia e Stratigrafia 111:375–392.

Retallack, G.J. 1997. Earliest Triassic origin of *Isoetes* and quillwort evolutionary radiation. Journal of Paleontology 71:500–521.

Ritchie, A. 1981. First complete specimen of the dipnoan *Gosfordia truncata* Woodward from the Triassic of New South Wales. Records of the Australian Museum 33:606–615.

Romano, C., I. Kogan, J. Jenks, I. Jerjen, and W. Brinkmann. 2012. *Saurichthys* and other fossil fishes from the late Smithian (Early Triassic) of Bear Lake County (Idaho, USA), with a discussion of saurichthyid palaeogeography and evolution. Swiss Bulletin of Geosciences 87:543–570.

Rusconi, C. 1948. Plesiosaurios del Jurasico de Mendoza. Anales de la Sociedad Científica 146:327–351.

Sander, P. M. 1989. The pachypleurosaurids (Reptilia, Nothosauria) from the Middle Triassic of Monte San Giorgio (Switzerland) with the description of a new species. Philosophical Transactions of the Royal Society London B 325:561–666.

Schaller, D. 1985. Wing evolution. Pp. 333–348 in M. K. Hecht, J. H. Ostrom, G. Viohl, and P. Wellnhofer (eds), The Beginnings of Birds. Freunde des Juramuseums, Eichstätt.

Schimper, W.P. 1869. Traité de Paléontologie végétale ou la flore du monde primitif dans ses rapports avec les formations géologiques et la flore du monde actuel. Paris: J.B. Baillière et Fils

Schneider, J. W., and W. Reichel. 1989. Chondrichthyer-Eikapseln aus dem Rotliegenden (Unterperm) Mitteleuropas – Schlussfolgerungen zur Paläobiologie paläozoischer Süßwasserhaie. Freiberger Forschungshefte C436:58–69.

Schoch, R., S. Voigt, and M. Buchwitz. 2010. A chroniosuchid from the Triassic of Kyrgyzstan and analysis of chroniosuchian relationships. Zoological Journal of the Linnean Society 160:515–530.

Schudack, U. 2008. Ostracodenexpertise Fossillagerstätte Madygen (SW Kirgistan, Zentralasien). Freie Universität Berlin. [Unpublished report]

Schultze, H.-P. 2004. Mesozoic sarcopterygians. Pp. 463–492 in G. Arratia and A. Tintori (eds), Mesozoic Fishes 3 – Systematics, Palaeoenvironments and Biodiversity. Verlag Dr. Friedrich Pfeil, Munich.

Schweitzer, M. H. 2011. Soft tissue preservation in terrestrial Mesozoic vertebrates. Annual Review of Earth and Planetary Sciences 39:187–216.

Scotese, C. R. 2000. PALEOMAP Project. University of Texas at Arlington.

Seilacher, A. 1970. Begriff und Bedeutung der Fossil-Lagerstätten. Neues Jahrbuch für Geologie und Paläontologie, Monatshefte 1970:34–39.

Seilacher, A., W.-E. Reif, and F. Westphal. 1985. Sedimentological, ecological and temporal patterns of fossil Lagerstätten. Philosophical Transactions of the Royal Society London B 311:5–23.

Sellwood, B. W., and P. J. Valdes. 2006. Mesozoic climates: general circulation models and the rock record. Sedimentary Geology 190:269–287.

Sennikov, A. G. 1996. Evolution of the Permian and Triassic tetrapod communities of Eastern Europe. Palaeogeography, Palaeoclimatology, Palaeoecology 120:331–351.

Senter, P. 2004. Phylogeny of Drepanosauridae (Reptilia: Diapsida). Journal of Systematic Paleontology 2:257–268.

Seward, A.C. (1919) Fossil Plants. Volume IV: Ginkgoales, Coniferales, Gnetales. Cambridge: Cambridge University Press.

Sharapova, E. G. 1948. [The ostracod fauna from Upper Permian deposits (Tatarian and Kazanian Stage) of oil-bearing regions of the USSR.] Trudy Vsesoyuznogo Neftyanogo Naukno-Issledovatelskogo Geologo-Razvednoknogo Instituta (VNIGRI) 31(1):21–48. [Russian]

Sharov, A. G. 1968. [Phylogeny of the Orthopteroidea.] Trudy Paleontologicheskogo Instituta Akademiya Nauk SSSR 118:1–217. [Russian]

Sharov, A. G. 1970. [A peculiar reptile from the Lower Triassic

of Fergana.] Paleontologicheskii Zhurnal 1970(1):127–130. [Russian]
Sharov, A. G. 1971a. [New flying reptiles from the Mesozoic of Kazakhstan and Kyrgyzstan.] Doklady Akademiya Nauk SSSR 130:104–113. [Russian]
Sharov, A. G. 1971b. [Unique paleontological finds.] Nauka i Zhizn 7:28–32. [Russian]
Shcherbakov, D. E. 1984. [System and phylogeny of Permian Cicadomorpha (Cimicida, Cicadina).] Paleontologicheskii Zhurnal 18(2):87–97. [Russian]
Shcherbakov, D. E. 2002. Order Forficulida Latreille, 1810. The earwigs and protelytropterans. Pp. 298–301 in A. P. Rasnitsyn and D. L. J. Quicke (eds), History of Insects. Kluwer Academic Publishers, Dordrecht.
Shcherbakov, D. E. 2008a. Madygen, Triassic Lagerstätte number one, before and after Sharov. Alavesia 2:113–124.
Shcherbakov, D. E. 2008b. Insect recovery after the Permian/Triassic crisis. Alavesia 2:125–31
Shcherbakov, D. E. 2008c. [On Permian and Triassic insect faunas in relation to biogeography and the Permian-Triassic crisis.] Paleontologicheskii Zhurnal 42(1):15–31.
Shcherbakov, D. E. 2011. New and little-known families of Hemiptera Cicadomorpha from the Triassic of Central Asia—early analogs of treehoppers and planthoppers. Zootaxa 2836:1–26.
Shcherbakov, D. E., and Y. A. Popov. 2002. Order Hemiptera Linne, 1758. Pp. 143–157 in A. P. Rasnitsyn and D. L. J. Quicke (eds), History of Insects. Kluwer Academic Publishers, Dordrecht.
Shcherbakov, D. E., E. D. Lukashevich, and V. A. Blagoderov. 1995. Triassic Diptera and initial radiation of the order. International Journal of Dipterological Research 6:75–115.
Shishkin, M. A., A. G. Sennikov, I. V. Novikov, and N. V. Ilyna. 2006. Differentiation of tetrapod communities and some aspects of biotic events in the Early Triassic of Eastern Europe. Paleontological Journal 40:1–10.
Sinitshenkova, N. D. 2000. A review of Triassic mayflies with description of new species from Western Siberia and Ukraine (Ephemerida = Ephemeroptera). Paleontological Journal 34:S275–S283.
Sinitshenkova, N. D. 2002. Ecological history of the aquatic insects. Pp. 388–417 in A. P. Rasnitsyn and D. L. J. Quicke (eds), History of Insects. Kluwer Academic Publishers, Dordrecht.
Sixtel, T. A. 1949. [Discovery of a Paleozoic flora with *Gigantopteris* in Fergana.] Doklady Akademiya Nauk SSSR 66:925–928. [Russian]
Sixtel, T. A. 1956. [*Prynadia, Madygenopteris, Madygenia, Kryschtofovichiella*.] Trudy Vsesoyuznogo Naucno-Issledovatelskogo Geologicheskogo Instituta (VSEGEI) 12: 219–220, 225–229, 251–253. [Russian]
Sixtel, T. A. 1960a. [On the presence of Upper Permian continental deposits in South Fergana.] Trudy Uzbekskogo Geologicheskogo Upravleniya 1960(1):29–38. [Russian]
Sixtel, T. A. 1960b. [Stratigraphy of continental deposits of the Upper Permian and Triassic of Middle Asia.] Trudy Tashkentskogo Gosudarstvennogo Universiteta Imeni V. I. Lenina 176:1–146. [Russian]
Sixtel, T. A. 1961. [Representatives of *Gigantopteris* and some accompanying plants from the Madygen Formation of Fergana.] *Paleontologicheskii Zhurnal* 1961(1):151–158. [Russian]
Sixtel, T. A. 1962. [Flora of the Late Permian and Early Triassic in Southern Fergana.] Pp. 271–414 in Anonymous (ed), [Stratigraphy and paleontology of Uzbekistan and adjacent areas.] Institut Geologii Akademii Nauk Uzbekskoi SSR, Tashkent. [Russian]
Storozhenko, S. Yu. 1992. [New Mesozoic grylloblattid insects (Grylloblattida) from Middle Asia.] Paleontologicheskii Zhurnal 1992(1):67–75. [Russian]
Sues, H.-D., and N. C. Fraser. 2010. Triassic Life on Land: The Great Transition. Columbia University Press, New York.
Sukatsheva, I. D. 1973. [New Trichoptera from the Mesozoic of Middle Asia.] Paleontologicheskii Zhurnal 1973(3):100–107. [Russian]
Sytchevskaya, E. K. 1999. Freshwater fish fauna from the Triassic of Northern Asia. Pp 445–468 in G. Arratia and H.-P. Schultze (eds), Mesozoic Fishes 2 – Systematics and Fossil Record. Verlag Dr. Friedrich Pfeil, Munich.
Tatarinov, L. P. 1989. [About the systematic position and the kind of life of the problematic late Triassic reptile *Sharovipteryx mirabilis*.] Paleontologiceskii Zhurnal 1989(2):110–112. [Russian]
Tatarinov, L. P. 2005. A new cynodont (Reptilia, Theriodontia) from the Madygen Formation (Triassic) of Fergana, Kyrgyzstan. Paleontological Journal 39:192–198.
Taylor, P. D., and O. Vinn. 2006. Convergent morphology in small spiral tubes ('*Spirorbis*') and its palaeoenvironmental implications. Journal of the Geological Society of London 163:225–228.
Thomas, H. 1913. On some new and rare Jurassic plants from Yorkshire – *Eretmophyllum*, a new type of Ginkgoalean leaf. Proceedings of the Cambridge Philosophical Society 17:256–262.
Troitsky, V. I., I. R. Ikramov, and V. V. Kurbatov. 1987. [Lake formations of the Late Permian and Triassic of Middle Asia.] Pp. 167–177 in Anonymous (ed), [History of Late Paleozoic and Mesozoic lakes.] Nauka, Leningrad. [Russian]
Tverdokhlebov, V. P., G. I. Tverdokhlebova, M. V. Surkov, and M. J. Benton. 2002. Tetrapod localities from the Triassic of the SE of European Russia. Earth-Science Reviews 60: 1–66.
Unwin, D. M., and M. J. Benton. 2001. *Longisquama* fossil and feather morphology. Science 291:1900–1901.
Unwin, D. M., V. R. Alifanov, and M. J. Benton. 2000. Enigmatic small reptiles from the Middle-Late Triassic of Kirgizstan. Pp. 177–186 in M. J. Benton, M. A. Shishkin, D. M. Unwin, and E. N. Kurochkin (eds), The Age of Dinosaurs in Russia and Mongolia. Cambridge University Press, Cambridge.
Vinn, O. 2010. Adaptive strategies in the evolution of encrusting tentaculitoid tubeworms. Palaeogeography, Palaeoclimatology, Palaeoecology 292:211–221.
Vinn, O., and H. Mutvei. 2009. Calcareous tubeworms of the Phanerozoic. Estonian Journal of Earth Sciences 58(4):286–296.
Vishnyakova, V. N. 1998. Cockroaches (Insecta, Blattodea) from the Triassic Madygen locality, Central Asia. Paleontological Journal 32:505–512.
Voigt, S., H. Haubold, S. Meng, D. Krause, J. Buchantschenko, K. Ruckwied, and A. E. Götz. 2006. Die Fossil-Lagerstätte

Madygen: Ein Beitrag zur Geologie und Paläontologie der Madygen-Formation (Mittel- bis Ober-Trias, SW-Kirgisistan, Zentralasien). Hallesches Jahrbuch für Geowissenschaften, Beiheft 22:85–119.

Voigt, S., M. Buchwitz, J. Fischer, D. Krause, and R. Georgi. 2009. Feather-like development of Triassic diapsid skin appendages. Naturwissenschaften 96:81–86.

Voigt, S., and D. Hoppe. 2010. Mass occurrence of penetrative trace fossils in Triassic lake deposits (Madygen fossil site, Kyrgyzstan, Central Asia). Ichnos 17:1–11.

Vorobyeva, E. I. 1967. [A Triassic ceratodontid from Southern Fergana and some remarks on the systematics and phylogeny of ceratodontids.] Paleontologicheskii Zhurnal 1967(4): 102–111. [Russian]

Vršanský, P., V. N. Vishniakova, and A. P. Rasnitsyn. 2002. Order Blattida Latreille, 1810. Pp. 263–70 in A. P. Rasnitsyn and D.L.J. Quicke (eds), History of Insects. Kluwer Academic Publishers, Dordrecht.

Wade, R.T. 1935. The Triassic Fishes of Brookvale, New South Wales. London: British Museum (Natural History).

Wood, T. S. 1983. General features of the class Phylactolaemata. Pp. 287–303 in R. A. Robinson (ed), Treatise on Invertebrate Paleontology, Part G: Bryozoa. University of Kansas Press, Lawrence.

Woodward, A.S. 1890. The fossil fishes of the Hawkesbury series at Gosford. Memoirs of the Geological Survey of New South Wales, Palaeontology 4: 1–55.

Woodward, S. 1830. A Synoptical Table of British Organic Remains, in which all the Edited British Fossils are Systematically and Stratigraphically Arranged. Longman, Rees, Orme, Brown and Green, London.

Wu, S.Q., M.N. Ye, and B.X. Li. 1980. Upper Triassic and Lower and Middle Jurassic plants from the Hsiangchi Group, Western Hubei. Memoirs of Nanjing Institute of Geology and Paleontology, Academia Sinica 14: 63–131. [Chinese]

Wu, X.-C. 1981. [The discovery of a new thecodont from north east Shanxi.] Vertebrata PalAsiatica 19:122–132. [Chinese]

Wu, X.-C., and A. P. Russell. 2001. Redescription of *Turfanosuchus dabanensis* (Archosauriformes) and new information on its phylogenetic relationships. Journal of Vertebrate Paleontology 21:40–50.

Xu, G.-H., and K.-Q. Gao. 2011. A new scanilepiform from the Lower Triassic of northern Gansu Province, China, and phylogenetic relationships of non-teleostean Actinopterygii. Zoological Journal of the Linnean Society 161:595–612.

Xu, X., X. Zheng, and H. You. 2010. Exceptional dinosaur fossils show ontogenetic development of early feathers. Nature 461:1338–1341.

Young, C. C. 1964. The pseudosuchians in China. Palaeontologica Sinica, N. S., C 19:1–205.

Young, C. C. 1973. [On the occurrence of *Vjushkovia* in Sinkiang.] Memoirs of the Institute of Vertebrate Paleontology and Paleoanthropology 10: 38–53. [Chinese]

Zhang, F.-K. 1975. [A new thecodont *Lotosaurus*, from Middle Triassic of Hunan.] Vertebrata PalAsiatica 13:144–147. [Chinese]

Zherikin, V. V. 2002. Insect trace fossils. Pp. 303–24 in A. P. Rasnitsyn and D. L. J. Quicke (eds), History of Insects. Kluwer Academic Publishers, Dordrecht.

Zherikhin, V. V. and V. G. Gratshev. 1993. [Obrieniidae, fam. nov., the oldest Mesozoic weevils (Coleoptera, Curculionidae).] Paleontologicheskii Zhurnal 27(1):50–69.

Zidek, J. 1976. A new shark egg capsule from the Pennsylvanian of Oklahoma, and remarks on the chondrichthyan egg capsules in general. Journal of Paleontology 50:907–915.

Chapter 4

The Solite Quarry – a window into life by a late Triassic lake margin

Nicholas C. Fraser[1], David A. Grimaldi[2], Brian J. Axsmith[3], Andrew B. Heckert[4], Cynthia Liutkus-Pierce[4], Dena Smith[5], Alton C. Dooley Jr.[6]

1 National Museums Scotland, Edinburgh, UK.
2 Division of Invertebrate Zoology, American Museum of Natural History, USA.
3 Biology Department, University of South Alabama, USA.
4 Department of Geological and Environmental Sciences, Appalachian State University, USA.
5 University of Colorado, Boulder, USA.
6 Western Science Center, Hemet, USA.

Abstract

The Virginia Solite Quarry, situated in the Dan River basin of the Newark Supergroup, preserves one of the few fossil assemblages that include intact Triassic insects. These are, for the most part, completely articulated and preserve exquisite anatomical detail including microtrichiae. The insects are very diverse, but the assemblage is significant for representatives of many modern orders and families including four extant families of Diptera, the oldest belostomatid water bugs and the oldest thysanopteran (thrips). There is considerable debate regarding the depositional environment, but evidence is presented to support deposition under relatively shallow conditions in a saline/alkaline lake. Plants and vertebrates are also present, but by contrast with the insects, the tetrapods represent archaic lineages that apparently did not survive the Triassic–Jurassic boundary.

Introduction

The early Mesozoic sediments of the Newark Supergroup extending along part of the eastern seaboard of North America (Fig 1A) are part of a much wider system of rift basins associated with the breakup of the Pangaean supercontinent during Triassic and Jurassic times. The sequences in North America have a long history of research dating back to the 1840s and 1850s, with William C. Redfield and Edward Hitchcock describing some of the classic fish and footprint localities in the more northerly basins in Connecticut and Massachusetts (e.g., Redfield, 1856; Hitchcock, 1858, 1865). Partly as a result of its proximity to New York City, the Newark Basin has been particularly well documented. For example, the former North Bergen Quarry (or Granton Quarry), New Jersey, just west of the Hudson River and with a direct view of Manhattan, is the site of several major discoveries including a juvenile rutiodontine phytosaur (Colbert, 1965), the gliding tetrapod, *Icarosaurus* (Colbert, 1966), the aquatic protorosaur *Tanytrachelos* and the rather unusual drepanosaur, *Hypuronector* (Colbert and Olsen, 2001), that was often referred to as the 'deep-tailed swimmer' (e.g. Olsen, 1980).

Basins further to the south have attracted attention for an equally long period of time. For example, Ebenezer Emmons (1856) working in the Deep River basin in North Carolina described 'mammalian' and phytosaur remains. No lesser giants of palaeontology than Charles Lyell, Othniel Charles Marsh, Edward Drinker Cope and Henry Fairfield Osborn also weighed in on various aspects of the palaeontology of the Newark Supergroup (e.g., Lyell, 1847; Cope, 1887; Marsh, 1896). The first specimen catalogued in the American Museum of Natural History's Fossil Reptile collection (AMNH FR.1) is the phytosaur, *Rutiodon*, collected by William Diller Matthew in 1895 at a coal mine in Chatham County, North Carolina (Colbert, 1947).

In the past decade field work in various brick quarries in the Deep River basin has yielded a number of

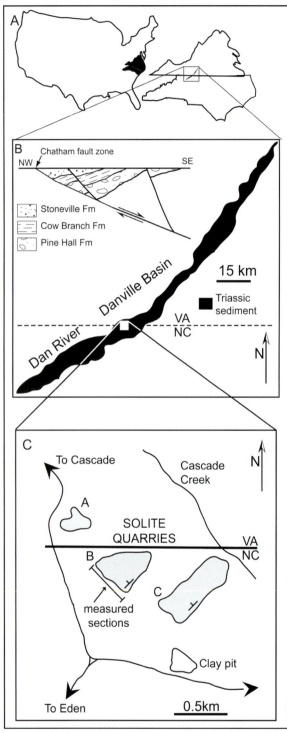

Figure 1 **A**, Location map of Virginia and North Carolina; **B**, Detail of (A) with cross-section of the Dan River/Danville Basin (after Olsen et al., 1991); **C**, map of the three separate Solite quarries.

remarkable specimens including the partial skeleton of a large rauisuchian, *Postosuchus alisonae* (Peyer et al., 2008). Although much of the skeleton of this large articulated specimen was missing as a result of quarrying operations, the body cavity was essentially intact. It was found to contain a variety of other tetrapods, which have been interpreted as gut contents. These include remains of a cynodont, a dicynodont and an aetosaur (Sues et al., 1999). Furthermore beneath the rauisuchian was a new crocodylomorph, *Dromicosuchus grallator* (Sues et al., 2003) that, based on tooth marks on the skull and neck, has also been regarded as an intended prey item for the rauisuchian. Remarkable new finds continue to be made in these southern basins and include the new aetosaur *Gorgetosuchus* (Heckert et al., 2015) and a new species of *Coahomasuchus* (Heckert et al., 2017). Despite these quite extraordinary discoveries elsewhere in the Newark Supergroup, one site deserves special attention, and that is the so-called Solite Quarry in the Dan River/Danville basin (Olsen et al., 1978; Olsen et al., 1989; Fraser et al., 1996). The preservation of the vertebrates makes study of the vertebrate fossils difficult, and the plant remains, although abundant, are usually rather fragmentary and lack preservation of cuticle (Fraser et al., 1996). Yet the Solite Quarry is still providing us with a completely new insight into terrestrial life in the Pangaean interior, and this is largely due to the remarkably preserved and very diverse insect fossils that continue to be recovered from the site.

Geological Context and History of the Quarry

Most rift basins of the Newark Supergroup are half-grabens that are bounded by major border faults. The sediments of these basins, together with basaltic lavas, record an extensive period of geological time probably ranging from the late Middle Triassic until well into the Early Jurassic. However, while the deposits in the northern basins span the Triassic–Jurassic boundary, sedimentation in the south seems to have ceased at some point during the Late Triassic (e.g., Olsen et al., 1989; Olsen and Johansson, 1994).

A series of lakes occupied the rift system in a similar manner to the African lakes in the modern Great Rift Valley, and the Newark deposits are well noted for their repetitive cycles of sedimentation reflecting the regular waxing and waning of water levels in these lakes (Olsen, 1986; Olsen et al., 1989). These are widely considered to have been directly linked to periodic variations in rainfall and temperature, which in turn were under the control of consistent variations in the earth's orbit around the sun known as Milankovitch cycles (Olsen, 1986; Olsen and Kent, 1999). The size of the lakes varied considerably, just as they do in today's African rift system, with some of the lakes (e.g., the Danville/Dan River basin fossil lake) being long and narrow and others much broader (e.g., the Newark basin palaeolake). There is some debate regarding their depth, with the traditional, and still widely held, view that periodically they all typically reached substantial depths (e.g., over 100 metres) (Olsen et al., 1978; Olsen et al., 1989; Olsen and Johansson, 1994). However, more recently one school of thought (Liutkus et al., 2010) has argued that at least 'Fossil Lake Danville' never reached particularly significant depths, and that it may have mirrored the present day situation in lakes such as Natron in northern Tanzania or Logipi in Kenya, which are shallow and highly alkaline.

The Dan River–Danville basin extends across parts of both Virginia and North Carolina (Fig 1A). The basin is an elongate half-graben with a southeast-dipping border fault system. The 4000m of tilted basin fill is entirely Triassic in age and unconformably overlies Proterozoic and Palaeozoic rocks primarily of the Smith River allochthon (Horton and McConnell, 1991 (Fig 1B)). The Virginia Solite Quarry is situated in the middle of the basin, where it straddles the North Carolina–Virginia state line at Cascade, VA (Fig 1B).

Today, three separate pits (Fig 1C) collectively constitute the 'Solite Quarry'. The first quarry (Quarry A of Fraser and Grimaldi, 2003) was opened in 1957 by the Virginia Solite Corporation on the Virginia side of the state line and was worked for its black 'shales' with high organic content. Work in this pit ceased many years ago, and it is now almost completely overgrown. A second pit (Quarry B) was also opened in the late 1950s a little to the south of the first, and today this pit is largely situated in North Carolina. This second pit was mined for over 30 years, but operations here largely ceased in the early 1980s when a third pit (Quarry C) was opened to the east of the second. Although originally operated by the Solite Corporation, Giant Cement owned the site for several years, working pit C in the late 1990s and early 2000s before Cemex continued quarrying primarily for road aggregate. Most recently, in 2014, Ararat Rock took ownership of the quarry and they have re-opened operations in pit B. In turn, all four owners have recognized the scientific importance of the site, and they are to be greatly commended for facilitating research over the years.

The original commercial value of the sediments lay in their properties of expansion when heated so that lightweight aggregate could be generated for use in concrete. The Virginia Solite Corporation even fuelled the kilns using a technique they developed, recycling lacquers from the local furniture industry. However, with a down-turning economy production of lightweight aggregate ceased and today it is quarried for straightforward gravel production.

Between the three quarries several hundred metres of the largely lacustrine upper member of the Cow Branch Formation (Meyertons, 1963; Olsen et al. 1991; Kent and Olsen, 1997) are exposed (Fig 2). Within the Solite Quarry section, bedded lacustrine shales, mudstones, and sandstones exhibit a distinct repetitive sequence interpreted in terms of Van Houten cyclicity (Olsen et al., 1989; Olsen and Johansson, 1994; Olsen, 1997). Seventeen such cycles are exposed in quarry B, and this is the quarry that has been the subject of much of the palaeontological research effort. Reflecting a predictable pattern of the waxing and waning in the level of an ancient lake, Olsen and Johansson (1994) suggested that the cycles reflect conditions ranging from complete desiccation to a lake 200m or more in depth. However, as already mentioned, some authors have argued for shallower lake conditions (Casey et al., 2007; Liutkus et al., 2010) and this is discussed in more detail below.

Parts of the quarry are arguably among the richest and most productive Triassic terrestrial sequences in the world (Fraser et al., 1996; Sues and Fraser, 2010). The diversity and remarkable preservation of the insects is particularly significant, but plants and vertebrates are also very well represented with soft tissue preservation in some of the tetrapod specimens.

One fossiliferous unit in quarry B (Fig 1C), designated as cycle 2 by Olsen (1979), is characterized by an approximately 330mm-thick micro-sequence at its base (Liutkus et al., 2010; Liutkus-Pierce et al., 2014). This particular unit has been the prime focus of fossil collecting at the quarry. The base of this measured (Fig 2E) micro-sequence comprises

Figure 2. Excavations in Solite Quarry B. **A**, Southwest wall of Quarry B as it was in October 2006; **B**, Outcrop along the north wall; **C**, The extension of Olsen's original pit as it was in February 2006; **D**, New trench along strike in August 2007. **E**, detail of the fossiliferous 34 cm described in Liutkus et al. (2010).

a boundary dolostone (Liutkus et al., 2010) on top of which lies a thin microlaminated unit bearing numerous soft-bodied insects, which was designated the 'insect layer' by Fraser and Grimaldi (2003). Although only 34 mm thick, this unit is the most productive in the entire quarry. In addition to the remarkable insect remains, a number of specimens of the small aquatic reptile *Tanytrachelos ahynis* (Olsen, 1979) have been recovered and here they also preserve soft tissues (e.g., Fraser et al., 1996). From our experience, fish are exceptionally rare in this particular unit and only two specimens have been collected from the 'insect layer' during the intensive collecting of the past 12 years. By contrast, Olsen *et al.* (1978) claim that fish are not uncommon in this unit. In any event, fish become more abundant and occur in approximately equal numbers to *Tanytrachelos* in the layers immediately above the insect layer, and they become even more abundant in a silty sand layer positioned approximately 21cm above the insect bed. This silty sand unit (imaginatively termed the 'fish bed') also preserves abundant plant material and coprolites, and it is also the only horizon that has yielded remains of the small gliding tetrapod, *Mecistotrachelos* (Fraser *et al.*, 2007). At the top of the micro-sequence is another thin bed that is notable for beautifully preserved *Tanytrachelos* specimens preserving some details of soft-part anatomy.

Elsewhere in the quarry, cycle 17 has also been the subject of some large-scale collecting, particularly in the early 1990s. Numerous fossils of *Tanytrachelos*, together with some fish, a diversity of plants, and a single horizon bearing numerous water bugs are known from this cycle. A tantalizing fragment of a slightly larger tetrapod was also recovered from the edge of the outcrop. Comprising vertebrae that are a little over 15mm long together with the remnants of what appear to be a row of osteoderms, nothing further can be said about its identity.

Some of the sandier units towards the top of certain cycles are notable for their trackways. These range from isolated dinosaur (e.g., *Grallator* and *Atreipus*) and phytosaur imprints to smaller *Rhynchosauroides*- and *Gwyneddichnium*-type fossils (e.g., Olsen *et al.*, 1989). Occasionally, the partial remains of tetrapods larger than *Tanytrachelos* and *Mecistotrachelos* are exposed, such as the fragmented rostrum of a small phytosaur. These are invariably discovered in hard dolomitized layers that are exceptionally difficult to prepare.

Age

Historically, the sediments of the Cow Branch Formation were regarded as Carnian in age (e.g., Olsen *et al.*, 1989; Huber *et al.*, 1993; Weems and Olsen, 1997; Kent and Olsen, 1999). However, following the recent work of Muttoni *et al.* (2004) and Furin *et al.* (2006) a Norian age has been proposed for much of the Newark Supergroup, including the Cow Branch Formation. Nevertheless, we note that Whiteside *et al.* (2011) provide an age estimate of around 226Ma for the Solite Quarry, which while 'Norian' on their timescale, would still be regarded as 'Carnian' in the usage advocated by Lucas *et al.* (2012).

The Cow Branch Formation correlates with the Chinle Formation of the southwestern United States (Olsen *et al.*, 1978; Fraser *et al.*, 1996), and palaeomagnetic data as well as the fishes and phytosaurs provide a strong correlation with the Lockatong Formation of Pennsylvania, New Jersey, and New York.

Preservation

Many of the insects are minute, and they are preserved as silvery carbonaceous films. In the field they can be very difficult to see with the naked eye, and may be easily mistaken for flecks of mica or organic detritus. By contrast, the vertebrate remains are typically covered by thin, tightly adhering layers of sediment that are practically impossible to remove. They therefore each present a significant challenge to their study. Mechanical preparation of the small tetrapod *Tanytrachelos* with pin vises or pneumatic tools (microjacks) is not possible, as there is no separation between bone and matrix. While some limited preparation has been possible using airbrasive techniques and sodium bicarbonate as the abrasive, almost invariably bone surfaces are lost (Fig 3A). As a consequence, details of features such as cranial sutures or muscle attachment scars on long bones cannot be determined. Natural weathering of vertebrate-rich layers has proved valuable in providing some understanding of basic skeletal morphology (Fig 3B), but CT scanning is essential for elucidating many skeletal characters. Even then, the extreme compression of the fossils still reduces the amount of detail that can be recovered (Fig 3C).

For study of the insects, the best results are obtained by immersing the specimens in ethanol, which increases the contrast between the matrix and the fossilized remains. Optimal observation requires either polarized light filters or an intense, diffuse nondirectional light source (e.g., a fibre-optic ring light). Specimens have also been examined using scanning electron microscopy, either in backscatter or secondary electron mode (e.g., Chatzimanolis *et al.*, 2012).

Figure 3 A, a weathered fish recovered from the main excavation trench, approximately 3 cm long; **B**, Partially weathered *Tanytrachelos ahynis* specimen, skeleton approximately 16 cm long; **C**, Posterior part of the skull and articulated anterior neck of *Tanytrachelos ahynis* that was partially mechanically prepared using airbrasive techniques.

Biodiversity

The layers of the Solite Quarry contain such an array of fossiliferous material that it is hard to understand how Meyertons (1963) considered them to contain few fossils. Even away from the main fossiliferous sequences, small fragments of plant foliage are extremely abundant, and on the bedding surfaces, exposing ripple marks and mud cracks, footprints are not uncommon. When taken together, the fossils provide a vivid picture of life around a freshwater lake in the centre of Pangaea.

Flora

Without doubt the most abundant fossils are plant remains, with numerous fragments of stems, conifer shoots and partial fronds readily apparent at many horizons throughout the quarries. Little shoots of cheirolepidaceous conifers, referred to the form taxa *Pagiophyllum* or *Brachyphyllum*, are the most ubiquitous fossils (Fig 4A). Sometimes these are found as branching clusters. Bennettitalean foliage is also quite common and includes forms such as *Pterophyllum* and *Zamites*. These tend to be most widely distributed in the units considered to represent the most humid part of the cycle. One particular bedding plane, situated just above the insect bed, extends for tens of metres and displays densely packed *Pterophyllum* fronds over its surface.

Ginkgophytes are represented by foliage that can be referred to *Eretmophyllum* and *Sphenobaeira*. *Eretmophyllum* is quite common in certain horizons low down in cycle 16. While *Sphenobaeira* is less abundant, most examples have been recovered from the beds above the insect layer in cycle 2. Sphenophyte remains are only rarely encountered, and they are typically rather fragmentary.

Ferns are rarer components of the foliage form taxa, but they include *Dictyophyllum* (Fig 4A) and other somewhat unusual specimens in which the terminal lobe of the fronds are consistently larger than the more proximal lobes.

Pannaulika triassica (Fig 4C) was first described, rather controversially, by Cornet (1993) as the oldest flowering plant. However, we have since collected new material of a dipteridaceous fern from Solite that also exhibits the characters used as evidence for angiosperm relationships of *Pannaulika* (Fraser and Grimaldi, 2003). Given that the generally accepted age of the angiosperm radiation is Early Cretaceous (Friis *et al*., 2011) *Pannaulika* should be considered a rather aberrant fragment of foliage from a dipteridaceous fern.

There are two intriguing occurrences of very distinct seeds that are adapted for wind dispersal. The first is *Fraxinopsis* (Axsmith *et al*., 1997), a distinct form taxon with a large, elongate 'wing'. This taxon was previously only known from Gondwana, and its association there with the foliage genus *Yabeiella* is indicative of gymnosperm affinities. The second is *Edenia villisperma* (Axsmith *et al*., 2013) and represents the earliest record of a seed with a pappus-like, parachute seed dispersal mechanism. These seeds typically occur as isolated specimens (Fig 4B) – most frequently in the insect bed. However, one specimen of axis fragment is known with a row of at least 10 seeds tightly packed along the axis (Fig 4D). The individual seeds are attached by their narrow (proximal) ends, with their dispersal hairs tightly folded up along the seed bodies. Its similarity to the achenes of derived angiosperms, particularly those of the basal eudicot *Platanus*, is striking. Moreover, its occurrence in a lake deposit is also interesting, in that extant *Platanus* achenes are secondarily dispersed via water. Nevertheless, despite these similarities to angiosperm fruits, it is very likely that *Edenia* represents a gymnosperm seed that independently evolved a pappus-like structure (Axsmith *et al*., 2013).

A variety of cone scales are relatively common. The majority of these are simple oval discs with a bluntly pointed base. Some of these can be referred either to Cheirolepidiaceae or Majonicaceae (Axsmith *et al*., 2004). Rarely, partial or almost complete cones are preserved.

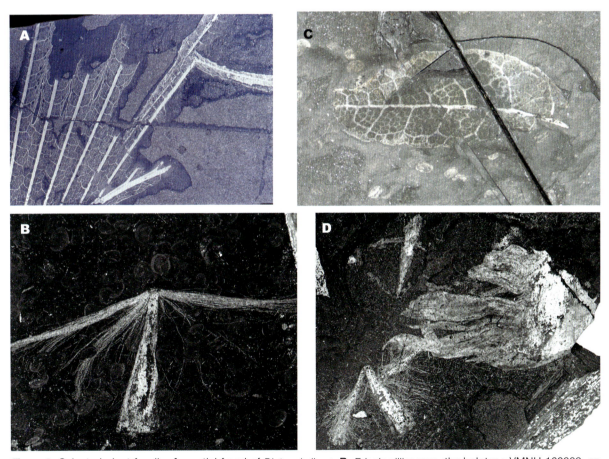

Figure 4. Selected plant fossils. **A**, partial frond of *Dictyophyllum*; **B**, *Edenia villisperma*, the holotype VMNH 160002, an isolated seed showing seed body with expanded dispersal hairs. Seed body approximately 1.4 mm; **C**, the holotype of *Pannaulika triassica*, approximately 32 mm long, VMNH 201; **D**, *E. villisperma*, VMNH 767, axis with closely spaced attached seeds visible along the left margin.

Invertebrates

Conchostracans are very abundant in some of the Solite beds, often completely covering the surfaces. In the insect bed they are so dense on some cleavage planes that they often partially obscure the insects. Conchostracans are considered to have potential biostratigraphic significance (Kozur and Weems, 2010) as they are globally widespread in Triassic freshwater deposits. Kozur and Weems (2010) assigned some to *Anyuanestheria*, which they consider to support a Carnian age assignment.

Spiders

Spiders are extremely rare in Triassic deposits, with one mygalomorph, *Rosamygale*, (Selden and Gall, 1992) from the Grès à Voltzia of eastern France and the araneomorph *Triassaraneus* from the Molteno Formation of South Africa (Selden et al., 1999). The four specimens of spider from the Solite quarry are therefore extremely important. All are likely to belong to the same taxon, *Argyrarachne solitus*, which is based on a specimen preserving all eight legs and much of the carapace but lacking the opisthosoma. It is considered likely to be a juvenile (Selden et al., 1999).

Insects

Unlike those in the Madygen Formation, the Solite Quarry insects are not overshadowed by the rather aberrant tetrapods. Indeed, it is the diversity of insect forms, coupled with the beautiful preservation of almost complete individuals, that makes the Solite Quarry arguably the world's most important single locality for Triassic insects. In all other Triassic deposits – from Europe, Central Asia, southern Africa, South America, and Australia – the insects are highly disarticulated, with usually just the wings preserved.

Insects occur sporadically within Newark Supergroup outcrops from Massachusetts to North Carolina (Huber et al., 2003), but only certain outcrops within the Solite Quarry yield abundant, diverse, and well-preserved insects (e.g., Fraser et al., 1996; Fraser and Grimaldi, 2003). The vast majority of insects originate in the rather obviously named 'insect bed' in cycle 2, which so far has produced well over 3000 individual specimens. Most occur as isolated individuals, but occasionally groups of insects occur together (Fig 5), sometimes associated with plant debris or on occasions, discrete clumps of fragmentary insects possibly representing regurgitated or faecal matter from an insectivorous vertebrate.

Preservation is on a microscopic scale: even fine, hair-like microtrichia ($c.5\mu m$ long, $<1\mu m$ width) on the wing membranes, antennae, and other parts of tiny insects have been preserved (Figs 5–12). In addition to the 'insect bed', one extremely thin, but persistent horizon in cycle 17 is known for a dense accumulation of water bugs (Nepomorpha), which are possibly notonectids or corixids (Fig 6.7). The insect accumulation on this bed is almost monospecific, but very occasionally there are hints of other taxa, including a fragmentary wing of Elcanidae (a Triassic to Cretaceous group of grasshopper-like insects). In Quarry C another horizon was found to contain fragmentary beetle elytra that resemble those referred to *Holcoptera* (Handlirsch, 1906-1908) which have a very distinctive pattern. These were originally described from the lower Lias of southwest England, but similar material has also been recovered from Early Jurassic strata of the Newark Supergroup (e.g., the Mount Toby Formation, Massachusetts) (Huber et al., 2003; Kelly et al., in press). Interestingly, larval-like forms referred to *Mormolucoides* are also common to both the Mount Toby Formation and the unit that yields the *Holcoptera*-like elytra in Quarry C at Solite (Huber et al., 2003).

The insect diversity includes at least 11 orders of pterygote insects: Auchenorrhyncha, Diptera (Fig 6), Heteroptera (Fig 7), Blattodea (Figs 8A, B), Thysanoptera (Fig 8C–F), Mecopterida (Fig 9E), Coleoptera (Figs 10, 11), Odonata, Sternorrhyncha, Orthoptera (Fig 13) and ?Plecoptera, and comprising perhaps 50 species known thus far (Fraser and Grimaldi, 1999, 2003).

Currently eight families of Dipteran (Fig 6) divided into eight genera and 16 species have been recorded (Blagoderov et al., 2007). The Nematocera are the predominant forms, but *Prosechamyia* appears to be a very early stem group to the extremely diverse infraorder Brachycera, the earliest definitive members of which appear in the Early Jurassic. Brachycera includes over 100 living families, with such familiar ones as horseflies (Tabanidae), robber flies (Asilidae), fruit flies (Drosophilidae), and house flies (Muscidae). A culicomorphan with a long proboscis is of particular interest, as it represents the earliest fossil record of a structure apparently specialized for blood feeding. This group includes mosquitoes, blackflies, biting midges and other blood feeders. Other families include

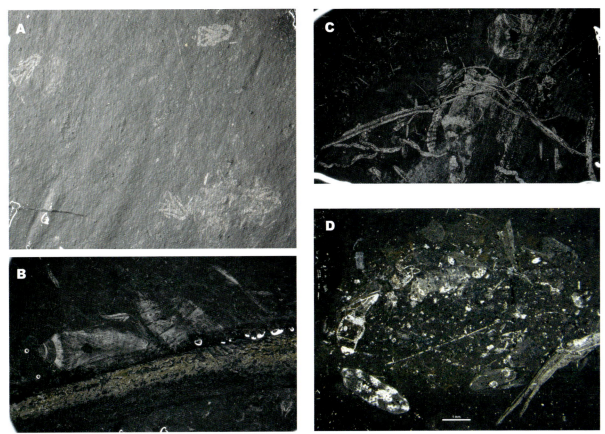

Figure 5. Assemblages of insects from the Solite shale. **A**, five individuals of an aquatic nepomorphan bug (*cf.* Corixidae?: Fig 7F), individual bodies approximately 2.5 mm long; **B**, Remains of four individuals of Belostomatidae, including bodies and the fringed swimming legs. Each specimen approximately 15 mm long; **C**, Accumulation of at least 11 insects amongst strands of plant remains. Identifiable insects include adult and nymphal Belostomatidae, Archescytinidae, and a dipteran. The belostomatid in the centre of the field of view is approximately 12 mm long; **D**, Remains of at least a dozen fragmentary insects, including isolated elytra and wings of Coleoptera and Diptera, and assorted other body parts (VMNH 934, scale bar represents 1.0 mm.). Fragmentation of the insects and their compaction into an oval shape readily indicates this is regurgitate or fecal matter of an insectivorous or scavenging vertebrate, possibly from *Tanytrachelos*. All of the insect specimens were photographed while wet with alcohol and using a ring light. Intensity of the silvery image in places probably corresponds to the darkness (melanization) of the original cuticle.

the Limoniidae (three species), Crosaphidae, Psychodidae, Procramptonomyiidae (basal bibionomorphs represented by two species. A new family, Prosechamyiidae, was erected for the two species of basal brachyceran (Blagoderov *et al.*, 2007).

Despite their relatively high diversity, Diptera only comprise some 1.5% of all insects recovered from Solite. This is comparable with the Ipswich Series of Australia, where only four out of 500 insects were dipterans (Anderson and Anderson, 1993) while Krzemiński and Krzemińska (2003) and Marchal-Papier *et al.* (2000) recorded only 59 adult dipterans from a total of 5300 other insects in the Grès à Voltzia of eastern France.

At some other Triassic localities dipterans are even rarer. For example, at the Madygen locality of Dzhailoucho, Kyrgyzstan, only 11 fly specimens were found among approximately 20,000 insects collected, or just about 0.05%.

Although coleopterans and water bugs are the most abundant insects, they have largely not been described in any detail to date. In the insect bed all the known water bugs can be referred to the extant family Belostomatidae (Fig 7). Today belostomatids can be up to 10 cm long and are voracious predators, even capable of feeding on small fishes. The Solite belostomatids are rather smaller, only attaining lengths of about 1.5 cm. Nevertheless with their

Figure 6 Diptera (true flies). The Solite shales preserve the most detailed and diverse record of flies from the Triassic; no other Triassic deposit has such delicate insects preserved fully articulated. **A**, *Metarchilimonia krzemiskoroum* (Tipulomorpha) (VMNH 3671); **B**, *Virginiptera certa* (Sciaroidea) (VMNH 731); **C**, *Triassopsychoda olseni* (Psychodomorpha) (VMNH 733, body 3.5 mm long); **D**, *Yalea rectimedia* (Procramptonomyiidae) (VMNH 1041b); **E**, Diptera indet. Scale bars A, C–E represent 1.0 mm, 0.5 mm in B.

'oar-like' fringed hind (swimming) legs and short raptorial forelegs they show a close resemblance to modern forms, and would still have been active predators. The adults have well-preserved wings and would have been capable of flight. Details such as the paired row of spiracles on the ventral surface of the abdomen are occasionally preserved, as is forewing venation. Belostomatids (Figs 7A–E) are most predominant in the bottom 1.0 mm, and instars can occasionally occur in densities as high as 75 per 0.01 m^2, although this is rare. There are also examples of instars and adults occurring in intimate association with sections of plant stems (Fig 5c). These potentially represent conditions whereby insects became entangled in floating plant debris, which then sank together to the bottom of the pool or lake margin, or simply that they lived amidst submerged and emergent vegetation at the lake margin, a habitat that belostomatids frequent today.

As noted above, a single horizon in cycle 16 of Quarry B is notable for the incredibly dense accumulation of relatively small water bugs that are considered to be either corixiids or notonectids (Fig 7F). Since they are invariably preserved with the ventral side exposed, they

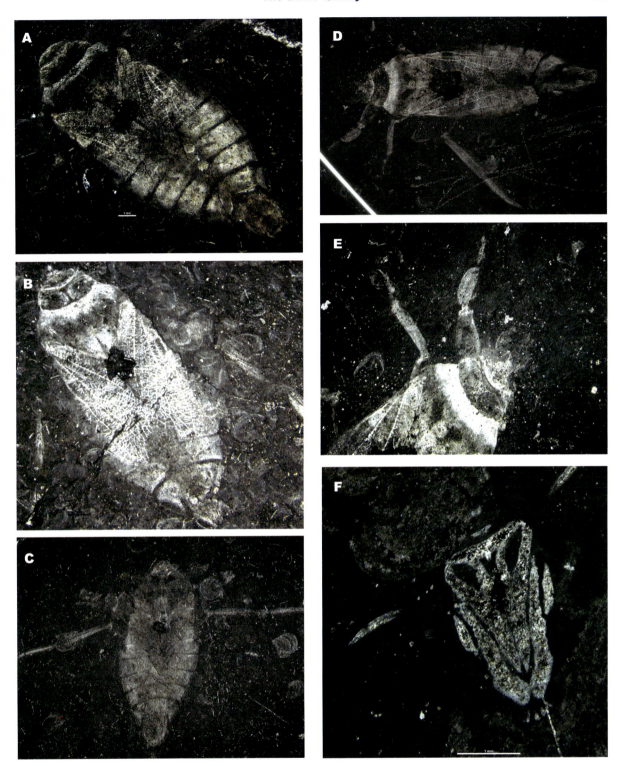

Figure 7 Predaceous aquatic Hemiptera (Nepomorpha). **A-E**, Adult Belostomatidae. These are the most common large insects in the Solite shale; nymphs are particularly common. Specimens in figs. 7.B, 7.C, and 7.D show the fringed (hind) swimming legs outstretched. Aggregations are frequently found (see Figure 5A, B). **A**, YPM 36437; **B**, VMNH 806; **C**, YPM 36437; **D**, AMNH; **E**, Detail of specimen in fig. 7.D. This is the only specimen showing the raptorial fore legs typical of the family; **F**, Family uncertain (near Corixidae or Notonectidae). The hemeltyra (modified forewings) have a distinctive colour pattern. Scale bars represent 1.0 mm.

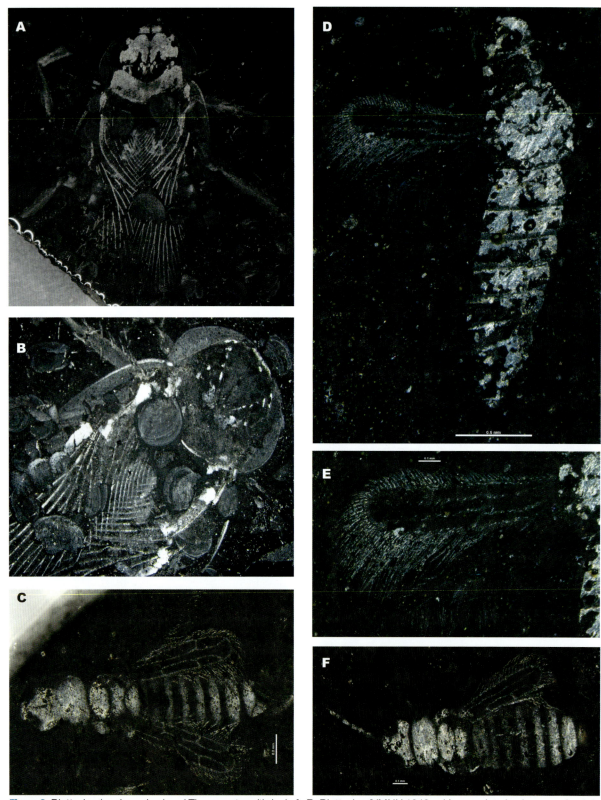

Figure 8 Blattodea (cockroaches) and Thysanoptera (thrips). **A**, **B**, Blattodea (VMNH 1043 a, b), complete body approximately 8.0 mm long. The colour pattern is preserved on the pronotum on the half shown in B. **C-F**, Thysanoptera (*Triassothrips*), with the fine fringe of hairs on the forewing margin well resolved (shown in detail in 8E). Specimens: C (VMNH 747); D, E (VMNH 836); F (VMNH 921A). Scale bar in C, D 0.5 mm and in E, F 0.1 mm.

are thought most likely to be notonectids or 'backswimmers'. This single layer has been sampled at many points across the quarry, and throughout it yields specimens at a density ranging from 10–100 per 0.01m². Therefore, despite being restricted to a single bedding surface, numerically these are the most well represented insects in the quarry.

The occurrence of thysanopterans (Figs 8C–8F) is of particular interest, not least for the fidelity of detail preserved in insects that are little more than 1.5mm long. Today the majority of thysanopterans, or thrips, are phytophagous with a reputation for being sometimes known as pests of various angiosperms, including fruit crops. Some flower-feeding forms probably aid in pollination, and one lineage is also known as cycad pollinators. However, there is also some fossil evidence for thysanopterans facilitating gymnosperm pollination (Peñalver et al., 2012). Nevertheless, given the apparent close relationship of thrips to lophioneurid psocopteroids, it seems likely that the Triassic thrips were also saprophagous, similar to the diets of the extant basal family Merothripidae. A number of small differences can be seen from one specimen to another, and it is unclear whether there may be more than the one species that has been described, *Triassothrips virginicus*. For example, there is quite a bit of variation in shape of the head, eyes and pronotum among the various specimens, and VMNH 921(Fig. 8F) is noticeably short-bodied with no apparent pronotum, so that it was not even included among the paratypes (Grimaldi et al., 2005). In addition, a silvery film on the forewings of one specimen, VMNH 3214, is suggestive of pigmentation, which points to a separate species. However, the asymmetry of the 'pigmentation' casts some doubt on this and it may simply be an artefact of preservation.

Mecopterans (scorpionflies) are rare with only three specimens of the single representative, *Pseudopolycentropodes virginicus*, so far recovered from the site (Fig 9E). Other Triassic members of the family occur in Eurasia, with no records from Gondwana. Certain Late Jurassic and Cretaceous scorpionflies are known to possess long, fine proboscides. Such a proboscis may have served to facilitate probing for ovular

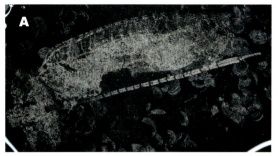

Figure 9 Assorted arthropods from the Solite Quarry. **A**, a Neuropteran. The filamentous, segmented structure is the antenna. **B**, **C**, Stem-group Amphiesmenoptera (Lepidoptera + Trichoptera). **D**, a common arthropod in the Solite shale of uncertain class (possibly Crustacean), showing fan-like structures of what are either terminal gills or uropods; **E**, scorpionfly (Mecopterida), of the extinct family Pseudopolycentropodidae (VMNH 944). Scale bars 1mm in B-D and 0.5mm in E.

secretions in gymnosperms, and as such it has been proposed that they could have played a role in gymnosperm pollination (Ren *et al.*, 2009). However, the Triassic form lacks such a long proboscis, and presumably therefore did not serve a specific gymnosperm pollinating role in the Triassic. *P. virginicus* is very similar to *P. triassicus* from the Middle Triassic of eastern France (Papier *et al.*, 1996). The broad triangular forewing, smaller hindwing, but large mesothorax suggest that *P. virginicus* was a strong, agile flier that mostly used its forewings for flight. The large tibial spurs and thick, multiarticulate antennae are also characteristic features not present in later Pseudopolycentropodidae (Grimaldi *et al.*, 2005).

Many of the records of Coleoptera (Figs 10, 11) are predominantly restricted to the bodies and elytra

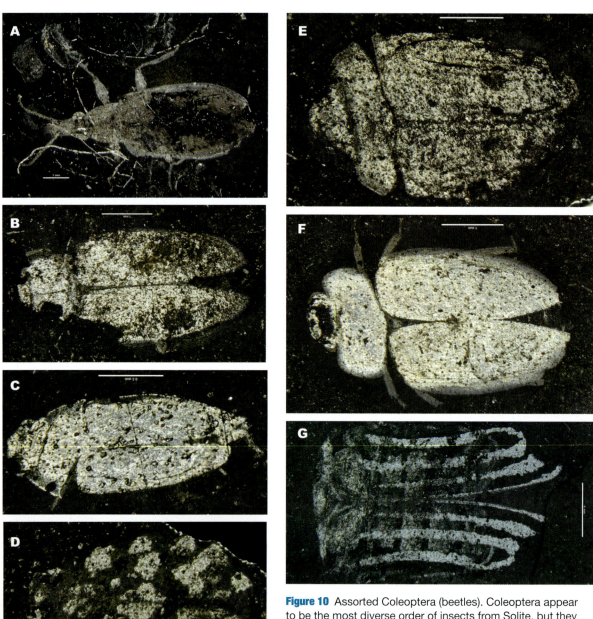

Figure 10 Assorted Coleoptera (beetles). Coleoptera appear to be the most diverse order of insects from Solite, but they can only rarely be identified to families since wing venation is not observable. Specimens in Figs 10B, D, and G have preserved colour patterns. **A**, VMNH 838A; **B**, VMNH 1359; **C**, AMNH 04-77; **D**, AMNH 04-70A, **E**, VMNH 129; **F**, VMNH 959; **G**, a specimen of Holcoptera sp. (Coptoclavidae). Scale bars represent 1 mm in A, B, D-G and 0.5 mm in C.

and are difficult to assign to specific families. Instead they will essentially remain as form taxa based on variation in the ornamentation, colour pattern, size and shape of the elytra. One of the few exceptions is the staphylinid (rove) beetle, *Leehermania prorova* (Chatzimanolis *et al.*, 2012) (Figs 11A–C). Interestingly, given the group's exceptional diversity today (with over 57,000 species) *L. prorova* is the only staphylinid from Solite, a statistic that contrasts dramatically with the remarkable diversity of Diptera from Solite (Blagoderov *et al.*, 2007).

Hemipterans are represented by a variety of forms (Fig 12). They include one of the most abundant groups of insects at Solite, the Archescytinidae (Figs 12C–E), which are minute, winged sternorrhynchans. Like other sternorrhynchans, the aphids and scale

Figure 11 Assorted Coleoptera. Figs. 11**A-C** are separate specimens of *Leehermania*, an early member of the rove beetle superfamily, Staphylinoidea, possessing one of the hallmarks of this group, the short elytra. **A**, Dorsal view (VMNH 1343); **B**, Dorsal view, with elytra slightly parted; **C**, Lateral view (VMNH 1244b); **D**, undetermined Coleoptera indet., dorsolateral view (VMNH 758); **E**, Coleoptera indet., lateral view; **F**, Coleoptera indet., dorsal view (VMNH 3113). Scale bars represent 0.5 mm in A-C and 1.0 mm in D and E.

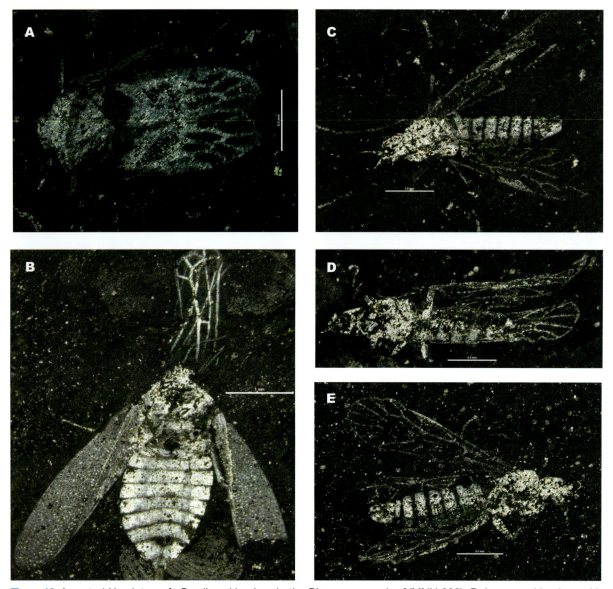

Figure 12 Assorted Hemiptera. **A**, Small sucking bug in the Dipsocoromorpha (VMNH 930); **B**, Large sucking bug with distinctive hemelytra having dense rows of dark spots; **C-E**, Archescytinidae, an early relative of aphids and scale insects. After the water bugs, this is the most common insect in the Solite shales. **C**, VMNH 3163; **D**, YPM 36436; **E**, YPM 36439. Scale bars represent 0.5 mm in A, C-E and 1mm in B.

insects, these were almost certainly phytophagous. The phytophagous insect fauna of the Solite beds is significant, consisting of the Elcanidae (Orthoptera) (Figs 13B–D), Auchenorrhyncha (plant hoppers), and perhaps *Triassothrips* and some of the beetles.

Although the insect bed is very uniform in thickness and composition across its entire exposure along the north face of the quarry, there is some preliminary data that indicates subtle variation in the relative abundance of different groups. Towards the east end of the exposure in the vicinity of Paul Olsen's initial excavation, belostomatids are very abundant and sometimes quite densely distributed across a bedding surface. For example, one small block measuring less than 30×20 cm exhibits over 70 nymphs, yet 100 metres to the west belostomatids, while still the most abundant forms, are fewer in number, and coleopterans are more abundant. Perhaps this reflects the presence of localized pools along the lake margins; but a more detailed study of the distribution of the insects over the entire extent of the quarry is clearly warranted.

Vertebrates

Fishes

Olsen and colleagues (e.g., Olsen *et al.*, 1989; Olsen and Johansson, 1994) have documented a range of fishes but these have not been described in detail, and illustrations are mostly limited to generic outline sketches and not detailed specimen drawings. This owes much to the difficulty in preparing these fossils. *Turseodus* is one of the most prevalent fishes, and the holostean *Semionotus* is also well documented. Coelacanths include *Diplurus* and some large examples of *Pariostegus*. Finally, the redfieldiid genera *Synorichthys* and *Cionichthys* have also been recorded. Neverteheless, detailed studies of all the fishes have still to be undertaken, and examination using CT scanning is likely to provide a great deal more information about the fish fauna.

Within cycle 2 the so-called 'fish bed' has, not surprisingly, been most productive for fish, but they have also been recovered from many of the beds yielding the small protorosaur, *Tanytrachelos*. However, the paucity of fish in the insect unit is striking. Although Olsen (pers. comm.) suggests that there are fish in the insect bed, in over 20 years' collecting by crews from VMNH and AMNH only two small fish have been recovered from the insect unit.

The 'fish layer' consists of interbedded siltstone, sandstone and dolostone and is rich in plant material, coprolites and fish, including some of the largest specimens known from the site. This bed was considered by Liutkus *et al.* (2010) to be representative of the deepest lake deposits in cycle 2.

Tetrapods

None of the tetrapods so far documented from the Solite Quarry could even be remotely considered as representatives of modern lineages. Indeed, the footprints (*Atreipus* and *Grallator*) and trackways of small non-avian dinosaurs are perhaps the closest records to living tetrapods.

Aside from the dinosaur trackways, perhaps the most conventional tetrapod records belong to phytosaurs. Isolated phytosaur teeth are not common, but have been recovered from various levels throughout the quarries. Occasionally, sections of more complete, sometimes articulated, material, including a partial rostrum, are found. However, preparation of this material is exceptionally difficult, and to date no identifications beyond *Rutiodon* sp. (e.g., Olsen and Johansson, 1994) have been made.

Figure 13 Assorted Orthoptera from the Solite Quarry. **A**, a specimen of Tetrigidae (?) showing a large hind femur and the very large pronotal shield that is characteristic of the family, body approximately 9 mm long; **B**, adult specimen of Elcanidae (VMNH 766). The legs have colour patterns; **C**, wing of an elcanid (AMNH 04-518A); **D**, specimen of an additional elcanid wing (VMNH 784B). Scale bars represent 1 mm.

The most common vertebrate is the small protorosaur, *Tanytrachelos ahynis* (Olsen, 1979) (Fig 14). It is now known from a few hundred articulated skeletons that have been collected from the quarry along with many associated remains and isolated elements. Cycles 16 and 2 are the most productive, although it has been recovered in other cycles in Quarry B (Olsen, 1978) and also in Quarry C. Some of the most complete specimens have been recovered from the insect bed, and typically there is some preservation of soft tissue in these specimens (e.g., Fraser *et al.*, 1996). On the basis of these specimens, it is probable that the skin was quite smooth and that at least the hind feet were webbed (Fraser, 2006). In addition, small footprints on some beds in cycle 16 closely match the pedal structure of *Tanytrachelos* and these trackways occasionally show clear impressions of webbing.

Casey *et al.* (2007) undertook a detailed quantitative taphonomic study of almost 100 specimens recovered from cycles 2 and 16. Apart from within the insect bed, the remains of *Tanytrachelos* are, surprisingly, rather disarticulated and, in the absence of bioturbation, suggest disruption by currents. Such taphonomic patterns support a low to moderate depth depositional environment, which corroborates the preliminary data based on the insect taphonomy.

While *Tanytrachelos* is also known from a handful of other localities in the Newark Supergroup, the small gliding tetrapod *Mecistotrachelos* (Fraser *et al.*, 2007) was for a long time known on the basis of just two specimens (Fig 15) that were both recovered from the unit termed the 'fish layer'. When discovered, the two specimens were recognized purely on the basis of poorly defined swellings on the bedding surface. In the case

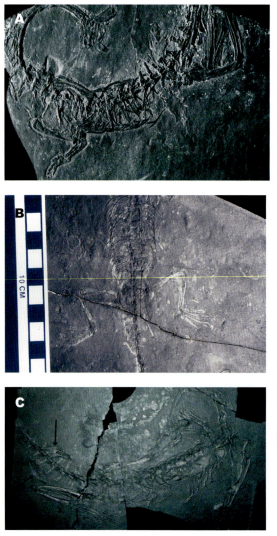

Figure 14 *Tanytrachelos ahynis*. **A**, the holotype, YPM 7496, skull approximately 2.6mm long; **B**, VMNH sexual dimorph A lacking the heterotopic bones; **C**, VMNH sexual dimorph B with heterotopic bones (indicated by arrows). **D**, illustration of restored *Tanytrachelos* by Karen Carr Studio.

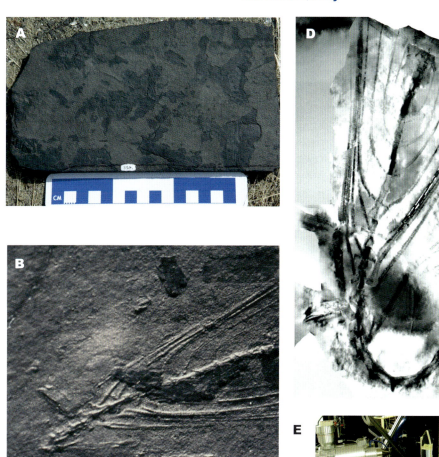
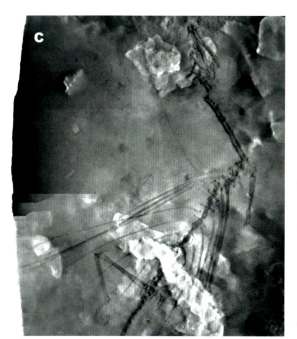

Figure 15. *Mecistotrachelos apeoros*. **A**, VMNH 3650 as recovered in the field; **B**, the holotype, VMNH 3649; Composite CT images of **C**, VMNH 3650 and; **D**, VMNH 3649; **E**, VMNH 3649 in the CT facility at Penn State University; **F**, illustration of restored *Mecistotrachelos* by Karen Carr Studio. The skull in both specimens is approximately 2.5cm long.

of VMNH 3650, the evidence for bone on the surface is very faint indeed, whereas in the holotype (VMNH 3649) at least the impressions of the elongate 'thoracic' ribs and the skull and neck are quite clear. Nevertheless, it has proved impossible to prepare either specimen by traditional means, and consequently *Mecistotrachelos* is described solely on the basis of CT scans. The elongate cervical vertebrae are reminiscent of protorosaurs including *Tanytrachelos*; however, elongate cervical ribs have not been discerned from the present scans, and resolution of details in the potentially diagnostic areas of the foot and skull are not sufficient to comment further on phylogenetic relationships. The elongate dorsal ribs are markedly similar to those of additional gliding tetrapods from equivalent-aged deposits of the Newark Basin in New Jersey and the Late Triassic British fissure fills. The forms from the UK are often referred to as two separate genera, *Kuehneosaurus* and *Kuehneosuchus* (Robinson, 1967), while the single specimen from New Jersey was described by Colbert (1966; 1970) as a third genus, *Icarosaurus*. However, Sues and Fraser (2010) regarded the differences between all three to be trivial and argued that all should be referred to *Icarosaurus*. *Mecistotrachelos* differs from *Icarosaurus* in markedly elongate cervical vertebrae and the lack of very pronounced transverse processes on the dorsal vertebrae that supported the ribs. In addition the pes, as seen in VMNH 3650, is noticeably short and is preserved in an unusual flexed posture that Fraser *et al.* (2007) considered as further evidence of an arboreal habit.

A recent phase of excavation has resulted in the discovery of two additional specimens of *Mecistotrachelos*. Like the two earlier discoveries, both originated in the fish layer, and both are apparently in complete articulation. It is hoped that CT scanning will yield more detailed information on the anatomy and help resolve phylogenetic relationships.

Depositional environment

As noted earlier, traditionally the most fossiliferous units at the Solite Quarry have been considered to represent deposition in a relatively deep lake (e.g., Olsen *et al.*, 1989; Kent and Olsen, 1997). Olsen and Johansson (1994) stated that abundant complete fish and tetrapods are only present in the microlaminated calcareous clay-rich siltstones lacking desiccation cracks, and the finely laminated beds, coupled with a lack of bioturbation, do tend to support such a view.

However, more recently this interpretation has been challenged. Firstly, Casey *et al.* (2007) noted that apart from the so-called insect bed, the numerous skeletons of the small aquatic reptile *Tanytrachelos* were unexpectedly incomplete for deposition within an unbioturbated deep lake. They attributed the incompleteness of the skeletons to currents, which in turn is suggestive of shallower water conditions.

Secondly, Liutkus *et al.* (2010) undertook a detailed examination of the lower part of cycle 2 and made a number of observations that are strongly suggestive of a rather unique depositional environment. They found that the basal part of the section, which they referred to as Unit 1, exhibits weak layering, high clay content, high siliciclastic content, and low carbonate content. One of the most significant features is a very distinct thin boundary layer (2–3mm) comprising a carbonaceous dolomite, and which defines the base of the insect layer above. This 'boundary dolostone' is broken at regular intervals and exhibits numerous downward-pointing wedge-shaped cracks that have been interpreted by Liutkus *et al.* (2010) as desiccation cracks. Liutkus *et al.* (2010) also noted that although the insect beds were very finely laminated, they considered the laminations to be discontinuous. The wavy clay laminites are reminiscent of algal mats, or at least appear to be biologically mediated. Liutkus *et al.* (2010) further argued that the presence of primary dolomite in these laminites indicates that the lake was saline, alkaline and probably shallow (Last, 1990). Certainly, similar interpretations have been made for other dolomite-bearing basin sediments (e.g., Gierlowski-Kordesch and Rust, 1994). Thus the lack of bioturbation and predation, as well as exquisite preservation of the soft tissue and microscopic detail of fully articulated, delicate insects could equally well be attributed to toxicity of lake-margin waters or to extraordinary depth.

A recent survey (Smith, 2012) of exceptional insect preservation in lacustrine environments suggests that the type and size of insects preserved and their completeness might provide insights into lake depth. Offshore-deepwater assemblages tend to have relatively low species richness, an abundance of terrestrial taxa that are good flyers, and there is often a greater quantity of large specimens, although the range in size is relatively great. Moreover, the quality of preservation is typically high, and specimens are frequently articulated. By contrast, nearshore assemblages tend to exhibit high species richness, a mixture of weak and strong flyers (the latter having been wafted back to shore), a tendency to be relatively small, comprise more disarticulated specimens, and variable preservation quality from high to low. Near-shore environments are also noted for their higher abundance of Coleoptera. It should be noted that, again, an abundance of phytophagous insects in the Solite beds suggests preservation of these near their habitat (i.e., near-shore or emergent vegetation).

When comparing the Solite insect assemblages against these criteria, they do contain many complete specimens, yet at the same time they are, for the most part, small (particularly the archescytinids, thysanopterans, dipterans and coleopterans) and coleopterans are relatively abundant. Moreover, the abundance of belostomatids, including abundant instars, is highly suggestive of a nearshore environment. Interestingly the greatest abundance of belostomatids occurs in the lower 5mm of the insect bed, while the greatest number of dipterans occurs in the upper 10mm of the insect bed. There is even some suggestion of a gradual deepening of the lake, moving up within the insect bed itself. However, such small, but completely preserved, dipterans, retaining their filamentous legs and antennae, are exceptionally rare in the fossil record. Where they are found, such as in the Bembridge Marls of the Isle of Wight (McCobb *et al.*, 1998; Ross and Self, 2014), then they are preserved in what are considered to be shallow-water environments, perhaps getting caught up in algal mats that eventually sink to the bottom.

Thus, while the high degree of articulation and high preservation quality of the specimens are indicators of an off-shore deposit, the high species richness, predominance of relatively small individuals, and the mixture of weak and strong flyers point to a near-shore deposit. This, coupled with the abundance of aquatic insects, (namely water bugs) typically associated with shallow water, favours the interpretation of the insect bed as a near-shore depositional environment.

The distribution and preservation of the two small Solite tetrapods perhaps offers some additional insight into the depositional environment. All four specimens of *Mecistotrachelos* were apparently preserved in complete articulation. While the posterior part of one specimen (VMNH 3649) is missing, this is entirely due to collection failure. At the same time, some specimens of *Tanytrachelos* are rather variable in their preservation, and outwith the insect bed many are incompletely articulated. It seems a little curious that these two similar-sized terrestrial vertebrates have yet to be found together within exactly the same units of the main fossiliferous sequence, but presumably they were occupying very different niches. The function of the elongate ribs of *Mecistotrachelos* does not seem to be in doubt, and it would seem most likely that it inhabited areas away from the shoreline with stands of trees. Presumably on occasions they were blown off course and out over the lake. Any individuals landing off-shore would inevitably have rapidly drowned. If the insect bed does indeed represent a very shallow marshy type of environment, then any *Mecistotrachelos* accidentally 'flying' into this environment may have been able to extricate itself and scramble back to land. This could explain why *Mecistotrachelos* is rare and only found in the units bearing the largest fish, and which we consider to represent some of the deeper lake sequences.

On the other hand, *Tanytrachelos* would seem to have lived and died in the area immediately surrounding the shoreline, which would certainly explain its abundance in the insect bed and the beds just above.

It is probably fair to say that neither of the two interpretations for the depositional environment is completely satisfactory, and there are various lines of evidence to support both competing arguments. Clearly these were very complex lake systems, and further detailed analyses of both the Solite and other Newark Supergroup sequences are necessary before a more definitive conclusion can be drawn.

Summary

The Norian Cow Branch Formation exposures at the Virginia Solite Quarry have yielded a remarkable assemblage of vertebrates, terrestrial arthropods and plants (Olsen *et al.*, 1978; Fraser *et al.*, 1996). Not surprisingly, the lacustrine sequences include a number of fishes, but they also contain equally abundant tetrapod remains, albeit low in diversity. Moreover, the body fossils so far recovered only represent lineages that died out at the end of the Triassic (Olsen, 1979; Fraser *et al.*, 2007). On the other hand, the insect assemblage is exceptionally diverse and includes representatives of living orders and families. It is also notable as the earliest known aquatic insect assemblage and the only Triassic locality in the world yielding abundant articulated and mostly complete insects.

The depositional environment and post-depositional conditions responsible for the remarkable preservation has been subject to some debate. Although this has been traditionally viewed as a deep lake environment, one group of workers has questioned this assumption. Research by Liutkus and colleagues (Liutkus et al., 2010, Liutkus-Pierce et al., 2014) is one of the first attempts to describe the sedimentology and geochemistry of a microsection surrounding a Lagerstätte. Their results provide support for a competing hypothesis where a shallow, yet toxic, lacustrine depositional environment was key in the successful preservation of detailed fossils and soft parts. The research on the Solite microsection provides an alternative, albeit unorthodox, scenario (shallow, toxic lake margin) in which Lagerstätten can be preserved, and could be a tool in the search for future terrestrial Lagerstätten.

Acknowledgements

We especially thank the various quarry workers of the Virginia Solite over the past 25 years for free access to the site and regular help with excavations. The owners of the Virginia Solite Corporation, Giant Cement, Cemex and Ararat Products have all enthusiastically supported our research and without their assistance this internationally important site would have languished. The numerous members of the volunteer field crews contributed greatly to the success of the research. We particularly recognize the contributions of Jim Beard, Christina Byrd, Alexa Chew, Anne Davis, Tim Dooley, Christa Hampton, Phil Huber, Milton Hundley, Julian MacCarthy, Miles Trail, and Ray Vodden, and all the many students from Virginia Tech and Appalachian State universities. We have benefited greatly from discussions with various colleagues including Jim Beard, Vladimir Blagoderov, Derek Briggs, Michele Casey, Bruce Cornet, Dennis Kent, Conrad Labandeira, Pete Le Tourneau, Steve Scheckler, Roy Schlische, Bill Shear, and Hans Sues. We are indebted to US National Science Foundation (NSF EAR 0106309 to NCF and DAG) for funding the research and the excavations were supported by National Geographic Society grants 5103-93 and 9517-14. Lastly, many thanks to Paul Olsen for the numerous insightful discussions, often held long into the night, and his unique field skills. Without Paul's involvement and enthusiasm none of these discoveries would have been made.

References

Anderson, J. M., and H. M. Anderson. 1993. Terrestrial flora and fauna of the Gondwana Triassic: 1, Occurrences. New Mexico Museum of Natural History and Science Bulletin 3:3–12.

Axsmith, B. J., F. M. Andrews, and N. C. Fraser. 2004. The structure and phylogenetic significance of the conifer *Psuedohirmerella delawarensis* nov. comb. from the Upper Triassic of North America. Review of Palaeobotany and Palynology 129:251–263.

Axsmith, B. J., N. C. Fraser, and T. Corso. 2013. A Triassic seed with an angiosperm-like wind dispersal mechanism. Palaeontology 56: 1173-1177.

Ayers, J. C., and L. Zhang. 2005. Zircon aqueous solubility and partitioning systematics. Goldschmidt conference abstracts, Moscow 15:A5.

Axsmith, B. J., T. N. Taylor, N. C. Fraser, and P. E. Olsen. 1997. An occurrence of the Gondwanan plant *Fraxinopsis* in the Upper Triassic of eastern North America. Modern Geology 21: 299–308.

Blagoderov, V., D. A. Grimaldi, and N. C. Fraser. 2007. How time flies for flies: diverse Diptera from the Triassic of Virginia and early radiation of the order Diptera. American Museum Novitates 3572:1–39.

Calvert, S. E., R. M. Bustin, and E. D. Ingall. 1996. Influence of water column anoxia and sediment supply on the burial and preservation of organic carbon in marine shales. Geochimica et Cosmochimica Acta 60:1577–1593.

Calvo, J. P., J. A. McKenzie, C. Vasconcelos. 2003. Microbially-mediated lacustrine dolomite formation: evidence and current research trends. Pp. 229–251 in B. L. Valero-Garcés (ed.), Limnogeology in Spain: a tribute to Kerry Kelts. Consejo Superior de Investigaciones Cientificas, Madrid, Spain.

Carroll, A. R., K. M. Bohacs. 1999. Stratigraphic classification of ancient lakes: balancing tectonic and climatic controls. Geology 27:99–102.

Casey, M. M., N. C. Fraser, and M. Kowalewski. 2007. Quantitative taphonomy of a Triassic reptile *Tanytrachelos ahynis* from the Cow Branch Formation, Dan River Basin, Solite Quarry, Virginia. Palaios 22:598–611.

Chatzimanolis, S., D. A. Grimaldi, M. S. Engel, and N. C. Fraser. 2012. *Leehermania prorova*, the earliest staphyliniform beetle, from the Late Triassic of Virginia (Coleoptera: Staphylinidae). American Museum Novitates 3761:1–28.

Colbert, E. H. 1947. Studies of the phytosaurs *Machaeroprosopus* and *Rutiodon*. Bulletin of the American Museum of Natural History 88:53–96.

Colbert, E. H. 1965. A phytosaur from North Bergen, New Jersey. American Museum Novitates 2230:1–25.

Colbert, E. H. 1966. A gliding reptile from the Triassic of New Jersey. American Museum Novitates 2246:1–23.

Colbert, E. H. 1970. The Triassic gliding reptile *Icarosaurus*. Bulletin of the American Museum of Natural History 143:85–142.

Colbert, E. H. and P. E. Olsen. 2001. A new and unusual aquatic reptile from the Lockatong Formation of New Jersey (Late Triassic, Newark Supergroup). American Museum Novitates 3334:1–44.

Cope, E. D. 1887. A contribution to the history of the Vertebrata of the Trias of North America. Proceedings of the American Philosophical Society 24:209–228.

Cornet, B. 1993. Dicot-like leaf and flowers from the Late Triassic tropical Newark Supergroup rift zone, USA. Modern Geology 19:81–99.

DeDeckker, P., and W. M. Last. 1989. Modern, non-marine dolomite in evaporitic playas of western Victoria, Australia. Sedimentary Geology 64:223–238.

deWet, C. B., C. I. Mora, P. J. W. Gore, E. Gierlowski-Kordesch, and S. J. Cucolo. 2002. Deposition and geochemistry of lacustrine and spring carbonates in Mesozoic rift basins, eastern North America. Pp. 309–325 in R. W. Renaut and G. M. Ashley, (eds), Sedimentation in Continental Rifts. SEPM, Tulsa, OK, Special Publication 73.

Duffy, C. J., and S. Al-Hassan. 1988. Groundwater circulation in a closed desert basin: topographic scaling and climate forcing. Water Resources Research 24:1675–1688.

El Tabakh, M., and B. C. Schreiber. 1998. Diagenesis of the Newark Rift Basin, eastern North America. Sedimentology 45:855–874.

Emmons, E. 1856. Geological Report of the Midland Counties of North Carolina. Henry D. Turner, Raleigh, 351pp.

Fraser, N. C., and D. A. Grimaldi. 1999. A significant Late Triassic Lagerstätte from Virginia, U.S.A. Revista del Museo Civico di Scienze Naturali 'Enrico Caffi' 20:79–84.

Fraser, N. C., and D. A. Grimaldi. 2003. Late Triassic continental faunal change: new perspectives on Triassic insect diversity as revealed by a locality in the Danville basin, Virginia, Newark Supergroup. Pp. 192–205 in P. M. LeTourneau and P. E. Olsen, (eds), The Great Rift Valleys of Pangaea in Eastern North America. Volume Two: Sedimentology, Stratigraphy, and Paleontology. Columbia University Press, New York.

Fraser, N. C. and D. Henderson. 2006. Dawn of the Dinosaurs: Life in the Triassic. Indiana University Press, Bloomington, Indiana, 328pp.

Fraser, N. C., D. A. Grimaldi, P. E. Olsen, and B. Axsmith. 1996. A Triassic Lagerstätte from eastern North America. Nature 380:615–619.

Fraser, N. C., A. B. Heckert, V. P. Schneider, and S. G. Lucas. 2006. The first record of *Coahomasuchus* (Archosauria: Stagonolepididae) from the Carnian of eastern North America. Journal of Vertebrate Paleontology 26(3, Supplement):63A.

Fraser, N. C., P. E. Olsen, A. C. Dooley Jr., and T. R. Ryan. 2007. A new gliding tetrapod (Diapsida: ?Archosauromorpha) from the Upper Triassic (Carnian) of Virginia. Journal of Vertebrate Paleontology 27:261–265.

Friis, E. M., P. R. Crane, and K. R. Pedersen. 2011. Early Flowers and Angiosperm Evolution. Cambridge University Press, Cambridge, UK, 585pp.

Furin, S., N. Preto, M. Rigo, G. Roghi, P. Gianolla, J. L. Crowley, and S. A. Bowring. 2006. High-precision U-Pb zircon age from the Triassic of Italy: implications for the Triassic time scale and the Carnian origin of calcareous nannoplankton and dinosaurs. Geology 34:1009–1012.

Gierlowski-Kordesch, E. 2010. Lacustrine carbonates. Pp. 1–101 in A. M. Alonso-Zarza and L. H. Tanner, (eds), Carbonates in Continental Settings, Elsevier, Volume 61.

Gierlowski-Kordesch, E., B. R. and Rust. 1994. The Jurassic East Berlin Formation, Hartford Basin, Newark Supergroup (Connecticut and Massachusetts): a saline lake–playa–alluvial plain system. Pp. 249–265 in R. W. Renaut and W. M. Last, (eds), Sedimentology and Geochemistry of Modern and Ancient Saline Lakes. SEPM, Tulsa, OK, Special Publication 50.

Grimaldi, D., and M. S. Engel. 2005. Evolution of the Insects, Cambridge University Press, New York, 755pp.

Grimaldi, D., A. Shmakov, and N. C. Fraser. 2004. Mesozoic thrips and early evolution of the order Thysanoptera (Insecta). Journal of Paleontology 78:941–952.

Grimaldi, D., Z. Junfeng, N. C. Fraser, and A. A. Rasnitsyn. 2005. Revision of the bizarre Mesozoic scorpionflies in the Pseudopolycentropodidae (Mecopteroidea). Insect Systematics and Evolution 36:443–458.

Heckert, A. B. *et al.* 2015. A new aetosaur (Archosauria, Suchia) from the Upper Triassic Pekin Formation, Deep River Basin, North Carolina, U.S.A., and its implications for early aetosaur evolution. Journal of Vertebrate Paleontology e881831. DOI: 10.1080/02724634.2014.881831

Hitchcock, E. 1858. Ichnology of New England. A Report on the Sandstone of the Connecticut Valley, Especially Its Fossiliferous Footmarks, Made to the Government of the Commonwealth of Massachusetts. William White, Boston.

Hitchcock, E. 1865. Supplement to the Ichnology of New England. With an Appendix by C. H. Hitchcock. Wright and Potter, Boston.

Horton, J. W., and K. J. McConnell. 1991. The western piedmont. Pp. 36–58 in J. W. Horton and V. A. Zullo (eds), The Geology of the Carolinas. The University of Tennessee Press, Knoxville.

Huber, P., S. G. Lucas, and A. P. Hunt. 1993. Vertebrate biochronology of the Newark Supergroup Triassic, eastern North America. New Mexico Museum of Natural History and Science Bulletin 3:179–185.

Huber, P., N. G. McDonald, and P. E. Olsen. 2003. Early Jurassic insects from the Newark Supergroup, northeastern United States. Pp. 206–222 in P. M. LeTourneau and P. E. Olsen, (eds), The Great Rift Valleys of Pangea in eastern North America. Volume Two: Sedimentology, Stratigraphy, and Paleontology. Columbia University Press, New York.

Kent, D. V., and P. E. Olsen. 1997. Paleomagnetism of Upper Triassic continental sedimentary rocks from the Dan River–Danville rift basin (eastern North America). Geological Society of America Bulletin 109:366–377.

Kent, D. V., and P. E. Olsen. 1999. Astronomically tuned geomagnetic polarity time scale for the Late Triassic. Journal of Geophysical Research 104:12,831–12,841.

Kilham, P., and R. E. Hecky. 1973. Fluoride: geochemical and ecological significance in east African waters and sediments. Limnology and Oceanography 18:932–945.

Kozur, H., and R. E. Weems. 2007. Upper Triassic conchostracan biostratigraphy of the continental rift basins of eastern North America: its importance for correlating Newark Supergroup events with the Germanic basin and the international geologic time scale. New Mexico Museum of Natural History and Science Bulletin 41:137–188.

Kozur, H.W., and R. E. Weems. 2010. The biostratigraphic importance of conchostracans in the continental Triassic of the northern hemisphere. Geological Society, London,

Special Publications 334:315–417.

Krzemiński, W., and E. Krzemińska. 2003. Triassic Diptera: descriptions, revisions and phylogenetic relations. Acta zoologica cracoviensia 46:153–184.

Larsen, C. P. S., R. Pienitz, J. P. Smol, K. A. Moser, B. F. Cumming, J. M. Blais, G. M. Macdonald, and R. I. Hall. 1998. Relations between lake morphometry and the presence of laminated lake sediments: a re-examination of Larsen and Macdonald (1993). Quaternary Science Reviews 17:711–717.

Last, W. M. 1990. Lacustrine dolomite: an overview of modern, Holocene, and Pleistocene occurrences. Earth Science Reviews 27:221–263.

Last, W. M., R. E. Vance, S. Wilson, and J. P. Smol. 1998. A multi-proxy limnologic record of rapid early Holocene hydrologic change on the northern Great Plains, southwestern Saskatchewan, Canada. The Holocene 8:503–520.

Litwin, R.J., and S. R. Ash. 1993. Revision of the biostratigraphy of the Chatham Group (Upper Triassic), Deep River basin, North Carolina, USA. Review of Palaeobotany and Palynology 77:75–95.

Liutkus, C. M., J. S. Beard, N. C. Fraser, and P. C. Ragland. 2010. Use of fine-scale stratigraphy and chemostratigraphy to evaluate conditions of deposition and preservation of a Triassic Lagerstätte, south-central Virginia. Journal of Paleolimnology 44:645–666.

Liutkus-Pierce, C.M., N. C. Fraser, and A. B. Heckert. 2014 Stratigraphy, sedimentology, and paleontology of the Upper Triassic Solite Quarry, North Carolina and Virginia. Pp. 255–269 in C. M. Bailey, and L. V. Coiner, (eds), Elevating Geoscience in the Southeastern United States: New Ideas about Old Terranes—Field Guides for the GSA Southeastern Section Meeting, Blacksburg, Virginia, 2014: Geological Society of America Field Guide No. 35.

Lucas, S. G., L. H. Tanner, H. W. Kozur, R. E. Weems, and A. B. Heckert. 2012. The Late Triassic timescale: age and correlation of the Carnian–Norian boundary. Earth-Science Reviews 114:1–18.

Marchal-Papier, F., A. Nel, and L. Grauvogel-Stamm. 2000. Nouveaux Orthoptéres (Ensifera, Insecta) du Trias des Vosges (France). Acta Geologica Hispanica 35:5–18.

Marsh, O. C. 1896. A new belodont reptile (*Stegomus*) from the Connecticut River Sandstone. American Journal of Science, series 4, 2:59–62.

McCobb, L. M. E., I. J. Duncan, E. A. Jarzembowski, B. A. Stankiewicz, M. A. Wills, and D. E. G. Briggs. 1998. Taphonomy of the insects from the Insect Bed (Bembridge Marls), late Eocene, Isle of Wight, England. Geological Magazine 135:553–563.

Meyertons, C. T. 1963. Triassic formations of the Danville basin. Virginia Division of Mineral Resources, Report of Investigations 6:1–65.

Muttoni, G., D. V. Kent, P. E. Olsen, P. D. Stefano, W. Lowrie, S. M. Bernasconi, and F. M. Hernández. 2004. Tethyan magnetostratigraphy from Pizzo Mondello (Sicily) and correlation to the Late Triassic Newark astrochronological polarity time scale. Geological Society of America Bulletin 116:1043–1058.

Olsen, P. E. 1978. On the use of the term Newark for Triassic and Early Jurassic rocks of eastern North America. Newsletters on Stratigraphy 7:90–95.

Olsen, P. E. 1979. A new aquatic eosuchian from the Newark Supergroup (Late Triassic–Early Jurassic) of North Carolina and Virginia. Postilla 176:1–14.

Olsen, P. E. 1980. A comparison of the vertebrate assemblages from the Newark and Hartford basins (Early Mesozoic, Newark Supergroup) of eastern North America. Pp. 35–53 in L. L. Jacobs (ed.), Aspects of Vertebrate History: Essays in Honour of Edwin Harris Colbert. Museum of Northern Arizona Press, Flagstaff.

Olsen, P. E. 1986. A 40-million year lake record of early Mesozoic orbital climatic forcing. Science 234:842–848.

Olsen, P. E. 1997. Stratigraphic record of the early Mesozoic breakup of Pangea in the Laurasia–Gondwana rift system. Annual Review of Earth and Planetary Sciences 25:337–401.

Olsen, P. E., and A. K. Johansson. 1994. Field guide to Late Triassic tetrapod sites in Virginia and North Carolina. Pp. 408–430 in N. C. Fraser and H.-D. Sues, (eds), In the Shadow of the Dinosaurs. Cambridge University Press, Cambridge.

Olsen, P. E., and D. V. Kent. 1999. Long-period Milankovitch cycles from the Late Triassic and Early Jurassic of eastern North America and their implications for the calibration of the Early Mesozoic time-scale and the long-term behaviour of the planets. Philosophical Transactions of the Royal Society of London, A 357:176–1786.

Olsen, P. E., A. R. McCune, and K. S. Thomson. 1982. Correlation of the early Mesozoic Newark Supergroup by vertebrates, principally fishes. American Journal of Science 282:1–44.

Olsen, P. E., C. L. Remington, B. Cornet, and K. S. Thomson. 1978. Cyclic change in Late Triassic lacustrine communities. Science 201:729–733.

Olsen, P. E., R. W. Schlische, and P. J. W. Gore, (eds). 1989. Tectonic, depositional, and paleoecological history of early Mesozoic rift basins, eastern North America. American Geophysical Union, Washington, D.C., 184pp.

Olsen, P. E., A. J. Froehlich, D. L. Daniels, J. P. Smoot, and P. J. W. Gore. 1991. Rift basins of early Mesozoic age. Pp. 142–170 in J. W. Horton, Jr., and V. A. Zullo, (eds), Geology of the Carolinas, Carolina Geological Society 50th Anniversary Volume. University of Tennessee Press, Knoxville.

Papier, F. A., A. Nel, and L. Grauvogel-Stamm. 1996. Deux nouveaux insectes Mecopteroidea du Buntsandstein Superieur (Trias) des Vosges (France)]. Paleontologia Lombardia 5:37–45.

Partey, F., S. Lev, R. Casey, E. Widom, V. W. Lueth, and J. Rakovan. 2009. Source of fluorine and petrogenesis of the Rio Grande Rift-type barite–fluorite–galena deposits. Economic Geology 104:505–520.

Peñalver, E., C. C. Labandeira, E. Barrón, X. Delclòs, P. Nel, A. Nel, P. Tafforeau, and C. Soriano. 2012. Thrips pollination of Mesozoic gymnosperms. Proceedings of the National Academy of Science 109:8623–8628.

Peyer, K., J. G. Carter, H.-D. Sues, S. E. Novak, and P. E. Olsen. 2008. A new suchian archosaur from the Upper Triassic of North Carolina. Journal of Vertebrate Paleontology 28:363–381.

Quinby-Hunt, M. S., and P. Wilde. 1996. Chemical depositional environments of calcic marine black shales. Economic Geology 91:4–13.

Redfield, W. C. 1856. On the relation of fossil fishes of the sandstone of the Connecticut and other Atlantic States to the Liassic and Oolitic periods. American Journal of Science, ser.2, 22:357–363.

Ren, D., C. C. Labandeira, J. A. Santiago-Blay, A. Rasnitsyn, C. K. Shih, A. Bashkuev, M. A. V. Logan, C. L. Hotton, and D. Dilcher. 2009. A probable pollination mode before angiosperms: Eurasian, long-proboscid scorpionflies. Science 326:840–847.

Robinson, P. L. 1962. Gliding lizards from the upper Keuper of Great Britain. Proceedings of the Geological Society, London 1601:137–146.

Robinson, P. L. 1967. Triassic vertebrates from lowland and upland. Science and Culture 33:169–173.

Ross, A. J. and A. Self. 2014. The fauna and flora of the Insect Limestone (late Eocene), Isle of Wight, UK: introduction, history and geology. Earth and Environmental Transactions of the Royal Society of Edinburgh 104:233–244.

Schlische, R.W. 2003. Progress in understanding the structural geology, basin evolution, and tectonic history of the eastern North American rift system. Pp. 21–64 in P. M. LeTourneau and P. E. Olsen, (eds), The Great Rift Valleys of Pangea in Eastern North America. Volume One: Tectonics, Structure, and Volcanism. Columbia University Press, New York.

Selden, P. A., and J.-C. Gall. 1992. A Triassic mygalomorph spider from the northern Vosges, France. Palaeontology 35:211–235.

Selden, P. A., H. M. Anderson, J. M. Anderson, and N. C. Fraser. 1999. Fossil araneomorph spiders from the Triassic of South Africa and Virginia. Journal of Arachnology 27:401–414.

Smith, D. M. 2012. Exceptional preservation of insects in lacustrine environments. Palaios 27:346–353.

Stocker, M. R., and R. J. Butler. 2013. Phytosauria. Geological Society, London, Special Publications 379:91–117.

Sues, H.-D., and N. C. Fraser. 2010. Triassic Life on Land – The Great Transition. Columbia University, New York, 236pp.

Sues, H.-D., P. E. Olsen, and J. G. Carter.1999. A Late Triassic traversodont cynodont from the Newark Supergroup of North Carolina. Journal of Vertebrate Paleontology 19:351–354.

Sues, H.-D., P. E. Olsen, J. G. Carter, and D. M. Scott. 2003. A new crocodylomorph reptile from the Upper Triassic of North Carolina. Journal of Vertebrate Paleontology 23:329–343.

Van Alstine, R. E. 1976. Continental rifts and lineaments associated with major fluorspar districts. Economic Geology 71:977–987.

Van de Kamp, P. C., and B. E. Leake. 1996. Petrology, geochemistry, and Na-metasomatism of Triassic–Jurassic non-marine clastic sediments in the Newark, Hartford, and Deerfield rift basins, northeastern USA. Chemical Geology 133:89–124.

Wright, V. P. 1999. The role of sulphate-reducing bacteria and cyanobacteria in dolomite formation in distal ephemeral lakes of the Coorong region, South Australia. Sedimentary Geology 126:147–157.

Weems, R. E., and P. E. Olsen. 1997. Synthesis and revision of groups within the Newark Supergroup, eastern North America. Geological Society of America Bulletin 109:195–209.

Whiteside, J. H., D. S. Grogan, P. E. Olsen, and D. V. Kent. 2011. Climatically driven biogeographic provinces of Late Triassic tropical Pangea. Proceedings of the National Academy of Sciences USA. 108:8972–8977.

Chapter 5

The Yanliao Biota: a trove of exceptionally preserved Middle-Late Jurassic terrestrial life forms

Xing Xu, Zhonghe Zhou, Corwin Sullivan, and Yuan Wang

Key Laboratory of Vertebrate Evolution and Human Origins, Institute of Vertebrate Paleontology and Paleoanthropology, Chinese Academy of Sciences, Beijing, China.

Abstract

The Yanliao Biota comprises fossil plants, invertebrates and vertebrates from Middle to Upper Jurassic sediments exposed in the same part of northeast China that is home to the better-known Early Cretaceous Jehol Biota. As in the Jehol Biota, many vertebrate specimens from the Yanliao Biota are exceptionally well preserved in fine-grained sediments, often as articulated skeletons accompanied by soft tissue. The Yanliao flora is rich in gymnosperms, but also includes plants that may be among the oldest angiosperms. The invertebrate fauna is dominated by insects, of which hundreds of species have been described. Among the notable vertebrates are the oldest known crown-group salamanders, pterosaurs that fall just outside the derived clade Pterodactyloidea, feathered non-avian theropod dinosaurs that shed light on the transition to birds, and an ecologically diverse array of mammaliaforms including a glider, a probable semi-aquatic form, and the oldest known member of the placental lineage.

Introduction

Over the last decade, numerous exceptionally preserved fossils have been recovered from Middle and Upper Jurassic terrestrial sediments exposed at several sites in southeastern Nei Mongol Autonomous Region (Inner Mongolia), western Liaoning Province, and northern Hebei Province, China (Fig 1). These fossils include plants that may be among the oldest known angiosperms (Wang, 2010a), in addition to the earliest known members of several insect clades (Huang et al., 2012, Ren et al., 2010, Gao et al., 2012), the earliest known crown-salamanders (Gao and Shubin, 2003, Gao and Shubin, 2012), transitional pterosaurs (Lü, Unwin et al., 2010, Wang, Kellner et al., 2010), the earliest known feathered dinosaurs (Hu et al., 2009, Xu and Zhang, 2005, Zhang, Zhou et al., 2008, Godefroit, Cau et al., 2013, Godefroit, Demuynck et al., 2013), the earliest known gliding and aquatic mammaliaforms (Meng et al., 2006, Ji et al., 2006), and the earliest known eutherian mammal (Luo et al., 2011), among others (Zhou et al., 2010). Many of the specimens are remarkably complete, even including preserved soft tissues such as setae, eyes, skin, feathers, and/or fur (Zhou et al., 2010). These fossils make up the Yanliao Biota, a diverse terrestrial assemblage that flourished during a short period close to the Middle–Late Jurassic boundary in northern China, and it has significant implications for our understanding of Middle-Late Jurassic life forms and ecosystems.

The Yanliao Biota takes its name from the Yanliao Area, which roughly coincides with the geographic range in which the biota occurs. The Yanliao Area is a large region north of the Yan Mountains, south of Daxinganling Mountain, east of the Hunshandak Sandland, and west of the Liao River (Fig 1). The Yanliao Area contains extensive exposures of Jurassic and Cretaceous terrestrial sediments that have produced spectacular fossil remains over the last twenty years. As well as the Middle-Late Jurassic Yanliao Biota, which is the focus of the present paper, the fossil assemblages preserved in this region include the famous Early Cretaceous Jehol Biota (Zhou et al., this volume). The Yanliao Biota includes the 'Daohugou Biota' of many previous authors (Ji and Juan, 2002, Zhang, 2002, Sullivan et al., 2014).

Figure 1 Map of 10 major Yanliao fossil localities (modified from Sullivan et al., 2014). Shaded area enlarged and shown in detail. **1**, Daohugou Locality; **2**, Daxishan Locality; **3**, Yaolugou Locality; **4**, Wubaiding Locality; **5**, Guancaishan Locality; **6**, Mutoudeng Locality, **7**, Nanshimen Locality; **8**, Fanshen Locality; **9**, Yujiagou Locality; **10**, Yipandaogou Locality; **11**, Liaotugou Locality.

The Yanliao Biota has received considerably less attention than the geographically coincident but considerably younger Jehol Biota (Zhou et al., 2010). Fossil insects were the first element of the former biota to be extensively studied, and in 1983 the term 'Yanliao Insect Fauna' was introduced to refer to what was then viewed as a rich Middle Jurassic insect assemblage from the Yanliao Area (Hong, 1983). Subsequently, Ren et al. (1995) proposed the term 'Yanliao Fauna' to cover a diverse terrestrial fossil assemblage, including not only insects but also conchostracans, bivalves, fish, reptiles, and mammaliaforms, from the Mentougou, Jiulongshan, Tiaojishan, and Tuchengzi formations exposed in the Yanliao Area.

These formations were considered at the time to have been deposited from the late Early Jurassic through to the middle Late Jurassic (Ren et al., 1995), but are now thought to range in age from the Middle Jurassic through to the Early Cretaceous (Xu et al., 2003, Ji et al., 2004). Because these formations span a long geological period and display striking differences in faunal content, sedimentology and taphonomy, the proposition that they contain a single coherent 'fauna' is, at best, debatable. Nevertheless, Ren et al. (1995) proposed that the stratigraphic extent of the 'Yanliao Fauna' could be defined in terms of a *Yanliaocorixa-Liaosteus-'Yabeinosaurus'* assemblage zone (Ren et al., 1995), though all three elements occur together only in the laterally equivalent Middle Jurassic Jiulongshan and Haifanggou formations (Zhou et al., 1991, Hoffstetter, 1964, Jin, 1999, Ren et al., 2010). It should be noted that '*Yabeinosaurus*' here refers to '*Yabeinosaurus*' *youngi*, which is known from the Jiulongshan/Haifanggou Formation of

Lingyuan, Liaoning. A recent study indicated that this species differs markedly in limb proportions from the type species *Yabeinosaurus tenuis* from the Lower Cretaceous Jehol Group, and its attribution to the genus is still equivocal (Evans and Wang, 2012). In practical terms, the Yanliao Fauna proposed by Ren *et al.* (1995) therefore refers to a unique fossil assemblage of essentially Middle Jurassic age in the Yanliao Area.

Major known Yanliao Fauna localities include the Fanshen Locality (Zhou *et al.*, 1991), the Yujiagou Locality (Lu, 1995, Jin, 1999), the Yipandaogou Locality, and the Liaotugou Locality (Wang *et al.*, 1989). The Yujiagou Locality is near Haifanggou Village in Sanbao Town, Beipiao City, western Liaoning Province, and exposes the Haifanggou Formation. Numerous specimens of insects and the acipenseriform fish *Liaosteus hongi* have been recovered from this locality (Hong, 1983, Jin, 1999). The Fangshen Locality is located near Fangshen Village, Songzhangzi Town, Lingyuan City, western Liaoning Province, and exposes the Middle Jurassic Jiulongshan Formation (Zhou *et al.*, 1991). The triconodont mammaliaform *Liaotherium gracile* and the lizard '*Yabeinosaurus*' *youngi* were probably both collected from this locality (Zhou *et al.*, 1991), although '*Yabeinosaurus*' *youngi* was originally reported to be from a site called Gezidong (Yang, 1958). The Yipandaogou Locality is located near Shajingou Village in Nuanchitang Town, Nanpiao District, Huludao City, western Liaoning, and exposes the Middle Jurassic Haifanggou Formation (Wang *et al.*, 1989). Known fossils from this locality include bivalves, fish, and plants (Wang *et al.*, 1989). The Liaotugou Locality is located in Batuyingzi Town, Beipiao County, western Liaoning. It exposes the Haifanggou Formation, and has yielded bivalves, insects, fish and plants (Wang *et al.*, 1989).

The Yanliao Fauna received little attention prior to the recent discovery of a controversial fossil site, the Daohugou Locality in Ningcheng County in the extreme southeast of Nei Mongol (Fig 2A,B). The strata exposed at the Daohugou Locality mainly comprise fine-grained lacustrine beds intercalated with

Figure 2 Two major Yanliao fossil localities. **A**, Daohugou Locality; **B**, Close-up of the Daohugou fossil-bearing beds; **C**, Daxishan Locality; **D**, Close-up of the Daxishan fossil-bearing beds.

volcanic ash (Liu, Kuang et al., 2012, Liu, Zhao et al., 2006, Sullivan et al., 2014) (Fig 3). Early finds from this locality included plants (Zhang, 2002), insects (Zhang, 2002, Ren and Yin, 2002), conchostracans (Zhang, 2002), salamanders (Wang, 2000, Ji and Yuan, 2002), and pterosaurs (Wang et al., 2002, Ji and Yuan, 2002).

Because the Daohugou strata are similar in terms of lithology and taphonomy to the Jehol Group, they were initially referred to the Yixian Formation, the lowest subdivision of the Jehol Group (Wang et al., 2000, Wang, 2000). However, later studies noted clear differences in fossil content between the Daohugou assemblage and the Jehol Biota, leading to suggestions

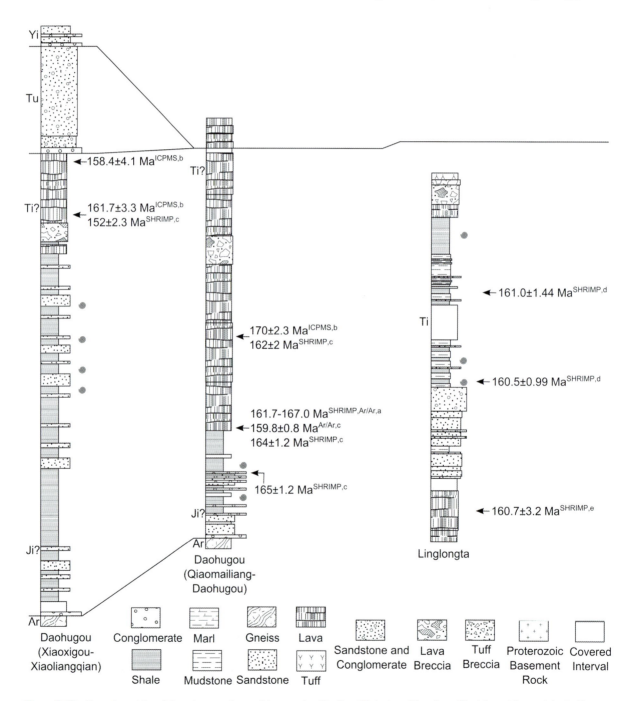

Figure 3 Stratigraphy of fossil-bearing sections at two major Yanliao Biota localities (modified from Figure 1 in Sullivan et al., 2014).

that the Daohugou strata were referable to either the Middle Jurassic Jiulongshan/Haifanggou Formation (Ren and Yin, 2002, Ji et al., 2006, Gao and Shubin, 2003) or an Upper Jurassic 'Daohugou Formation' (Zhang, 2002). The term 'Daohugou Formation' was informally introduced by Zhang (2002), who considered the formation to be comparable to the Tuchengzi Formation. Some authors proposed that the Daohugou strata were Middle-Late Jurassic (Xu and Zhang, 2005), Late Jurassic (Ji and Yuan, 2002) or Late Jurassic-Early Cretaceous (Wang et al., 2005) in age without assigning them to a formal stratigraphic unit.

Following the early discoveries, more significant fossils have been recovered from the Daohugou Locality, including the possible angiosperm *Solaranthus daohugouensis* (Zheng and Wang, 2010), the crown-salamander *Chunerpeton tianyiensis* (Gao and Shubin, 2003), the feathered dinosaurs *Pedopenna daohugouensis*, *Epidendrosaurus ningchengensis* and *Epidexipteryx hui* (Xu and Zhang, 2005, Zhang, Zhou et al., 2008, Zhang et al., 2002), the gliding mammaliaform *Volaticotherium antiquum*, the aquatic mammaliaform *Castorocauda lutrasimilis* (Meng et al., 2006, Ji et al., 2006), and the herbivorous haramiyidan *Megaconus mammaliaformis* (Zhou et al., 2013). It is noteworthy that most of the individual vertebrate specimens known from the locality are fully articulated skeletons with soft tissue traces (Fig 4; Zhou et al., 2010). Ji and Yuan (2002) introduced the term 'Daohugou Biota' for the fossil assemblage from the Daohugou Locality, but did not provide an explicit definition. Subsequently, Zhang (2002) clearly defined the 'Daohugou Biota' as a distinct pre-Jehol Biota represented by the fossil assemblage at the Daohugou Locality. It is now clear that the Daohugou Biota overlaps extensively in geographic and stratigraphic terms with the Yanliao Fauna of Ren et al. (1995) and the associated flora, and in this paper both are considered to be part of a single Yanliao Biota (see below for a more detailed analysis).

Several other localities in northern Hebei Province and western Liaoning Province have yielded fossils that are similar to those recovered from the Daohugou Locality, and were regarded as part of the Daohugou Biota by Sullivan et al. (2014). The Daxishan Locality is near Daxishan Village in Linglongta Town, Jianchang County, western Liaoning Province (Fig 2B,C). The fossil-bearing beds of the Daxishan Locality have generally been referred to the Tiaojishan/Lanqi Formation (Jiang et al., 2010, Hu et al., 2009, Lü and Fucha, 2010, Lü, Fucha et al., 2010, Lü, Unwin et al., 2010, Liu, Kuang et al., 2012), but see Wang et al., 2009 and Wang, Kellner et al., 2010 for a different opinion. The Daxishan section consists mainly of andesites, rhythmic tuffs, and tuffaceous sediments along with lavas (Fig 3). The muddy shale and mudstone or tuffaceous layers have produced numerous fossils, including the conchostracan *Euestheria* sp., the ostracods *Darwinula sarytirmenensis*, *D. impudica*, *D. magna*, *D. submagna*, *D. changxinensis*, *D. xiaofenzhangziensi* and *Timiriasevia catenularia*, the bivalve *Shaanxiconcha cliovata*, the plants *Neocalamites carcinoides*, *N. nathorsti*, *Coniopteris hymenophylloides*, *C. margaretae*, *Zamites* cf. *gigas*, *Czekanowskia rifida*, *C. ketova*, *Phoenicopsis speciosa* and *Sphenobaiera kazachstanica*, and ptycholepid fish (Duan et al., 2009). Described taxa from this locality also include the feathered paravian theropods *Anchiornis huxleyi* and *Xiaotingia zhengi* (Xu, Zhao et al., 2009, Hu et al., 2009, Xu et al., 2011), various wukongopterid pterosaurs such as *Wukongopterus lii* and *Darwinopterus modularis* (Lü, Unwin et al., 2010, Lü, Fucha et al., 2010, Wang, Kellner et al., 2010, Lü et al., 2011, Wang et al., 2009, Lü and Fucha, 2010, Zhou and Schoch, 2011), the salamander *Chunerpeton tianyiensis* (Hu et al., 2009, Jiang et al., 2010, Sullivan et al., 2014), and the earliest known eutherian *Juramaia sinensis* (Luo et al., 2011). In the original report, the holotype specimen of *Juramaia sinensis* was stated to be from the 'Daxigou Locality'. We have established (Zhexi Luo, pers. comm.) that this is the same site as the Daxishan Locality, a name commonly used in the published literature.

The second Daohugou Biota locality in Jianchang County is the Yaolugou Locality, near Yaolugou Town. It exposes the Tiaojishan Formation/Lanqi Formation (Liu, Kuang et al., 2012). Fossil remains of the feathered ornithischian *Tianyulong confuciusi*, and the holotype specimens of the paravians *Anchiornis huxleyi*, *Eosinopteryx brevipenna*, and *Aurornis xui*, have been described as originating from this locality (Liu, Kuang et al., 2012, Xu, Zhao et al., 2009, Godefroit, Cau et al., 2013, Godefroit, Demuynck et al., 2013,). However, the provenance of these specimens requires confirmation by additional evidence, given that they were all acquired from fossil dealers.

The Wubaiding Locality is located near Wubaiding Village in Reshuitang Town, Lingyuan City, western Liaoning Province (Fig 2C). The Wubaiding Locality has produced specimens of two salamander taxa,

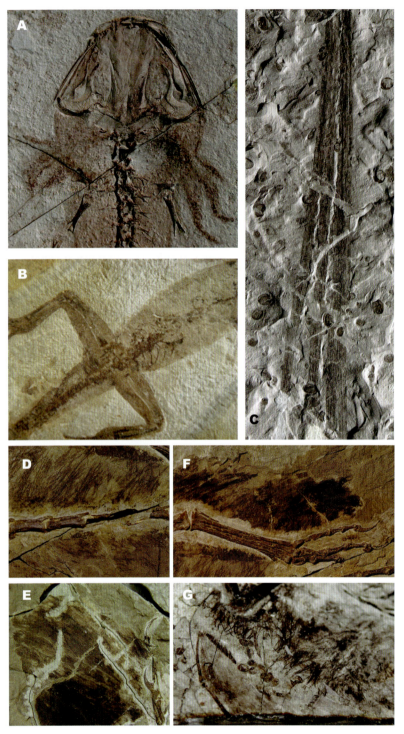

Figure 4 Close-up photographs of soft tissues in selected Yanliao vertebrates. **A**, external gills of the salamander *Chunerpeton tianyiensis*; **B**, skin impression of a juvenile lizard; **C**, tail feathers of the maniraptoran *Epidexipteryx hui*; feathers along the tail (**D**), manus (**E**), and pes (**F**) in the paravian *Anchiornis huxleyi*; **G**, filamentous integumentary structures in the anurognathid pterosaur *Jeholopterus ningchengensis*. Not to scale.

Pangerpeton sinensis and *Chunerpeton tianyiensis*, in addition to insects and conchostracans comparable to those from the Daohugou Locality (Wang et al., 2005, Zhang and Wang, 2004, Wang and Evans, 2006). *C. tianyiensis* also occurs at other Daohugou Biota localities (Wang, Dong et al., 2010b, Sullivan et al., 2014).

The Guancaishan Locality is near Muyingzi Village in Shahai Town, Jianping County, western Liaoning Province, and the fossil-bearing beds at this site have been referred to the Tiaojishan/Lanqi Formation (Liu, Liu et al., 2006, Liu, Kuang et al., 2012). In addition to fossils of the conchostracan *Euestheria* (Liu, Kuang et al., 2012) and the salamander *Chunerpeton tianyiensis* (Wang, Dong *et al.*, 2010a), this locality has recently produced specimens of the crown-salamander *Beiyanerpeton jianpingensis* (Gao and Shubin, 2012).

The Mutoudeng Locality is located in Mutoudeng Town, Qinglong County, northern Hebei Province (Fig 2D). The Mutoudeng strata have been referred to the Tiaojishan Formation and have recently produced the 'rhamphorhynchoid' (i.e. non-pterodactyloid) pterosaurs *Changchengopterus pani* (Lü, 2009) and *Qinglongopterus guoi* (Lü *et al.*, 2012) and the haramiyidan *Arboroharamiya jenkinsi* (Zheng *et al.*, 2013). This locality has also yielded the salamander *Chunerpeton tianyiensis*, which is also known from the other Daohugou Biota localities (Sullivan *et al.*, 2014). The Nanshimen Locality is also located in Qinglong County, near Nanshimen Village in Gangou Town. This locality has produced fossils of *Chunerpeton tianyiensis* and several other Daohugou taxa (Sullivan *et al.*, 2014).

The various Daohugou Biota localities, including the Daohugou Locality itself, are broadly similar in terms of lithology, taphonomy, and fossil content (Sullivan *et al.*, 2014). Several taxa, such as the conchostracan *Euestheria* and the salamander *Chunerpeton tianyiensis*, are known from all these sites. The available evidence thus suggests that the fossil assemblages from these localities can be considered to belong to a single regionally widespread biota.

A Middle-Late Jurassic age for the Daohugou Biota is biostratigraphically supported by evidence from fossil plants (Chen, 2003, Zhou and Zhang, 1992, Zhou *et al.*, 2007), invertebrates (Shen *et al.*, 2003, Chen, 2003, Chen *et al.*, 2003, Jiang, 2006, Zhang, 2010, Wang and Ren, 2007, Ren *et al.*, 2010, Rasnitsyn and Zhang, 2004, Zhang, 2002, Wang *et al.*, 2006) and vertebrates (Ji and Yuan, 2002, Lü and Fucha, 2010, Lü, Fucha *et al.*, 2010, Lü, 2009, Wang, Kellner *et al.*, 2010, Zhou *et al.*, 2010), and also suggested by radiometric dating (Peng *et al.*, 2012, Liu, Liu *et al.*, 2006, Chen *et al.*, 2004, Yang and Li, 2004, He *et al.*, 2005, Liu, Kuang *et al.*, 2012, Wang *et al.*, 2013). He *et al.* (2005) reported an age of 159 ± 0.8 Ma based on samples overlying the fossil-bearing beds collected at the Daohugou Locality, but suggested that the Daohugou strata were inverted and that the date represented a maximum rather than a minimum age for the fossil-bearing beds. We agree with Liu, Liu *et al.* (2006) and others that the date in question represents a minimum age for the beds, because there is no convincing evidence for any inversion of the stratigraphy in the area. Estimates of the youngest possible age for the Daohugou fossil assemblages range from 157 ± 3 Ma to 166 ± 1.1 Ma, and estimates of the oldest possible age range from 160.7 ± 3.2 Ma to 168 ± 4.4 Ma (see below for a more detailed consideration of both the biostratigraphic and the radiometric evidence). Radiometric data thus suggest that the Daohugou Biota essentially coincided in space and time with the previously described Yanliao Fauna. Furthermore, some typical elements of the Yanliao Fauna (e.g., *Euestheria*, *Yanliaocorixa* and '*Yabeinosaurus*' *youngi*) are also known from Daohugou Locality and the other Daohugou Biota sites. Consequently, the Daohugou Biota should be united with the Yanliao Fauna as the Yanliao Biota, a term used by a few previous authors including Sun *et al.* (2009) and Chang *et al.* (2009).

Sun *et al.* (2009) defined the Yanliao Biota as the entire Jurassic flora and fauna of western Liaoning Province and neighbouring areas (Sun *et al.*, 2009, Sun *et al.*, 2011), a very broad definition that few later studies have followed. Chang *et al.* (2009) defined the Yanliao Biota much more narrowly as the fossil content of the Haifanggou Formation of Liaoning Province and the laterally equivalent Jiulongshan Formation of Hebei Province.

Several other authors, such as Zhou *et al.* (2010), have considered the Yanliao Biota to include all fossils from not only the Middle Jurassic Jiulongshan Formation and its equivalents (such as the Haifanggou Formation) but also the Upper Jurassic Lanqi Formation and its equivalents (such as the Tiaojishan Formation). The present study follows Zhou *et al.*'s (2010) definition based on the lithological resemblances between the Jiulongshan and Tiaojishan formations and the strong similarities in both composition and preservational mode between their fossil assemblages.

The Yanliao Biota represents a terrestrial biota that existed from the late Middle Jurassic through the early Late Jurassic in the Yanliao Area. The best-known elements of this biota include the plants *Yimaia* and *Schmeissneria*, the conchostracan *Euestheria*, the bivalve *Ferganoconcha sibirica*, the insect *Yanliaocorixa*, the fish *Liaosteus*, the salamander *Chunerpeton*, the pterosaur *Darwinopterus*, and the dinosaur *Anchiornis*, among others. Although several reviews of various aspects of the Yanliao Biota have been published recently (Guo *et al.*, 2012, Zhou *et al.*, 2010, Liu *et al.*, 2010, Sullivan *et al.*, 2014, Zheng, pers. comm.), no comprehensive overview of the entire biota has been presented. This chapter summarizes the composition, palaeoecology and geological context of the Yanliao Biota, and reviews some recent advances in research on this rich and distinctive fossil assemblage.

Geological context of the Yanliao Biota

During the middle-late Mesozoic, the Yanliao Area was located at a palaeolatitude of approximately 45° N in the northeastern part of the North China Craton (Smith et al., 1994). The craton is composed of Archaean to Palaeoproterozoic basement rocks, overlain by unmetamorphosed terrestrial Mesoproterozoic to Cenozoic sedimentary and volcanic rocks (Chang et al., 2009). The Middle Jurassic–Early Cretaceous terrestrial deposits in the Yanliao Area are as follows, in ascending order: the Haifanggou Formation (called the Jiulongshan Formation in Hebei Province and Nei Mongol), the Lanqi Formation (called the Tiaojishan Formation in Hebei Province and Nei Mongol), the Tuchengzi Formation (called the Houcheng Formation in Hebei Province), the Zhangjiakou Formation, the Dabeigou Formation, the Yixian Formation (the Dadianzi and Huajiying formations in Hebei Province may correlate with at least the lower part of the Yixian Formation), and the Jiufoutang Formation. Among these stratigraphic units, the Haifanggou/Jiulongshan and Lanqi/Tiaojishan formations preserve the Yanliao Biota, and the Dabeigou, Yixian and Jiufoutang formations preserve the Jehol Biota (see Zhou et al., this volume, for an explicit definition of the Jehol Biota).

The Middle Jurassic–Early Cretaceous terrestrial deposits in the Yanliao Area were laid down in the course of three major volcanic and sedimentary cycles: one that produced the Haifanggou/Jiulongshan and Lanqi/Tiaojishan formations, a second that produced the Tuchengzi/Houcheng and Zhangjiakou formations, and a third that produced the Dabeigou, Yixian and Jiufotang formations. The cycles differed substantially from one another in the nature of both the erupting magma and the non-volcanic sediments that were deposited, and represent three distinct geological episodes (Xu et al., 2003, Ren et al., 1995). The Yanliao fossil assemblages formed during the first volcanic and sedimentary cycle, which, it has been suggested, may have taken place in the context of a new tectonic regime centred on the North China Craton (Dong, 2010). Other plates converged on the North China Craton from the north, east and southwest, resulting in intracontinental subduction, orogeny, and considerable thickening of the crustal lithosphere of the eastern part of the Asian continent (Dong, 2010).

The Haifanggou/Jiulongshan Formation has a thickness of more than 200 metres, and comprises sandstones, conglomerates, shales, and interbedded tuffs. The Haifanggou/Jiulongshan Formation has been widely accepted as being Middle Jurassic in age, based on palaeontological data (Shen et al., 2003, Ren et al., 1995, Xu et al., 2003), though the possibility that it extends into the Late Jurassic in some areas cannot be excluded (Liu, Zhao et al., 2006, Zhang, Wang et al., 2008). Numerous radiometric dates have been reported for the Haifanggou/Jiulongshan Formation based on samples from various stratigraphic sections, ranging from 146.3 ± 8.5 to 188 ± 43.7 Ma (Chen et al., 1997, Wu et al., 2004, Xu et al., 2003). The Jiulongshan Formation in Chengde Basin, northern Hebei Province may even be partially Late Jurassic in age (Liu, Zhao et al., 2006, Zhang, Wang et al., 2008). However, the accuracy of these results is questionable (Chang et al., 2009). A more recent analysis resulted in a $^{40}Ar/^{39}Ar$ age of $166.7 + 1.0$ Ma, based on samples collected from around the middle of the Haifanggou Formation of Beipiao, western Liaoning Province (Chang et al., 2013). The possible early angiosperm *Schmeissneria sinensis* is from the same beds (Chang et al., 2013).

The Lanqi/Tiaojishan Formation ranges in thickness from 200 to 800 metres, and is mainly composed of acid volcanic rocks (lavas and pyroclastics) and tuffaceous sediments. Biostratigraphic age estimates for the Lanqi/Tiaojishan Formation range from Early Jurassic to Late Jurassic (Cheng and Li, 2007, Xu et al., 2003), and reported radiometric ages range from 188 ± 19 Ma to 148.9 ± 3.0 Ma (Chang et al., 2009, Zhang et al., 2005, Yang and Li, 2008, Zhang, Wang et al., 2008, Liu, Zhao et al., 2006). Two recent dates for the Lanqi Formation of western Liaoning are a $^{40}Ar/^{39}Ar$ age of 159.5 ± 0.6 Ma based on samples from the top of the formation (Chang et al., 2013) and an age of 161.8 ± 0.4 Ma based on samples from the basal portion of the formation (Chang et al., 2009, Chang et al., 2013). These results would imply that the Lanqi formation was deposited during a short episode close to the Middle-Late Jurassic boundary.

The deposits that contain the Yanliao Biota are assignable to either the Haifanggou/Jiulongshan Formation or the Lanqi/Tiaojishan Formation, and the Yanliao fossils are consistent with a Middle-Late Jurassic age for the strata in which they occur. The plant *Neocalamites* is known from the Triassic through Middle Jurassic, and *Coniopteris margaretae* is a Middle Jurassic species known from China and Yorkshire, UK (Chen, 2003, Harris, 1961). *Yimaia* is a ginkgoalean previously reported from the early Middle Jurassic of northern China (Zhou and Zhang, 1992). Almost identical fossils have also been recovered from Early Jurassic sediments in Germany (Kirchner, 1992) and Middle Jurassic sediments in England (Black, 1929). This combination of plant fossils suggests a Middle Jurassic age for the Yanliao Biota (Zhou et al., 2007).

Invertebrate fossils provide further support for a Middle Jurassic age. The conchostracan *Euestheria* is widely distributed in the early-middle Middle Jurassic sediments of China (Shen et al., 2003), and the *Darwinula sarytirmenensis-magna* assemblage is the most widespread Middle Jurassic ostracode assemblage (Chen, 2003). The bivalve *Shaanxiconcha* is a common element in the Late Triassic of northern China, although it is also seen in Middle Triassic and Early Jurassic sediments (Chen et al., 2003). The bivalve *Ferganoconcha* is widely distributed in the Lower and Middle Jurassic of northern China, central Asia, Siberia and the Ural region, and its presence at the Daohugou Locality supports referral of the strata to the Haifanggou Formation/Jiulongshan Formation (Jiang, 2006). The insect fossil assemblage from the Daohugou Locality generally supports a Middle Jurassic age (Zhang, 2010, Wang and Ren, 2007, Ren et al., 2010). However, some studies have posited strong similarities to the entomofauna of the Karabastau Formation of Kazakhstan, which is considered to be Callovian-Oxfordian in age (Rasnitsyn and Zhang, 2004, Zhang, 2002), and other studies have supported an Early-Middle Jurassic age (Wang et al., 2006).

Yanliao vertebrate fossils are also consistent with a Middle-Late Jurassic age. Ptycholepidae is a group of palaeonisciform fish known mainly from the Triassic. The youngest previously known ptycholepids are from lower Middle Jurassic sediments in western China. For example, *Yuchoulepis* is from the Middle Jurassic Lower Shaximiao Formation of Sichuan (Su, 1974) and the Yaojie Formation of Gansu (Zhang et al., 2003). Discoveries of ptycholepids in the Yanliao Biota favour a Middle rather than a Late Jurassic age. Pterosaur fossils recovered from the Yanliao Biota sites are clearly Jurassic in character, but they are not helpful in determining whether a Middle Jurassic or Late Jurassic age is more probable (Ji and Yuan, 2002, Lü and Fucha, 2010). To date all of the known pterosaurs are non-pterodactyloids, including long-tailed rhamphorhynchids and scaphognathids, short-tailed anurognathids, and the unusual wukongopterids (Lü, Fucha et al., 2010, Lü, 2009, Wang, Kellner et al., 2010, Zhou et al., 2010). Long-tailed pterosaurs are known only from pre-Cretaceous strata, whereas anurognathids are known from the Middle and Late Jurassic, with a single exception from the Early Cretaceous (Ji and Ji, 1998, Ji et al., 1999, Unwin, 2006).

Several radiometric dates have been reported based on samples collected from strata overlying the fossil-bearing layers at the Daohugou Locality (Fig 3). They include two ^{206}Pb/^{238}U SHRIMP ages of 162 ± 2 Ma and 152 ± 2.3 Ma (Liu, Liu et al., 2006), two ^{206}Pb/^{238}U SHRIMP ages of 166 ± 1.5 Ma and 165 ± 2.4 Ma, a ^{40}Ar/^{39}Ar age of 164 ± 2.5 Ma (Chen et al., 2004), two ^{206}Pb/^{238}U SHRIMP ages of 164 ± 1.2 Ma and 165 ± 1.2 Ma (Yang and Li, 2004), and a ^{40}Ar/^{39}Ar age of 159.8 ± 0.8 Ma (He et al., 2004, Liu, Liu et al., 2006). Estimates of the youngest possible age for the fossiliferous strata at the Daohugou Locality thus range from 166 ± 1.1 Ma to 159.8 ± 0.8 Ma. Because the samples analysed in the radiometric studies were taken from varying stratigraphic levels and geographic locations (the Daohugou Locality is actually a sizeable region containing multiple areas of stratigraphic exposure), the differing dates probably reflect the occurrence of multiple volcanic events after the formation of the Daohugou fossil assemblage.

Several radiometric dates have also been obtained based on samples collected from both below and above the *Anchiornis*-bearing beds at the Daxishan Locality (Liu, Kuang et al., 2012, Wang et al., 2013, Peng et al., 2012). A sample from a layer of clay-rich tuff 30 cm above the *Anchiornis*-bearing beds produced a U/Pb SHRIMP age of 160.5 ± 0.99 Ma, and a second sample from a tuff layer 100 m above the first sample produced a U/Pb SHRIMP age of 161.0 ± 1.44 Ma (Liu, Kuang et al., 2012). An andesite lava about 220 m below the *Anchiornis*-bearing beds produced a U/Pb SHRIMP age of 160.7 ± 3.2 Ma (Peng et al., 2012). Wang and colleagues also dated samples from the *Anchiornis*-bearing beds at the Daxishan Locality, obtaining ages of 160.7 ± 1.7 Ma, 158.9 ± 1.7 Ma, and 159.5 ± 2.3 Ma (Wang et al., 2013). It should be noted that the dates based on tuffaceous samples from the fossil-bearing beds provide the most direct evidence for the age of the fossil assemblage. These dates suggest that the *Anchiornis*-bearing beds are very close in age to the Middle/Late Jurassic boundary.

Radiometric dating results are also available from several other Yanliao Biota localities. A date of 158.5 ± 1.6 Ma has been reported based on a sample collected from volcanic rocks overlying the *Tianyulong*-bearing beds at the Yaolugou Locality (Liu, Kuang et al., 2012), representing the youngest possible age of the fossil-bearing beds (Liu, Kuang et al., 2012). Samples collected from lavas below the salamander-fossil-bearing beds and above the Jiulongshan Formation at the Wubaiding Locality (Liu, Liu et al., 2006) yielded a $^{206}Pb/^{238}U$ SHRIMP age of 164 ± 4 Ma, representing the oldest possible age of the fossil-bearing beds at this site (Liu, Liu et al., 2006). The same sample also produced a $^{206}Pb/^{238}U$ TIMS age of 168 ± 4.4 Ma (Liu Yongqing, personal communication). A $^{206}Pb/^{238}U$ age of 157 ± 3 Ma provides a minimum age constraint for the salamander-bearing beds at the Guancaishan Locality (Liu, Liu et al., 2006).

In summary, the youngest age constraints for the Yanliao fossil assemblages range from 157 ± 3 Ma (Guancaishan Locality) to 166 ± 1.1 Ma (Daohugou Locality) and the oldest age constraints range from 160.7 ± 3.2 Ma (Daxishan Locality) to 168 ± 4.4 Ma (Wubaiding Locality). All these dates fall within the chronological range in which the Haifanggou/Jiulongshan Formation and Lanqi/Tiaojishan Formation were deposited. The available radiometric data suggest that the Yanliao fossil assemblages formed during a short period near the Middle-Late Jurassic boundary, like some other Chinese Jurassic assemblages including the Shishugou Fauna from the Junggar Basin of northwestern China (Eberth et al., 2010, He et al., 2013).

The Yanliao Biota is currently known from several localities across a relatively large geographical area (Fig 1), but in general these localities are very similar in terms of sedimentology and taphonomy. The Yanliao fossils are preserved in laminated tuffaceous mudstones or tuffaceous shales, and they consistently display certain preservational characteristics.

First, most individual animal fossils are completely preserved. In particular, most vertebrate fossils retain nearly complete skeletons. Second, nearly all specimens are preserved in full articulation. Some skeletons are only semi-articulated, but this mode of preservation is rare. Third, many fine-scale structures are preserved in Yanliao fossils. For example, insect fossils include preserved veins and mouthparts (Ren et al., 2010, Wang and Zhang, 2010, Zhang and Kluge, 2007), and the preservation of tiny branchiopod eggs has been reported (Shen and Huang, 2008). Fourth, most specimens preserve soft tissue, including setae and hairs in insects (Gao et al., 2012, Zhang, Ren et al., 2008, Ren et al., 2010), skin impressions in many salamanders (Wang, Dong et al., 2010a, Wang and Evans, 2006) and lizards (Wang, Dong et al., 2010a, Evans and Wang, 2007), filamentous structures and membranes in pterosaurs (Ji and Yuan, 2002, Wang et al., 2002), feathers in dinosaurs (Hu et al., 2009, Zhang, Zhou et al., 2008), and fur in mammaliaforms (Meng et al., 2006). Some fossils retain preserved stomach contents, such as conchostracans or corixid insects preserved within the guts of salamanders (Dong, Huang et al., 2012). Some rarely preserved organisms are also known from the Yanliao Biota; for example, fossil lichens have been documented at the Daohugou Locality (Wang, Krings et al., 2010). Finally, the recovered fossils show a high level of taxonomic diversity. These levels of taxonomic richness and preservational quality indicate that the Yanliao Biota can be regarded as a Konservat-Lagerstätte.

It is worth pointing out that the preservational patterns reported above are inferred from our qualitative observations of Yanliao specimens housed in several major repositories (e.g., the Shandong Tianyu Museum of Nature, the Institute of Vertebrate Palaeontology and Palaeoanthropology of the Chinese Academy of Sciences, Capital Normal University, the Beijing Natural History Museum, and Shenyang Normal University), and we cannot exclude the possibility of biased sampling (e.g., private collectors tend to collect small fossils). Quantitative surveys of museum collections, and taphonomic investigations in the field, will be required to confirm these results.

Recurrent volcanism is probably responsible for the exceptional preservation of the Yanliao fossils (Chang et al., 2009), as has also been suggested for the Early Cretaceous Jehol Biota (Fürsich et al., 2007). The mode of fossil preservation and its relationship to local volcanic activity has been well studied in the Jehol Group (Jiang et al., 2011, Fürsich et al., 2007, Guo et al., 2003, Guo et al., 2008, Wang et al., 1999), but little comparable research has been done on Yanliao fossils. Given the extreme similarity between the two biotas in terms of taphonomy and the sedimentology of the strata in which they occur, some results derived from the biogeochemical, sedimentological and taphonomic studies in the Jehol Group may also be helpful in understanding the preservation of the Yanliao Biota.

The Yanliao Biota seems to be similar to the Jehol Biota in having occupied an area where volcanic eruptions frequently took place in the vicinity of volcanic lakes (Chang et al., 2009), potentially explaining the apparent sudden death, rapid burial, and excellent preservation in large numbers of many individual plants and animals in both biotas (Jiang et al., 2011, Fürsich et al., 2007, Guo et al., 2003, Guo et al., 2008, Wang et al., 1999, Zhou et al., 2010). However, a recent study suggested that volcanic activity might not have been a direct driver of mortality and burial in the case of the Jehol Biota, and instead might have exerted an indirect influence by causing serious environmental and climatic changes and repeatedly inducing collapse of the aquatic ecosystem (Pan, Sha et al., 2011).

Nevertheless, several recent studies have elucidated mechanisms of exceptional fossil preservation that depend on volcanism. For example, microbial processes may have been the key to preserving extremely high-quality fossils of insects and plants in the Eocene Florissant Formation at Florissant, Colorado, USA (O'Brien et al., 2002). More specifically, the well-preserved fossils occur within diatomite layers and are covered with a Mucous film that may have been produced by diatoms, implying that the mucuous film retarded decomposition of the bodies of the plants and insects and was instrumental in their preservation (O'Brien et al., 2002). Volcanic lakes are optimal places for abundant diatom growth and associated bacterial activity, which might explain the frequent occurrence of well-preserved soft tissues in Jehol fossils (Jiang et al., 2011). If so, the same factor is likely to have played a role in the similarly spectacular soft-tissue preservation seen in Yanliao Biota specimens, though more investigation will be needed to confirm this.

However, there are some salient differences between the modes of volcanism that appear to have been characteristic of Yanliao and Jehol times (Guo et al., 2008). Geochemical analysis indicates that de-gassing associated with andesitic and trachyandesitic volcanism during the deposition of the Tiaojishan Formation featured high levels of H_2O and CO_2, which might have led to high air temperatures. By contrast, Jehol volcanism probably resulted in high atmospheric concentrations of S, F, and Cl. It has been suggested that the Jehol type of volcanism led to acid rain and low air temperatures, occurring at least occasionally during the time when the Jehol Biota existed, as well as releasing highly poisonous gases that directly resulted in the deaths of birds (Guo et al., 2008). There is some independent isotopic evidence for lowered air temperatures (Amiot et al., 2011).

It has been demonstrated that clay minerals played a key role in soft-tissue preservation in some fossil Lagerstätten (Briggs, 2003, Martínez-Delclòs et al., 2004). However, they do not appear to have been important in the case of the Daohugou Locality. The only highly concentrated elements in insect fossils from this site are C and Ca, and the high concentration of the latter element has been linked to calcium carbonate minerals (Wang et al., 2008). Wang et al. (2008) also noticed pyritization, a key mode of soft tissue preservation, in Daohugou insect fossils (Briggs, 2003, Zhang and Shu, 2001, Leng and Yang, 2003). Few sedimentological and taphonomical data are available for other Yanliao localities.

Biodiversity and Palaeoecology of the Yanliao Biota

Although not comparable to that of the Jehol Biota, the documented diversity of the Yanliao Biota is high by the standards of Mid-Late Jurassic biotas worldwide, despite the short history of research on the Yanliao Biota. Among the reported taxa from the Yanliao Biota are 36 vertebrate species including fish, salamanders, lizards, pterosaurs, dinosaurs, and mammaliaforms; 454 insect species, and at least 25 other invertebrate species; and 216 plant species including bryophytes, ferns, cycads, ginkgopsians, conifers, and possible angiosperms (Jiang et al., 2008, Wang, 2010a, Zheng, pers. comm.).

Floral diversity

The Yanliao flora comprises a diverse array of taxa from several major plant groups (Fig 5), and even includes plants that may be among the earliest known angiosperms (Pan, 1983, Wang, 2010a, Wang and Wang, 2010). In an internal report, Zheng (pers. comm.) recently presented a comprehensive review of the plant fossils from the Jiulongshan/Haifanggou and Tiaojishan/Lanqi formations in the Yanliao Area, all of which are referable to the Yanliao Biota by definition.

Figure 5 Selected Yanliao plant fossils. The possible angiosperms *Schmeisseria* (**A**) and *Xingxueanthus* (**B**); **C**, the bennettitalean *Anomozamites* sp.; **D**, the filicalean *Coniopteris* sp.; the cycads *Nilssonia* sp. (**E**) and *Cycadolepis* sp. (**F**); **G**, the conifer *Yanliaoa* sp.; the ginkgoaleans *Ginkgo* sp. (**H**), *Czekanowskia setacea* (**I**), and *Yimaia capituliformis* (**J**). Not to scale.

Known plant fossils from the Yanliao Biota (Fig 5) include wood and spores, as well as leaves (Zheng, pers. comm.). Reported plants include lichens, horsetails, lycopsids, filicins, bennettitaleans, cycads, ginkgopsians, czekanowskialeans, conifers, osmundaceans, dipterids, dicksonians, and possible angiosperms (Pan, 1983, Wang, 2010a, Wang and Wang, 2010, Zheng, pers. comm.).

There are 124 known fossil plant species from the Haifanggou Formation of Liaoning Province, including 2 lichens, 1 articulatan, 6 lycopsidans, 29 filicines, 19 bennettitaleans, 13 cycads, 12 ginkgopsidans, 11 czekanowskialeans, and 11 conifers (Zheng, pers. comm.). Exposures of the Haifanggou Formation in neighbouring areas such as Ningcheng, Nei Mongol have also produced some well-preserved plant fossils, such as reproductive organs of the bennettitaleans *Anomozamites haifanggouensis* (Zheng *et al.*, 2003) and *Weltrichia daohugouensis* (Li *et al.*, 2004) and the ginkgoalean *Yimaia capituliformis* (Zhou *et al.*, 2007). The potentially most important plant species of the Yanliao Biota are *Solaranthus daohugouensis* (Zheng and Wang, 2010), *Xingxueanthus sinensis* (Wang and Wang, 2010), and *Schmeissneria sinensis* (Wang, 2010a), which might be among the earliest known angiosperms (Wang, 2010a, Wang, 2010b) (Fig 5). The oldest undoubted angiosperms, such as *Archaefructus* (Sun *et al.*, 1998), date from the Early Cretaceous, and the possible angiosperms from the Yanliao Biota are among a small number of fossils worldwide that may document the existence of this clade prior to the Jurassic/Cretaceous boundary.

The Tiaojishan/Lanqi Formation of western Liaoning Province has produced at least 92 plant species, including lichens, articulatans, filicines, osmundaceans, dipteridaceans, dicksoniaceans, bennettitaleans, cycads, ginkgopsians, and conifers (Zheng, pers. comm, Duan *et al.*, 2009, Jiang *et al.*, 2008, Zhang and Zheng, 1987, Zhang and Zheng, 1991, Cheng and Li, 2007, Cheng *et al.*, 2007). This assemblage includes such important fossils as stems of the cycad *Lioxylon liaoningense* (Zhang *et al.*, 2006) and reproductive organs of *Araucaria beipiaoensis* (Zheng *et al.*, 2008).

Faunal diversity

The most taxonomically diverse group in the Yanliao Biota is Insecta (Fig 6). Some 350 insect species had been reported from the Yanliao Biota when a recent review by Liu *et al.* (2010) was completed, and an additional 104 species have been reported since. These fossils represent 24 major groups of insects, with coleopterans, dipterans, homopterans, hemipterans and hymenopterans being especially high in specific diversity. The species numbers of these groups are as follows: 13 species of Blattaria (Liu *et al.*, 2010, Wei, Shih *et al.*, 2012), 6 species of Blattida (Guo and Ren, 2011, Vrsansky *et al.*, 2012, Liang, Vrsansky *et al.*, 2012, Liang, Huang *et al.*, 2012, Wei, Liang *et al.*, 2012), 1 species of Chresmodidae, 55 species of Coleoptera (Liu *et al.*, 2010, Dong and Huang, 2011, Wang, Zhang *et al.*, 2012, Jarzembowski *et al.*, 2012, Wang and Zhang, 2011, Nikolajev *et al.*, 2011, Tan, Wang *et al.*, 2012, Tan, Ren *et al.*, 2012, Yu *et al.*, 2012, Bai *et al.*, 2012, Pan, Chang *et al.*, 2011), 3 species of Dermaptera (Liu *et al.*, 2010, Zhao *et al.*, 2011), 80 species of Diptera (Liu *et al.*, 2010, Liu, Shih *et al.*, 2012, Huang, Nel *et al.*, 2013, Wang, Zhao *et al.*, 2012), 8 species of Ephemeroptera (Liu *et al.*, 2010), 5 species of Grylloblattodea (Liu *et al.*, 2010, Cui *et al.*, 2012, Ren and Aristov, 2011), 36 species of Hemiptera (Liu *et al.*, 2010, Dong, Yao *et al.*, 2012a, Dong, Yao *et al.*, 2012b, Wang, Shih, Szwedo *et al.*, 2012, Li, Wang *et al.*, 2013, Li, Szwedo *et al.*, 2011, Wang, Szwedo *et al.*, 2012, Yang, Yao *et al.*, 2012, Yang, Yao *et al.*, 2013, Wang *et al.*, 2011, Lu *et al.*, 2011), 4 species of Heteroptera (Hou *et al.*, 2012), 43 species of Homoptera (Liu *et al.*, 2010), 39 species of Hymenoptera (Liu *et al.*, 2010, Wang, Shih *et al.*, 2012, Ding *et al.*, 2013, Zhang, Zheng, Zhang *et al.*, 2013, Shih, Feng *et al.*, 2011), 28 species of Mecoptera (Liu *et al.*, 2010, Zhang *et al.*, 2011, Shih, Yang *et al.*, 2011, Qiao *et al.*, 2012, Petrulevičius and Ren, 2012, Yang, Shih *et al.*, 2012, Wang, Labandeira *et al.*, 2012), 2 species of Megaloptera (Liu, Wang *et al.*, 2012), 39 species of Neuroptera (Engel *et al.*, 2011, Jepson *et al.*, 2013, Makarkin *et al.*, 2011, Shi *et al.*, 2012, Yang, Makarkin *et al.*, 2013, Liu *et al.*, 2010, Yang *et al.*, 2011, Shi *et al.*, 2011, Liu, Shi *et al.*, 2011), 18 species of Odonata (Liu *et al.*, 2010, Li, Nel, Shih *et al.*, 2013, Li, Nel *et al.*, 2013, Li, Nel *et al.*, 2011, Li *et al.*, 2012, Petrulevičius *et al.*, 2011, Zhang, Zheng *et al.*, 2013), 32 species of Orthoptera (Liu *et al.*, 2010, Gu *et al.*, 2012, Gu *et al.*, 2013, Gu *et al.*, 2011), 1 species of Phasmatoptera (Shang *et al.*, 2011), 14 species of Plecoptera (Liu *et al.*, 2010, Liu, Sinitshenkova *et al.*, 2011), 5 species of Psocoptera (Liu *et al.*, 2010), 3 species of Raphidioptera (Liu *et al.*, 2010), 4 species of Siphonaptera (Huang, Engel *et al.*, 2013, Gao

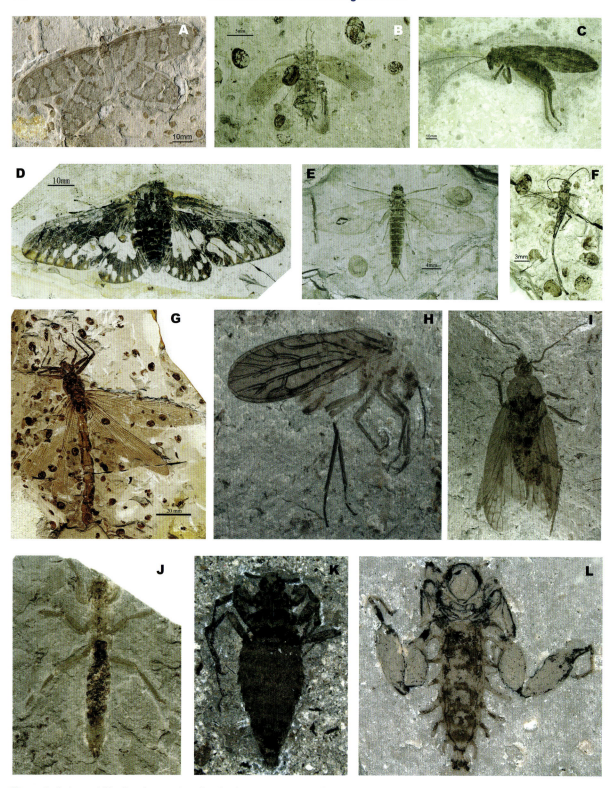

Figure 6 Selected Yanliao insect fossils. **A**, the neuropteran *Grammolingia boi*; **B**, the blattarian *Fuzi dadao*; **C**, the orthopteran *Allaboilus gigantus*; **D**, the palaeontinid *Daohugoucossus shii*; **E**, the ephemeropteran *Jurassonurus amoenus*; **F**, the hymenopteran *Archaeopelecinus tebbei*; **G**, the odonatan *Amnifleckia guttata*; **H**, the corrodentian *Archipsylla sinica*; **I**, the grylloblattid *Sinosepididontus chifengensis*; **J**, the mantophasmatod *Juramantophasma sinica*; the siphonapterans *Pseudopulex wangi* (**K**) and *Strashila daohugouensis* (**L**). Not to scale.

et al., 2012, Gao, Shih *et al.*, 2013), and 5 species of Trichoptera (Liu *et al.*, 2010, Gao, Yao *et al.*, 2013, Wu and Huang, 2012).

Other invertebrate groups (Fig 7) include gastropods of uncertain systematic position (Chang and Sun, 1997, Jiang, 2006), the bivalves *Ferganoconcha sibirica* (Chang and Sun, 1997, Jiang, 2006), *Ferganoconcha curta*, *Ferganoconcha elongata*, *Ferganoconcha subcentralis*, *Ferganoconcha tomiensis*, *Ferganoconcha haifanggouensis*, *Sibireconcha sp.* (Wang *et al.*, 1989), and *Shaanxiconcha cliovata* (Duan *et al.*, 2009), the conchostracans *Euestheria ziliujingensis*, *Euestheria haifanggouensis*, *Euestheria luanpinensis*, and *Euestheria jingyuanensis* (Shen *et al.*, 2003), the ostracods *Darwinula sarytirmenensis*, *Darwinula impudica*, *Darwinula magna*, *Darwinula submagna*, *Darwinula changxinensis*, *Darwinula xiaofenzhangziensi*, and *Timiriasevia catenularia* (Duan *et al.*, 2009), and the arachnids *Patarchaea muralis*, *Sinaranea metaxyostraca* (Selden *et al.*, 2008), *Mesobunus martensi* (Fig 7A) (Huang *et al.*, 2009), *Eoplectreurys gertschi* (Selden and Huang, 2010), and *Nephila jurassica* (Selden *et al.*, 2011).

The taxonomic diversity of the Yanliao vertebrate fauna is not particularly high, though most of the known vertebrate fossils are exceptionally well preserved (Figs 8–12). Reported fish include the acipenseriform *Liaosteus hongi* (Lu, 1995, Jin, 1999) and some undescribed ptycholepids (Duan *et al.*, 2009). Caudates (Fig 8) include *Jeholotriton paradoxus* (Wang, 2000, Wang and Rose, 2005) and *Pangerpeton siensis* (Wang and Evans, 2006), both of uncertain systematic position, as well as the cryptobranchid *Chunerpton tianyiensis* (Gao and Shubin, 2003), the hynobiid *Liaoxitriton daohugouensis* (Wang, 2004), and the salamandroid *Beiyanerpeton jianpingensis* (Gao and Shubin, 2012). The final three taxa represent the earliest known crown-group caudates. One enigmatic specimen has been claimed to represent an anuran tadpole (Yuan *et al.*, 2004), but it has been recently interpreted as an incomplete insect (Huang, 2013). Perhaps the most problematic lizard of the Yanliao Biota is '*Yabeinosaurus*' *youngi* (Ji, 2004, Estes, 1983), a species whose holotype was originally referred to *Yabeinosaurus tenuis* (Yang, 1958) but whose placement in the genus '*Yabeinosaurus*' is still equivocal (Evans and Wang, 2012). Also present are another lizard of uncertain systematic position (Fig 9C), and one possible scleroglossan (Evans and Wang, 2007, Evans and Wang, 2009) (Fig 9D).

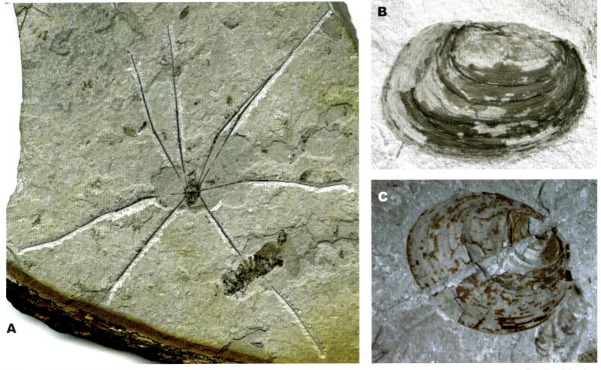

Figure 7 Selected Yanliao non-insect invertebrate fossils. **A**, the harvestman *Mesobunus martensi*; **B**, the bivalve *Ferganoconcha sibirica*; **C**, the conchostracan *Euestheria luanpinensis*. Not to scale.

Figure 8 Selected Yanliao caudate fossils. The basal caudates *Pangerpeton siensis* (**A**) and *Jeholotriton paradoxu* (**B**); **C**, the cryptobranchid *Chunerpton tianyiensis*; **D**, the hynobiid *Liaoxitriton daohugouensis*; **E**, the salamandroid *Beiyanerpeton jianpingensis*. Not to scale.

Figure 9 Selected Yanliao fish and lizard fossils. **A**, a ptycholepid fish; **B**, the lizard *'Yabeinosaurus' youngi*; **C**, A juvenile lizard with uncertain systematic position; **D**, a possible scleroglossan lizard. Not to scale.

The Yanliao Biota contains a diverse assemblage of pterosaurs, including forms that are closely related to pterodactyloids and provide important information regarding the pattern of character evolution that resulted in the highly distinctive and specialized pterodactyloid body plan (Fig 10). However, no undoubted pterodactyloids are known from the Yanliao Biota, in marked contrast to the situation in the somewhat younger Solnhofen assemblage from Germany. Pterosaur species reported in the Yanliao Biota include the anurognathids *Jeholopterus ningchengensis* (Wang et al., 2002) and *Dendrorhynchus mutoudengensis* (Lü and Hone, 2012), the rhamphorhynchine *Qinglongopterus guoi* (Lü et al., 2012), the scaphognathines *Pterorhynchus wellnhoferi* (Czerkas and Ji, 2002), *Fenghuangopterus lii* (Lü, Fucha et al., 2010),

Figure 10 Selected Yanliao pterosaur fossils. **A**, the anurognathid *Jeholopterus ningchengensis*; **B**, the rhamphorhynchine *Qinglongopterus guoi*, the scaphognathines *Jianchangopterus zhaoianus* (**C**) and *Jianchangnathus robustus* (**D**); the wukongopterids *Wukongopterus lii* (**E**), *Darwinopterus robustodens* (**F**), and *Darwinopterus linglongtaensi* (**G**). Not to scale.

Jianchangopterus zhaoianus (Lü and Bo, 2011) and *Jianchangnathus robustus* (Cheng et al., 2012), the wukongopterids *Wukongopterus lii* (Wang et al., 2009), *Darwinopterus modularis* (Lü, Unwin et al., 2010), *Darwinopterus linglongtaensis* (Wang, Kellner et al., 2010), *Darwinopterus robustodens* (Lü et al., 2011), *Kunpengopterus sinensis* (Wang, Kellner et al., 2010) and *Changchengopterus pani* (Lü, 2009, Wang, Kellner et al., 2010), and the possible istiodactylid *Archaeoistiodactylus linglongtaensis* (Lü and Fucha, 2010, but see Martill and Etches, 2012). The poor preservation of the holotype of *Archaeoistiodactylus linlongtaensis* renders its proposed istiodactylid affinities questionable, and a recent examination indicates that the holotype might actually be a poorly preserved specimen of *Darwinopterus*.

The dinosaur fossils of the Yanliao Biota all represent maniraptoran theropods (Fig 11), apart from a few specimens that represent a single heterodontosaurid ornithischian species. Furthermore, all of the Yanliao maniraptorans are either paravians or closely related species. They include the bizarre theropods *Epidendrosaurus ningchengensis* (Zhang et al., 2002) and *Epidexipteryx hui* (Zhang, Zhou et al., 2008) (=*Scansoriopteryx heilmanni*) (Czerkas and Yuan, 2002) and the paravians *Pedopenna daohugouensis* (Xu and Zhang, 2005), *Anchiornis huxleyi* (Xu, Zhao et al., 2009), *Xiaotingia zhengi* (Xu et al., 2011), *Aurornis xui* (Godefroit, Cau et al., 2013), and *Eosinopteryx brevipenna* (Godefroit, Demuynck et al., 2013). The sole heterodontosaurid species is *Tianyulong confuciusi* (Zheng et al., 2009). When originally reported, the

Figure 11 Selected Yanliao theropod dinosaur fossils. **A**, the maniraptoran *Epidexipteryx hui*; the paravians *Anchiornis huxleyi* (**B**), *Xiaotingia zhengi* (**C**), *Aurornis xui* (**D**), and *Eosinopteryx brevipenna* (**E**). Not to scale.

holotype was claimed to be from the Lower Cretaceous Yixian Formation, but later studies have considered the taxon to belong to the Yanliao Biota. Both proposals await the publication of solid supporting evidence.

The Yanliao mammaliaforms are relatively diverse (Fig 12). Reported taxa include the triconodonts *Liaotherium gracile* (Zhou et al., 1991) and *Manchurodon simplicidens* (Yabe and Shikama, 1938), the yinotherian *Pseudotribos robustus* (Luo et al., 2007), the basal mammaliaform *Volaticotherium antiquus* (Meng et al., 2006), the docodont *Castorocauda lutrasimiliss* (Ji et al., 2006), the haramiyidans *Arboroharamiya jenkinsi* (Zheng et al., 2013) and *Megaconus mammaliaformis* (Zhou et al., 2013) and the eutherian *Juramaia sinensis* (Luo et al., 2011).

Palaeoecology of the Yanliao Biota

Palaeoecological investigation of the Yanliao Biota is hampered by the almost unquestionable presence of a taphonomic filter that has facilitated the preservation of some kinds of taxa and excluded the preservation of others. Taking the fossil record at face value, for example, would imply a total absence of vertebrates larger than a few kilograms, a situation that seems unlikely in any but the most unusual and/or isolated Mesozoic environments. Also unusual is the complete

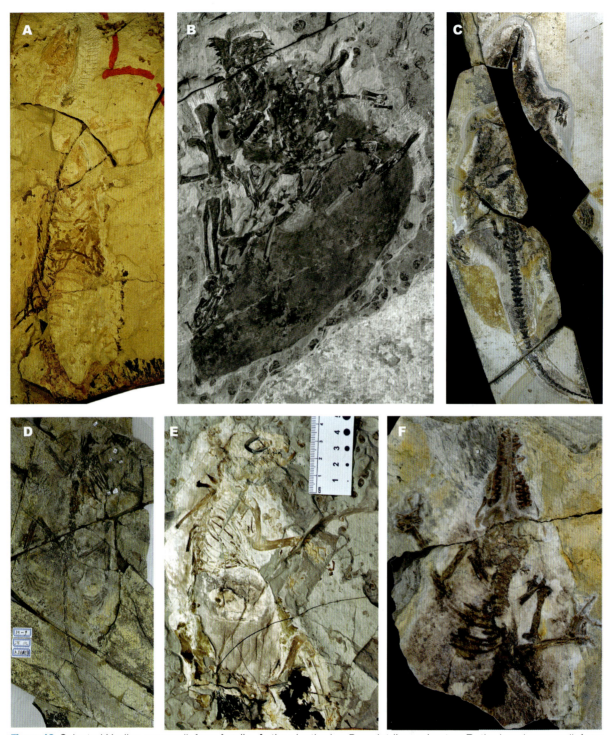

Figure 12 Selected Yanliao mammaliaform fossils. **A**, the yinotherian *Pseudotribos robustus*; **B**, the basal mammaliaform *Volaticotherium antiquum*; **C**, the docodont *Castrocauda lustrasinilis*; the haramiyidans *Arboroharamiya jenkinsi* (**D**) and *Megaconus mammaliaformis* (**E**); **F**, the eutherian *Juramaia sinensis*. Not to scale.

absence of definitively aquatic or semi-aquatic reptiles, although the possibility that the known Yanliao lizards spent some time in the water cannot be ruled out. The probable semi-aquatic mammaliaform *Castorocauda* and even the Yanliao fish are relatively rare, leaving *Chunerpeton tianyiensis* and, to some extent, other salamanders as the numerically dominant aquatic vertebrates preserved in the record. The Yanliao Biota resembles the Jehol Biota in being rich in salamanders (and aquatic insects). Among the dinosaurs, theropods are much more abundant than ornithischians (no sauropodomorphs have been reported), in contrast to most other dinosaur faunas.

Nevertheless, several studies have attempted to reconstruct the palaeoenvironmental and climatic conditions in which the Yanliao Biota existed, and these studies have found evidence for temporal and spatial variation in the Yanliao palaeoenvironment and climate. A taphonomic study of the insect fossils from the Daohugou Locality discerned multiple preservational modes, suggesting the presence of several different microenvironments in the Daohugou lakes (Wang *et al.*, 2008). Tan and Ren (2002) conjectured that the Daohugou area might have been a near-shore shallow lacustrine basin with a humid and warm climate, diverse and abundant vegetation, and highly aquiferous soil that would have been beneficial to many plants (Tan and Ren, 2002). Subsequent discoveries at the Daohugou Locality of a few fossil snakeflies (Raphidioptera) that would have inhabited mountains (Ren, 2003, Engel and Ren, 2008), and of a salamander (Wang, 2004) that would have preferred mountain streams, suggests that there might have been high-altitude streams near the lake in which the strata exposed at the Daohugou locality were deposited, or even that their deposition took place within a montane lacustrine basin. The composition of the Daohugou Locality insect assemblage implies a warm and humid climate (Ren *et al.*, 2010, Selden *et al.*, 2011), consistent with palaeobotanical data from the Haifanggou Formation of western Liaoning Province that suggest warm but seasonal and/or humid conditions (Sun *et al.*, 2008). By contrast, most palaeobotanical data suggest a hot, dry climate in the Yanliao Area during the depositional period of the Lanqi Formation (Zheng, pers. comm.), though some silicified coniferous wood fossils suggest a subtropical, humid and seasonal climate in western Liaoning Province (Jiang *et al.*, 2008).

Dietary information is available for some Yanliao animals, and in some cases it is even possible to deduce trophic relationships. It has been demonstrated that the Yanliao insects had diverse diets. The insects, such as scorpionflies, may have fed on pollination drops of early angiosperms and gymnosperms in the Yanliao Biota (Ren, 1998, Ren *et al.*, 2009). The pseudopolycentropodid *Sinopolycentropus rasnitsyni* has a tubular proboscis, which, it is suggested, may have been used in feeding on ovulate fluids of Mesozoic gymnosperms (Shih, Yang *et al.*, 2011). The prophalangopsid *Bacharboilus lii* has stout mandibles with well-defined molar dentes, indicating an herbivorous feeding habit (Gu *et al.*, 2011). The Yanliao geinitziids have been interpreted as aboveground diurnal predators, as they combine a flat body with powerful legs characterized by long tarsomeres and small arolia (Cui *et al.*, 2012). The raphidiomimid cockroach *Graciliblatta bella* is inferred to have been a carnivore based on its sharp, prognathous mandibles (Liang, Huang *et al.*, 2012). The damsel-dragonfly *Zygokaratawia reni* is said to be similar to modern calopterygid damselflies in its wing proportions, suggesting that it may have resembled calopterygids in being 'a predator of small insects along riverbanks overhung by trees' (Nel *et al.*, 2008). The large nephilid spider *Nephila jurassica*, known from the Daohugou locality, presumably resembled extant members of the same genus in being a web-weaving predator of medium-sized to large insects (Selden *et al.*, 2011). Several stem-fleas from the Yanliao Biota are likely to have ectoparasitized dinosaurs and pterosaurs (Gao *et al.*, 2012, Huang *et al.*, 2012). Two species of the genus *Pumilanthocoris* represent the oldest known flower bugs (Anthocoridae *sensu lato*) and are thought to have lived on gymnosperms as predators of the small arthropods that were abundant in the Yanliao Biota (Hou *et al.*, 2012). Based on the morphology of their mouthparts, head and legs it has been suggested that the alloioscarabaeids of the Yanliao Biota fed on decaying organic materials (Bai *et al.*, 2012).

Some interesting data have been published regarding interactions between insects and plants in the Yanliao Biota. The bennettitalean *Anomozamites villosus* in the Daohugou Biota frequently shows signs of insect damage, and the leaves bear 'hairs' that may have acted as a defence against the insects by disrupting their movements (Pott *et al.*, 2011). Damage tends to be restricted to the distal part of each leaf, suggesting that the hairs successfully protected the proximal part (Pott *et al.*, 2011). Mimesis between insect wings and angiosperm leaves was known previously, but the Yanliao Biota is unique among modern and fossil biotas in offering examples of mimesis between insects and gymnosperms. There is evidence of pinnate cycadophyte leaf mimesis in lacewings (Wang, Labandeira *et al.*, 2010), and of mimesis between a hangingfly from the Daohugou Locality and a type of ginkgo leaf known from the same site (Wang, Labandeira *et al.*, 2012). In the latter case, the wings and abdomen of the hangingfly are nearly identical in general morphology to the multilobed leaf of the ginkgo, representing a degree of leaf mimesis that is likely to have offered some protection to both the insect and the plant (Wang, Labandeira et al., 2012).

The most important inference concerning insect–plant interactions in the Yanliao Biota is probably that of a mutualistic relationship between certain gymnosperms (Ren et al., 2009) and the Yanliao scorpionflies, based largely on the anatomy of both the reproductive organs of the former and the mouthparts of the latter. The scorpionflies would have used their siphon-like proboscAes to feed on the gymnosperms' pollination drops, helping the gymnosperms achieve pollination in the process. This Jurassic mutualism between plants and insects greatly preceded 'the similar and independent coevolution of nectar-feeding flies, moths, and beetles on angiosperms' (Ren et al., 2009).

Some vertebrate fossils preserve direct evidence of dietary habits. Some specimens of the salamander *Jeholotriton paradoxus* contain undigested shells of the conchostracan *Euestheria luanpingensis* inside the belly, and at least nine specimens of the salamander *Chunerpeton tianyiensis* similarly contain remains of the corixid insect *Yanliaocorixa chinensis* (Dong, Huang et al., 2012). In both cases the stomach contents provide direct evidence of the last meals of the salamanders in question, and of selective feeding in Jurassic caudates (Dong, Huang et al., 2012). Dietary inferences, albeit less precise, can be drawn for some other vertebrate taxa based on morphological data. Like other anurognathid pterosaurs, *Jeholopterus ningchengensis* was probably an aerial insectivore based on its wide head, simple teeth, and low-aspect-ratio wings (Bennett, 2007; Sullivan et al., 2014). Several Yanliao pterosaurs have been interpreted as piscivorous based on their cranial and dental morphology (Sullivan et al., 2014), and the pterosaur *Darwinopterus robustodens* may have used its unusually stout teeth to feed on the hard-carapaced coleopterans that are known from the Daxishan Locality (Lü et al., 2011). The heterodontosaurid dinosaur *Tianyulong* was probably primarily herbivorous, based on its heavily worn teeth. The diets of the Yanliao theropods, such as *Epidexipteryx* and *Anchiornis*, are difficult to infer. However, tooth morphology suggests that *Anchiornis* was carnivorous and/or insectivorous. Most Yanliao mammaliaforms were probably insectivorous, like most other Mesozoic mammaliaforms (Luo, 2007), but the haramiyidan *Megaconus mammaliaformis* was probably herbivorous (Zhou et al., 2013).

Data are available on the habitat preferences and locomotion styles of some Yanliao animals. Tan and Ren (2002) recognized three distinct insect communities at the Daohugou locality: an aquatic insect community, a soil insect community and a forest and vegetation insect community. The siphlonurid ephemeropteran *Multramificans ovalis* is represented by a single imago from Daohugou (Huang et al., 2007), which occurs on a slab along with freshwater conchostracans. The aquatic nymphs of extant siphlonurids prefer cool water and adults 'do not move far from water' (Huang et al., 2007), and the habits of *M. ovalis* were presumably similar. The chresmodid *Jurachresmoda gaskelli* has fringing hairs on its legs and other specialized features suggestive of 'water-skiing' locomotion (Zhang, Ren et al., 2008).

Among Yanliao vertebrates, the external gills of the salamanders *Jeholotriton* and *Chunerpeton* (Fig 4A) clearly imply an aquatic lifestyle, the latter taxon may have been benthic given its flat skull morphology (Dong, Huang et al., 2012). Interestingly, the mammaliaform *Castorocauda lutrasimilis* also displays some morphological features (e.g. a wide and flat otter-like tail) suggesting aquatic habits (Ji et al., 2006). However, most of the Yanliao mammaliaforms were probably fossorial (Luo et al., 2007), and even *Castorocauda* appears to have been a capable digger as well as a capable swimmer (Sullivan et al., 2014). Unlike the majority of Mesozoic mammaliaforms, which are inferred to have been ground-dwellers (Luo, 2007), several Yanliao mammaliaforms show scansorial (Luo et al., 2011) or even arboreal (Meng et al., 2006, Zheng et al., 2013) adaptations. Similar adaptations are also likely to be present in several Yanliao theropod dinosaurs (Zhang, Zhou et al., 2008) and lizards (Evans and Wang, 2009). For example, the highly specialized manual digit IV in the theropod *Epidendrosaurus* (Zhang et al., 2002) and the elongated penultimate phalanges and curved pedal claws in several paravians such as *Anchiornis* (Xu, Zhao et al., 2009, Hu et al., 2009) have been interpreted as correlates of arboreality, and the relatively long manus and forelimbs of one of the lizards from the Daohugou Locality suggest that this animal may have been a climber in life (Evans and Wang, 2009).

The Yanliao Biota pterosaurs vary in their limb proportions, and presumably also in their modes of flight. For example, *Fenghuangopterus lii* has proportionally extremely long legs, implying the presence of very broad wings, given that the pterosaurian brachiopatagium attaches to the leg and extends to the ankle (Sullivan *et al.*, 2014). This type of wing profile is uncommon among Jurassic pterosaurs (Sullivan *et al.*, 2014). Some capacity for aerial locomotion was also clearly present in the mammaliaform *Volaticotherium*, which has a large membranous patagium superficially resembling that of a flying squirrel (Meng *et al.*, 2006). The presence of large pennaceous feathers (Fig 4) in the theropods *Anchiornis huxleyi* and *Xiaotingia zhengi*, and their relatively long, robust forelimbs, suggest that these dinosaurs were capable of aerial locomotion (Xu *et al.*, 2011). This may also have been true of *Pedopenna daohugouensis*, which is known only from a feathered hindlimb associated with an additional patch of pennaceous feathers that may be from the forelimb (Sullivan *et al.*, 2014). Pennaceous feathers are unknown in the scansoriopterygids *Epidendrosaurus* and *Epidexipteryx*, but both had relatively long, robust forelimbs as in other basal paravians (Xu *et al.*, 2011). However, it is impossible to rule out the possibility that the absence of pennaceous feathers in the known specimens of *Epidendrosaurus* and *Epidexipteryx* represents a preservational artefact.

In a few cases, the exceptional preservation that is characteristic of the Yanliao Biota offers a surprisingly clear window into some aspect of the biology of a particular species. Male–female pairs of the pachymeridiid *Peregrinpachymeridium comitcola* are occasionally preserved in a mating position that is also seen in some living bugs, with the copulating individuals facing each other but aligned so that the head of the male is near the tail of the female and *vice versa* (Lu *et al.*, 2011). The exceptional preservation of some Yanliao insect fossils has even allowed the reconstruction of acoustic communication (Gu *et al.*, 2012). In katydids, the morphology of the stridulatory structures can be used to predict the type of song that stridulation would produce. For example, the exceptionally well-preserved stridulatory structures in a particular basal katydid allowed Gu *et al.* (2012) to reconstruct its low-frequency song. This suggests that such low-frequency 'chirping' may have been a normal part of the auditory background in the Yanliao Biota ecosystem (Gu *et al.*, 2012).

Significance of the Yanliao Biota

The Yanliao fossils are informative regarding the origins and early evolution of several major groups of organisms, whether viewed from the perspective of phylogeny, structural and functional transformation, or ecological diversification. The possible angiosperms from the biota may, if their status as flowering plants is confirmed, shed new light on the sequence in which major angiosperm features evolved (Wang *et al.*, 2007). Even if these plants prove not to be angiosperms, analysis of their angiosperm-like characteristics may lead to insights into wider patterns of character evolution. Numerous exquisitely preserved insect fossils from the Yanliao Biota, particularly the earliest known members of several insect clades, have provided useful evidence regarding the systematic positions of the groups to which they belong, the initial assembly of their highly specialized body-plans, and the interactions between insects and other organisms such as plants (Huang *et al.*, 2012, Ren *et al.*, 2010, Ren *et al.*, 2009, Gao *et al.*, 2012). The Yanliao Biota contains the earliest known crown-group salamanders, which provide important data on the temporal and biogeographical framework of salamander evolution (Gao and Shubin, 2003, Gao and Shubin, 2012, Wang, 2004). The transitional pterosaurs from the biota provide insights into the process of mosaic and even modular evolution that resulted in the body plan of pterodactyloids (Lü, Unwin *et al.*, 2010). Among other key vertebrate fossils from the Yanliao Biota, the earliest known feathered dinosaurs have implications for the phylogenetic relationships of theropods and patterns of character evolution in this group, and particularly for the origin of birds and related issues such as the origin of avian flight and the origin and early evolution of feathers (Hu *et al.*, 2009, Xu and Zhang, 2005, Zhang, Zhou *et al.*, 2008, Godefroit, Demuynck *et al.*, 2013, Godefroit, Cau *et al.*, 2013). The earliest known gliding and semi-aquatic mammaliaforms highlight the extent of early ecological diversification in mammaliaform evolution (Meng *et al.*, 2006, Ji *et al.*, 2006), and the earliest known eutherian mammal help to establish the temporal framework of the eutherian–metatherian divergence and facilitate reconstruction of the primitive condition of the clade Eutheria (Luo *et al.*, 2011).

Several major clades are inferred to have originated in the Jurassic, and their Jurassic representatives inevitably play an important role in efforts to reconstruct phylogenetic relationships and analyses of character evolution within these groups (Clark *et al.*, 2006). Many taxa from the Yanliao Biota occupy otherwise sparsely sampled areas of phylogenetic trees, breaking up long branches that often lead to incorrect phylogenetic reconstructions, and are helpful in understanding the early evolutionary histories of their respective groups.

The origin and early evolution of flowering plants is a contentious subject in evolutionary biology (Davies et al., 2004, Frohlich and Chase, 2007). Molecular systematics posits that crown-group angiosperms originated in the Jurassic at the latest (Sanderson et al., 2004), and perhaps even in the Triassic (Smith et al., 2010). Stem-angiosperms, by definition, would have appeared even earlier than crown ones. However, the earliest confirmed angiosperm fossils are of Early Cretaceous age (Friis et al., 2005, Sun et al., 2002). The lack of pre-Cretaceous angiosperm fossils has been attributed to the physiology and ecological preferences of the early angiosperms, which might have restricted their distribution and diversification (Feild et al., 2009). However, several plant fossils from the Yanliao Biota might be helpful in understanding the origin and early evolution of the flowering plants (Wang, 2010a, Wang et al., 2007, Wang and Wang, 2010, Wang, 2010b). It has been suggested that the Early-Middle Jurassic plant *Schmeissneria*, from the Yanliao Area and Europe (Wang et al., 2007), and *Xingxueanthus* and *Solaranthus*, known only from the Yanliao Biota, may either represent the earliest known angiosperms or be close angiosperm relatives. These discoveries help to reduce the discrepancy between molecular and palaeontological estimates of the time at which angiosperms appeared (Wang, 2010a), and potentially provide important information on patterns of character change during early angiosperm evolution (Wang, 2010a). On the other hand, *Problematospermum* and some other plant fossils from the Yanliao Biota appear to represent seed plants that are not closely related to angiosperms but have independently evolved some angiosperm-like features, demonstrating the presence of unusual modes of pollen dispersal and pollination in certain Jurassic seed plants (Wang, Zheng et al., 2010).

Fleas are among the major ectoparasitic insect lineages, but their phylogenetic relationships to other insect groups remain uncertain. Many studies have suggested that fleas are most closely related to the Boreidae (Grimaldi and Engel, 2005), a group of scorpionflies with strong mandibles, relatively short lacinia and labial palps, or to true flies (Friedrich and Beutel, 2010). A recent study proposed a close relationship between fleas and Nannochoristidae, a scorpionfly group characterized by tube-like labial palps with elongate lacinia and very small mandibles (Schneeberg and Beutel, 2011). Several stem-fleas from the Yanliao Biota have siphon-like mouthparts similar to those seen in Nannochoristidae, supporting the hypothesis of a phylogenetic link to this group (Huang et al., 2012). The stem-fleas from the Yanliao Biota also provide significant new information on the evolution of parasite–host relationships in fleas and their extinct relatives (Gao et al., 2012, Huang et al., 2012). Mammals have been widely regarded as the primary hosts of fleas, based on neontological data (Whiting et al., 2008), but the evidence now available from the Yanliao Biota suggests that ectoparasitic fleas might have originally targeted members of the dinosaur–bird lineage as hosts and only later transitioned to targeting mammals. Early mammals were small, and the large size of the Mesozoic fleas and the robustness of their mouthparts suggests that they must have been feeding on larger, presumably dinosaurian, hosts (Gao et al., 2012, Huang et al., 2012).

Yanliao insect fossils also provide significant information regarding other aspects of insect evolution. The Yanliao Biota contains the earliest representatives of several insect groups, including flower bugs (Hou et al., 2012), libelluloid dragonflies (Huang and Nel, 2007), sclerosomatids (Huang et al., 2009), webspinners (Huang and Nel, 2009), taeniopterygids (Liu et al., 2007), fishflies (Wang and Zhang, 2010), tenebrionoids (Wang and Zhang, 2011), axymyiids (Zhang, 2004), blephariceriids (Zhang and Lukashevich, 2007), earwigs (Zhao et al., 2011), and rhophlids (Yao et al., 2006), among others. Along with other insect fossils that have been recovered from the Yanliao Biota, these specimens provide significant new information on phylogenetic relationships and patterns of character evolution (Huang and Nel, 2009, Tan and Ren, 2007, Vršanský et al., 2009, Yang, Makarkin et al., 2012). Interestingly, they reveal a high degree of evolutionary stasis in many insect groups (Huang et al., 2009, Ren et al., 2010).

The Yanliao Biota contains examples of apparent mimesis between lacewing and hangingfly insects and gymnosperms (Wang, Liu et al., 2010, Wang, Labandeira et al., 2012) and evidence for 'musical' stridulation in katydid insects (Gu et al., 2012), suggesting that many adaptations seen in the modern entomofauna extend back in some form to the Jurassic. Most significantly, a mutualistic relationship revolving around pollination may have existed between Yanliao scorpionflies and gymnosperms, well before 'the similar and independent coevolution of nectar-feeding flies, moths, and beetles on angiosperms' (Ren et al., 2009). If this conclusion is correct, scorpionflies must have undergone a major change in feeding mode between the Jurassic and the Recent, as modern scorpionflies are typically predators and scavengers (Ren et al., 2009).

Jurassic arachnid fossils are globally rare, but hundreds have been recovered from the Daohugou Locality (Selden et al., 2008). These discoveries provide significant new information on spider evolution during the Jurassic (Selden et al., 2008). For example, a golden orb-weaver spider from Daohugou not only represents the largest known fossil spider but also extends the fossil record of the family Nephilidae back in time by approximately 35 Ma and that of the genus *Nephila* by approximately 130 Ma, making *Nephila* the spider genus with the longest known temporal range (Selden et al., 2011). Similarly, haplogyne spiders from Daohugou extend the temporal range of the family by at least 120 Ma (Selden and Huang, 2010). These discoveries demonstrate suprisingly strong evolutionary conservatism in some spider lineages (Selden and Huang, 2010, Selden et al., 2008).

The early fossil record of caudates was poor prior to the discovery of numerous fossils representing this group in the Yanliao Biota. The five species of caudates recovered from the Biota are all known from at least some articulated skeletons (Wang, 2004, Wang, 2000, Wang and Evans, 2006, Gao and Shubin, 2003, Gao and Shubin, 2012), and have gone a long way towards filling the large temporal and evolutionary gap between formerly known Middle Jurassic taxa from England (*Marmorerpeton*) and Kyrgyzstan (*Kokartus*) on the one hand and Cretaceous forms on the other. In particular, *Marmorerpeton* and *Kokartus* are represented by isolated materials and can not be attributed to any living group. The Yanliao caudates include not only probable stem taxa (*Jeholotriton* and *Pangerpeton*) but also the earliest known crown members of the Caudata from anywhere in the world. Furthermore, the Yanliao crown-caudates show considerable phylogenetic diversity in that they include the earliest known members of three modern clades: cryptobranchids (*Chuerpeton*), hynobiids (*Liaoxitriton*), and salamandroids (*Beiyanerpeton*) (Wang, 2004, Wang and Evans, 2006, Gao and Shubin, 2003, Gao and Shubin, 2012).

Pterosaurs are a group of extinct flying reptiles whose evolutionary history extends from the Late Triassic to the end of the Cretaceous (Unwin, 2006). Pterodactyloids, the derived members of the group, have a highly specialized body-plan and include the largest known aerial animals in Earth history. They were isolated from all other known pterosaurs by a huge morphological gap until the recent discoveries of several transitional pterosaurs from the Yanliao Biota (Lü, Unwin et al., 2010, Wang, Kellner et al., 2010). These new pterosaurs are closely related to pterodactyloids, falling just outside the group in phylogenetic terms. They possess the 'modular' combination of a derived skull and neck and a primitive post-cervical skeleton, which supports the broader idea that modularity can play a key role in major evolutionary transformations (Lü, Unwin et al., 2010).

Coelurosaurian theropod phylogeny has been the subject of many recent studies, partly due to the intense level of interest in the problem of bird origins (Witmer, 2002, Zhou, 2004). Because of the high incidence of reversals and convergences in coelurosaurian evolution (Holtz, 2001), basal members of coelurosaurian subgroups are particularly important in reconstructing coelurosaurian phylogeny. However, the fossil record of Jurassic coelurosaurian theropods is relatively sparse. Denser sampling of Jurassic coelurosaurians will be needed in order to build an accurate coelurosaurian phylogeny.

Several theropods from the Yanliao Biota occupy sufficiently basal positions within particular derived coelurosaurian clades that they are helpful in reducing the problem of limited sampling, and new information from these species is already leading to novel phylogenetic hypotheses. On the one hand, the scansoriopterygids *Epidendrosaurus* (Fucheng et al., 2002) and *Epidexipteryx* (Zhang, Zhou et al., 2008) from the Yanliao Biota are more similar in many respects to basal birds such as *Jeholornis* and *Sapeornis* (Zhou and Zhang, 2003a, Zhou and Zhang, 2003b) than to *Archaeopteryx*, despite the traditional view of *Archaeopteryx* as the basal-most bird. On the other, the discoveries of the deinonychosaurians *Anchiornis* and *Xiaotingia* from the Yanliao Biota have significantly reduced the morphological gap between known deinonychosaurians and *Archaeopteryx* (Hu et al., 2009, Xu, Zhao et al., 2009, Xu et al., 2011), providing fresh evidence for a close relationship between these taxa (Xu et al., 2010). Indeed, a recent numerical phylogenetic analysis (Xu *et al.*, 2011) disrupted a long-standing consensus by placing *Archaeopteryx* within the Deinonychosauria, rather than within the Avialae. This result potentially carries great significance for our understanding of bird origins, but has been questioned in other recent studies (Turner *et al.*, 2012, Lee and Worthy, 2012, Agnolín and Novas, 2013, Godefroit, Cau *et al.*, 2013).

The Yanliao theropods are particularly important in reconstructing the phylogenetic and temporal distributions of salient avian features, such as flight and feathers. First, these taxa strongly suggest that pennaceous feathers and the four-winged condition are primitive for the Paraves and appeared around the Middle Jurassic (Hu *et al.*, 2009). The Yanliao theropods are uniformly characterized by long and robust arms, most plausibly interpreted as an adaptation for withstanding the forces involved in aerial locomotion, and by a number of pedal features similar to those in extant arboreal animals (Zhang, Zhou *et al.*, 2008, Xu and Zhang, 2005, Zhang *et al.*, 2002). These characteristics suggest that both flight and arboreality were primitively present in paravians and had appeared, at the latest, by the earliest part of the Late Jurassic.

The several mammaliaform fossils known from the Yanliao Biota have greatly improved our understanding of the early evolution of mammaliaforms and, in particular, of mammaliaform ecological diversity in the Jurassic. *Castorocauda lutrasimilis* and *Pseudotribos robustus* were probably fossorial, as indicated in both taxa by the wide distal end of the humerus with its hypertrophied epicondyles and massive, widely separated ulnar and radial condyles, the massive and asymmetrical olecranon process of the ulna, and the block-like carpals (Luo *et al.*, 2007). Most interestingly, *Castorocauda lutrasimilis* was probably also a semi-aquatic, otter-like animal, as indicated by the presence of caudal vertebrae with dorsoventrally compressed centra and bifurcated transverse processes (Ji *et al.*, 2006). Fossorial habits appear to have been widely present in Jurassic mammaliaforms, and this may also have been true of aquatic habits (Sullivan *et al.*, 2014). For example, the basal mammaliaform *Fruitafossor* from the United States (Luo and Wible, 2005) and the docodont *Haldanodon* from Portugal (Martin, 2005) also display digging adaptations in the forelimb and shoulder girdle, and aquatic habits have also been suggested for *Haldanodon* (Martin, 2005). *Volaticotherium antiquum* is a squirrel-sized mammaliaform, with a broad, furry patagium (gliding membrane) and proportionally long limbs, features suggestive of gliding capability (Meng *et al.*, 2006). *Volaticotherium* is the earliest known aerially capable mammaliaform, and its discovery extends the record of gliding mammaliaforms back by about 110 million years (Meng *et al.*, 2006). *Juramaia sinensis* represents the earliest known eutherian, predating the previous record by about 35 Ma (Luo *et al.*, 2011), and significantly reduces the discrepancy between molecular and fossil-based estimates of when this clade originated (Kitazoe, 2007). Information from *Juramaia sinensis* is also helpful in showing that scansorial adaptation was characteristic of ancestral eutherians and may have contributed to the success of the group by facilitating exploration of arboreal niches (Luo *et al.*, 2011). Two recently described haramiyidan mammals provide further new information on early mammaliaform ecological diversity (Zheng *et al.*, 2013, Zhou *et al.*, 2013). *Arboroharamiya jenkinsi* displays highly specialized features for arboreal adaptation, which are rarely seen in other Jurassic mammals (Zheng *et al.*, 2013). *Megaconus mammaliaformis* was clearly herbivorous and different from other known Jurassic mammals (Zhou *et al.*, 2013). However, these two studies disagree on the placement of the haramiyidans, and thus suggest different temporal frameworks for mammaliaform diversification. Zheng *et al.* (2013) placed haramiyidans within crown-Mammalia based on new information from *Arboroharamiya jenkinsi* and therefore inferred the origin of crown-mammals in the Late Triassic Period, but Zhou *et al.* (2013) posited the haramiyidans outside crown-Mammalia. The latter phylogenetic hypothesis, in combination with the presence of fur in *Megaconus mammaliaformis* (and also in *Castorocauda lutrasimilis*), would imply that mammalian integument originated well before the appearance of crown-Mammalia (Zhou *et al.*, 2013).

On a larger scale, the Yanliao Biota may reveal some major evolutionary events among terrestrial organisms, and the relationships between these events and contemporaneous environmental changes. One of the most poorly known intervals in the terrestrial fossil record is the Middle Jurassic, which represents a transitional period between the low diversity, cosmopolitan faunas of the Early Jurassic and the more diverse and somewhat endemic faunas in the Late Jurassic (Behrensmeyer *et al.*, 1992, Fraser and Sues, 1994, Upchurch *et al.*, 2002). The Yanliao Biota and other assemblages, including the Middle-Late Jurassic Shishugou and Shaximiao faunas of western China, show that considerable diversity was already in place by approximately the end of the Middle Jurassic. The fossil record suggests that basic patterns of divergence within many groups, including several insect clades, crown-salamanders, crown-squamates, pterosaurs, major dinosaur clades, and mammaliaforms, were established during the poorly known Middle Jurassic.

Recently discovered Jurassic coelurosaurians, particularly basal paravians, support the hypothesis that avialans and other major coelurosaurian groups originated and underwent considerable diversification prior to the beginning of the Late Jurassic. A stratigraphically calibrated theropod phylogeny based only on well-corroborated fossil occurrences suggests that all coelurosaurian groups, including Avialae, diversified rapidly in the Middle Jurassic (Hu et al., 2009). The same pattern appears to hold for other major groups of organisms. Most groups are less diverse, in terms of species richness, in the Yanliao Biota than in the Jehol Biota. However, the earliest known members of many animal groups are from the Yanliao Biota, indicating that the groups in question originated in the Middle Jurassic. The Yanliao Biota also appears to contain a relatively high diversity of smaller clades within these animal groups. In combination with other Middle-Late Jurassic assemblages, the Yanliao Biota shows that rapid divergence must have occurred in order to result in the appearance of these higher taxa. Although this interesting pattern is likely to be a sampling artefact, it nevertheless deserves further investigation.

It is perhaps significant that this interval in Earth's history also witnessed drastic palaeogeographic and climatic changes. Rifting within Pangaea, which had begun in the Late Triassic, continued through the Middle Jurassic (Manspeizer, 1988). By the latter epoch South America and Africa had already separated from each other, and from Laurasia. Volcanic activity was common along the rifts. Sea levels were relatively high, creating epicontinental seaways in North America, Europe, and Asia. For example, Central and East Asia were among the first regions to be separated by epicontinental seas from adjacent parts of Pangaea. In other cases, local climatic conditions resulted in geographic barriers, such as the large central Gondwanan desert that existed during the Middle and Late Jurassic (Rees et al., 2000, Sellwood and Valdes, 2008, Myers et al., 2011). The global geography of the Middle and Late Jurassic was characterized by more separate land masses, and more coastlines, than were present during the Triassic and Early Jurassic. Endemic faunas began to develop in various areas by the Middle Jurassic (Russell, 1993, Luo, 1999, Pol and Rauhut, 2012).

Although the Jurassic climate was generally warm and moist, there is evidence for the presence of some arid regions (Rees et al., 2000, Sellwood and Valdes, 2008, Myers et al., 2011) and cold periods (Royer, 2006). In particular, global temperatures appear to have been cool during the Bathonian (167.7–164.7 Ma) (Price, 1999) and from the late Callovian to middle Oxfordian (162–159 Ma), based on a survey of marine temperatures (Dromart, Garcia et al., 2003, Dromart, Garcia, Picard et al., 2003, Lécuyer et al., 2003), a putative drop in sea level (Dromart, Garcia et al., 2003, Dromart, Garcia, Picard et al., 2003), and the distribution of reefs (Cecca et al., 2005). These climatic changes must have had an effect on the evolution of terrestrial life forms.

If rapid divergence affecting multiple major groups of organisms did indeed occur around the Middle Jurassic, it will be desirable to investigate whether this change was linked to the drastic palaeogeographic and climatic changes that were occurring at about the same time. It is possible that continental breakup and the appearance of geographic barriers might have driven an increase in endemism, which in turn permitted diversification through allopatric speciation. It is also possible that climatic changes might have caused occasional episodes of extinction, which in turn opened up ecological space for fresh diversification. Admittedly, the available data are far from sufficient to confirm the presence of such evolutionary events, let alone any linkage to environmental changes. However, rapid advances in our understanding of climatic and geographic patterns during the Mesozoic are beginning to yield finer-scale information about environmental conditions. For example, contrary to the traditional reconstruction of a consistent greenhouse climate for the entire Mesozoic, multiple intervals of cool or even cold conditions are now thought to have occurred during the Mesozoic. Two such intervals appear to have occurred during the Middle Jurassic. Meanwhile, the presence of filamentous feathers in Yanliao dinosaurs such as *Tianyulong* and *Anchiornis*, and pterosaurs such as the Yanliao taxon *Jeholopterus*, suggests that a feathery covering may represent the ancestral condition for a group that contains both the Pterosauria and Dinosauria and accounts for the majority of non-marine tetrapods during the Jurassic and Cretaceous (Xu, Zheng et al., 2009, Zheng et al., 2009, Brusatte, 2012, Xu and Guo, 2009). If feathers were indeed ancestral for both pterosaurs and dinosaurs, their insulating properties may have helped both groups survive and flourish in the changing climatic conditions of the Mesozoic.

Acknowledgements

The authors thank Nick Fraser and Hans Sues for inviting us to contribute to this volume, and Ge Sun, Shaolin Zheng, Dong Ren, Diying Huang, Yongqing Liu, Keqin Gao, Changfu Zhou, Lijun Zhang, Xin Wang, Haichun Zhang, David Hone, and Zhiyan Zhou for discussion and providing references and photographs. This project was supported by the National Natural Science Foundation of China and the National Basic Research Program of China (91514302, 41688103, and 41120124002).

Note added in proof

Since the first submission of this chapter many significant new discoveries have been made, improving our understanding of the Yanliao Biota in various respects. For example, the composition of the biota has now been suggested to comprise two distinct phases: the Bathonian-Callovian Daohugou phase (about 168–164 million years ago) and the Oxfordian Linglongta phase (164–159 million years ago) (Huang 2015, Xu, Zhou et al., 2016). Furthermore, the systematic positions and evolutionary implications of certain key taxa are better understood. For additional information, readers may wish to refer to (Xu, Zhou et al., 2016) and references therein.

References

Agnolín, F. L., and F. E. Novas. 2013. Avian Ancestors – A Review of the Phylogenetic Relationships of the Theropods Unenlagiidae, Microraptoria, Anchiornis and Scansoriopterygidae. Springer, Dordrecht and Heidelberg, 96 pp.

Amiot, R., X. Wang, Z. Zhou, X. Wang, E. Buffetaut, C. L'Ecuyer, Z. Ding, F. Fluteau, T. Hibino, N. Kusuhashi, J. Mo, V. Suteethorn, Y. Wang, X. Xu, and F. Zhang. 2011. Oxygen isotopes of East Asian dinosaurs reveal exceptionally cold Early Cretaceous climates. Proceedings of the National Academy of Sciences of the United States of America 108:5179–5183.

Bai, M., D. Ahrens, X. Yang, and D. Ren. 2012. New fossil evidence of the early diversification of scarabs: *Alloioscarabaeus cheni* (Coleoptera: Scarabaeoidea) from the Middle Jurassic of Inner Mongolia, China. Insect Science 19:159–171.

Behrensmeyer, A. K., J. D. Damuth, W. A. DiMichele, R. Potts, H.-D. Sues, and S. L. Wing (eds). 1992. Terrestrial Ecosystems Through Time. The University of Chicago Press, Chicago, 568 pp.

Bennett, S. C. 2007. A second specimen of the pterosaur *Anurognathus ammoni*. Paläontologische Zeitschrift 81:376–398.

Black, M. 1929. Drifted plant-beds of the Upper Estuarine Series of Yorkshire. Quarterly Journal of the Geological Survey of London 85:389–437.

Briggs, D. E. G. 2003. The role of decay and mineralization in the preservation of soft-bodied fossils. Annual Review of Earth and Planetary Sciences 31:275–301.

Brusatte, S. L. 2012. Dinosaur Paleobiology. Wiley-Blackwell, Chichester, 322 pp.

Cecca, F., B. Martingarin, D. Marchand, B. Lathuiliere, and A. Bartolini. 2005. Palaeoclimatic control of biogeographic and sedimentary events in Tethyan and peri-Tethyan areas during the Oxfordian (Late Jurassic). Palaeogeography, Palaeoclimatology, Palaeoecology 222:10–32.

Chang, J.-P., and Y.-W. Sun. 1997. [Aquatic biocoenose of the Haifanggou Formation during the Middle Jurassic in Beipiao Basin, Liaoning Province.] Journal of Changchun University of Earth Sciences 27:241–245. [Chinese]

Chang, S.-C., H.-C. Zhang, P. R. Renne, and Y. Fang. 2009. High-precision $^{40}Ar/^{39}Ar$ age constraints on the basal Lanqi Formation and its implications for the origin of angiosperm plants. Earth and Planetary Science Letters 279:212–221.

Chang, S.-C., H. Zhang, S. R. Hemming, G. T. Mesko, and Y. Fang, 2013. $^{40}Ar/^{39}Ar$ age constraints on the Haifanggou and Lanqi formations: when did the first flowers bloom? Pp. 277–284 in F. Jourdan, D. F. Mark, and C. Verati (eds), Advances in $^{40}Ar/^{39}Ar$ Dating: From Archaeology to Planetary Science. Geological Society of London, Special Publication 378.

Chen, J.-H., M.-Z. Cao, and T. Komatsu. 2003. A discovery of the North China provincial fauna in the Triassic–Jurassic boundary beds in western Fujian, South China. Science Technology and Engineering 3:230–231.

Chen, P.-J. 2003. Jurassic biostratigraphy of China: pp. 423–464 in W.-T. Zhang, P.-J. Chen, and A. R. Palmer (eds), Biostratigraphy of China. Science Press, Beijing.

Chen, W., Q. Ji, D. Liu, Y. Zhang, B. Song, and X. Liu. 2004. Isotope geochronology of the fossil-bearing beds in the Daohugou area, Ningcheng, Inner Mongolia. Geological Bulletin of China 23:1165–1169.

Chen, Y., W. Chen, X. Zhou, Z. Li, H. Liang, and Q. Li. 1997. [Liaoxi and Adjacent Mesozoic Volcanic Rocks: Chronology, Geochemistry and Tectonic Settings.]. The Seismological Press, Beijing, 279 pp. [Chinese]

Cheng, X., X. Wang, S. Jiang, and A. W. A. Kellner. 2012. A new scaphognathid pterosaur from western Liaoning, China. Historical Biology 24:101–111.

Cheng, Y., and C. Li. 2007. A new species of *Millerocaulis* (Osmundaceae, Filicales) from the Middle Jurassic of China. Review of Palaeobotany and Palynology 144:249–259.

Cheng, Y.-M., Y.-F. Wang, and C.-S. Li. 2007. A new species of *Millerocaulis* (Osmundaceae) from the Middle Jurassic of China and its implication for evolution of *Osmunda*. International Journal of Plant Science 168:1351–1358.

Clark, J. M., X. Xu, D. Eberth, C. A. Forster, M. Malkus, S. Hemming, and R. Hernandez. 2006. The Mid-Late Jurassic terrestrial transition: new discoveries form the Shishugou Formation, China. Pp. 26–28 in P. M. Barrett, D. J. Batten, S. E. Evans, J. Nudds, P. A. Selden, and A. Ross (eds), Ninth International Symposium on Mesozoic Terrestrial Ecosystems and Biota. Manchester.

Cui, Y., Y. S. Storozhenko, and D. Ren. 2012. New and little-known species of Geinitziidae (Insecta: Grylloblattida) from the Middle Jurassic of China, with notes on taxonomy, habitus and habitat of these insects. Alcheringa 36:251–261.

Czerkas, S. A., and Q. Ji. 2002. A new rhamphorhynchoid with a head crest and complex integumentary structures. Pp. 15–41 in S. J. Czerkas (ed), Feathered Dinosaurs and the Origin of Flight. The Dinosaur Museum, Blanding, Utah.

Czerkas, S. A., and C.-X. Yuan. 2002. An arboreal maniraptoran from Northeast China. Pp. 63–95 in S. J. Czerkas (ed), Feathered Dinosaurs and the Origin of Flight. The Dinosaur Museum, Blanding, Utah.

Davies, T. J., T. G. Barraclough, M. W. Chase, P. S. Soltis, D. E. Soltis, and V. Savolainen. 2004. Darwin's abominable mystery: insights from a supertree of the angiosperms. Proceedings of the National Academy of Sciences of the United States of America 101:1904–1909.

Ding, M., D.-R. Zheng, Q. Zhang, and H.-C. Zhang. 2013. A new species of Ephialtitidae (Insecta: Hymenoptera: Stephanoidea) from the Middle Jurassic of Inner Mongolia, China. Acta Palaeontologica Sinica 52:51–56.

Dong, F., and D.-Y. Huang. 2011. A new elaterid from the Middle Jurassic Daohugou Biota (Coleoptera: Elateridae: Progagrypninae). Acta Geologica Sinica (English Edition) 85:1224–1230.

Dong, L.-P., D.-Y. Huang, and Y. Wang. 2012) Two Jurassic salamanders with stomach contents from Inner Mongolia, China. Chinese Science Bulletin 57:72–76.

Dong, Q.-P., Y.-Z. Yao, and D. Ren. 2012a. A new species of Progonocimicidae (Hemiptera, Coleorrhyncha) from the Middle Jurassic of China. Alcheringa 37:31–37.

Dong, Q.-P., Y.-Z. Yao, and D. Ren. 2012b. A new species of Progonocimicidae (Hemiptera, Coleorrhyncha) from northeastern China. Zootaxa 3495:73–78.

Dong, S.-W. 2010. Jurassic tectonic revolution in China and new interpretation of the 'Yanshan Movement'. Acta Geological Sinica (English Edition) 82:334–347.

Dromart, G., J. P. Garcia, S. Picard, F. Atrops, C. Lécuyer, and S. M. F. Sheppard. 2003. Ice age at the Middle–Late Jurassic transition. Earth and Planetary Science Letters 213:205–220.

Dromart, G., J. P. Garcia, F. Gaumet, S. Picard, M. Rousseau, F. Atrops, C. Lécuyer, and S. M. F. Sheppard. 2003. Perturbations of the carbon cycle at the Middle/Late Jurassic transition: geological and geochemical evidence. American Journal of Science 303:667–707.

Duan, Y., S.-L. Zheng, D.-Y. Hu, L.-J. Zhang, and W.-L. Wang. 2009. Preliminary report on Middle Jurassic strata and fossils from Linglongta area of Jianchang, Liaoning. Global Geology 28:143–147.

Eberth, D. A., X. Xu, and J. M. Clark. 2010. Dinosaur death pits from the Jurassic of China. Palaios 25:112–125.

Engel, M. S., and D. Ren. 2008. New snakeflies from the Jiulongshan Formation of Inner Mongolia, China (Raphidioptera). Journal of the Kansas Entomological Society 81:188–193.

Engel, M. S., D.-Y. Huang, and Q. Lin. 2011. A new genus and species of Aetheogrammathidae from the Jurassic of Inner Mongolia, China (Neuroptera). Journal of the Kansas Entomological Society 84:315–319.

Estes, R. 1983. Sauria terrestria, Amphisbaenia. Pp. 1–249 in P. Wellnhofer (ed), Handbuch der Paläoherpetologie, Vol. 10A. Gustav Fischer Verlag, Stuttgart.

Evans, S. E., and Y. Wang. 2007. A juvenile lizard specimen with well-preserved skin impressions from the Upper Jurassic/Lower Cretaceous of Daohugou, Inner Mongolia, China. Naturwissenschaften 94:431–439.

Evans, S. E., and Y. Wang. 2009. A long-limbed lizard from the Upper Jurassic/Lower Cretaceous of Daohugou, Ningcheng, Nei Mongol, China. Vertebrata PalAsiatica 47:21–34.

Evans, S. E., and Y. Wang. 2012. New material of the Early Cretaceous lizard *Yabeinosurus* from China. Cretaceous Research 34:48–60.

Feild, T. S., D. S. Chatelet,. and T. J. Brodribb. 2009. Ancestral xerophobia: a hypothesis on the whole plant ecophysiology of early angiosperm. Geobiology 7:237–264.

Fraser, N. C., and H.-D. Sues (eds). 1994. In the Shadow of the Dinosaurs: Early Mesozoic Tetrapods. Cambridge University Press, New York.

Friedrich, F. and R. G. Beutel. 2010. The thoracic morphology of *Nannochorista* (Nannochoristidae) and its implications for the phylogeny of Mecoptera and Antliophora. Journal of Zoological Systematics and Evolutionary Research 48:50–74.

Friis, E. M., K. R. Pedersen, and P. R. Crane. 2005. When Earth started blooming: insights from the fossil record. Current Opinion of Plant Biology 8:5–12.

Frohlich, M. W., and M. W. Chase. 2007. After a dozen years of progress the origin of angiosperms is still a great mystery. Nature 450:1184–1189.

Fürsich, F. T., J. Sha, B. Jiang, and Y. Pan. 2007. High resolution palaeoecological and taphonomic analysis of Early Cretaceous lake biota, western Liaoning (NE China). Palaeogeography, Palaeoclimatology, Palaeoecology 253:434–457.

Gao, K.-Q., and N. H. Shubin. 2003. Earliest known crown-group salamanders. Nature 422:424–428.

Gao, K.-Q., and N. Shubin. 2012. Late Jurassic salamandroid from western Liaoning, China. Proceedings of the National Academy of Sciences of the United States of America 109:5767–5772.

Gao, Y.-H., Y.-Z. Yao, and D. Ren. 2013. A new Middle Jurassic caddisfly (Trichoptera, Hydrobiosidae) from China. Fossil Record 16:111–116.

Gao, T.-P., C.-K. Shih, X. Xu, S. Wang, and D. Ren. 2012. Mid-Mesozoic flea-like ectoparasites of feathered or haired vertebrates. Current Biology 22:732–735.

Gao, T.-P., C.-K. Shih, A. P. Rasnitsyn, X. Xing, S. Wang, and D. Ren. 2013. New transitional fleas from China highlighting diversity of Early Cretaceous ectoparasitic insects. Current Biology 22:732–735.

Gao, Y.-H., Y.-Z. Yao, and D. Ren. 2013. A new Middle Jurassic caddisfly (Trichoptera, Hydrobiosidae) from China. Fossil Record 16:111–116.

Godefroit, P., A. Cau, D.-Y. Hu, F. Escuillié, W.-H. Wu, and G.

Dyke. 2013. A Jurassic avialan dinosaur from China resolves the early phylogenetic history of birds. Nature 498:359–362.

Godefroit, P., H. Demuynck, G. Dyke, D. Hu, F. Escuillié, and P. Claeys. 2013. Reduced plumage and flight ability of a new Jurassic paravian theropod from China. Nature Communications 4:1394. doi:10.1038/ncomms2389

Grimaldi, D., and M. S. Engel. 2005. Evolution of the Insects. Cambridge University Press, Cambridge, 772 pp.

Gu, J.-J., G.-X. Qiao, and D. Ren. 2011. An exceptionally preserved new species of *Barchaboilus* (Orthoptera: Prophalangopsidae) from the Middle Jurassic of Daohugou, China. Zootaxa 2909:64–68.

Gu, J.-J., G.-X. Qiao, and D. Ren. 2013. The first discovery of Cyrtophyllitinae (Orthoptera, Haglidae) from the Middle Jurassic and its morphological implications. Alcheringa 36:27–34.

Gu, J.-J., F. Montealegre-Z, D. Robert, M. S. Engel, G.-X. Qiao, and D. Ren. 2012. Wing stridulation in a Jurassic katydid (Insecta, Orthoptera) produced low-pitched musical calls to attract females. Proceedings of the National Academy of Sciences of the United States of America 109:3868–3873.

Guo, X.-Q., J.-G. Han, and S.-A. Ji. 2012. [Advances in the study of vertebrate fossils of the Middle Jurassic Yanliao biota in western Liaoing Province and adjacent areas.] Geological Bulletin of China 31:928–935. [Chinese]

Guo, Y.-X., and D. Ren. 2011. A new cockroach genus of the family Fuziidae from northeastern China (Insecta: Blattida). Acta Geologica Sinica (English Edition) 85:501–506.

Guo, Z., J. Liu, and X. Wang. 2003. Effect of Mesozoic volcanic eruptions in the western Liaoning Province, China on paleoclimate and paleoenvironment. Science in China D 46:1261–1272.

Guo, Z.-F., J.-Q. Liu, X.-Y. Chen, and X.-H. Li. 2008. [Effect of volatiles erupted from Mesozoic activities in Yanliao area of North China on paleooenvironmental changes.] Acta Petrologica Sinica 24:2595–2603. [Chinese]

Harris, T. M. 1961. The Yorkshire Jurassic Flora I. Thallophyta–Pteridophyta. British Museum (Natural History), London, 212 pp.

He, H.-Y., X.-L. Wang, Z.-H. Zhou, R.-X. Zhu, F. Jin, F. Wang, X. Ding, and A. Boven. 2004. $^{40}Ar/^{39}Ar$ dating of ignimbrite from Inner Mongolia, northeastern China, indicates a post-Middle Jurassic age for the overlying Daohugou Bed. Geophysical Research Letters 31:1–4. doi:10.1029/2004GL020792.

He, Y.-M., J. M. Clark, and X. Xu. 2013. A large theropod metatarsal from the upper part of the Jurassic Shishugou Formation in Junggar Basin, Xinjiang, China. Vertebrata PalAsiatica 51:29–42.

Hoffstetter, R. 1964. Les Sauria du Jurassique supérieur et spécialement les Gekkota de Bavière et de Mandchourie. Senckenbergiana Biologica 45:281–324.

Holtz, T. R., Jr. 2001. Arctometatarsalia revisited: the problem of homoplasy in reconstructing theropod phylogeny. Pp. 99–124 in J. A. Gauthier and L. F. Gall (eds), New Perspectives on the Origin and Early Evolution of Birds. Peabody Museum of Natural History, Yale University, New Haven, Connecticut.

Hong, Y.-C. 1983. Middle Jurassic Fossil Insects in North China. Geological Publishing House, Beijing.

Hou, W.-J., Y.-Z. Yao, W.-T. Zhang, and D. Ren. 2012. The earliest fossil flower bugs (Heteroptera: Cimicomorpha: Cimicoidea: Vetanthocoridae) from the Middle Jurassic of Inner Mongolia, China. European Journal of Entomology 109:281–288.

Hu, D.-Y., L.-H. Hou, L.-J. Zhang, and X. Xu. 2009. A pre-*Archaeopteryx* troodontid from China with long feathers on the metatarsus. Nature 461:640–643.

Huang, D.-Y. 2013. Discussions on the fossil "tadpole" from the Daohugou Biota. Acta Palaeontologica Sinica 52:141–145.

Huang, D. Y. 2015. [Yanliao Biota and Yanshan Movement.] Acta Palaeontologica Sinica 54(4): 501-546. [Chinese]

Huang, D.-Y., and A. Nel. 2007. Oldest "libelluloid" dragonfly from the Middle Jurassic of China (Odonata: Anisoptera: Cavilabiata). Neues Jahrbuch für Geologie und Paläontologie Abhandlungen 246:63-68.

Huang, D.-Y., and A. Nel. 2009. Oldest webspinners from the Middle Jurassic of Inner Mongolia, China (Insecta: Embiodea). Zoological Journal of the Linnean Society 156:889–895.

Huang, D.-Y., P. A. Selden, and J. A. Dunlop. 2009. Harvestmen (Arachnida: Opiliones) from the Middle Jurassic of China. Naturwissenschaften 96:955–962.

Huang, D.-Y., M. S. Engel, C.-Y. Cai, and A. Nel. 2013. Mesozoic giant fleas from northeastern China (Siphonaptera): taxonomy and implications for palaeodiversity. Chinese Science Bulletin 58:1682–1690.

Huang, D.-Y., A. Nel, C.-Y. Cai, and M. S. Engel. 2013. Amphibious flies and paedomorphism in the Jurassic period. Nature 495:94–97.

Huang, D.-Y., M. S. Engel, C.-Y. Cai, H. Wu, and A. Nel. 2012. Diverse transitional giant fleas from the Mesozoic era of China. Nature 483:201–204.

Huang, J.-D., Y. Liu, N. D. Sinitshenkova, and D. Ren. 2007. A new fossil genus of Siphlonuridae (Insecta: Ephemeroptera) from the Daohugou, Inner Mongolia, China. Annales Zoologici (Warszawa) 57:221–225.

Jarzembowski, E. A., E.-V. Yan, B. Wang, and H. Zhang. 2012. A new flying water beetle (Coleoptera: Schizophoridae) from the Jurassic Daohugou lagerstätte. Palaeoworld 21:160–166.

Jepson, J. E., S. W. Heads, V. N. Makarkin, and D. Ren. 2013. New fossil mantidflies (Insecta, Neuroptera, Mantispidae) from the Mesozoic of northeastern China. Palaeontology 56:603–613.

Ji, Q., and C.-X. Yuan. 2002. [Discovery of two kinds of protofeathered pterosaurs in the Mesozoic Daohugou Biota in the Ningcheng region and its stratigraphic and biologic significances.] Geological Review 48:221–224. [Chinese]

Ji, Q., Z.-X. Luo, C.-X. Yuan, and A. R. Tabrum. 2006. A swimming mammaliaform from the Middle Jurassic and ecomorphological diversification of early mammals. Science 311:1123–1127.

Ji, Q., W. Chen, W.-L. Wang, X.-C. Jin, J.-P. Zhang, Y.-X. Liu, H. Zhang, P.-Y. Yao, S.-A. Ji, C.-X. Yuan, Y. Zhang, and H.-L. You. 2004. [Mesozoic Jehol Biota of Western Liaoning, China.] Geological Publishing House, Beijing, 375 pp. [Chinese]

Ji, S. 2004. [Late Mesozoic fossil lizards (Reptilia, Squamata) from western Liaoning and northern Hebei, China.] Science Technology and Engineering 4:756–767. [Chinese]

Ji, S.-A., and Q. Ji. 1998. A new fossil pterosaur (Rhamphorhynchoidea) from Liaoning. Jiansu Geology 22:199–206.

Ji, S.-A., Q. Ji, and K. Padian. 1999. Biostratigraphy of new pterosaurs from China. Nature 398:573–574.

Jiang, B.-Y. 2006. [Non-marine *Ferganoconcha* (Bivalvia) from the Middle Jurassic in Daohugou Area, Ningcheng County, Inner Mongolia, China.] Acta Palaeontologica Sinica 45:259–264. [Chinese]

Jiang, B.-Y., F. T. Fürsich, J.-G. Shang, B.-X. Wang, and Y. Niu. 2011. Early Cretaceous volcanism and its impact on fossil preservation in western Liaoning, NE China. Palaeogeography, Palaeoclimatology, Palaeoecology 302:255–269.

Jiang, H.-E., D. K. Ferguson, C.-S. Li, and Y.-M. Cheng. 2008. Fossil coniferous wood from the Middle Jurassic of Liaoning Province, China. Review of Palaeobotany and Palynology 150:37–47.

Jiang, X.-J., Y.-Q. Liu, N. Peng, and J.-C. Lü. 2010. [Middle Jurassic pterosaur and feathered dinosaur-bearing strata at Linglongta, Jianchang County, western Liaoning.] Acta Geoscientica Sinica 31:33–35. [Chinese]

Jin, F. 1999. Middle and late Mesozoic acipenseriforms from northern Hebei and western Liaoning, China. Palaeoworld 11:188–280.

Kirchner, M. 1992. Untersuchungen an einigen Gymnospermen der fränkischen Rhät-Lias-Grenzschichten. Palaeontographica B 224:17–61.

Kitazoe, Y., H. Kishino, P. J. Waddell, N. Nakajima, T. Okabayashi, T. Watabe, and Y. Okuhara 2007. Robust time estimation reconciles views of the antiquity of placental mammals. PLoS ONE 2:e384. doi:10.1371/journal.pone.0000384

Lécuyer, C., S. Picard, J.-P. Garcia, S. M. F. Sheppard, P. Grandjean, and G. Dromart. 2003. Thermal evolution of Tethyan surface waters during the Middle-Late Jurassic: evidence from $\delta^{18}O$ values of marine fish teeth. Paleoceanography 1:1076. doi:10.1029/2002PA000863

Lee, M. S. Y., and T. H. Worthy. 2012. Likelihood reinstates *Archaeopteryx* as a primitive bird. Biology Letters 8:299–303.

Leng, Q., and H. Yang. 2003. Pyrite framboids associated with the Mesozoic Jehol Biota in northeastern China: implications for microenvironment during early fossilization. Progress in Natural Science 12:206–212.

Li, N., Y. Li, L.-X. Wang, S.-L. Zheng, and W. Zhang. 2004. A new species of *Weltrichia* Braun in North China with a special bennettitalean male reproductive organ. Acta Botanica Sinica 46: 1269–1275.

Li, S., J. Szwedo, D. Ren, and H. Pang. 2011. *Fenghuangor imperator* gen. et sp. nov. of Fulgoridiidae from the Middle Jurassic of Daohugou Biota (Hemiptera: Fulgoromorpha). Zootaxa 3094:52–62.

Li, S., Y. Wang, D. Ren, and H. Pang. 2013. Revision of the genus *Sunotettigarcta* Hong, 1983 (Hemiptera: Tettigarctidae), with a new species from Daohugou, Inner Mongolia, China. Alcheringa 36:501–507.

Li, Y., A. Nel, D. Ren, and H. Pang. 2011. A new genus and species of hawker dragonfly of uncertain affinities from the Middle Jurassic of China (Odonata: Aeshnoptera). Zootaxa 2927:57–62.

Li, Y.-J., A. Nel, D. Ren, and H. Pang. 2013. A new damsel-dragonfly from the Mesozoic of China with a hook-like male anal angle (Odonata: Isophlebioptera: Campterophlebiidae). Journal of Natural History 47:1953–1958.

Li, Y.-J., A. Nel, D. Ren, B.-L. Zhang, and H. Pang. 2012. Reassessment of the Jurassic damsel-dragonfly genus *Karatawia* (Odonata: Campterophlebiidae). Zootaxa 3417:64–68.

Li, Y.-J., A. Nel, C.-K. Shih, D. Ren, and H. Pang. 2013. The first euthemistid damsel-dragonfly from the Middle Jurassic of China (Odonata, Epiproctophora, Isophlebioptera). Zookeys 261:41–50.

Liang, J.-H., W. Huang, and D. Ren. 2012. *Graciliblatta bella* gen. et sp. n. – a rare carnivorous cockroach (Insecta, Blattida, Raphidiomimidae) from the Middle Jurassic sediments of Daohugou in Inner Mongolia, China. Zootaxa 3449:62–68.

Liang, J.-H., P. Vršanský, and D. Ren. 2012. Variability and symmetry of a Jurassic nocturnal predatory cockroach (Blattida: Raphidiomimidae). Revista Mexicana de Ciencias Geologicas 29:411–421.

Liu, J., Y. Zhao, and X. Liu. 2006. [Age of the Tiaojishan Formation volcanics in the Chengde Basin, northern Hebei province.] Acta Petrologica Sinica 22:2617–2630. [Chinese]

Liu, L.-X., C.-K. Shih, and D. Ren. 2012. Two new species of Ptychopteridae and Trichoceridae from the Middle Jurassic of northeastern China (Insecta: Diptera: Nematocera). Zootaxa 3501:55–62.

Liu, P.-J., J.-D. Huang, and D. Ren. 2010. [Palaeoecology of the Middle Jurassic Yanliao entomofauna.] Acta Zootaxonomica Sinica 35:568–584. [Chinese]

Liu, X., Y. Wang, C. Shih, D. Ren, and D. Yang. 2012. Early evolution and historical biogeography of fishflies (Megaloptera: Chauliodinae): implications from a phylogeny combining fossil and extant taxa. PLoS ONE 7:e40345. doi:10.1371/journal.pone.0040345

Liu, Y., D. Ren, N.D. Sinitshenkova, and C.-K. Shih. 2007. The oldest known record of Taeniopterygidae in the Middle Jurassic of Daohugou, Inner Mongolia, China (Insecta: Plecoptera). Zootaxa 1521:1–8.

Liu, Y.-Q., H. Kuang, X. Jiang, N. Peng, H. Xu, and H. Sun. 2012. Timing of the earliest known feathered dinosaurs and transitional pterosaurs older than the Jehol Biota. Palaeogeography, Palaeoclimatology, Palaeoecology 323–325:1–12.

Liu, Y.-S., C.-F. Shi, and D. Ren. 2011. A new lacewing (Insecta: Neuroptera: Grammolingiidae) from the Middle Jurassic of Inner Mongolia, China. Zootaxa 2897:51–56.

Liu, Y.-S., N. D. Sinitshenkova, D. Ren, and C.-K. Shih, C. K. 2011. Pronemouridae fam. nov. (Insecta: Plecoptera), the stem group of Nemouridae and Notonemouridae, from the Middle Jurassic of Inner Mongolia, China. Palaeontology 54:923–933.

Liu, Y., Y. Liu, and H. Zhang. 2006. LA-ICPMS zircon U-Pb dating in the Jurassic Daohugou beds and correlative strata in Ningcheng of Inner Mongolia. Acta Geologica Sinica (English Edition) 80:733–742.

Lu, L.-W. 1995. [Fish fossils.] Pp. 121–140 in D. Ren, L.-W. Lu, and Z. G. Guo (eds), [Jurassic–Cretaceous Faunas and Stratigraphy of Beijing and Neighbouring Areas.] Geological

Publishing House, Beijing. [Chinese]

Lu, Y., Y.-Z. Yao, and D. Ren. 2011. Two new genera and species of fossil true bugs (Hemiptera: Heteroptera: Pachymeridiidae) from northeastern China. Zootaxa 2835:41–52

Lü, J.-C., and X. Bo. 2011. A new rhamphorhynchid pterosaur (Pterosauria) from the Middle Jurassic Tiaojishan Formation of western Liaoning, China. Acta Geologica Sinica 85:977–983.

Lü, J.-C., and X.-H. Fucha. 2010. A new pterosaur (Pterosauria) from Middle Jurassic Tiaojishan Formation of western Liaoning, China. Global Geology 13:113–118.

Lü, J.-C., and D. W. E. Hone. 2012. A new Chinese anurognathid pterosaur and the evolution of pterosaurian tail lengths. Acta Geologica Sinica (English Edition) 86:1317–1325.

Lü, J.-C., L. Xu, H.-L. Chang, and X.-L. Zhang. 2011. A new darwinopterid pterosaur from the Middle Jurassic of western Liaoning, northeastern China and its ecological implications. Acta Geologica Sinica (English Edition) 85:507–514.

Lü, J.-C., D. M. Unwin, X.-S. Jin, Y.-Q. Liu, and Q. Ji. 2010. Evidence for modular evolution in a long-tailed pterosaur with a pterodactyloid skull. Proceedings of the Royal Society B 277:383–389.

Lü, J.-C., D. M. Unwin, B. Zhao, C. Gao, and C. Shen. 2012. A new rhamphorhynchid (Pterosauria: Rhamphorhynchidae) from the Middle/Upper Jurassic of Qinglong, Hebei Province, China. Zootaxa 3158:1–19.

Lü, J.-C. 2009. A new non-pterodactyloid pterosaur from Qinglong County, Hebei Province of China. Acta Geologica Sinica (English edition) 83:89–199.

Lü, J.-C., X.-H. Fucha, and J.-M. Chen. 2010. A new scaphognathine pterosaur from the Middle Jurassic of western Liaoning, China. Acta Geoscientica Sinica 31:263–266.

Luo, Z.-X. 1999. A refugium for relicts. Nature 400:23–24.

Luo, Z.-X. 2007. Transformation and diversification in early mammal evolution. Nature 450:1011–1019.

Luo, Z.-X., and J. R. Wible. 2005. A Late Jurassic digging mammal and early mammalian diversification. Science 308:103–107.

Luo, Z.-X., Q. Ji, and C.-X. Yuan. 2007. Convergent dental evolution in pseudotribosphenic and tribosphenic mammals. Nature (London) 450:93–97.

Luo, Z.-X., C.-X. Yuan, Q.-J. Meng, and Q. Ji. 2011. A Jurassic eutherian mammal and divergence of marsupials and placentals. Nature 476:442–445.

Makarkin, V. N., Q. Yang, and D. Ren. 2011. Two new species of *Sinosmylites* Hong (Neuroptera, Berothidae) from the Middle Jurassic of China, with notes on Mesoberothidae. Zookeys 130:199–215.

Manspeizer, W. (ed). 1988. Triassic–Jurassic Rifting: Continental Breakup and the Origin of the Atlantic Ocean and Passive Margins. Two volumes. (Developments in Geotectonics 22.) Elsevier, Amsterdam.

Martínez-Delclòs, X., D. E. G. Briggs, and E. Peñalver. 2004. Taphonomy of insects in carbonates and amber. Palaeogeography, Palaeoclimatology, Palaeoecology 203:19–64.

Martill, D. M., and S. Etches. 2012. A new monofenestratan pterosaur from the Kimmeridge Clay Formation (Upper Jurassic, Kimmeridgian) of Dorset, England. Acta Palaeontologica Polonica 58:285–294.

Martin, T. 2005. Postcranial anatomy of *Haldanodon exspectatus* (Mammalia, Docodonta) from the Late Jurassic (Kimmeridgian) of Portugal and its bearing for mammalian evolution. Zoological Journal of the Linnean Society 145:219–248.

Meng, J., Y.-M. Hu, Y.-Q. Wang, X.-L. Wang, and C.-K. Li. 2006. A Mesozoic gliding mammal from northeastern China. Nature 444:889–893.

Myers, T. S., N. J. Tabor, and L. L. Jacobs. 2011. Late Jurassic paleoclimate of central Africa. Palaeogeography, Palaeoclimatology, Palaeoecology 311:111–125.

Nel, A., D. Y. Huang, and Q. Lin. 2008. A new genus of isophlebioid damsel-dragonflies with "calopterygid"-like wing shape from the Middle Jurassic of China (Odonata: Isophlebioidea: Campterophlebiidae). European Journal of Entomology 105:783–787.

Nikolajev, G. V., B. Wang, Y. Liu, and H. Zhang. 2011. Stag beetles from the Mesozoic of Inner Mongolia, China (Scarabaeoidea: Lucanidae). Acta Palaeontologica Sinica 50:41–47.

O'Brien, N. R., H. W. Meyer, K. Reilly A. M. Ross, and S. Maguire. 2002. Microbial taphonomic processes in the fossilization of insects and plants in the late Eocene Florissant Formation, Colorado. Rocky Mountain Geology 37:1–11.

Pan, G. 1983. The Jurassic precursors of angiosperms from Yanliao region of North China and the origin of angiosperms. Chinese Science Bulletin 28:15–20.

Pan, X., H. Chang, , D. Ren, and C. Shih. 2011. The first fossil buprestids from the Middle Jurassic Jiulongshan Formation of China (Coleoptera: Buprestidae). Zootaxa 2745:53–62.

Pan, Y., J. Sha, F. T. Fürsich, Y. Wang, X. Zhang, and X. Yao. 2011. Dynamics of the lacustrine fauna from the Early Cretaceous Yixian Formation, China: implications of volcanic and climatic factors. Lethaia 45:299–314.

Peng, N., Y. Liu, H. Kuang, X. Jiang, and H. Xu. 2012. Stratigraphy and geochronology of vertebrate fossil-bearing Jurassic strata from Linglongta, Jianchang County, western Liaoning, northeastern China. Acta Geologica Sinica (English Edition) 86:1326–1339.

Petrulevicius, J. F., D.-Y. Huang, and A. Nel. 2011. A new genus and species of damsel-dragonfly (Odonata: Isophlebioidea: Campterophlebiidae) in the Middle Jurassoc of Inner Mongolai, China. Acta Geological Sinica (English Edition) 85:733–738.

Petrulevičius, J. F., and D. Ren. 2012. A new species of "Orthophlebiidae" (Insecta: Mecoptera) from the Middle Jurassic of Inner Mongolia, China. Revue de Paléobiologie 11:311–315.

Pol, D. and O. W. M. Rauhut. 2012. A Middle Jurassic abelisaurid from Patagonia and the early diversification of theropod dinosaurs. Proceedings of the Royal Society B 279:3170–3175.

Pott, C., S. McLoughlin, S.-Q. Wu, and E. M. Friis. 2011. Trichomes on the leaves of *Anomozamites villosus* sp. nov. (Bennettitales) from the Daohugou beds (Middle Jurassic), Inner Mongolia, China: mechanical defence against herbivorous arthropods. Review of Palaeobotany and Palynology 169:48–60.

Price, G. D. 1999. The evidence and implications of polar ice during the Mesozoic. Earth-Science Reviews 48:183–210.

Qiao, X., C.-K. Shih, and D. Ren. 2012. Two new Middle Jurassic species of orthophlebiids (Insecta: Mecoptera) from Inner Mongolia, China. Alcheringa 36:467–472.

Rasnitsyn, A. P., and H.-C. Zhang. 2004. Composition and age of the Daohugou hymenopteran (Insecta, Hymenoptera = Vespida) assemblage from Inner Mongolia, China. Palaeontology 47:1507–1517.

Rees, P. M., A. M. Ziegler, and P. J. Valdes. 2000. Jurassic phytogeography and climates: new data and model comparisons. Pp. 297–318 in B. T. Huber, K. G. MacLeod, and S. L. Wing (eds), Warm Climates in Earth History. Cambridge University Press, Cambridge.

Ren, D. 1998. Flower-associated Brachycera flies as fossil evidence for Jurassic angiosperm origins. Science 280:85–88.

Ren, D. 2003. Study on the Jurassic Fossil Raphidioptera and Neuroptera from Northeast China. PhD dissertation. Academy of Forestry, Beijing Forestry University, Beijing.

Ren, D., and D. S. Aristov. 2011. A new species of *Plesioblattogryllus* Huang, Nel et Petrulevičius (Grylloblattida: Plesioblattogryllidae) from the Middle Jurassic of China. Paleontological Journal 45:273–278.

Ren, D., and J. C. Yin. 2002. A new Middle Jurassic species of *Epiosmylus* (Neuroptera: Osmylidae) from Inner Mongolia, China. Acta Zootaxonomica Sinica 27:274–277.

Ren, D., L.-W. Lu, Z.-G. Guo, and S.-A. Ji. 1995. [Faunas and Stratigraphy of Jurassic-Cretaceous in Beijing and the Adjacent Areas.] Seismological Press, Beijing, 222 pp. [Chinese]

Ren, D., C.-K. Shih, T.-P. Gao, Y.-Z. Yao, and Y.-Y. Zhao. 2010. Silent Stories – Insect Fossil Treasures from Dinosaur Era of Northeastern China. Science Press, Beijing, 332 pp.

Ren, D., C. C. Labandeira, J. A. Santiago-Blay, A. Rasnitsyn, C. Shih, A. Bashkuev, M. A. V. Logan, C. L. Hotton, and D. Dilcher. 2009. A probable pollination mode before angiosperms: Eurasian, long-proboscid scorpionflies. Science 326:840–847.

Royer, D. L. 2006. CO_2-forced climate thresholds during the Phanerozoic. Geochimica et Cosmochimica Acta 70:5665–5675.

Russell, D. A. 1994. The role of Central Asia in dinosaurian biogeography. Canadian Journal of Earth Sciences 30:2002–2012.

Sanderson, M. J., J. L. Thorne, N. Wilk, and K. G. Bremer. 2004. Molecular evidence on plant divergence times. American Journal of Botany 91:1656–1665.

Schneeberg, K., and R. G. Beutel. 2011. The adult head structures of Tipulomorpha (Diptera, Insecta) and their phylogenetic implications. Acta Zoology 92:316–343.

Selden, P. A., and D.-Y. Huang. 2010. The oldest haplogyne spider (Araneae: Plectreuridae) from the Middle Jurassic of China. Naturwissenschaften 97:449–459.

Selden, P. A., D.-Y. Huang, and D. Ren. 2008. Palpimanoid spiders from the Jurassic of China. The Journal of Arachnology 36:306–321.

Selden, P. A., C.-K. Shih, and D. Ren. 2011. A golden orb-weaver spider (Araneae: Nephilidae: *Nephila*) from the Middle Jurassic of China. Biology Letters 7:775–778.

Sellwood, B. W., and P. J. Valdes. 2008. Jurassic climates. Proceedings of Geological Association 119:5–17.

Shang, L.-J., O. Béthoux, and D. Ren. 2011. New stem-Phasmatodea from the Middle Jurassic of China. European Journal of Entomology 108:677–685.

Shen, Y.-B., and D.-Y. Huang. 2008. Extant clam shrimp egg morphology: taxonomy and comparison with other fossil branchiopod eggs. Journal of Crustacean Biology 28:352–360.

Shen, Y.-B., P.-J. Chen, and D.-Y. Huang. 2003. Age of the fossil conchostracans from Daohugou of Ningcheng, Inner Mongolia. Journal of Stratigraphy 27:311–313.

Shi, C.-F., Q. Yang, and D. Ren. 2011. Two new fossil lacewing species from the Middle Jurassic of Inner Mongolia, China (Neuroptera: Grammolingiidae). Acta Geologica Sinica (English Edition) 85:482–489.

Shi, C.-F., Y.-J. Wang, Q. Yang, and D. Ren. 2012. *Chorilingia* (Neuroptera: Grammolingiidae): a new genus of lacewings with four species from the Middle Jurassic of Inner Mongolia, China. Alcheringa 36:309–318.

Shih, C.-K., H. Feng, and D. Ren. 2011. New fossil Heloridae and Mesoserphidae wasps (Insecta, Hymenoptera, Proctotrupoidea) from the Middle Jurassic of China. Annals of the Entomological Society of America 104:1334–1348.

Shih, C.-K., X.-G. Yang, C. C. Labandeira, and D. Ren. 2011. A new long-proboscid genus of Pseudopolycentropodidae (Mecoptera) from the Middle Jurassic of China and its plant-host specializations. Zookeys 130:281–297.

Smith, S. A., J. M. Beaulieu, and M. J. Donoghue. 2010. An uncorrelated relaxed-clock analysis suggests an earlier origin for flowering plants. Proceedings of the National Academy of Sciences of the United States of America 107:5897–5902.

Smith, A. G., D. G. Smith, and B. M. Funnell. 1994. Atlas of Mesozoic and Cenozoic Coastlines. Cambridge University Press, Cambridge, 99 pp.

Su, D.-Z. 1974. New Jurassic ptycholepid fishes from Szechuan, S. W. China. Vertebrata PalAsiatica 12:1–15.

Sullivan, C., Y. Wang, D. W. E. Hone, Y. Wang, X. Xu, and F. Zhang. 2014. The vertebrates of the Jurassic Daohugou Biota of northeastern China. Journal of Vertebrate Paleontology 34:243-280.

Sun, C.-L., D. L. Dilcher, H.-S. Wang, G. Sun, and Y.-H. Ge. 2008. A study of *Ginkgo* leaves from the Middle Jurassic of Inner Mongolia, China. International Journal of Plant Science 169:1128–1139.

Sun, G., D. L. Dilcher, S. Zheng, and Z. Zhou. 1998. In search of the first flower: a Jurassic angiosperm, *Archaefructus*, from northeast China. Science 82:1692–1695.

Sun, G., Q. Ji, D. L. Dilcher, S.-L. Zheng, and K. C. Nixon. 2002. Archaefructaceae, a new basal angiosperm family. Science 296: 899–904.

Sun, G., L. Zhang, C. Zhou, Y. Duan, and D. Hu. 2009. A General Outline for the Liaoning Paleontological Museum. Liaoning Paleontological Museum, Shenyang.

Sun, G., L. Zhang, C. Zhou, Y. Duan, and D. Hu. 2011. The Fossils over Last 300 Myr in Liaoning Province. Shanghai Science & Technology Press, Shanghai.

Tan, J.-J., and D. Ren. 2002. Palaeoecology of insect community from Middle Jurassic Jiulongshan Formation in Ningcheng County, Inner Mongolia, China. Acta Zootaxonomica Sinica 27:428–434.

Tan, J.-J., and D. Ren. 2007. Two exceptionally well-preserved catiniids (Coleoptera: Archostemata: Catiniidae) from the late Mesozoic of Northeastern China. Annals of the Entomological Society of America 100:666–672.

Tan, J.-J., D. Ren, C. Shih, and X. Yang. 2012. New schizophorid fossils from China and possible evolutionary scenarios for Jurassic archostematan beetles. Journal of Systematic Palaeontology 11:47–62.

Tan, J.-J., Y. J. Wang, D. Ren, and X.-K. Yang. 2012. New fossil species of ommatids (Coleoptera: Archostemata) from the Middle Mesozoic of China illuminating the phylogeny of Ommatidae. BMC Evolutionary Biology 12:113–132.

Turner, A. H., P. J. Makovicky, and M. A. Norell. 2012. A review of dromaeosaurid systematics and paravian phylogeny. Bulletin of the American Museum of Natural History 371:1–206.

Unwin, D. M. 2006. The Pterosaurs From Deep Time. Pi Press, New York, 347 pp.

Upchurch, P., C. A. Hunn, and D. B. Norman. 2002. An analysis of dinosaurian biogeography: evidence for the existence of vicariance and dispersal patterns caused by geological events. Proceedings of the Royal Society of London B 269:613–621.

Vršanský, P., J.-H. Liang, and D. Ren. 2009. Advanced morphology and behaviour of extinct earwig-like cockroaches (Blattida: Fuziidae fam. nov.). Geologica Carpathica 60:449–462.

Vršanský, P., J.-H. Liang, and D. Ren. 2012. Malformed cockroach (Blattida: Liberiblattinidae) in the Middle Jurassic sediments from China. Oriental Insects 46:12–18.

Wang, B., J. Szwedo, and H. Zhang. 2012. New Jurassic Cercopoidea from China and their evolutionary significance (Insecta: Hemiptera). Palaeontology 55:1223–1243.

Wang, B., and H.-C. Zhang. 2010. Earliest evidence of fishflies (Megaloptera: Corydalidae): an exquisitely preserved larva from the Middle Jurassic of China. Journal of Paleontology 84:774–780.

Wang, B., and H. Zhang. 2011. The oldest Tenebrionoidea (Coleoptera) from the Middle Jurassic of China. Journal of Paleontology 85:266–270.

Wang, B., H.-C. Zhang, and Y. Fang. 2006. Some Jurassic Palaeontinidae (Insecta, Hemiptera) from Daohugou, Inner Mongolia, China. Palaeoworld 15:115–125.

Wang, B., H. Zhang, and A. G. Ponomarenko. 2012. Mesozoic Trachypachidae (Insecta: Coleoptera) from China. Palaeontology 55:341–353.

Wang, B.-X., J.-F. Li, Y. Fang, and H.-C. Zhang. 2008. Preliminary elemental analysis of fossil insects from the Middle Jurassic of Daohugou, Inner Mongolia and its taphonomic implications. Chinese Science Bulletin 54:783–787.

Wang, L., D. Hu, L. Zhang, S. Zheng, H. He, C. Deng, X.-L. Wang, Z.-H. Zhou, and R.-X. Zhu. 2013. [SIMS U-Pb zircon age of Jurassic sediments in Linglongta, Jianchang, western Liaoning: constraint on the age of oldest feathered dinosaurs.] Chinese Science Bulletin 58:1–8. [Chinese]

Wang, M., C.-K. Shih, and D. Ren. 2012. *Platyxyela* gen. nov. (Hymenoptera, Xyelidae, Macroxyelinae) from the Middle Jurassic of China. Zootaxa 3456:82–88.

Wang, Q., Y. Zhao, and D. Ren. 2012. Two new species of Mesosciophilidae (Insecta: Diptera: Nematocera) from the Yanliao biota of Inner Mongolia, China. Alcheringa: An Australasian Journal of Palaeontology 36:509–514.

Wang, W.-L., S.-L. Zheng, L.-J. Zhang, R.-G. Pu, W. Zhang, H.-Z. Wu, R.-H. Ju, G.-Y. Dong, and H. Yuan. 1989. [Mesozoic Stratigraphy and Palaeontology of Western Liaoning 1.] Geological Publishing House, Beijing, 168 pp. [Chinese]

Wang, X. 2010a. Flower-related fossils from the Jurassic. Pp. 91–153 in X. Wang (ed), The Dawn Angiosperms: Uncovering the Origin of Flowering Plants. (Lecture Notes in Earth Sciences 121.) Springer, Heidelberg and Dordrecht

Wang, X. 2010b. *Schmeissneria*: an angiosperm from the Early Jurassic. Journal of Systematics and Evolution 48:326–335.

Wang, X., and S.-J. Wang. 2010. *Xingxueanthus*: an enigmatic Jurassic seed plant and its implications for the origin of angiospermy. Acta Geologica Sinica (English Edition) 84:47–55.

Wang, X., M. Krings, and T. N. Taylor. 2010. A thalloid organism with possible lichen affinity from the Jurassic of northeastern China. Review of Palaeobotany and Palynology 162:591–598.

Wang, X., S.-L. Zheng, and J.-H. Jin. 2010. Structure and relationships of *Problematospermum*, an enigmatic seed from the Jurassic of China. International Journal of Plant Science 171:447–456.

Wang, X.-F., S. Duan, B. Geng, J. Cui, and Y. Yang. 2007. *Schmeissneria*: a missing link to angiosperms? BMC Evolutionary Biology 7:1–13.

Wang, X.-L., A. W. A. Kellner, S. Jiang, and X. Meng. 2009. An unusual long-tailed pterosaur with elongated neck from western Liaoning, China. Anais da Academia Brasileira de Ciências 81:793–812.

Wang, X., Z. Zhou, F. Zhang, and X. Xu. 2002. A nearly completely articulated rhamphorhynchoid pterosaur with exceptionally well-preserved wing membranes and 'hairs' from Inner Mongolia, northeast China. Chinese Science Bulletin 47:226–230.

Wang, X., A. W. A. Kellner, S. Jiang, X. Cheng, X. Meng, and T. Rodrigues. 2010. New long-tailed pterosaurs (Wukongopteridae) from western Liaoning, China. Anais da Academia Brasileira de Ciências 82:1045–1062.

Wang, X.-L., Y.-Q. Wang, X. Xu, Y. Wang, J.-Y. Zhang, F.-C. Zhang, F. Jin, and G. Gu. 1999. Record of the Sihetun vertebrate mass mortality events, western Liaoning, China: caused by volcanic eruptions. Geological Review 45(Suppl.):458–467.

Wang, X.-L., Y.-Q. Wang, F.-C. Zhang, Zhang, Z.-H. Zhou, F. Jin, Y.-M. Hu, G. Gu, and H.-C. Zhang. 2000. [Vertebrate biostratigraphy of the Lower Cretaceous Yixian Formation in Linghuan, western Liaoning and its neighboring southern Nei Mongol (Inner Mongolia), China.] Vertebrata PalAsiatica 38:81–99. [Chinese]

Wang, X., Z. Zhou, H. He, F. Jin, Y. Wang, J. Zhang, Y. Wang, X. Xu, and F. Zhang. 2005. Stratigraphy and age of the Daohugou Bed in Ningcheng, Inner Mongolia. Chinese Science Bulletin 50:2369–2376.

Wang, Y. 2000. [A new salamander (Amphibia: Caudata) from the Early Cretaceous Jehol Biota.] Vertebrata PalAsiatica 38:100–103. [Chinese]

Wang, Y. 2004. A new Mesozoic caudate (*Liaoxitriton daohugouensis* sp. nov.) from Inner Mongolia, China. Chinese Science Bulletin 49:858–860.

Wang, Y., and S. E. Evans. 2006. A new short-bodied salamander from the Upper Jurassic/Lower Cretaceous of China. Acta Palaeontologica Polonica 51:127–130.

Wang, Y., and D. Ren. 2007. Revision of the genus *Suljuktocossus* Becker-Migdisova, 1949 (Hemiptera, Palaeontinidae), with description of a new species from Daohugou, Inner Mongolia, China. Zootaxa 1576:57–62.

Wang, Y., and C. Rose. 2005. *Jeholotriton paradoxus* (Amphibia: Caudata) from the Lower Cretaceous of southeastern Inner Mongolia, China. Journal of Vertebrate Paleontology 25:523–532.

Wang, Y., L. Dong, and S. E. Evans. 2010a. Herpetofaunas from the Jehol associated strata in NE China: evolutionary and ecological implications. Bulletin of the Chinese Academy of Sciences 24:76–79.

Wang, Y., L. Dong, and S. E. Evans. 2010b. Polydactyly in a Mesozoic salamander from China. Journal of Vertebrate Paleontology Program and Abstracts 30:183A–184A.

Wang, Y., Z. Liu, D. Ren, and C. Shih. 2011. New Middle Jurassic kempynin osmylid lacewings from China. Acta Palaeontologica Polonica 56:865–869.

Wang, Y., C. Shih, J. Szwedo, and D. Ren. 2012. New fossil palaeontinids (Hemiptera, Cicadomorpha, Palaeontinidae) from the Middle Jurassic of Daohugou, China. Alcheringa 37:19–30.

Wang, Y., C. C. Labandeira, C. Shih, Q. Ding, C. Wang, Y. Zhao, and D. Ren. 2012. Jurassic mimicry between a hangingfly and a ginkgo from China. Proceedings of the National Academy of Sciences of the United States of America 109:20514–20519.

Wang, Y., Z. Liu, X. Wang, C. Shih, Y. Zhao, M. S. Engel, and D. Ren. 2010. Ancient pinnate leaf mimesis among lacewings. Proceedings of the National Academy of Sciences of the United States of America 107:16212–16215.

Wei, D., J. Liang, and D. Ren. 2012. A new species of Fuziidae (Insecta, Blattida) from Inner Mongolia, China. Zookeys 217:53–61.

Wei, D., C. Shih, and D. Ren. 2012. *Arcofuzia cana* gen. et sp. nov. (Insecta, Blattaria, Fuziidae) from the Middle Jurassic sediments of Inner Mongolia, China. Zootaxa 3597:25–32.

Whiting, M. F., A. S. Whiting, M. W. Hastriter, and K. Dittmar. 2008. A molecular phylogeny of fleas (Insecta: Siphonaptera): origins and host associations. Cladistics 24:1–31.

Witmer, L. M. 2002. The debate on avian ancestry: phylogeny, function, and fossils. Pp. 3–30 in L. M. Chiappe and L. M. Witmer (eds), Mesozoic Birds: Above the Heads of Dinosaurs. University of California Press, Berkeley, California.

Wu, G., C. Li, and W. Wang. 2004. [Geochemical features and geological implications of Middle Mesozoic Haifanggou Formation in West Liaoning.] Acta Petrologica et Mineralogica 23:97–105. [Chinese]

Wu, H. and D.-Y. Huang. 2012. A new species of *Liaodotaulius* (Insecta: Trichoptera) from the Middle Jurassic of Daohugou, Inner Mongolia. Acta Geological Sinica (English Edition) 86:320–324.

Xu, K., J. Yang, M. Tao, H. Liang, C. Zhao, R. Li, H. Kong, Y. Li, C. Wan, and W. Peng. 2003. [Jurassic System in the North of China. Vol. VII: The Stratigraphic Region of Northeast China.] Petroleum Industry Press, Beijing. [Chinese]

Xu, X., and Y. Guo. 2009. The origin and early evolution of feathers: insights from recent paleontological and neontological data. Vertebrata PalAsiatica 47:311–329.

Xu, X., and F. Zhang. 2005. A new maniraptoran dinosaur from China with long feathers on the metatarsus. Naturwissenschaften 92:173–177.

Xu, X., Q. Y. Ma, and D. Y. Hu. 2010. Pre-*Archaeopteryx* coelurosaurian dinosaurs and their implications for understanding avian origins. Chinese Science Bulletin 55:3971–3977.

Xu, X., X. Zheng, and H. You. 2009. A new feather type in a nonavian theropod and the early evolution of feathers. Proceedings of the Natural Academy of Sciences of the United States of America 106:832–834.

Xu, X., H. You, K. Du, and F. Han. 2011. An *Archaeopteryx*-like theropod from China and the origin of Avialae. Nature 475:465–470.

Xu, X., Q. Zhao, M. Norell, C. Sullivan, D. Hone, G. Erickson, X. Wang, F. Han, and Y. Guo. 2009. A new feathered maniraptoran dinosaur fossil that fills a morphological gap in avian origin. Chinese Science Bulletin 54:430–435.

Xu, X., Z. H. Zhou, C. Sullivan, Y. Wang, and D. Ren 2016. An updated review of the Middle-Late Jurassic Yanliao Biota: Chronology, Taphonomy, Paleontology, and Paleoecology. Acta Geologica Sinica (English Edition) 90(6): 1801–1840.

Yabe, H., and T. Shikama. 1938. A new Jurassic Mammalia from South Manchuria. Proceedings of the Imperial Academy (Japan) 14:353–357.

Yang, G., Y. Yao, and D. Ren. 2012. A new species of Protopsyllidiidae (Hemiptera, Sternorrhyncha) from the Middle Jurassic of China. Zootaxa 3274:36–42.

Yang, G., Y. Yao, and D. Ren. 2013. *Poljanka strigosa*, a new species of Protopsyllidiidae (Hemiptera, Sternorrhyncha) from the Middle Jurassic of China. Alcheringa 37:125–130.

Yang, Q., V. N. Makarkin, and D. Ren. 2011. Two interesting new genera of Kalligrammatidae (Neuroptera) from the Middle Jurassic of Daohugou, China. Zootaxa 2873:60–68.

Yang, Q., V. N. Makarkin, and D. Ren. 2013. A new genus of the family Panfiloviidae (Insecta, Neuroptera) from the Middle Jurassic of China. Palaeontology 56:49–59.

Yang, Q., V. N. Makarkin, S. L. Winterton, A. V. Khramov, and D. Ren. 2012. A remarkable new family of Jurassic insects (Neuroptera) with primitive wing venation and its phylogenetic position in Neuropterida. PLoS ONE 7(9):e44762. doi:10.1371/journal.pone.0044762

Yang, W., and S.-G. Li. 2004. [The chronological framework of the Mesozoic volcanic rocks of western Liaoning and its implications for the Mesozoic lithosphere thinning in eastern China.] Abstracts of Symposium on National Petrology and Continental Geodynamics. Petroleum Committee of Chinese Society for Geology. Haikou, China. [Chinese]

Yang, W., and S. Li. 2008. Geochronology and geochemistry of the Mesozoic volcanic rocks in western Liaoning: implications for lithospheric thinning of the North China Craton. Lithos 102:88–117.

Yang, X.-G., C.-K. Shih, D. Ren, and J. F. Petrulevičius. 2012. New Middle Jurassic hangingflies (Insecta: Mecoptera) from Inner Mongolia, China. Alcheringa 36:195–201.

Yao, Y., W. Cai, and D. Ren. 2006. The first discovery of fossil rhopalids (Heteroptera: Coreoidea) from Middle Jurassic of

Inner Mongolia, China. Zootaxa 1269:57–68.

Young, C.-C. 1958. A new *Yabeinosaurus* locality and its stratigraphical implication. Vertebrata PalAsiatica 2:151–156.

Yu, Y., R. A. B. Leschen, A. Slipinski, D. Ren, and H. Pang. 2012. The first fossil bark-gnawing beetle from the Middle Jurassic of Inner Mongolia, China (Coleoptera: Trogossitidae). Annales Zoologici 62:245–252.

Yuan, C.-X., H.. Zhang, M. Li, and X. Ji. 2004. Discovery of a Middle Jurassic fossil tadpole from Daohugou Regon, Ningcheng, Inner Mongolia, China. Acta Geologica Sinica 78:145–148.

Zhang, F., Z. Zhou, X. Xu, and X. Wang. 2002. A juvenile coelurosaurian theropod from China indicates arboreal habits. Naturwissenschaften 89:394–398.

Zhang, F., Z. Zhou, X. Xu, X. Wang, and C. Sullivan. 2008. A bizarre Jurassic maniraptoran from China with elongate ribbon-like feathers. Nature 455:1105–1108.

Zhang, H., M. Wang, and X. Liu. 2008. Constraint on the upper boundary age of the Tiaojishan Formation volcanic rocks in western Liaoning–northern Hebei by LA-ICP-MS. Chinese Science Bulletin 53:3574–3584.

Zhang, H., H. Yuan, Z. Hu, X. Liu, and C.-R. Diwu. 2005. [U-Pb zircon dating of the Mesozoic volcanic strata in Luanping of north Hebei and its significance.] Earth Science Journal of China University of Geosciences 30:707–720. [Chinese]

Zhang, H., D. Zheng, B. Wang, Y. Fang, and E. A. Jarzembowski. 2013. The largest known odonate in China: *Hsiufua chaoi* Zhang et Wang, gen. et sp. nov. from the Middle Jurassic of Inner Mongolia. Chinese Science Bulletin 58:1579–1584.

Zhang, H.-C., D.-R. Zheng, Q. Zhang, E. A. Jarzembowski, and M. Ding. 2013. Re-description and systematics of *Paraulacus sinicus* Ping, 1928 (Insecta, Hymenoptera). Palaeoworld 22:32–35.

Zhang, J.-F. 2002. [Discovery of Dahugou Biota (Pre-Jehol Biota) with a discussion on its geological age.] Journal of Stratigraphy 26:173–177. [Chinese]

Zhang, J.-F. 2004. First description of axymyiid fossils (Insecta: Diptera: Axymyiidae). Geobios 37:687–694.

Zhang, J. F. 2010. Records of bizarre Jurassic brachycerans in the Daohugou Biota, China (Diptera, Brachycera, Archisargidae and Rhagionemestriidae). Palaeontology 53:307–317.

Zhang, J.-F., and N. J. Kluge. 2007. Jurassic larvae of mayflies (Ephemeroptera) from the Daohugou Formation in Inner Mongolia, China. Oriental Insects 41:351–366.

Zhang, J., and E. D. Lukashevich. 2007. The oldest known net-winged midges (Insecta: Diptera: Blephariceridae) from the late Mesozoic of northeast China. Cretaceous Research 28:302–309.

Zhang, J. X., C. Shih, J. F. Petrulevičius, and D. Ren. 2011. A new fossil eomeropid (Insecta, Mecoptera) from the Jiulongshan Formation, Inner Mongolia, China. Zoosystema 33:443–450.

Zhang, L.-J., and L.-X. Wang. 2004. [Mesozoic salamander fossil-bearing strata in Reshuitang area near Lingyuan, western Liaoning.] Geology and Resources 13:202–206. [Chinese]

Zhang, W., and S.-L. Zheng. 1987. [Early Mesozoic fossil plants from western Liaoning.] Pp. 239–368 in X.-H. Yu, W.-L. Wang, X.-T. Liu, W. Zhang, S.-L. Zheng, Z.-C. Zhang, J. S. Yu, F. Z. Ma, G. Y. Dong, and P. Y. Yao (eds), [Mesozoic Stratigraphy and Palaeontology of Western Liaoning.] Geological Publishing House, Beijing. [Chinese]

Zhang, W., and S.-L. Zheng. 1991. [A new species of osmundaceous rhizome from Middle Jurassic of Liaoning, China.] Acta Palaeontologica Sinica 30:714–727. [Chinese]

Zhang, W., Y. Wang, K. I. Saiki, N. Li, and S. Zheng. 2006. A structurally preserved cycad-like stem, *Lioxylon liaoningense* gen. et sp. nov., from the Jurassic in western Liaoning, China. Progress in Natural Science 26(Suppl. 1):236–248.

Zhang, X., D. Yan, and S. Bainian. 2003. The discovery of Jurassic *Yuchoulepis szechuanensis* Su from Yaojie, Gansu Province and its geological significance. Journal of Gansu Sciences 15:45–48.

Zhang, X.-L., and D.-G. Shu. 2001. [Preservation mechanisms of nonbiomineralized animal tissue.] Acta Sedimentologica Sinica 19:13–19. [Chinese]

Zhang, X.-W., D. Ren, H. Pang, and C.-K. Shih. 2008. A water-skiiing chresmodid from the Middle Jurassic in Daohugou, Inner Mongolia, China (Polyneoptera: Ornthopterida). Zootaxa 1762:53–62.

Zhao, J., C. Shih, D. Ren, and Y. Zhao. 2011. New primitive fossil earwig from Daohugou, Inner Mongolia, China (Insecta: Dermaptera: Archidermaptera). Acta Geologica Sinica (English Edition) 85:75–80.

Zheng, S., and X. Wang. 2010. An undercover angiosperm from the Jurassic of China. Acta Geologica Sinica (English Edition) 84:895–902.

Zheng, S.-L., L.-J. Zhang, and E.-P. Gong. 2003. A discovery of *Anomozamites* with reproductive organs. Acta Botanica Sinica 45:667–672.

Zheng, S., L. Zhang, W. Zhang, and Y. Yang. 2008. A new female cone, *Araucaria beipiaoensis* sp. nov. from the Middle Jurassic Tiaojishan Formation, Beipiao, western Liaoning, China and its evolutionary significance. Acta Geologica Sinica (English Edition) 82:266–282.

Zheng, X., S. Bi, X. Wang, and J. Meng. 2013. A new arboreal haramiyid shows the diversity of crown mammals in the Jurassic period. Nature 500:199–202.

Zheng, X.-T., H.-L. You, X. Xu, and Z.-M. Dong. 2009. A heterodontosaurid dinosaur from the Early Cretaceous of China with filamentous integumentary structures. Nature 458:333–336.

Zhou, C.-F., and R. R. Schoch. 2011. New material of the non-pterodactyloid pterosaur *Changchengopterus pani* Lü, 2009 from the Late Jurassic Tiaojishan Formation of western Liaoning. Neues Jahrbuch für Geologie und Paläontologie Abhandlungen 260:265–275.

Zhou, C.-F., S. Wu, T. Martin, and Z.-X. Luo. 2013. A Jurassic mammaliaform and the earliest mammalian evolutionary adaptations. Nature 500:163–167.

Zhou, M.-Z., Z.-W. Cheng, and Y.-Q. Wang. 1991. A mammalian lower jaw from the Jurassic of Lingyuan, Liaoning. Vertebrata PalAsiatica 29:165–175.

Zhou, Z. 2004. The origin and early evolution of birds: discoveries, disputes, and perspectives from fossil evidence. Naturwissenschaften 91:455–471.

Zhou, Z., and F. Zhang. 2003a. Anatomy of the primitive bird *Sapeornis chaoyangensis* from the Early Cretaceous of Liaoning, China. Canadian Journal of Earth Sciences 40:731–747.

Zhou, Z., and F. Zhang. 2003b. *Jeholornis* compared to *Archaeopteryx*, with a new understanding of the earliest avian evolution. Naturwissenschaften 90:220–225.

Zhou, Z.-H., F. Jin, and Y. Wang. 2010. Vertebrate assemblages from the Middle-Late Jurassic Yanliao Biota in northeast China. Earth Science Frontiers 17:252–254.

Zhou, Z., and B. Zhang. 1992. *Baiera hallei* Sze and associated ovule-bearing organs from the Middle Jurassic of Henan, China. Palaeontographica B 224:151–169.

Zhou, Z.-Y., S.-L. Zheng, and L.-J. Zhang. 2007. Morphology and age of *Yimaia* (Ginkgoales) from Daohugu village, Nincheng, Inner Mongolia, China. Cretaceous Research 28:348–362.

Chapter 6

The Jehol Biota: an exceptional window into Early Cretaceous terrestrial ecosystems

Zhonghe Zhou[1], Yuan Wang[1], Xing Xu[1], and Dong Ren[2]

1 Key Laboratory of Vertebrate Evolution and Human Origins, Institute of Vertebrate Paleontology and Paleoanthropology, Chinese Academy of Sciences, Beijing, China.
2 Beijing Capital Normal University, Beijing, China.

Abstract

The Early Cretaceous Jehol Biota (approximately 131Ma to 120Ma), mainly distributed in northeastern China, comprises assemblages of diverse lacustrine and land-dwelling animals and plants. It is particularly well-known for the numerous exceptionally preserved feathered non-avian dinosaurs, early birds and mammals. Discoveries in the last two decades have greatly enriched the known diversity and provided significant new data concerning critical issues with respect to evolution in the Mesozoic, particularly on the origin and early evolution of birds, mammals and flowering plants. Understanding of the ancient global and local tectonics, characterized by frequent volcanic eruptions, and the unique fossil preservation together provide a rare chance to assess the interaction between the evolution of the biota and the geological and palaeoenvironmental background, adding to our understanding of this Early Cretaceous terrestrial ecosystem.

Introduction

The Jehol Biota represents an Early Cretaceous terrestrial ecosystem that is mainly distributed in northeastern China, particularly in western Liaoning, northern Hebei and southeastern Inner Mongolia, an area once called 'Jehol'. It is best known now for yielding fossils of early birds, feathered non-avian dinosaurs and early mammals, as well as many other exceptionally preserved animals (pterosaurs, insects etc.) and plants (e.g., flowering plants).

The study of fossils from the Jehol Biota has a long history. Fossil fishes from the biota have been known to farmers in the region possibly for hundreds of years. Scientific collecting of these fossils dates back to the efforts of the French missionary Père David in the 1860s, and the first formally named vertebrate fossil from the Jehol Biota, *Lycoptera davidi*, has its species name dedicated to this researcher.

The first systematic palaeontological and stratigraphic studies on the Jehol fossils were conducted by Amadeus William Grabau, a German-American palaeontologist who began teaching at Peking University in 1920 and mentored the first generation of Chinese palaeontologists. In 1923, Grabau coined the name 'Jehol Series' for the fossil fish-bearing lacustrine deposits in Linyuan, Liaoning Province. In 1928, he gave the name Jehol Fauna to the fossil community from the Jehol Series.

During the 1930s and 1940s additional vertebrate (reptile) fossils were studied by Japanese palaeontologists. They reported and named the lizard *Yabeinosaurus*, the choristoderan *Monjurosuchus*, and the turtle *Manchurochelys*.

In 1962, the Chinese palaeontologist Zhiwei Gu first proposed the terms 'Jehol Group' and 'Jehol Biota'. The Jehol Biota was proposed for the characteristic '*Eosestheria* (conchostracan)–*Ephemeropsis* (insect)–*Lycoptera* (fish)' (EEL) fossil assemblage (e.g., Grabau, 1928; Gu, 1962; Chen, 1988). All three taxa are lacustrine (Fig 1). Liu *et al*. (1963) published the first monograph on the fossil fish *Lycoptera*.

Despite much geological survey work on the Mesozoic deposits alongside biostratigraphical work in the region during the 1970s and 1980s, few additional vertebrates apart from the ornithischian dinosaur *Psittacosaurus* were discovered during this period. The Jehol Biota did not earn its international reputation

Figure 1 Traditional three representative fossils (E-E-L assemblage) of the Jehol Biota. **A.** *Eosestheria*; **B.** *Ephemeropsis*; **C.** *Lycoptera*. Scale bars equal 5 mm (A) and 1 cm (B-C).

until early in the 1990s when remarkable discoveries of new fossils were made. In particular, specimens of early birds, feathered non-avian dinosaurs, mammals, pterosaurs, amphibians, flowering plants and insects had a major impact on palaeontological research (Sereno and Rao, 1992; Chen *et al.*, 1998; Sun *et al.*, 1998; Chang *et al.*, 2003; Zhou *et al.*, 2003).

The rapid accumulation of the fossils from the biota in western Liaoning and neighbouring areas also raised the question of a precise definition of the Jehol Biota. Traditionally, the Jehol Biota was defined based on the EEL fossil assemblage mainly distributed in western Liaoning. The discoveries of typical Jehol fossils from neighbouring regions such as northern Hebei and southeastern Inner Mongolia expanded the distribution of the biota beyond Liaoning Province, and this wider distribution has become increasingly accepted by many researchers (e.g., Zhou, 2006; Zhou and Wang, 2010; Zhang *et al.*, 2008). However, some workers argue for an even more inclusive palaeogeographical distribution of the biota, comprising fossil assemblages from the entire Lower Cretaceous Jehol Group (Yixian and Jiufotang formations) or equivalent strata of adjacent areas in eastern and central Asia, including northern China, the Korean Peninsula, Japan, Mongolia, Kazakhstan, and Siberia (e.g., Gu, 1962; Chen, 1988, 1999; Chang *et al.*, 2003; Zhou *et al.*, 2003; Zhou, 2006).

Recently, the Jehol Biota was defined as a large-scale tectonic-sedimentary cycle in Liaoning (Jin *et al.*, 2008). However, this definition seems to lack a definite geographic restriction as well as any biological context.

Considering the current controversy over the use of the term 'Jehol Biota', Zhou and Wang (2010) proposed a more precise concept of the Jehol Biota *sensu stricto*, with deposits distributed in northeastern China, including only northern Hebei, western Liaoning, and southeastern Inner Mongolia, and the Jehol Biota *sensu lato*, which is known from northern China, the Korean Peninsula, Japan, Mongolia, Kazakhstan, and Siberia.

Obviously, the traditional representatives of the Jehol Biota, i.e., the *Eosestheria–Ephemeropsis–Lycoptera* (EEL) assemblage can no longer appropriately represent the composition of the entire Jehol Biota. Thus, we believe that the traditional definition of the biota based primarily on the EEL assemblage in western Liaoning is outdated and should no longer be used.

The lack of a precise definition of the Jehol Biota has hindered our precise understanding of such important issues as its palaeodiversity, its spatial and temporal distribution, and the pattern of the evolutionary radiation of the Jehol Biota. Pan *et al.* (2013) recently suggested that the Jehol Biota can be defined using a palaeoecological concept, and by combining ecological and taphonomic aspects – organisms that lived in the Early Cretaceous volcanic-influenced environments of northeastern China, and were buried in lacustrine and rarely fluvial sediments, where most turned into exceptionally preserved fossils. This definition restricts the composition of the biota to fossil assemblages from the Yixian and Jiufotang formations of western Liaoning and neighbouring Inner Mongolia and Hebei, and the Huajiying Formation of northern Hebei. Thus, the distribution of the biota is confined to a comparatively small area (similar to the Jehol Biota *sensu stricto* by Zhou and Wang, 2010), ranging from the Barremian to the Aptian, for about 10Ma (Swisher *et al.*, 1999, 2002; He *et al.*, 2004, 2006, 2008; Zhou, 2006; Zhu *et al.*, 2007; Chang *et al.*, 2009). We follow this definition in the description and discussion of the Jehol Biota in this chapter.

Finally, it is also necessary to discuss briefly the relationship between the Jehol Biota and the Jurassic Yanliao (or Daohugou) Biota (see previous chapter) from approximately the same region in northeastern China. Previously, due to a lack of detailed biostratigraphic correlation, insufficient sampling of fossils and in particular the lack of data providing absolute ages of the fossil-bearing deposits, some elements of the Yanliao Biota were incorrectly referred to the Jehol Biota. However, it has now become clear that the Yanliao Biota is much older, with an age of about 160Ma (Liu *et al.*, 2012; Wang *et al.*, 2013; chapter 5), which predates the Jehol Biota by about 30Ma. Thus, there exists a significant temporal gap between the two biotas.

Geological background

During the Early Cretaceous significant global tectonic activity (Ingle and Coffin, 2004), particularly the subduction of the West Pacific Plate, most likely accounted for frequent volcanic eruptions, local crustal thinning and the breakup of the North China Craton (Wu *et al.*, 2008), which, in turn, resulted in the extensive formation of fault basins and freshwater lakes in northeastern China.

A series of basins of various sizes, with lacustrine deposits up to 3000 metres thick, are widely distributed and well exposed in western Liaoning, northern Hebei and southeastern Inner Mongolia. The fossil-bearing sediments comprising the Yixian and the overlying Jiufotang formations produce the majority of the components of the Jehol Biota, in particular feathered non-avian dinosaurs, birds, pterosaurs, mammals, insects and flowering plants. The slightly older Huajiying Formation, which is mainly distributed in northern Hebei, has yielded the oldest known elements of the biota such as the basalmost enantiornithine bird *Protopteryx*. (The Dabeigou and Dadianzi formations have also been referred to the same deposits in the region, and more work is needed to clarify the stratigraphic relations among these different stratigraphic units in northern Hebei). Deposits of these formations mainly comprise shales, mudstones and fine-grained sandstones (Figs 2–4), with interbedded ashes and volcanic rocks (andesites and basalts). The Jehol Biota probably first appears in northern Hebei in a small area with lower biodiversity and later increasingly expands its distribution and increases its biodiversity during the time of the deposition of the Yixian and Jiufotang formations.

The exceptional preservation of fossils in the Jehol Biota has generally led the Jehol Biota-bearing deposits to be regarded as an exceptional conservation fossil Lagerstätte, characterized by the preservation of soft tissue remains and fully articulated skeletons and providing an unparalleled opportunity for the reconstruction of Early Cretaceous ecosystems (Pan *et al.*, 2013).

Following Pan *et al.* (2013), the Jehol fossil-bearing deposits can be broadly divided into two types. The most common represents normal lake deposits (Figs 2–4) consisting of finely laminated sediments interbedded with volcanic ash beds. The fossils from these deposits are characterized by exceptional preservation of soft tissues, such as body outlines, impressions of skin and wing membranes as well as integumentary structures and colour patterns (e.g., Ji *et al.*, 2001,

Figure 2 One of the best known Jehol sites – the Sihetun locality (Yixian Formation) – in Beipiao, Liaoning Province.

Figure 3 Close-up photo of the Sihetun locality, showing the volcanic rock intruding into the lacustrine sediments. Yellow bands in the section represent volcanic ash beds.

Figure 4 One excavation site in Dapingfang (Jiufotang Formation) in Chaoyang, Liaoning Province. Yellow bands in the section represent volcanic ash beds.

2004; Xu and Norell, 2006; Evans and Wang, 2010; Zhang et al., 2010; Li et al., 2012; Huang et al., 2012). Vertebrates are typically completely preserved with little postmortem transportation. Mass mortality has been invoked to explain the exceptional preservation of fossils such as birds and dinosaurs that are not preserved in such quantity and quality in other localities. The exact causes for the mass mortality are unclear although it is generally considered to be related to volcanic eruptions, releasing poisonous gases as well as significant airborne particulate matter that was finally deposited in the lakes (Wang, 1991; Wang et al., 1999; Guo and Wang, 2002; Guo et al., 2003). However, some studies suggest that the mass mortality layers may be more indirectly related to volcanic activity. They propose that frequent, often severe volcanic activity adversely influenced water quality, which, in turn, caused repeated collapse of the aquatic ecosystem. Fluctuations in oxygen level related to climate probably controlled the shallow eutrophic lake system (Pan et al., 2012).

The remarkable preservation of the vertebrates in the first type of deposit provides interesting information concerning the reproductive behaviour of vertebrates. For instance, specimens of *Monjurosuchus* (Wang, Miao et al., 2005) and *Yabeinosaurus* (Wang and Evans, 2011) preserve embryos in their body cavities. There are also examples of *Sinosauropteryx* with eggs (Chen et al., 1998) and early birds with follicles in the uterus (Zheng, O'Connor et al., 2013). These deposits are also known for the embryos of a choristoderan (Ji et al., 2006), a pterosaur (Wang and Zhou, 2004), and the earliest known bird embryo (Zhou and Zhang, 2004).

Zhang and Sha (2012) carefully studied the fine sedimentary laminations in the Yixian Formation in an excavated section of the Sihetun Fossil Museum in Liaoning. They reported that the section consists mainly of mudstones and shales that are composed of siliciclastic and organic-rich laminae and layers of volcanic ash. Most of the laminations represent varves that record seasonal climatic changes. Zhang and Sha also estimated that the sedimentation rate for the majority of the mudstones and shales was 0.2 to 0.7mm/year and inferred that quiet, anoxic lacustrine bottom waters were critical for the preservation of the laminations.

The second type of Jehol fossil-bearing deposit is relatively rare and only occurs in the Yixian Formation. It consists of massive tuffaceous, pebbly sandstones (Fig 5), which yield three-dimensionally preserved, articulated vertebrate skeletons (Fig 6), rarely with soft-tissue traces (e.g., Hu et al., 2005; Evans et al., 2007; Roček et al., 2012), along with occasional isolated teeth and fragmentary postcranial remains (Xu et al., 2004), a few plant fragments, but without fossils of invertebrates or flying vertebrates (birds and pterosaurs). This type of deposit is generally believed to have resulted from volcanic ash flow or volcanically generated mudslides, clearly representing mass-mortality events (Zhou et al., 2003; Evans et al., 2007; Zhao et al., 2007). Some intriguing information on the lifestyle of various vertebrates is preserved, including resting behaviour of the troodontid dinosaur *Mei long* (Xu and Norell, 2004) and the post-nestling gregarious behaviour (Zhao et al., 2007) and possible parental care in the ornithischian dinosaur *Psittacosaurus* (Meng et al., 2004).

The preservation of stomach contents in vertebrate fossils is quite common in specimens from both types of deposit (e.g., Chen et al., 1998; Zhou and Zhang, 2002b; Zhou, 2004; Zhou et al., 2002, 2004; Hu et al., 2005; Ji S. et al., 2007; Zheng et al., 2011; O'Connor et al., 2011). For instance, a mammaliaform (*Repenomamus*) preserved remains of a juvenile dinosaur (*Psittacosaurus*) in its stomach (Hu et al., 2005), and specimens of *Sinosauropteryx* have remains of lizards and mammals preserved in their stomachs (Chen et al., 1998). *Caudipteryx* has gastroliths (Ji et al., 1998), *Jeholornis* has seeds (Zhou and Zhang, 2002a), skeletons of *Microraptor* contain remains of birds or fish (Xing et al., 2013), ornithurine birds such as *Yanornis* contain fish, and *Archaeorhynchus* are found with gastroliths in their stomachs (Zhou et al., 2004; Zhou et al., 2013). Recently, seeds have been reported in the crop of several birds (*Sapeornis*, *Hongshanornis*), suggesting a digestive system similar to that in extant birds (Zheng et al., 2011).

Jiang et al. (2011) studied the volcanic succession in the Sihetun area where the Jehol Biota has the best fossil record. They suggested that the volcanic succession in this region comprises four complexes, each with distinct

Figure 5 Section of the sediments at Lujiatun Locality (Yixian Formation), showing deposits of ash flow.

Figure 6 Aggregation of lizards from the Luojiatun locality in Beipiao, Liaoning, suggesting a mass mortality event.

products; a shield volcano, an intermediate multi-vent centre, a volcanic lake (caldera), and finally lava domes. They argued that frequent volcanic activity and the widespread existence of volcanic lakes during the Early Cretaceous could account for the exceptional preservation of invertebrates and vertebrates of the Jehol Biota in western Liaoning. Recently it was proposed that phreatomagmatic eruptions were probably responsible for major casualties and for transporting most of the terrestrial vertebrates, such as lizards, birds, non-avian dinosaurs and mammals, into the lacustrine environment for burial (Jiang *et al.*, 2014).

East Asia was almost completely isolated from the rest of Laurasia from the Middle Jurassic until the Early Cretaceous (Enkin *et al.*, 1992; Russell, 1994; Upchurch, 1995; Barrett *et al.*, 2002; Zhou *et al.*, 2003). Consequently, a number of endemic forms are known from the Jehol Biota (Chang *et al.*, 2003). East Asia has even been proposed as a 'refugium for relics' (Luo, 1999); however, this refugium hypothesis can, at best, only partially explain the observed features of the biota.

Current evidence suggests instead that the Jehol Biota was a cradle or a centre for the diversification of many groups of animals and plants (Manabe *et al.*, 2000; Zhou *et al.*, 2003; Wang, Kellner *et al.*, 2005). With the Aptian-Albian regression of the Turgai Sea resulting in the formation of a Europe-Asia 'land bridge' (Upchurch *et al.*, 2002), no significant palaeogeographical barrier existed between Asia and Europe at the time of the second and third phase of the Jehol Biota. Other palaeogeographical changes during the Early Cretaceous, such as the final formation of the Eurasian continent and the disappearance of several palaeogeographical barriers to biotic exchanges between the Jehol Biota and those from other continents are reflected by the presence of various globally distributed animals and plants in the Jehol Biota. For vertebrates such as birds and pterosaurs, dispersal to Europe and America would probably have been easier than for other, non-flying animals (Zhou, 2006; Wang *et al.*, 2012). In addition, many dinosaurian, mammalian and amphibian groups in the Jehol Biota also show cosmopolitan distributions

(Li *et al.*, 2003; Zhou *et al.*, 2003). Therefore, despite the fact that the Jehol Biota had retained some primitive forms that may have had an earlier origin in Europe, by the second and particularly the third phase of the Jehol Biota, the communities must have experienced widespread dispersal to other regions. The presence of many of the most basal representatives of various dinosaurian, mammalian, lizard, pterosaur and avian groups supports this scenario (Manabe *et al.*, 2000; Luo *et al.*, 2003; Zhou, 2004, 2006; Wang, Kellner *et al.*, 2005; Evans and Wang, 2005).

Admittedly, some fossils typical of the Jehol Biota have also been reported from other areas in northern China and neighbouring areas of East and Central Asia. For instance, the ornithischian dinosaur *Psittacosaurus* is also known from the Junggar Basin in Xinjiang, China, and from Siberia, Russia (Averianov *et al.*, 2006). A confuciusornithid bird was recently reported from North Korea (Gao *et al.*, 2009). The choristoderan *Monjurosuchus* has also been reported from Japan, and there are probably additional shared vertebrate taxa between the Jehol Biota and the Tetori Group in Japan (S.E. Evans, pers. comm.). *Lycoptera*, one of the EEL elements of the Jehol Biota, and other Jehol fishes are also known from Siberia (Sychevskaya, 2001). Such distribution of the typical Jehol elements highlights the dispersal of Jehol vertebrates throughout East and Central Asia.

Sha *et al.* (2012) summarized the non-marine and marine correlations of Early Cretaceous deposits in northeastern China, southeastern Korea, and southwestern Japan and discussed the palaeogeographical implications based on non-marine molluscan biochronology. They concluded that the Jehol Group of NE China is Hauterivian-Barremian to Aptian in age, the Sindong Group and the Hayang Group (with the exception of the Jindong Formation) in SE Korea is Aptian to Albian, and the Myogog Formation, which unconformably underlies the Sindong Group, is mainly Hauterivian. The Tetori Group and Monobegawa Group in Japan both are Hauterivian to Albian in age. The close resemblances between Hauterivian to Albian non-marine bivalves in NE China, SE Korea, and SW Japan indicate that all three areas were connected and contained a single fluvial system during the Early Cretaceous (Sha *et al.*, 2012).

The palaeoclimate of the Early Cretaceous in the Jehol region is still not well understood, although forests were flourishing. Recently, Amiot *et al.* (2011) suggested that cooler climates occurred during the existence of the Jehol Biota, with mean air temperatures of about $12\pm3°C$ in Liaoning Province, which corresponds to present-day mid-latitude cool temperate climates, based on the oxygen isotope composition of apatites from vertebrate bone. They argued that such low temperatures are in agreement with previous reports of global icehouse intervals (Steuber *et al.*, 2005), reflected by the occurrence of the temperate to cool-temperate wood taxon *Xenoxylon* in northeastern China and the absence of crocodyliforms. The integuments of various Jehol reptiles may have provided insulation and helped these animals to maintain sustained activity during cold winters when other forms hibernated. The recent discovery of the large feathered tyrannosauroid *Yutyrannus* suggests that this dinosaur was well adapted for an unusually cold environment (Xu *et al.*, 2012).

Biodiversity

The discussion of the biodiversity of the Jehol Biota is based on data of the fossil assemblages from the Yixian and Jiufotang formations in western Liaoning, southeastern Inner Mongolia and the Huajiying Formation in Hebei (Pan *et al.*, 2013).

The Jehol Biota covers a temporal range from about 131 to 120Ma, recorded in three major stratigraphic units distributed in various regions. Thus, before we discuss the general biodiversity of the biota (in particular, insects, vertebrates and plants), we present a brief introduction of the evolutionary history of the Jehol Biota.

It has been suggested that the evolutionary radiation of the biota can be broadly divided into three phases, with the first phase limited to a small area in Hebei (Huajiying Formation), the second phase (Yixian Formation) representing the peak in biodiversity, and the third phase (Jiufotang Formation) representing the widest distribution. The typical and traditional fossil assemblages of the first phase include the *Nestoria–Keratestheria* conchostrachan assemblage (Chen, 2003), the *Luanpingella–Eoparacypris–Darwinula* ostracod assemblage (Cao and Hu, 2003), the *Arguniella* bivalve assemblage, the *Lymnaea websteri* gastropod assemblage (Pan and Zhu, 2003), and the *Peipiaosteus fengningensis–Yanosteus longidorsalis* acipenseriform fish assemblage. The assemblages are characterized by the presence of relatively few taxa and the lack of some of the most typical elements of the later stage of the Jehol Biota, such as the fish *Lycoptera* and the ostracod

Cypridea. The best-known birds from this assemblage are *Protopteryx fengningensis*, the most basal enantiornithine (Zhang and Zhou, 2000), and *Eoconfuciusornis* (Zhang et al., 2008), the most basal confuciusornithid. These two taxa represent the earliest birds known from China. Another noteworthy taxon from this interval is the basal cryptobranchiid salamander, *Regalerpeton weichangensis* (Zhang et al., 2009), of which relatively large numbers of specimens are preserved.

The fossil assemblages from the second phase of the Jehol Biota include the characteristic *Eosestheria–Lycoptera–Ephemeropsis trisetalis* assemblage, the *Cypridea (Cypridea) liaoningensis–C. (Ulwellia) muriculata–Djungarica camarata* and *Cypridea (C.) veridica orguata–C. (C.) jingangshanensis–C. (C.) zaocishanensis* ostracod assemblages, the *Arguniella–Sphaerium* bivalve assemblage, the *Probaicalia vitimensis–Reesidella robusta* gastropod assemblage, and the *Aeschnidium–Manlayamia dabeigouensis* insect assemblage (Chen, 1999, 2003). This phase also includes the most important and largest radiation of the Jehol Biota. Many of these evolutionary radiations occurred during the time of the lower Yixian Formation (Jianshangou Bed) at an age of about 125Ma. In recent years, a diverse vertebrate assemblage has been recovered from the lower Yixian Formation, including many non-avian dinosaurs, birds, pterosaurs, amphibians and mammals, which will be discussed later. The Jehol flora also witnessed its greatest diversity at this time, with representatives of nearly all Mesozoic plant groups including early angiosperms (Sun et al., 1998, 2001, 2002; Friis et al., 2003; Leng, 2006; Leng and Friis, 2003, 2006).

The fossil assemblages of the third phase of the Jehol Biota are characterized by the appearance of the *Yanjiestheria* conchostrachan assemblage, the *Mengyinaia–Nakamuranai–Sphaerium* bivalve assemblage, and the *Cypridea (Cypridea) veridica veridica–C. (C.) trispinosa–C. (Yumenia) acutiuscula* and *C. (Ulwellia) koskulensis–C. (Yumenia) casta–Limmocypridea abscondida–Dijungarica* ostracod assemblages. Both bird and pterosaur assemblages are distinct from those recorded in the Yixian Formation and are characterized by the presence of more cosmopolitan taxa (Unwin et al., 2000; Wang and Zhou, 2006; Zhou et al., 2003; Zhou, 2006).

Zhang et al. (2010) have summarized the diversity of the Jehol insects. There are three distinct insect assemblages, approximately corresponding to the early, middle, and late stages of the Jehol Biota. Zhang et al. (2010) counted approximately 150 species in about 40 families and 11 orders during the early phase, followed by a great increase in the middle phase - up to about 500 species in about 100 families and 16 orders. Finally, the third phase saw a decline to about 300 species in about 80 families and 14 orders.

On the other hand, Ren (unpublished data) only recognizes 470 species, belonging to 327 genera, 144 families and 15 orders of insects in the Jehol Biota in its entirety. In addition, there is 1 indeterminate genus, 6 indeterminate families and 2 indeterminate orders (Table 1).

Zhou and Wang (2010) reviewed the vertebrate diversity of the Jehol Biota most recently. An updated list is provided here based on rapidly increasing discoveries in the past three years (Table 2). To date, the Jehol vertebrate assemblage comprises at least 145 genera and 171 species. Among them are 15 genera and 17 species of mammaliaforms and mammals, 47 genera and 53 species of birds, 32 genera and 38 species of non-avian dinosaurs, 24 genera and species of pterosaurs, 5 genera and species of squamates, 5 genera and 7 species of choristoderans, 3 genera and species of turtles, 5 genera and 8 species of amphibians, 7 genera and 14 species of fishes, and 1 genus and species of lamprey. Only a small percentage of the 144 genera can be referred to extant families. The Jehol vertebrate diversity already exceeds those of other, more or less contemporaneous conservation fossil Lagerstätten such as the Santana Formation from Brazil and the Las Hoyas Formation from Spain, and is nearly as great as that of the Upper Jurassic Solnhofen Formation from Germany (Zhou and Wang, 2010). We note that many recently proposed taxa may ultimately prove to be synonymous with previously described forms, so the list provided here represents only our current estimate of the vertebrate diversity. We provide a brief introduction to the Jehol vertebrate assemblage below.

Mesomyzon (Fig 7) is the earliest known representative of freshwater lampreys (Chang et al., 2006, 2014). The best-known fishes from the biota are acipenseriforms and basal teleosts. The former comprise basal taxa such as the short-snouted *Peipiaosteus* and the long-snouted basal polyodontids including *Protopsephurus* (Jin, 1995, 1999; Grande et al., 2002; Zhou et al., 2010b; Fig 8). Basal teleosts such as *Lycoptera* are the most common fish in the Jehol Biota and are generally referred to the Osteoglossomorpha. Some other less well-known fishes from the biota include

Figure 7 *Mesomyzon*, the earliest known freshwater lamprey, preserved in association with a specimen of *Lycoptera*.

Figure 8 *Protopsephurus*, a basal paddlefish.

basal amiids such as *Sinamia*.

Amphibians were not known from the Jehol Biota until quite recently (Wang, 2006; Wang and Evans, 2006). In addition to four species of the basal crown-anuran, *Liaobatrachus* (Dong et al., 2013; Fig 9), salamanders (Gao and Shubin, 2001; Zhang et al., 2009) are also diverse and sometimes are represented by hundreds of individuals (e.g., *Liaoxitriton* (Fig 10), *Laccotriton* and *Regalerpeton*).

Turtles from the Jehol are often abundant but are less diverse. Currently only four genera are known. Three of them, *Ordosemys*, *Manchurochelys*, and

Liaochelys, represent basal taxa of the crown-group Cryptodira (Zhou, 2010a,b). *Perochelys* is referred to the Trionychidae (Li *et al.,* 2015).

Choristoderans represent the top predators of the Jehol lake ecosystem and comprise 3 families, 5 genera and 7 species. Reaching up to 2 metres in length, *Ikechosaurus* is the largest, and most closely resembled crocodiles in both morphology and dietary habits (Liu and Wang, 2003; Liu, 2004). The long-necked

Figure 9 *Liaobatrachus*, a basal crown-group anuran.

Figure 10 *Liaoxitriton*, a primitive hynobiid salamander.

Figure 11 *Hyphalosaurus*, a choristederan. Length: 1.16 m.

Hyphalosaurus (Gao *et al.*, 1999; Fig 11) is often preserved in great abundance. Another family is represented by *Monjurosuchus* (Gao *et al.*, 2000; Gao and Fox, 2005), which has a relatively stout body and is generally believed to have had a semi-aquatic life.

Lizards are also well known from the Jehol Biota and are quite diverse. The most common taxa are *Yabeinosaurus* (Fig 12) and *Dalinghosaurus* (Evans *et al.*, 2005; Evans and Wang, 2005). *Xianglong* represents the earliest known gliding lizard (Li *et al.*, 2007).

Pterosaurs from the Jehol Biota now represent the most diverse pterosaur assemblages anywhere. They comprise at least 10 families and 24 genera and species. The Jehol pterosaurs also show a great variation in size and skull morphology. The largest ones are estimated to have a wingspan of over 5 metres whereas the smallest is the size of a sparrow (Wang, Kellner *et al.*, 2005, 2008). Practically all taxa belong to the pterodactyloids. The sole exception is a short-tailed anurognathid 'rhamphorhynchoid' (Wang *et al.*, 2002; Wang and Zhou, 2006), but this record must be considered questionable and may actually come from the older Yanliao Biota. The Jehol pterosaur assemblage already exceeds that of the famous Upper Jurassic Solnhofen Formation in both abundance and diversity of genera.

Non-avian dinosaurs from the Jehol Biota are not only known for the presence of many feathered taxa but also for their great diversity. Among the 32 genera and 38 species of dinosaurs recorded to date, 22 genera and species are theropods, representing nearly all major lineages of Cretaceous coelurosaurs. The Jehol non-avian dinosaurs show great differentiation in terms of dietary adaptations, body size and locomotion (Xu *et al.*, 2000, 2002, 2003; Xu and Norell, 2006).

The birds from the Jehol Biota now comprise 47 genera and 53 species. Most of the Jehol birds are recognized as arboreal forms. However, some ornithurine birds probably fed mainly on fishes and other aquatic animals. They demonstrate significant differentiation in morphology, size, flight capability, diet, and habitat, representing the first major radiation in the evolutionary history of birds. Among the Jehol birds, enantiornithines are most diverse, comprising 25 genera, whereas ornithurines comprise 16 genera. In addition, 6 genera belong to basal avian lineages, including jeholornithiforms, sapeornithiforms and confuciusornithiforms (Zhou and Zhang, 2006a; O'Connor *et al.*, 2009, 2013).

The Jehol mammaliaforms/mammals comprise 15 genera and 17 species. As is the case for birds and non-avian dinosaurs, they are often represented by complete skeletons, sometimes with the exceptional preservation of integumentary structures. The Jehol mammals can be referred to five major groups of Mesozoic mammaliaforms/mammals: Triconodonta, Multituberculata, Symmetrodonta, Metatheria, and Eutheria (Wang *et al.*, 2001; Ji *et al.*, 2002; Luo *et al.*, 2003; Hu *et al.*, 2005; Ji *et al.*, 2009; Meng *et al.*, 2011).

The flora of the Jehol Biota is less well studied than the fauna. Fossils of Jehol plants have mainly been recovered from the lower Yixian Formation. Wu (1999) provided an estimate of the diversity of the Jehol Flora, with 31 genera and 51 species. They include mosses, lycopsids, ferns, gymnosperms and angiosperms.

Figure 12 *Yabeinosaurus*, a lizard missing part of its tail.

Most recent attention has been paid to the angiosperms, although the flora is dominated by gymnosperms. Several genera and species of angiosperms have been reported, including *Archaeofructus*, which most likely occupied an aquatic habitat, and *Sinocarpus*, a basal eudicot (Leng and Friis, 2006). Representatives of Gymnospermae include cycadophytes, ginkgophytes, conifers, and gnetaleans. Zhou and Zheng (2003) described a new species of the genus *Ginkgo* with transitional features between the Jurassic and Cenozoic species. Recently, Pott *et al.* (2012) described two fossil seed plants, *Baikalophyllum* and *Rehezamites*, with 'cycadophyte' foliage from the Yixian Formation. The Jehol is an important source of additional information on the Gnetales, an enigmatic group of seed plants. *Siphonospermum* was reported as an Early Cretaceous relative of Gnetales (Rydin and Friis, 2010), and Yang *et al.* (2013) recently described a new ephedroid plant, *Chengia laxispicata*. Yang and Wang (2013) reported a fleshy cone of the genus *Ephedra*, indicating an even greater diversity of gnetaleans in the Jehol Flora.

Evolutionary, ecological and biological implications

In addition to the great diversity of the Jehol Biota, its remains are generally well preserved and often contain high-quality information that is generally not available in other fossil deposits. Vertebrate fossils have attracted more attention than the other fossils. Here we will focus and provide a brief discussion of the evolutionary implications of major groups of vertebrates (non-avian dinosaurs, birds, mammals, pterosaurs, lizards, choristoderans, turtles, amphibians and fishes). With regard to invertebrates, we will focus on insects, and, for plants, we will concentrate on seed plants in general and angiosperms in particular. We will also provide a brief summary of trophic interactions among various vertebrate groups (e.g., birds–fishes, pterosaurs–fishes, mammal—dinosaurs, etc.), as well as between plants and animals (e.g., co-evolution between angiosperms and pollinating insects and between herbivorous vertebrates and plants), which will help reconstruct the food webs from the Jehol Biota.

The earliest known freshwater lamprey, *Mesomyzon* (Fig 7) has a long snout, a well-developed oral sucking disc, a relatively long branchial apparatus exhibiting the branchial basket, seven gill pouches, gill arches and impressions of gill filaments (Chang *et al.*, 2006). Its discovery confirms the group's invasion into a freshwater environment by the Early Cretaceous. It also bridges the gap between the Carboniferous representatives of the group and their extant relatives. A recent study of the larva of this fish also confirms that it has a three-phased life cycle, as do its extant relatives (Chang *et al.*, 2014).

The fish assemblage of the Jehol represents the major predators of the Jehol freshwater ecosystem. *Protopsephurus* represents the most basal member of the Polyodontidae, and its snout is short compared to later, more derived taxa (Jin, 1999; Fig 8). The basal teleosts (such as *Lycoptera*), referable to the Osteoglossomorpha (Zhang *et al.*, 1994), are generally believed to be endemic to East and Central Asia, indicating palaeogeographic isolation of these regions from the rest of Laurasia from the Middle Jurassic to the early part of the Early Cretaceous. It is notable that the Jehol fishes, especially the teleosts, are the most common prey for several groups of tetrapods such as choristoderans, lizards, birds, pterosaurs and even dinosaurs (Zhou *et al.*, 2002; Dalsätt *et al.*, 2006; Zhou, Zhang and Wang, 2010; Zhou and Martin, 2011; Evans and Wang, 2012; Xing *et al.*, 2013; see below for more detailed discussion), thus playing an important role in the Jehol ecosystem (Zhou *et al.*, 2010b).

Amphibians from the Jehol are important for studying the early evolution of frogs and salamanders. *Liaobatrachus* represents a basal clade of crown-group anurans (Dong *et al.*, 2013; Fig 9), with only Leiopelmatidae being more basal. The Jehol salamanders have provided much information for understanding the current, somewhat confusing phylogenetic relationships of caudates. It has been suggested that *Liaoxitriton* (Fig 10) is a crown-group salamander, similar to extant hynobiids, based on the morphology of the vomerine teeth row and hyobranchial apparatus (Wang and Evans, 2006). Other salamanders such as *Laccotriton* and *Sinerpeton* are also believed to be similar to hynobiids (Wang and Evans, 2006).

Turtles are abundant in the Jehol Biota, although they are less diversified compared to other reptilian groups. *Manchurochelys* and *Ordosemys*, the two best-known turtles from the Jehol, have been referred to the family Sinemydidae. Zhou (2010a) described a third eucryptodiran genus, *Liaochelys*, from the Jehol Biota and recognized the Sinemydidae as a monophyletic group comprising *Sinemys*, *Manchurochelys* and *Dracochelys*. Their phylogenetic analysis places *Liaochelys* along the phylogenetic stem of Cryptodira in a position more derived than Manchurochelys and *Ordosemys*. Tong and Brinkman (2012) described a new species of *Sinemys* (Cryptodira: Sinemydidae) from the Ordos Basin in Inner Mongolia, which is not part of the Jehol Biota, but it adds to our understanding of the relationship between Jehol turtles and contemporaneous turtles from other regions. Recently, a new basal eucryptodiran turtle has been described from the Early Cretaceous of Spain, which seems to share a close relationship to those from the Jehol and the Ordos Basin, further indicating intercontinental exchanges among vertebrates at that time. The most recent discovery is *Perochelys lamadongensis*, the oldest known morphologically modern soft-shelled turtle, which has been referred to the Trionychidae (Li *et al.*, 2015).

Choristoderans are a group of reptiles that became extinct only in the Miocene. They represent the largest animals in the Jehol lake ecosystem. Gao and Fox (2005) erected a new genus, *Philydrosaurus*, in the family Monjurosuchidae, and their phylogenetic analysis placed the Monjurosuchidae outside the Neochoristodera, but in a more derived position than the Cteniogenidae. Choristoderans from the Jehol also include the long-necked *Hyphalosaurus* (Gao *et al.*, 1999; Fig 11), a close relative of *Shokawa* from the Early Cretaceous of Japan (Evans and Manabe, 1999). There is evidence that *Hyphalosaurus* was oviparous (Wang and Evans, 2011), whereas *Monjurosuchus* was viviparous, as one female individual preserved seven embryos in the body cavity. (We reject the original explanation of cannibalism by Wang, Miao *et al*. [2005].)

Lizards from the Jehol Biota are often very well preserved, sometimes complete with skin impressions. The gliding lizard *Xianglong* represents the only known fossil lizard with a number of morphological features associated with gliding, including a patagium supported by the dorsal ribs (Li *et al.*, 2007). It has been suggested that the skull of *Dalinghosaurus* resembles those of the extant *Xenosaurus* and *Shinisaurus* as well as *Carusia* from the Late Cretaceous of Mongolia and China (Evans and Wang, 2005). Phylogenetic analysis also provides some support for this relationship. *Dalinghosaurus* was regarded as a climber on the basis of its delicate long pes and slender claws (Evans and Wang, 2005). Furthermore, evidence for aggregation of multiple individuals was reported from the ash deposits of the Yixian Formation (Evans *et al.*, 2007). Wang and Evans (2011) recently reported on a gravid lizard of the genus *Yabeinosaurus* bearing more than 15 embryos. The embryos exhibited a level of skeletal development corresponding to that of late embryos in extant viviparous lizards, thereby documenting the first occurrence of viviparity in a fossil reptile that was essentially land-dwelling. This record extends the first occurrence of viviparity in squamates back in time by at least 30Ma.

Yabeinosaurus was one of the first tetrapods described from the Jehol Biota (Fig 12). It was long misidentified as a gekkotan based on juvenile specimens, but is now recognized as a large stem-scleroglossan with an extended period of skeletal maturation. Newly discovered fossils also preserve gut contents, showing that large individuals took vertebrate prey, including fish (Evans and Wang, 2012). Their phylogenetic analysis placed *Yabeinosaurus* on the stem of Scleroglossa, as the sister-taxon of *Sakurasaurus* from the Early Cretaceous of Japan.

Pterosaurs from the Jehol Biota include Anhangueridae, Ctenochasmatidae, Ornithocheiridae, Gallodactylidae, Pteranodontidae, Tapejaridae, and Istiodactylidae (Wang, Kellner *et al.*, 2005; Wang and Zhou, 2006; Wang, Kellner *et al.*, 2007, 2008; Wang, de A. Campos *et al.*, 2008; Jiang *et al.*, 2011; Wang *et al.*, 2012). Many of these pterosaurs are piscivorous. For instance, *Liaoningopterus*, which has an estimated wing span of three to five metres, has the proportionately largest teeth known for any pterosaur, but the teeth are restricted to the anterior half of the upper and lower jaws. *Guidraco* has anterior teeth that are considerably larger than the posterior teeth (Fig 13). In addition, some toothless pterosaurs such as *Chaoyangopterus* are also regarded as piscivorous. Other pterosaurs probably fed on insects or plants. The small size of the tapejarid *Sinopterus* (Fig 14) enabled it to move easily through the dense forest, and its strong claws were an adaptation for climbing. Several pterosaurs were adapted for filter feeding; examples are *Gegepterus* and *Pterofiltrus* (Ctenochasmatidae) (Jiang and Wang, 2011). The skull of *Pterofiltrus* is greatly elongated, with an estimated length of 208 millimetres, and its dentition consists of exceptionally elongate pointed teeth, partly directed sideways.

The Jehol pterosaurs have implications for palaeobiogeography. Many appear to be closely related to cosmopolitan groups reported in South America and

Figure 13 *Guidraco*, piscivorous pterosaur.

Figure 14 *Sinopterus*, a toothless, possibly seed-eating pterosaur. Wing span: c.1.2 m.

Europe, suggesting that there were few palaeogeographical barriers to their distribution. This is also indicative of strong flight capability in these animals. Furthermore, phylogenetic studies strongly suggest that the origin and early diversification centre for Cretaceous pterosaur radiation was in northeastern China (Wang, Kellner *et al.*, 2005; Wang and Zhou, 2006; Wang *et al.*, 2012).

The discovery of the first pterosaurian egg (Fig 15) demonstrated that pterosaurs had a precocial mode of development, as in dinosaurs and early birds (Wang and Zhou, 2004) and that their eggs were most likely soft-shelled (Ji *et al.*, 2004).

The Jehol pterosaurs included the largest flying animals of their time and likely competed with birds for food. Indeed, it has been suggested that the competition

Figure 15 Fossil embryo of a pterodactyloid pterosaur. Scale bar equals 1 cm.

between these two equally diversified groups led to the dispersal of some pterosaur groups elsewhere (Wang, Kellner *et al.*, 2005). Although the Jehol pterosaurs probably fed mainly on fishes, the co-evolution between plants and pterosaurs in the Jehol Biota might also have played a significant role in the pterosaurian radiation.

The highlights among the vertebrates in the Jehol Biota are the feathered non-avian dinosaurs and certain basal birds that represent key stages in the evolutionary transition from non-avian theropod dinosaurs (e.g., *Sinosauropteryx*; Fig 16) to birds. The discovery of more than a dozen feathered dinosaurs has provided solid evidence in support of the presence of feathers at various evolutionary stages in different lineages of dinosaurs, including many theropods, as well as some ornithischian dinosaurs (Zheng

Figure 16 *Sinosauropteryx*, the first known feathered non-avian dinosaur.

et al., 2011) This implies that protofeathers probably appeared very early in dinosaurian evolution, or even outside Dinosauria. Many of the feathered non-avian theropods from the Jehol Biota are regarded as the closest relatives of birds: dromaeosaurids, oviraptorids, and troodontids (Ji *et al.*, 1998, 2001; Xu and Norell, 2006; Xu *et al.*, 1999, 2000, 2002, 2003, 2004, 2009).

Like other vertebrates, the Jehol dinosaurs are often preserved as complete skeletons and thus provide a significant amount of data for assessing their phylogenetic relationships. The study of the Jehol feathered theropods has provided some of the most important evidence for the dinosaurian origin of birds. The discovery of feathers on the legs of the dromaeosaurid *Microraptor* (Fig 17), as well as arboreal adaptations inferred from the skeletal characteristics, provides strong evidence in support of the arboreal hypothesis for the origin of bird flight and suggests that the ancestors of birds passed through a four-winged stage in the early development of flight (Xu *et al.*, 2003; Zheng, Zhou *et al.*, 2013).

Many Jehol dinosaurs also preserve stomach contents providing direct evidence of their diet. For instance, skeletons of several theropods, such as the oviraptorosaurian *Caudipteryx* (Zhou and Wang, 2000), the ornithomimosaurian *Shenzhousaurus* (Ji *et al.*, 2003), and the ornithischian *Psittacosaurus* frequently contain gastroliths, indicative of a herbivorous diet. The oviraptorosaurian *Incisivosaurus* has a pair of premaxillary teeth resembling rodent incisors and small, lanceolate 'cheek' teeth with large wear facets, suggesting a specialized herbivorous adaptation previously unknown in theropods (Xu *et al.*, 2002). The best-known carnivorous dinosaur from the Jehol Biota is probably the feathered *Sinosauropteryx*, which is usually found with mammal or lizard remains preserved in its stomach (Chen *et al.*, 1998; pers. obs.). Some skeletons of *Microraptor* preserve remains of birds in their stomach (O'Connor *et al.*, 2011). A recent study reported evidence of *Microraptor* also preying on fishes (Xing *et al.*, 2013).

Although most of the Jehol dinosaurs are small-sized, there were also some large-sized forms. Titanosaurs have been discovered in the Jehol Biota (Barrett *et al.*, 2007; Barrett and Wang, 2007), and the iguanodontid *Jinzhousaurus* is another large-sized dinosaur (Wang and Xu, 2001), with the largest known individual attaining a length of about seven metres. Tyrannosauroids include the small *Dilong* (Xu *et al.*, 2004), but *Yutyrannus* currently constitutes the largest known feathered non-avian dinosaur (Xu *et al.*, 2012). This, in turn, suggests relatively cold temperatures, which is consistent with some geochemical evidence. Fowler *et al.* (2011) argued that the small tyrannosaurid *Raptorex* actually represents a juvenile *Tarbosaurus* and is probably not even part of the Jehol Biota.

Figure 17 Microraptor gui, a four-winged dromaeosaur dinosaur. Total length of skeleton: 77 cm.

The exceptional preservation of animals from the Jehol Biota also provided the first evidence of melanosomes (colour-bearing organelles) for the colour of dinosaurian feathers. The discovery of both eumelanosomes and phaeomelanosomes in *Sinosauropteryx* not only confirmed the homology of the integumentary covering of this dinosaur to the feathers of extant birds, but also provided evidence for reconstructing the colours and colour patterning of this dinosaur (Fig 18). For example, the dark-coloured stripes on the tail of *Sinosauropteryx* can reasonably be inferred to have exhibited chestnut to reddish-brown tones (Zhang *et al.*, 2010). *Microraptor* is thought to have possessed iridescent feather colours based on distinctly narrow melanosomes that are comparable of those of extant birds with iridescent feathers. The reconstruction of the feather distribution in *Microraptor* is consistent with an ornamental function of the tail. This, in turn, suggests that the early evolution of plumage and their colours primarily served a display function (Li *et al.*, 2012).

Birds from the Jehol Biota document the greatest known radiation in the Mesozoic evolutionary history of the group. They provide key information for understanding early avian evolution, showing that the Jehol represents an important area for the origin and early evolution of the major groups of early birds. The biota includes not only the most basal enantiornithines (e.g., *Protopteryx*), and ornithurines, but also many stem-taxa that are evolutionarily placed between *Archaeopteryx* and the dominant Mesozoic groups Enantiornithes and Ornithurae. For example, *Jeholornis* is considered the most basal bird aside from *Archaeopteryx* and possesses a long tail with as many as 27 vertebrae. *Eoconfuciusornis* represents the earliest known bird with a horny beak (Zhang *et al.*, 2008). The recent report of a hindwing in *Sapeornis* further supports the four-winged stage in the evolution of bird flight (Zheng, Zhou *et al.*, 2013). The loss of teeth in early birds is thought to have occurred multiple times (Hou *et al.*, 1995; Zhou and Zhang, 2005, 2006b; Zhang *et al.*, 2008; Zhou *et al.*, 2010a).

With only a few exceptions (represented mostly by ornithurine birds), the Jehol birds are arboreal forms. Stem-avians such as *Jeholornis*, *Sapeornis* and *Confuciusornis* (Fig 19) probably only had limited flight capability whereas ornithurines, such as *Yixianornis*, already possessed a powerful flight, as indicated by an elongated keeled sternum and the derived structure of the pectoral girdle and wing. Enantiornithines are generally considered to lie between stem-avians and ornithurines in terms of flight capability. The appearance of the alula in enantiornithines, including the basal *Protopteryx* (Fig 20), suggests they possessed the ability for slow flight during takeoff and landing.

Figure 18 Phaeomelanosomes, fossil evidence for colour of the feathers of *Sinosauropteryx*.

Figure 19 A pair of *Confuciusornis*, one of the earliest beaked birds.

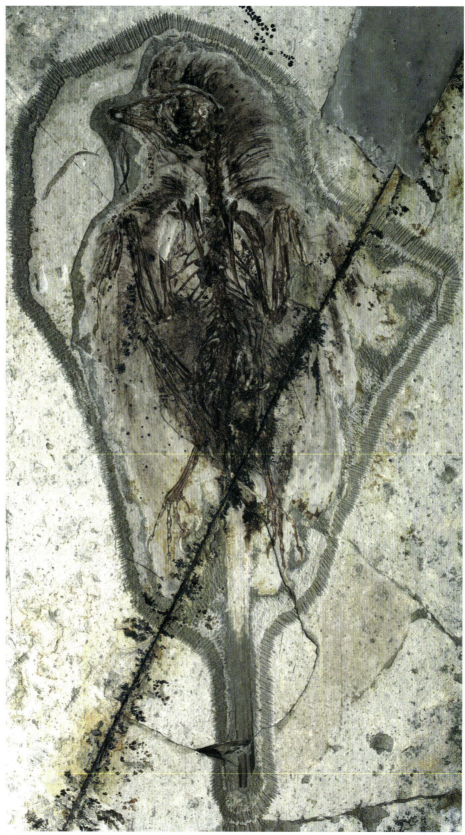

Figure 20 *Protopteryx*, a basal enantiornithine bird.

Jehol birds demonstrate a great variety of dietary adaptations. Again, their exceptional preservation provides several direct examples of their diet. For instance, *Jeholornis* is a specialized seed-eater. *Sapeornis* (Zhou and Zhang, 2002b; Fig 21) and *Hongshanornis* had a crop and a digestive system similar to that in extant birds (Zheng *et al.*, 2011). The ornithurine *Yanornis* was undoubtedly a piscivore, with many specimens containing fish remains in their stomachs or crops. Although there is, as yet, no direct evidence, we believe that many enantiornithines were insectivores whereas some forms such as *Longirostravis* with a long, slender rostrum and mandible and a small number of anterior teeth possibly employed a probing feeding behaviour (Hou *et al.*, 2004).

The Jehol birds also show a great degree of size differentiation. For instance, *Jeholornis* and *Sapeornis* are usually large (Zhou and Zhang, 2002a,b, 2003), exceeding the size of some of the smaller non-avian dinosaurs. They probably occupied a more open forest habitat, whereas small-sized enantiornithines might have lived in denser forests. Ornithurine birds living near the shore were larger than enantiornithines and adapted for a piscivorous habit (Zhou and Zhang, 2001; Zhou *et al.*, 2002; Zhou, 2006).

The most incredible bird fossils known from the Jehol Biota are a complete embryo (Fig 22), preserving an articulated skeleton and impression of feathers and indicating precocial or super-precocial development (Zhou and Zhang, 2004) and several specimens

Figure 21 *Sapeornis*, a basal bird with preservation of seeds in the crop.

Figure 22 Fossil embryo of an enantiornithine bird.

(*Jeholornis* and enantiornithines) that preserve follicles in their ovaries, showing that early birds had a single functional ovary and oviduct, much as in extant birds but unlike non-avian dinosaurs and extant reptiles (Zheng, O'Connor *et al.*, 2013). Furthermore, the discovery of many juveniles provides evidence for assessing the development of many flight-related characters such as the sternum and wing bones (Zheng *et al.*, 2012; Zhou *et al.*, 2013).

The Jehol mammaliaforms and mammals illustrate the diversification of mammals during the Early Cretaceous, which includes representatives of many of the major lineages such as Symmetrodonta, Triconodonta, Multituberculata, Eutheria and Metatheria (Hu *et al.*, 1997; Luo *et al.*, 2003; Hu *et al.*, 2005; Ji *et al.*, 2009; Meng *et al.*, 2011).

The discovery of the first ossified Meckel's cartilage in mammals provides support for the hypothesis of a single origin for the definitive mammalian middle ear (Wang *et al.*, 2001). The transfer of postdentary bones from the mandible to the cranium in mammals to serve as auditory ossicles is a classic example of an evolutionary transformation in vertebrates (Meng *et al.*, 2011), and the Jehol mammals have provided much critical fossil evidence. Based on an exceptionally preserved skeleton, Meng *et al.* (2011) reported the first unambiguous ectotympanic (angular), malleus (articular and prearticular) and incus (quadrate) in an Early Cretaceous eutriconodont (Fig 23). They determined that the ectotympanic and malleus lost their direct contact with the dentary but still remained connected to the ossified Meckel's cartilage.

Although we know that mammals were the prey of many dinosaurs (e.g., *Sinosauropteryx*), the discovery of one of the largest known mammals from the Mesozoic has great palaeoecological implications (Fig 24). One specimen of *Repenomamus* even contained remains of a juvenile dinosaur (*Psittacosaurus*), thus providing the first direct evidence of early mammals preying on dinosaurs (Hu *et al.*, 2005).

Insects represent some of the most important discoveries among the Jehol invertebrates. In addition to their remarkable diversity, exceptional preservation and implication for evolution of many insect lineages, they also provide some important evidence for the co-evolution between insects and plants (Figs 25–28). Ren (1998) reported on the discovery of pollinating orthorrhaphous brachycerans

Figure 23 *Liaoconodon*, an early mammal.

Figure 24 *Repenomamus*, one of the largest mammals in the Mesozoic. Scale bar: 3 cm.

Figure 26 *Vitimopsyche*, a pollinating scorpionfly with siphonate proboscis.

Figure 27 A male individual of *Saurophthirus exquisitus*, a blood-feeding flea. Body length up to 1.2 cm.

Figure 25 *Protonemestrius*, a pollinating brachyceran fly.

Figure 28 *Sophogramma*, a butterfly-like kalligrammatid lacewing.

(short-horned flies), suggesting that pollinating insects may have played a decisive role in the origin and early evolution of the angiosperms. By comparison with present-day related taxa, they also demonstrate that the orthorrhaphous brachycerans were some of the most ancient pollinators.

Recently Huang *et al.* (2012, 2013) and Gao *et al.* (2012, 2013) reported on the discovery of the largest definitive Mesozoic fleas from the Jehol Biota and the Jurassic Yanliao Biota respectively, highlighting the diversity of Early Cretaceous ectoparasitic insects (Fig 27). Despite the retention of primitive traits such as non-jumping hind legs, they all have stout and elongate sucking siphons for piercing the skins of their hosts. The specimens also indicate that fleas may be rooted among the pollinating 'long siphonate' scorpionflies of the Mesozoic. They further suggested that their hosts were probably feathered tetrapods.

A review by Zhang *et al.* (2010) divided the insect assemblages into four communities based on habitats, or five groups based on feeding habits. Of the four communities, the highest species diversity occurred in the forest community, followed by the aquatic, soil, and alpine communities. Of the five feeding groups, the highest species diversity appeared in the phytophagous group, followed by the carnivorous, parasitic, saprophagous, and heterophagous groups. More importantly, the insects played a key role in the whole Jehol ecosystem as they constituted prey for nearly all major groups of vertebrates: fishes, amphibians, choristoderans, turtles, lizards, birds, and pterosaurs. They may also have provided food for some dinosaurs and mammals.

The Jehol Flora is best known for its preservation of several early angiosperms, solving the long-standing 'abominable mystery' – the sudden appearance of angiosperms during the Cretaceous – that had concerned Charles Darwin. *Archaeofructus* represents a stem-group flowering plant (Sun *et al.*, 2001; Zhou *et al.*, 2003). Its 'flowers' contain male and female reproductive organs on the same shoot. It lacked petals, sepals and other organs associated with stamens and carpels (Fig 29) and is regarded as an aquatic plant (Leng and Friis, 2003). *Sinocarpus* (Fig 30) was described as a basal eudicot closest to Ranunculales. Its reproductive organs are associated with typical angiosperm dicot leaves, which have distinct petioles and laminae (Leng and Friis, 2006).

Figure 29 *Archaeofructus*, a stem angiosperm.

Figure 30 *Sinocarpus*, a stem eudicot angiosperm.

Ephedra and ephedroid fossils are relatively abundant in the Jehol Flora and provide important information concerning the early evolution of this group (Fig 31). Yang and Wang (2013) recently reported on the earliest fleshy cone of *Ephedra* from the Jehol. The ephedroid *Chengia* (Yang *et al.*, 2013) represents part of a leafy shooting system with reproductive organs attached, providing a 'missing link' between archetypal fertile organs in the crown-lineage of the Gnetales and compound female cones of the extant Ephedraceae. Yang *et al.* (2013) demonstrated that the Ephedraceae including *Ephedra* had attained considerable diversity by the Early Cretaceous. The diversity of well-preserved macrofossils of *Ephedra* and ephedroids from the Jehol Biota and adjacent areas illustrates different evolutionary stages, including primitive and derived characters of Ephedraceae. Northeastern China and adjacent areas may represent either the or one of the centres for the early diversification of the family (Yang *et al.*, 2013).

Summary

The Jehol Biota represents an exceptional window into Early Cretaceous continental ecosystems. A remarkable array of fossils has been discovered, enabling us to gain a good estimate of its biodiversity as well its implications for understanding the evolution of many lineages of vertebrates, insects and plants. However, much more information has yet to come to light, since the majority of fossils have only recently been found, and for the most part documented only in a preliminary fashion. Although a chronostratigraphic framework for the evolution of the biota is now in place, and comparisons have been made with other more or less contemporaneous biotas in neighbouring areas and on other continents, much remains unknown regarding the palaeoenvironments. This will require multidisciplinary collaborations by palaeontologists, geochemists and geologists.

Acknowledgements

We would like to thank Susan Evans and Makoto Manabe for providing useful information. We are indebted to Xiaolin Wang, Fan Jin and Yuanqing Wang for providing photos. The research was supported by the National Natural Science Foundation of China (91514302 and 41688103) and the Chinese Academy of Sciences (XDB18000000).

Figure 31 Fertile shoot of the gnetalean *Ephedrites*.

References

Amiot, R., X. Wang, Z. Zhou, X. Wang, E. Buffetaut, C. Lécuyer, Z. Ding, F. Fluteau, T. Hibino, N. Kusuhashi, J. Mo, V. Suteethorn, Y. Wang, X. Xu, and F. Zhang. 2011. Oxygen isotopes of East Asian dinosaurs reveal exceptionally cold Early Cretaceous climates. Proceedings of the National Academy of Sciences of the United States of America 108:5179–5183.

Averianov, A. O., A. V. Voronkevich, S. V. Leshchinskiy, and A. V. Fayngertz. 2006. A ceratopsian dinosaur *Psittacosaurus sibiricus* from the Early Cretaceous of West Siberia, Russia and its phylogenetic relationships. Journal of Systematic Palaeontology 4:359–395.

Barrett, P. M., and X. Wang. 2007. Basal titanosauriform (Dinosauria, Sauropoda) teeth from the Lower Cretaceous

Yixian Formation of Liaoning Province, China. Palaeoworld 16:265–271.

Barrett, P. M., Y. Hasegawa, M. Manabe, S. Isaji, and H. Matsuoka. 2002. Sauropod dinosaurs from the Lower Cretaceous of Eastern Asia: taxonomic and biogeographical implications. Palaeontology 45:1197–1217.

Cao, M.-Z., and Y.-X. Hu. 2003. Ostracods. Pp. 48–52 in M.-M. Chang, P.-J. Chen, Y.-Q. Wang, and Y. Wang (eds), The Jehol Biota: The Emergence of Feathered Dinosaurs, Beaked Birds and Flowering Plants. Shanghai Scientific & Technical Publishers, Shanghai.

Chang, M., D. Miao, and J. Zhang. 2006. A lamprey from the Cretaceous Jehol biota of China. Nature 441:972–974.

Chang, M.-M., P.-J. Chen, Y.-Q. Wang, and Y. Wang (eds). 2003. The Jehol Biota: The Emergence of Feathered Dinosaurs, Beaked Birds and Flowering Plants. Shanghai: Shanghai Scientific & Technical Publishers, 208 pp.

Chang, M.-M., F. Wu, D. Miao, and J. Zhang. 2014. Discovery of fossil lamprey larva from the Lower Cretaceous reveals its three-phased life cycle. Proceedings of the National Academy of Sciences of the United States of America 111:15486–15490.

Chang, S.-C., H. Zhang, P. R. Renne, and Y. Fang. 2009. High-precision $^{40}Ar/^{39}Ar$ age for the Jehol Biota. Palaeogeography, Palaeoclimatology, Palaeoecology 280:94–104.

Chen, J.-H. 1999. [A study of nonmarine bivalve assemblage succession from the Jehol Group (U. Jurassic and L. Cretaceous).] Pp. 92–113 in P.-J. Chen and F. Jin (eds), The Jehol Biota. Palaeoworld 11. [Chinese]

Chen, P.-J. 1988. [Distribution and migration of the Jehol fauna with reference to non-marine Jurassic-Cretaceous boundary in China.] Acta Palaeontologica Sinica 27:659–683. [Chinese]

Chen, P.-J. 1999. Fossil conchostracans from the Yixian Formation of western Liaoning, China. Pp. 114–130 in P.-J. Chen and F. Jin (eds), The Jehol Biota. Palaeoworld 11. [Chinese]

Chen, P.-J. 2003. Conchostracans. Pp. 44–47 in M.-M. Chang, P.-J. Chen, Y.-Q. Wang and Y. Wang (eds), The Jehol Biota: The Emergence of Feathered Dinosaurs, Beaked Birds and Flowering Plants. Shanghai Scientific & Technical Publishers: Shanghai.

Chen, P.-J., Z.-M. Dong, and S.-N. Zhen. 1998. An exceptionally well-preserved theropod dinosaur from the Yixian Formation of China. Nature 391:147–152.

Dalsätt, J., Z. Zhou, F. Zhang, and P. G. P. Ericson. 2006. Food remains in *Confuciusornis sanctus* suggest a fish diet. Naturwissenschaften 93:444–446.

Dong, L., Y. Wang, Z. Roček, and M. E. H. Jones. 2013. Anurans from the Lower Cretaceous Jehol Group of western Liaoning, China. PLoS ONE 8(7):e69723. doi:10.1371/journal.pone.0069723

Enkin, R. J., Z. Yang, Y. Chen, and V. Courtillot. 1992. Paleomagnetic constraints on the geodynamic history of the major blocks of China from the Permian to the Present. Journal of Geophysical Research 97(B10): 13953–13989.

Evans, S. E, and M. Manabe. 1999. A choristoderan reptile from the Lower Cretaceous of Japan. *Special Papers in Palaeontology* 60:101–119.

Evans, S. E., and Y. Wang. 2005. The Early Cretaceous lizard *Dalinghosaurus* from China. Acta Palaeontologica Polonica 50:725–742.

Evans, S. E., and Y. Wang. 2010. A new lizard (Reptilia: Squamata) with exquisite preservation of soft tissue from the Lower Cretaceous of Inner Mongolia, China. Journal of Systematic Palaeontology 8:81–95.

Evans S. E., and Y. Wang. 2012. New material of the Early Cretaceous lizard *Yabeinosaurus* from China. Cretaceous Research 34:48–60.

Evans, S. E., Y. Wang, and M. E. H. Jones. 2007. An aggregation of lizard skeletons from the Lower Cretaceous of China. Senckenbergiana lethaea 87:109–118.

Evans, S. E., Y. Wang, and C. Li. 2005. The Early Cretaceous lizard genus *Yabeinosaurus* from China: resolving an enigma. Journal of Systematic Palaeontology 3:319–335.

Fowler, D. W., H. N. Woodward, E. A. Freedman, P. L. Larson, and J. R. Horner. 2011. Reanalysis of "*Raptorex kriegsteini*": a juvenile tyrannosaurid from Mongolia. PLoS ONE 6(6):e21376. doi:10.1371/journal.pone.0021376

Friis, E. M., J. A. Doyle, P. K. Endress, and Q. Leng. 2003. *Archaefructus* – angiosperm precursor or specialized early angiosperm? Trends in Plant Science 8: 369–373.

Gao, K.-Q., and R. C. Fox. 2005. A new choristodere (Reptilia: Diapsida) from the Lower Cretaceous of western Liaoning Province, China, and phylogenetic relationships of Monjurosuchidae. Zoological Journal of the Linnean Society 145:427–444.

Gao, K.-Q., and N. H. Shubin. 2001. Late Jurassic salamanders from northern China. Nature 410:574–577.

Gao, K.-Q., Z.-L. Tang, and X.-L. Wang. 1999. A long-necked diapsid reptile from the Upper Jurassic/Lower Cretaceous of Liaoning Province, northeastern China. Vertebrata PalAsiatica 37:1–8.

Gao, K., S. E. Evans, Q. Ji, M. A. Norell, and S. Ji. 2000. Exceptional fossil material of a semi-aquatic reptile from China: the resolution of an enigma. Journal of Vertebrate Paleontology 20:417–421.

Gao, K.-Q., Q. Li, M. Wei, H. Pak, and I. Pak. 2009. Early Cretaceous birds and pterosaurs from the Sinuijui Series, and geographic extension of the Jehol Biota into the Korean Peninsula. Journal of the Paleontological Society of Korea 25: 57–61.

Gao, T., C. Shih, X. Xu, S. Wang, and D. Ren. 2012. Mid-Mesozoic flea-like ectoparasites of feathered or haired vertebrates. Current Biology 22:732–735.

Gao, T., C. Shih, A. Rasnitsyn, X. Xu, S. Wang and D. Ren. 2013. New transitional fleas from China highlighting diversity of Early Cretaceous ectoparasitic insects. Current Biology 23:1–6.

Grabau, A. W. 1923. Cretaceous Mollusca from north China. Bulletin of the Geological Survey of China 5:183–198.

Grabau A. W. 1928. Stratigraphy of China. Part 2. Mesozoic. Geological Survey of China, Beijing, 774 pp.

Grande, L., F. Jin, Y. Yabumoto, and W. E. Bemis. 2002. *Protopsephurus liui*, a well-preserved primitive paddlefish (Acipenseriformes: Polyodontidae) from the Lower Cretaceous of China. Journal of Vertebrate Paleontology 22:209–237.

Gu, Z.-W. 1962. [Jurassic and Cretaceous of China.] Science Press, Beijing. 84 pp. [Chinese]

Guo, Z.-G., and X. Wang. 2002. [A study on the relationship between volcanic activities and mass mortalities of the Jehol vertebrate fauna from Sihetun, western Liaoning, China.] Acta Petrologica Sinica 18:117–125. [Chinese]

Guo, Z., J. Liu, and X. Wang. 2003. Effect of Mesozoic volcanic eruptions in the western Liaoning Province, China on paleoclimate and paleoenvironment. Science in China 46:1261–1272.

He, H., Y. Pan, L. Tauxe, H. Qin, and R. Zhu. 2008. Toward age determination of the M0r (Barremian–Aptian boundary) of the Early Cretaceous. Physics of the Earth and Planetary Interiors 169:41–48.

He, H. Y., X. L. Wang, Z. H. Zhou, F. Wang, A. Boven, G. H. Shi, and R. X. Zhu. 2004. Timing of the Jiufotang Formation (Jehol Group) in Liaoning, northeastern China and its implications. Geophysical Research Letters 31:L12605. doi:10.1029/2004GL019790

He, H. Y., X. L. Wang, F. Jin, Z. H. Zhou, F. Wang, L. K. Yang, X. Ding, A. Boven, and R. X. Zhu. 2006. The $^{40}Ar/^{39}Ar$ dating of the early Jehol Biota from Fengning, Hebei Province, northern China. Geochemistry, Geophysics, Geosystems 7:Q04001. doi:10.1029/2005SGC001083

Hou, L., L. M. Chiappe, F. Zhang, and C. Chuong. 2004. New Early Cretaceous fossil from China documents a novel trophic specialization for Mesozoic birds. Naturwissenschaften 91:22–25.

Hou, L.-H., Z. Zhou, L. D. Martin, and A. Feduccia. 1995. A beaked bird from the Jurassic of China. Nature 377:616–618.

Hu, Y., J. Meng, Y. Wang and C. Li. 2005. Large Mesozoic mammals fed on young dinosaurs. Nature 433:149–152.

Hu, Y, Y..Wang, Z. Luo, and C. Li. 1997. A new symmetrodont mammal from China and its implications for mammalian evolution. Nature 390:137–142.

Huang, D., M. S. Engel, C. Cai, H. Wu, and A. Nel. 2012. Diverse transitional giant fleas from the Mesozoic era of China. Nature 483:201–204.

Huang, D., M. S. Engel, C. Cai, H. Wu, and A. Nel. 2013. Mesozoic giant fleas from northeastern China (Siphonaptera): taxonomy and implications for palaeodiversity. Chinese Science Bulletin 58:1682–1690.

Ingle, S., and M. F. Coffin. 2004. Impact origin for the greater Ontong Java Plateau? Earth and Planetary Science Letters 218:123–134.

Ji, Q., P. J. Currie, M. A. Norell, and S.-A. Ji. 1998. Two feathered dinosaurs from northeastern China. Nature 393:753–761.

Ji, Q., S. Ji, J. Lu, H. You, and C. Yuan. 2006. Embryo of Early Cretaceous choristodere (Reptilia) from the Jehol Biota in western Liaoning, China. Journal of the Paleontological Society of Korea 22:111–118.

Ji, Q., Z.-X. Luo, C.-Y. Yuan, J. R. Wible, J.-P. Zhang, and J.A. Georgi. 2002. The earliest known eutherian mammal. Nature 416:816–822.

Ji, Q., M. A. Norell, K.-Q. Gao, S.-A. Ji, and D. Ren. 2001. The distribution of integumentary structures in a feathered dinosaur. Nature 410:1084–1088.

Ji, Q., M. A. Norell, P. J. Makovicky, K.-Q. Gao, S. Ji, and C. Yuan. 2003. An early ostrich dinosaur and implications for ornithomimosaur phylogeny. American Museum Novitates 3420:1–19.

Ji, Q., S.-A. Ji, Y.-N. Cheng, H.-L. You, J.-C. Lü, Y.-Q. Liu, and C.-X. Yuan. 2004. Pterosaur egg with a leathery shell. Nature 432:572.

Ji, S., Q. Ji, J. Lü, and C. Yuan. 2007. A new giant compsognathid dinosaur with long filamentous integuments from Lower Cretaceous of northeastern China. Acta Geologica Sinica 81:8–15.

Jiang, S., and X. Wang. 2011. A new ctenochasmatid pterosaur from the Lower Cretaceous, western Liaoning, China. Anais da Academia Brasileira de Ciências 83:1243–1249.

Jiang, B., F. T. Fürsich, B. Wang, and Y. Niu. 2011. Early Cretaceous volcanism and its impact on fossil preservation in western Liaoning, NE China. Palaeogeography, Palaeoclimatology, Palaeoecology 302:255–269.

Jiang, B., Harlow, G. E., Wohletz, K., Zhou, Z. and J. Meng. 2014. New evidence suggests pyroclastic flows are responsible for the remarkable preservation of the Jehol Biota. Nature Communications 5, 3151, doi: 10.1038/ncomms4151.

Jin, F. 1995. [Late Mesozoic fish fauna from Western Liaoning, China.] Vertebrata PalAsiatica 33:169–193. [Chinese]

Jin, F. 1999. [Middle and late Mesozoic acipenseriforms from northern Hebei and western Liaoning, China.] Palaeoworld 11:188–280. [Chinese]

Jin, F., F. Zhang, Z. Li, J. Zhang, C. Li, and Z. Zhou. 2008. On the horizon of *Protopteryx* and the early vertebrate fossil assemblages of the Jehol Biota. Chinese Science Bulletin 53:2820–2827.

Larson, R. L., and E. Erba. 1999. Onset of the mid-Cretaceous greenhouse in the Barremian–Aptian: igneous events and the biological, sedimentary, and geochemical responses. Palaeoceanography 14:663–678.

Leng, Q. 2006. [Palaeobotanical data — the direct and decisive evidence for solving the 'Abominable Mystery' of angiosperm origin.] Pp. 593–609 and 917–921 in J.-Y. Rong, Z.-J. Fang, Z.-H. Zhou, R.-B. Zhan, X.-D. Wang, and X.-L. Yuan (eds), [Originations, Radiations and Biodiversity Changes – Evidences from the Chinese Fossil Record.] Science Press, Beijing. [Chinese]

Leng, Q., and E. M. Friis. 2003. *Sinocarpus decussatus* gen. et sp. nov., a new angiosperm with basally syncarpous fruits from the Yixian Formation of Northeast China. Plant Systematics and Evolution 241:77–88.

Leng, Q., and E. M. Friis. 2006. Angiosperm leaves associated with *Sinocarpus* infructescences from the Yixian Formation (mid-Early Cretaceous) of NE China. Plant Systematics and Evolution 262:173–187.

Li, C., Y. Wang, Y. Hu, and J. Meng. 2003. A new species of *Gobiconodon* (Triconodonta, Mammalia) and its implication for the age of Jehol Biota. Chinese Science Bulletin 48:1129–1134.

Li, P.-P., K.-Q. Gao, L.-H. Hou, and X. Xu. 2007. A gliding lizard from the Early Cretaceous of China. Proceedings of the National Academy of Sciences of the United States of America 104:5507–5509.

Li, Q., K.-Q. Gao, Q. Meng, J. A. Clarke, M. D. Shawkey, L. D'Alba, R. Pei, M. Ellison, M. A. Norell, and J. Vinther. 2012. Reconstruction of *Microraptor* and the evolution of iridescent plumage. Science 335:1215–1219.

Li, L., W. G. Joyce, and J. Liu. 2015. The first soft-shelled turtle from the Jehol Biota of China. Journal of Vertebrate Paleontology 35:e909450. doi:10.1080/02724634.2014.90

9450

Liu, H.-T., T.-T. Su, W.-L. Huang, and K.-J. Chang. 1963. [Lycopterid fishes from North China.] Memoirs of the Institute of Vertebrate Palaeontology and Palaeoanthropology, Academia Sinica 6:1–53. [Chinese]

Liu, J. 2004. A nearly complete skeleton of *Ikechosaurus pijiagouensis* sp. nov. (Reptilia: Choristodera) from the Jiufotang Formation (Lower Cretaceous) of Liaoning, China. Vertebrata PalAsiatica 42:120–129.

Liu, J., and X.-L. Wang. 2003. Choristoderes. Pp. 89–96 in M.-M. Chang, P.-J. Chen, Y.-Q. Wang, and Y. Wang (eds), The Jehol Biota: The Emergence of Feathered Dinosaurs, Beaked Birds and Flowering Plants. Shanghai Scientific & Technical Publishers: Shanghai.

Liu, Y.-Q., H.-W. Kuang, X.-J. Jiang, N. Peng, H. Xu, and H.-Y. Sun. 2012. Timing of the earliest known feathered dinosaurs and transitional pterosaurs older than the Jehol Biota. Palaeogeography, Palaeoclimatology, Palaeoecology 323–325:1–12.

Luo, Z. 1999. A refugium for relicts. Nature 400:24–25.

Luo, Z.-X., Q. Ji, J. R. Wible, and C.-X. Yuan. 2003. An Early Cretaceous tribosphenic mammal and metatherian evolution. Science 302:1934–1940.

Manabe, M., P. M. Barrett, and S. Isaji. 2000. A refugium for relicts? Nature 404:953.

Meng, J., Y. Wang, and C. Li. 2011. Transitional mammalian middle ear from a new Cretaceous Jehol eutriconodont. Nature 472:181–185.

Meng, Q., J. Liu, D. J. Varricchio, T. Huang, and C. Gao. 2004. Parental care in an ornithischian dinosaur. Nature 431:145–146.

O'Connor, J., Z. Zhou, and X. Xu. 2011. Additional specimen of *Microraptor* provides unique evidence of dinosaurs preying on birds. Proceedings of the National Academy of Sciences of the United States of America 108:19662–19665.

O'Connor, J. K., X.Wang, L. M. Chiappe, C. Gao, Q. Meng, X. Cheng, and J. Liu. 2009. Phylogenetic support for a specialized clade of Cretaceous enantiornithine birds with information from a new species. Journal of Vertebrate Paleontology 29:188–204.

O'Connor, J., X. Wang, C. Sullivan, X. Zheng, P. Tubaro, X. Zhang, and Z. Zhou. 2013. Unique caudal plumage of *Jeholornis* and complex tail evolution in early birds. Proceedings of the National Academy of Sciences of the United States of America 110: 17404–17408.

Pan, H.-Z., and X.-G. Zhu. 2003. Gastropods. Pp. 37–39 in M.-M. Chang, P.-J. Chen, Y.-Q. Wang, and Y. Wang (eds), The Jehol Biota: The Emergence of Feathered Dinosaurs, Beaked Birds and Flowering Plants. Shanghai Scientific & Technical Publishers, Shanghai.

Pan, Y., J. Sha, Z. Zhou, and F. T. Fürsich. 2013. The Jehol Biota: definition and distribution of exceptionally preserved relicts of a continental Early Cretaceous ecosystem. Cretaceous Research 44:30–38.

Pan, Y., J. Sha, F. T. Fürsich, Y. Wang, X. Zhang, and X. Yao. 2012. Dynamics of the lacustrine fauna from the Early Cretaceous Yixian Formation, China: implications of volcanic and climatic factors. Lethaia 45:299–314.

Pott, C., S. McLoughlin, A. Lindstrom, S. Wu, and E. M. Friis. 2012. *Baikalophyllum lobatum* and *Rehezamites anisolobus*: two seed plants with "cycadophyte" foliage from the Early Cretaceous of eastern Asia reviewed. International Journal of Plant Science 173:192–208.

Ren, D. 1998. Flower-associated Brachycera flies as fossil evidence for Jurassic angiosperm origins. Science 280:85–88.

Roček, Z., Y. Wang, and L. Dong. 2012. Post-metamorphic development of Early Cretaceous frogs as a tool for taxonomic comparisons. Journal of Vertebrate Paleontology 32:1285–1292.

Russell, D.A. 1994. The role of Central Asia in dinosaurian biogeography. Canadian Journal of Earth Sciences 30:2002–2012.

Rydin, C., and E. M. Friis. 2010. A new Early Cretaceous relative of Gnetales: *Siphonospermum simplex* gen. et sp. nov. from the Yixian Formation of Northeast China. BMC Evolutionary Biology 10:183.

Sereno, P. C., and C. Rao. 1992. Early evolution of avian flight and perching: new evidence from the Lower Cretaceous of China. Science 255:845–848.

Sha, J.-G., Y.-H. Pan, Y.-Q. Wang, X.-L. Zhang, and X. Rao. 2012. Non-marine and marine stratigraphic correlation of Early Cretaceous deposits in NE China, SE Korea and SW Japan, non-marine Molluscan biochronology, and palaeogeographic implications. Journal of Stratigraphy 36:357–381.

Steuber, T., M. Rauch, J.-P. Masse, J. Graaf, and M. Malkoc. 2005. Low-latitude seasonality of Cretaceous temperatures in warm and cold episodes. Nature 437:1341–1344.

Sun, G., D. L. Dilcher, S. Zheng, and Z. Zhou. 1998. In search of the first flower: a Jurassic angiosperm, *Archaefructus*, from northeast China. Science 282:1692–1695.

Sun, G., S.-L. Zheng, D. L. Dilcher, Y.-D. Wang, and S.-W. Mei. 2001. [Early Angiosperms and Their Associated Plants from Western Liaoning, China.] Shanghai Scientific and Technological Education Publishing House, Shanghai, 227 pp. [Chinese]

Sun, G., Q. Ji, D. L. Dilcher, S. Zheng, K. C. Nixon, and X. Wang. 2002. Archaefructaceae, a new basal angiosperm family. Science 296:899–904.

Swisher, C. C. III, Y.-Q. Wang, X.-L. Wang, X. Xu, and Y. Wang. 1999. Cretaceous age for the feathered dinosaurs of Liaoning, China. Nature 400:58–61.

Swisher, C. C. III, X. Wang, Z. Zhou, Y. Wang, F. Jin, J. Zhang, X. Xu, and Y. Wang. 2002. Further support for a Cretaceous age for the feathered-dinosaur beds of Liaoning, China: new $^{40}Ar/^{39}Ar$ dating of the Yixian and Tuchengzi formations. Chinese Science Bulletin 47:135–138.

Sychevskaya, E. K. 2001. New data on the Jurassic freshwater fish fauna of northern Eurasia. P. 60 in A. Tintori (ed), Meeting on Mesozoic Fishes, Systematics, Paleoenvironments and Biodiversity, Abstracts. Serpiano, Switzerland.

Tong, H., and D. Brinkman. 2012. A new species of *Sinemys* (Testudines: Cryptodira: Sinemydidae) from the Early Cretaceous of Inner Mongolia, China. Palaeobiodiversity and Palaeoenvironments 93:355–366.

Unwin, D. M., J. Lü, and N. N. Bakhurina. 2000. On the systematic and stratigraphical significance of pterosaurs from the Lower Cretaceous Yixian Formation (Jehol Group) of Liaoning, China. Mitteilungen aus dem Museum für Naturkunde Berlin, Geowissenschaftliche Reihe 3:181–206.

Upchurch, P. 1995. The evolutionary history of the sauropod dinosaurs. Philosophical Transactions of the Royal Society of London B 349:365–390.

Upchurch, P., C. A. Hunn, and D. B. Norman. 2002. An analysis of dinosaurian biogeography: evidence for the existence of vicariance and dispersal patterns caused by geological events. Proceedings of the Royal Society of London B 269:613–621.

Wang, L.-L., D.-Y. Hu, L.-J. Zhang, S.-L. Zheng, H.-Y. He, C.-L. Deng, X.-L. Wang, Z.-H. Zhou, and R.-X. Zhu. 2013. SIMS U-Pb zircon age of Jurassic sediments in Linglongta, Jianchang, western Liaoning: constraint on the age of oldest feathered dinosaurs. Chinese Science Bulletin 58:1346-1353.

Wang, S. 1991. Origin, evolution and mechanism of the Jehol fauna. Acta Geologica Sinica 64:203–215.

Wang, X., and X. Xu. 2001. A new iguanodontid (*Jinzhousaurus yangi* gen. et sp. nov.) from the Yixian Formation of western Liaoning, China. Chinese Science Bulletin 46:1669–1672.

Wang, X., and Z. Zhou. 2004. Pterosaur embryo from the Early Cretaceous of China. Nature 429:621.

Wang, X., and Z. Zhou. 2006. Pterosaur assemblages of the Jehol Biota and their implication for the Early Cretaceous pterosaur radiation. Geological Journal 41:405–418.

Wang, X., D. Miao, and Y. Zhang. 2005. Cannibalism in a semi-aquatic reptile from the Early Cretaceous of China. Chinese Science Bulletin 50:281–283.

Wang, X., D. de A. Campos, Z. Zhou, and A. W. A. Kellner. 2008. A primitive istiodactylid pterosaur (Pterodactyloidea) from the Jiufotang Formation (Early Cretaceous), northeast China. Zootaxa 1813:1–18.

Wang, X., A. W. A. Kellner, S. Jiang, and X. Cheng. 2012. New toothed flying reptile from Asia: close similarities between Early Cretaceous pterosaur faunas from China and Brazil. Naturwissenschaften 99:249–257.

Wang, X., A. W. A. Kellner, Z. Zhou, and D. de A. Campos. 2005. Pterosaur diversity in Cretaceous terrestrial ecosystems in China. Nature 437:875–879.

Wang, X., A. W. A. Kellner, Z. Zhou, and D. de A. Campos. 2007. A new pterosaur (Ctenochasmatidae, Archaeopterodactyloidea) from the Lower Cretaceous Yixian Formation of China. Cretaceous Research 28: 245–260.

Wang, X., A. W. A. Kellner, Z. Zhou, and D. de A. Campos. 2008. Discovery of a rare arboreal forest-dwelling flying reptile (Pterosauria, Pterodactyloidea) from China. Proceedings of the National Academy of Sciences of the United States of America 105:1983–1987.

Wang, X., Z. Zhou, Z. Zhang, and X. Xu. 2002. A nearly completely articulated rhamphorhynchoid pterosaur with exceptionally well-preserved wing membranes and 'hairs' from Inner Mongolia, northeast China. Chinese Science Bulletin 47:226–230.

Wang, X., H. You, Q. Meng, C. Gao, X. Chang, and J. Liu. 2007. *Dongbeititan dongi*, the first sauropod dinosaur from the Lower Cretaceous Jehol Group of western Liaoning Province, China. Acta Geologica Sinica 81:911–916.

Wang, X.-L., Y.-Q. Wang, X. Xu, Y. Wang, J.-Y. Zhang, F.-C. Zhang, F. Jin, and G. Gu. 1999. [Record of the Sihetun vertebrate mass mortality events western Liaoning, China: caused by volcanic eruptions.] Geological Review 45(Suppl.):458–467. [Chinese]

Wang, Y. 2006. [Phylogeny and early radiation of Mesozoic lissamphibians from East Asia.] Pp. 643–663 and 931–936 in J.-Y. Rong, Z.-J. Fang, Z.-H. Zhou, R.-B. Zhan, X.-D. Wang, and X.-L. Yuan (eds), [Originations, Radiations and Biodiversity Changes – Evidences from the Chinese Fossil Record.] Science Press, Beijing. [Chinese]

Wang, Y., and S. E. Evans. 2006. Advances in the study of fossil amphibians and squamates from China: the past fifteen years. Vertebrata PalAsiatica 44:60–73.

Wang, Y., and S. E. Evans. 2011. A gravid lizard from the Cretaceous of China and the early history of squamate viviparity. Naturwissenschaften 98:739–743.

Wang, Y., K. Gao, and X. Xu. 2000. Early evolution of discoglossid frogs: new evidence from the Mesozoic of China. Naturwissenschaften 87:417–420.

Wang, Y., Y. Hu, J. Meng, and C. Li. 2001. An ossified Meckel's cartilage in two Cretaceous mammals and origin of the mammalian middle ear. Science 294:257–361.

Wu, F.-Y., Y.-G. Xu, S. Gao, and J.-P. Zheng. 2008. [Lithospheric thinning and destruction of the North China Craton.] Acta Petrologica Sinica 24:1145–1174. [Chinese]

Wu, S.-Q. 1999. [A preliminary study of the Jehol flora from western Liaoning.] Pp. 7–57 in P.-J. Chen and F. Jin (eds), [The Jehol Biota.] Palaeoworld 11. [Chinese]

Xing, L., W. S. Persons, P. R. Bell, X. Xu, J. Zhang, T. Miyashita, F. Wang, and P. J. Currie. 2013. Piscivory in the feathered dinosaur *Microraptor*. Evolution 67:2441–2445.

Xu, X., and M.A. Norell. 2004. A new troodontid dinosaur from China with avian-like sleeping posture. Nature 431:838–841.

Xu, X., and M. A. Norell. 2006. Non-avian dinosaur fossils from the Lower Cretaceous Jehol Group of western Liaoning, China. Geological Journal 41:419–437.

Xu, X., X.-L. Wang, and X.-C. Wu. 1999. A dromaeosaurid dinosaur with a filamentous integument from the Yixian Formation of China. Nature 401:262–266.

Xu, X., X. Zheng, and H. You. 2009. A new feather type in a non-avian theropod and the early evolution of feathers. Proceedings of the National Academy of Sciences of the United States of America 106:832–834.

Xu, X., Z. Zhou, and X. Wang. 2000. The smallest known non-avian theropod dinosaur. Nature 408:705–708.

Xu, X., Y.-N. Cheng, X.-L. Wang, and C.-H. Chang. 2002. An unusual oviraptorosaurian dinosaur from China. Nature 419:291–293.

Xu, X., M. A. Norell, X. Kuang, X. Wang, Q. Zhao, and C. Jia. 2004. Basal tyrannosauroids from China and evidence for protofeathers in tyrannosauroids. Nature 431:680–684.

Xu, X., Z. Zhou, X. Wang, X. Kuang, F. Zhang, and X. Du. 2003. Four-winged dinosaurs from China. Nature 421:335–340.

Xu, X., K. Wang, K. Zhang, Q. Ma, L. Xing, C. Sullivan, D. Hu, S. Cheng, and S. Wang. 2012. A gigantic feathered dinosaur from the Lower Cretaceous of China. Nature 484:92–95.

Yang, Y., and Q. Wang. 2013. The earliest fleshy cone of *Ephedra* from the Early Cretaceous Yixian Formation of Northeast China. PLoS ONE 8(1):e53652. doi:10.1371/journal.pone.0053652

Yang, Y., L.-B. Lin, and Q. Wang. 2013. *Chengia laxispicata* gen. et sp. nov., a new ephedroid plant from the Early Cretaceous Yixian Formation of western Liaoning, Northeast China: evolutionary, taxonomic, and biogeographic implications. BMC Evolutionary Biology 13:72.

doi:0.1186/1471-2148-13-72

Zhang, F., and Z. Zhou. 2000. A primitive enantiornithine bird and the origin of feathers. Science 290:1955–1959.

Zhang, F., Z. Zhou, and M. J. Benton. 2008. A primitive confuciusornithid bird from China and its implications for early avian flight. Science in China, Earth Sciences 51:625–639.

Zhang, F., S. L. Kearns, P. J. Orr, M. J. Benton, Z. Zhou, D. Johnson, X. Xu, and X. Wang. 2010. Fossilized melanosomes and the colour of Cretaceous dinosaurs and birds. Nature 463:1075–1078.

Zhang, G., Y. Wang, M. E. H. Jones, and S. E. Evans. 2009. A new Early Cretaceous salamander (*Regalerpeton weichangensis* gen. et sp. nov.) from the Huajiying Formation of northeastern China. Cretaceous Research 30:551–558.

Zhang, H., B. Wang, and Y. Fang. 2010. Evolution of insect diversity in the Jehol Biota. Science China, Earth Sciences 53:1908–1917.

Zhang, J., F. Jin, and Z. Zhou. 1994. [A review of Mesozoic osteoglossomorph fish *Lycoptera longicephalus*.] Vertebrata PalAsiatica 32:41–59. [Chinese]

Zhang, X., and J. Sha. 2012. Sedimentary laminations in the lacustrine Jianshangou Bed of the Yixian Formation at Sihetun, western Liaoning, China. Cretaceous Research 36:96–105.

Zhao, Q., P. M. Barrett, and D. A. Eberth. 2007. Social behaviour and mass mortality in the basal ceratopsian dinosaur *Psittacosaurus* (Early Cretaceous, People's Republic of China). Palaeontology 50:1023–1029.

Zheng, X., X. Wang, J. O'Connor, and Z. Zhou. 2012. Insight into the early evolution of the avian sternum from juvenile enantiornithines. Nature Communications 3:1116. doi:10.1038/ncomms2104.

Zheng, X., L. D. Martin, Z. Zhou, D. Burnham, F. Zhang, and D. Miao. 2011. Fossil evidence of avian crops from the Early Cretaceous of China. Proceedings of the National Academy of Sciences of the United States of America 108:5904–15907.

Zheng, X., J. O'Connor, F. Huchzermeyer, X. Wang, Y. Wang, M. Wang, and Z. Zhou. 2013. Preservation of ovarian follicles reveals early evolution of avian reproductive behaviour. Nature 495:507–511.

Zheng, X., Z. Zhou, X. Wang, F. Zhang, X. Zhang, Y. Wang, G. Wei, S. Wang, and X. Xu. 2013. Hind wings in basal birds and the evolution of leg feathers. Science 339:1309–1312.

Zhou, C.-F. 2010a. A new eucryptodiran turtle from the Early Cretaceous Jiufotang Formation of western Liaoning, China. Zootaxa 2676:45–56.

Zhou, C.-F. 2010b. A second specimen of *Manchurochelys manchoukuoensis* Ento and Shikama, 1942 (Testudines: Eucryptodira) from the Early Cretaceous Yixian Formation of western Liaoning, China. Zootaxa 2534:57–66.

Zhou, S., Z. Zhou, and J. K. O'Connor. 2013. Anatomy of the basal ornithuromorph bird *Archaeorhynchus spathula* from the Early Cretaceous of Liaoning, China. Journal of Vertebrate Paleontology 33:141–152.

Zhou, Z. 2004. Vertebrate radiations of the Jehol Biota and their environmental background. Chinese Science Bulletin 49:754–756.

Zhou, Z. 2006. Evolutionary radiation of the Jehol Biota: chronological and ecological perspectives. Geological Journal 41:377–393.

Zhou, Z., and L. D. Martin. 2011. Distribution of the predentary bone in Mesozoic ornithurine birds. Journal of Systematic Palaeontology 9:25–31.

Zhou, Z.-H., and X.-L. Wang. 2000. A new species of *Caudipteryx* from the Yixian Formation of Liaoning, Northeast China. Vertebrata PalAsiatica 38:111–127.

Zhou, Z., and Y. Wang. 2010. Vertebrate diversity of the Jehol Biota as compared with other lagerstätten. Science China, Earth Sciences 53:1894–1907.

Zhou, Z., and F. Zhang. 2001. Two new ornithurine birds from the Early Cretaceous of western Liaoning, China. Chinese Science Bulletin 46:1258–1264.

Zhou, Z., and F. Zhang. 2002a. Largest bird from the Early Cretaceous and its implications for the earliest avian ecological diversification. Naturwissenschaften 89:34–38.

Zhou, Z., and F. Zhang. 2002b. A long-tailed, seed-eating bird from the Early Cretaceous of China. Nature 418:405–409.

Zhou, Z., and F. Zhang. 2003. *Jeholornis* compared to *Archaeopteryx*, with a new understanding of the earliest avian evolution. Naturwissenschaften 90:220–225.

Zhou, Z., and F. Zhang. 2004. A precocial avian embryo from the Lower Cretaceous of China. Science 306:653.

Zhou, Z., and F. Zhang. 2005. Discovery of a new ornithurine bird and its implication for Early Cretaceous avian radiation. Proceedings of the National Academy of Sciences of the United States of America 102:18998–19002.

Zhou, Z., and F. Zhang. 2006a. Mesozoic birds of China – a synoptic review. Vertebrata PalAsiatica 44:74–98.

Zhou, Z., and F. Zhang. 2006b. A beaked basal ornithurine bird (Aves, Ornithurae) from the Lower Cretaceous of China. Zoologica Scripta 35:363–373.

Zhou, Z., and S. Zheng. 2003. The missing link in *Ginkgo* evolution. Nature 423:821–822.

Zhou, Z, P. M. Barrett, and J. Hilton. 2003. An exceptionally preserved Lower Cretaceous ecosystem. Nature 421:807–814.

Zhou, Z., J. A. Clarke, and F. Zhang. 2002. *Archaeoraptor*'s better half. Nature 420:285.

Zhou, Z., F. Zhang, and Z. Li. 2010a. A new Lower Cretaceous bird from China and tooth reduction in early avian evolution. Proceedings of the Royal Society B 277:219–227.

Zhou, Z., J. Zhang, and X. Wang. 2010b. The Jehol fish fauna: ecological interaction and paleogeographic distribution. Pp. 419–431 in D. K. Elliott, J. G. Maisey, X. Yu, and D. Miao (eds), Morphology, Phylogeny and Paleobiogeography of Fossil Fishes. Verlag Dr. Friedrich Pfeil, Munich.

Zhou, Z., J. A. Clarke, F. Zhang, and O. Wings. 2004. Gastroliths in *Yanornis* – an indication of the earliest radical diet switching and gizzard plasticity in the lineage leading to living birds? Naturwissenschaften 91:571–574.

Zhu, R., Y. Pan, R. Shi, Q. Liu, and D. Li. 2007. Palaeomagnetic and $^{40}Ar/^{39}Ar$ dating constraints on the age of the Jehol Biota and the duration of deposition of the Sihetun fossil bearing lake sediments, northeastern China. Cretaceous Research 8:171–176.

Tables

Table 1: Taxonomic list of insect fossils of the Jehol Biota.

Order	Family	Genus	Species
Ephemeroptera	Hexagenitidae	*Ephemeropsis*	*E. trisetalis*
		Epicharmeropsis	*E. hexavenulosus*
			E. Quadrivenulosus
	Family indet.	*Caenoephemera*	*C. shangyuanensis*
Odonata	Aeschnidiidae	*Linaeschnidium*	*L. sinensis*
		Dracontaeschnidium	*D. orientale*
		Hebeiaeschnidia	*H. fengningensis*
		Sinaeschnidia	*S. heishankowensis*
			S. cancellosa
	Araripelibellulidae	*Sopholibellula*	*S. amoena*
			S. eleganta
	Corduliidae	*Mesocordulia*	*M. boreala*
	Gomphaeschnidae	*Sophoaeschna*	*S. frigida*
		Falsisophoaeschna	*F. generalis*
		Sinojagoria	*S. imperfecta*
			S. cancellosa
			S. magna
	Liupanshaniidae	*Protoliupanshania*	*P. wangi*
	Progobiaeshnidae	*Progobiaeshna*	*P. liaoningensis*
		Mongoliaeshna	*M. sinica*
			M. hadrens
			M. exiguusens
		Decoraeshna	*D. preciosa*
	Rudiaeschnidae	*Rudiaeschna*	*R. limnobia*
	Gomphidae	*Liaoninglanthus*	*L. latus*
		Jibeigomphus	*J. xinboensis*
		Sinogomphus	*S. taushanensis*
		Liogomphus	*L. yixianensis*
	Hemeroscopidae	*Abrohemeroscopus*	*A. mengi*
		Hemeroscopus	*H. baissicus*
	Nodalulaidae	*Nodalula*	*N. dalinghensis*
	Nannogomphidae	*Sinahemeroscopus*	*S. magnificus*

	Aktassiidae	*Sinaktassia*	*S. tangi*
		Pseudocymatophlebia	*P. boda*
	Proterogomphidae	*Lingomphus*	*L. magnificus*
	Campterophlebiidae	*Parafleckium*	*P. senjituense*
	Stenophlebiidae	*Sinostenophlebia*	*S. zhanjiakouensis*
	Tarsophlebiidae	*Turanophlebia*	*T. sinica*
	Family indet.	*Parapetala*	*P. liaoningensis*
	Family indet.	*Telmaeshna*	*T. paradoxica*
Plecoptera	Siberioperlidae	*Sinosharaperla*	*S. zhaoi*
	Perlidae	*Archaeoperla*	*A. rarissimus*
	Mesoleuctridae	*Mesoleuctra*	*M. peipiaoensis*
	Perlariopseidae	*Sinoperla*	*S. abdominalis*
		Perlariopsis	*P. peipiaoensis*
	Baleyopterygidae	*Aristoleuctra*	*A. yehae*
	Taeniopterygidae	*Liaotaenionema*	*L. tenuitibia*
	Nemouridae	*Parvinemoura*	*P. parvus*
Blattida	Mesoblattinidae	*Rhipidoblattina*	*R. laternoforma*
		Nipponoblatta	*N. acerba*
	Blattulidae	*Blattula*	*B. rudis*
		Macaroblattula	*M. ellipsoids*
		Habroblattula	*H. drepaniodes*
			H. eumeura
		Elisama	*E. cuboids*
Orthoptera	Haglidae	*Alloma*	*A. huanghuachunensis*
			A. faciata
		Yenshania	*Y. hebeiensis*
		Sinohagla	*S. anthoides*
	Prophalangopsidae	*Aethehagla*	*A. hongi*
		Allaboilus	*A. hani*
		Barcharaboilus	*B. jurassicus*
		Chifengia	*C. amans*
			C. angustata
			C. lata
			C. mosaica
		Grammohagla	*G. striata*

			H. yixianensis
	Protapioceridae	Protapiocera	P. convergens
			P. megista
			P. ischyra
	Empididae		
		Protempis	P. minuta
	Rhagionidae	Basilorhagio	B. venustus
		Longhuaia	L. orientalis
		Oiobrachyceron	O. limnogenus
		Orsobrachyceron	O. chinensis
		Pauromyia	P. oresbia
	Stratiomyidae	Gigantoberis	G. liaoningensis
	Tabanidae	Allomyia	A. ruderalis
		Eopangonius	E. pletus
		Palaepangonius	P. eupterus
Tricoptera	Vitimotauliidae	Macropteryx	M. xiaoshetaiensis
		Multimodus	M. dissitus
			M. elongatus
			M. stigmaeus
		Sinomodus	S. macilentus
			S. peltayus
			S. spatiosus
		Gen. indet.	Gen.et sp. indet
Hymenoptera			
	Anaxyelidae	Brachysyntexis	B. robusta
		Sinanxyela	S. dorunensis
			S. continentalis
	Paroryssidae	Parorysus	P. suspectus
	Sepulcidae	Trematothoracoides	T. liaoningensis
		Sinosepulca	S. gigathoracalis
	Sincidae	Sinosirex	S. gigantea
	Xyelotomidae	Davidsmithia	D. suni
		Liaotoma	L. linearis
		Synaptotoma	S. imi
	Xyelidae	Angaridyela	A. endemica
			A. exculpta
			A. robusta
			A. suspecta
			A. orientalis
		Ceratoxyela	C. decorosa
		Heteroxyela	H. ignota

	Isoxyela	I. rudis
	Lethoxyela	L. excurva
		L. vulgata
		L. antiqua
	Sinoxyela	S. viriosa
	Liadoxyela	L. chengdeensis
	Alloxyelula	A. lingyuanensis
	Xyelites	X. lingyuanensis
	Brachyoxyela	B. brevinodia
		B. gracilenta
Praesiricidae	Rudisiricius	R. belli
Gasteruptiidae	Manlaya	M. lexuosus
Angarospheiidae	Baissodes	B. grabaui
Bethylonymidae	Allogaster	A. ovate
Ephialtitidae	Crephanogaster	C. rara
	Tuphephialtites	T. zherikhini
Evaniidae	Procretevania	P. pristina
		P. exquisite
		P. vesca
Heloridae	Protocyrtus	P. validus
	Spherogaster	S. coronata
		S. saltatrix
	Sinohelorus	S. elegans
Ichneumonidae	Tanychora	T. beipiaoensis
		T. exquisita
		T. spinata
	Tanychorella	T. dubia
	Ovigaster	O. cephalotus
Megalyridae	Yanocleistogaster	Y. canaliculata
	Mesaulacinus	M. rasnitsyni
Mesoserphidae	Beipiaoserphus	B. elegans
Pelecinidae	Eopelecinus	E. shangyuanensis
		E. similaris
		E. vicinu
		E. huangi
		E. tumidus
	Scorpiopelecinus	S. versatilis
	Shoushida	S. regilla
	Sinopelecinus	S. delicatus
		S. epigaeus
		S. magicus
		S. viriosus
	Abropelecinus	A. annulatus
	Azygopelecinus	A. clavatus
	Megapelecinus	M. changi
		M. nashi
Praeaulacidae	Sinowestratia	S. communicata
Praeichneumonidae	Scolichneumon	S. rectivenius
Roproniidae	Jeholoropronia	J. pingi

		Habrohagla	*H. curtivenata*
		Huabeius	*H. suni*
		Shanxius	*S. meileyingziensis*
		Trachohagla	*T. jeholia*
	Trigonidiidae	*Liaonemobius*	*L. tanae*
Phasmatodea	Chresmodidae	*Chresmoda*	*C. orientalis*
			C. multinervis
			C. shihi
		Sinochresmoda	*S. magnicornia*
	Susamaniidae	*Hagiphasma*	*H. paradoxa*
		Aethephasma	*A. megista*
		Orephasma	*O. eumorpha*
Hemiptera	Oviparosiphidae	*Sinoviparosiphum*	*S. lini*
	Procercopidae	*Anomoscytina*	*A. anomala*
		Anthoscytina	*A. aphthosa*
		Cretocercopis	*C. yii*
	Scytinopteridae	*Sunoscytinopteris*	*S. lushangfenensis*
	Tettigarctidae	*Luanpingia*	*L. senjituensis*
	Cixiidae	*Cathaycixius*	*C. pustulosus*
			C. trinervus
		Lapicixius	*L. decorus*
		Yanducixius	*Y. yihi*
			Y. pardalinus
		Cretocixius	*C. stigmatosus*
	Pereboriidae	*Jiphara*	*J. wangi*
			J. reticulata
	Palaeontinidae	*Ilerdocossus*	*I. beipiaoensis*
			I. exiguus
			I. hui
			I. fengningensis
			I. pingquanensis
			I. ningchengensis
		Miracossus	*M. ingentius*
		Yanocossus	*Y. guoi*

Mesoveliidae	*Sinovelia*	S. mega
		S. popovi
Ochteridae	*Angulochterus*	A. quatrimaculatus
	Floricaudus	F. multilocellus
	Pristinochterus	P. ovatus
		P. zhangi
Corixidae	*Karataviella*	K. pontoforma
Naucoridae	*Exilcrus*	E. cameriferus
	Miroculus	M. laticephlus
Notonectidae	*Notonecta*	N. xyphiale
Saldidae	*Brevrimatus*	B. pulchalifer
	Venustsalda	V. locella
Archegocimicidae	*Longianteclypea*	L. tibialis
	Mesolygaeus	M. laiyangensis
		M. naevius
	Propritergum	P. opimum
Vetanthocoridae	*Byssoidecerus*	B. levigata
	Collivetanthocoris	C. rapax
	Curticerus	C. venustus
	Crassicerus	C. furtivus
	Curvicaudus	C. ciliatus
	Mecopodus	M. xanthos
	Pustulithoracalis	P. gloriosus
	Vetanthocoris	V. decorus
		V. longispicus
Pachymeridiidae	*Beipiaocoris*	B. multifurcus
	Bellicoris	B. mirabilis
Coreidae	*Bibiticen*	B. hebeiensis
Cydnidae	*Cilicydnus*	C. robustispinus
	Orienicydnus	O. hongi
Venicoridae	*Venicoris*	V. solaris
	Clavaticoris	C. zhengi
Primipentatomidae	*Primipentatoma*	P. peregrine
		P. fangi
	Breviscutum	B. lunatum
	Oropentatoma	O. epichara
	Quadrocoris	Q. radius

Raphidioptera	Alloraphidiidae	*Alloraphidia*	A. longistigmosa
			A. anomala
			A. obliquivenatica
		Caloraphidia	C. glossophylla
		Xynoraphidia	X. shangyuanensis
			X. polyphlebia
		Yanoraphidia	Y. gaoi
	Baissopteridae	*Baissoptera*	B. euneura
			B. grandis
		Rudiraphidia	R. liaoningensis
		Siboptera	S. fornicata
	Mesoraphidiidae	*Jilinoraphidia*	J. dalaziensis
		Mesoraphidia	M. amoena
			M. heteroneura
			M. sinica
		Mioraphidia	M. furcivenata
		Phiradia	P. myrioneura
		Xuraphidia	X. kezuoensis
			X. liaoxiensis
Neuroptera	Aetheogrammatidae	*Aetheogramma*	A. speciosum
	Ascalochrysidae	*Ascalochrysa*	A. megaptera
	Berothidae	*Oloberotha*	O. sinica
	Chrysopidae	*Lembochrysa*	L. miniscula
			L. polyneura
		Mesypochrysa	M. cf. chrysopoides
		Paralembochrysa	P. splendida
	Dipteromantispidae	*Dipteromantispa*	D. brevisubcosta
	Ithonidae	*Lasiosmylus*	L. newi
	Kalligrammatidae	*Kalligramma*	K. liaoningense
		Limnogramma	L. mirum
		Oregramma	O. gloriosum
		Sophogramma	S. eucallum
			S. lii
			S. papilionaceum
			S. plecophlebium
	Mesochrysopidae	*Kareninoides*	K. lii
		Longicellochrysa	L. yixiana
		Mesascalaphus	M. yangi
		Tachinymphes	T. delicatus
			T. magnificus
	Myrmeleontidae	*Choromyrmeleon*	C. aspoeckorum
			C. othneius
	Palaeoleontidae	*Guyiling*	G. jianboni
	Psychopsidae	*Alloepipsychopsis*	A. lata
		Angaropsychops	A. sinicus

		Undulopsychopsis	*U. alexi*
	Family indet.	*Liaoximyia*	*L. sinica*
	Family indet.	*Mesohemerobius*	*M. jeholensis*
	Family indet.	*Yanosmylus*	*Y. rarivenatus*
Coleoptera			
	Cupedidae	*Priacma*	*P. latidentata*
			P. subtilis
			P. tuberculosa
			P. clavata
			P. renaria
		Furcicupes	*F. raucus*
		Latocupes	*L. fortis*
			L. bellus
		Ovatocupes	*O. alienus*
		Diluticupes	*D. impressus*
		Lupicupes	*L. trachylaenus*
	Ommatidae	*Tetraphalerus*	*T. lentus*
			T. laetus
			T. latus
			T. curtinervis
		Sinocupes	*S. validus*
		Amblomma	*A. psilata*
			A. rudis
			A. epicharis
			A. stabilis
			A. cyclodonta
			A. miniscula
			A. porrecta
			A. eumeura
			A. protensa
		Euryomma	*E. tylodes*
		Odontomma	*O. trachylaena*
		Brochocoleus	*B. sulcatus*
			B. angustus
			B. magnus
			B. validus
			B. applanatus
		Monticupes	*M. surrectus*
			M. fengtainsis
		Cionocoleus	*C. magicus*
			C. planiusculus
			C. cervicalis
	Rhombocoleidae	*Sinorhombocoleus*	*S. papposus*
	Taldycupedidae	*Penecupes*	*P. rapax*
		Pulchicupes	*P. jiensis*
		Longaevicupes	*L. macilentus*
	Ademosynidae	*Atalosyne*	*A. sinuolata*
	Triaplidae	*Mesaplus*	*M. beipiaoensis*

Dytiscidae	Sinoporus	S. lineatus
	Mesoderus	M. magnus
		M. ventralis
	Liadytiscus	L. cretaceus
		L. longitibialis
		L. latus
		L. elegans
Carabidae	Aethocarabus	A. levigata
	Denudirabus	D. exstrius
	Fangshania	F. punctata
	Protorabus	P. polyphlebius
Trachypachidae	Beipiaocarabus	B. oblonga
	Fortiseode	F. pervalimand
	Undo	U. pandurata
Parandrexidae	Parandrexis	P. beipiaoensis
Odemeridae	Glypta	G. qingshilaensis
Hydrophilidae	Cretohelophorus	C. yanensis
	Sinosperchopsis	S. silinae
	Hydrophilopsia	H. shatrovskiyi
		H. hydraenoides
		H. gracilis
Lasiosynidae	Lasiosyne	L. gratiosa
		L. quadricollis
		L. fedorenkoi
		L. daohugouensis
Elateridae	Cryptocoelus	C. major
		C. buffoni
		C. giganteus
	Lithomerus	L. buyssoni
	Desmatinus	D. cognatus
	Bilineariselater	B. foveatus
	Curtelater	C. wui
	Paralithomerus	P. exquisitus
		P. parallelus
	Paradesmatus	P. dilatatus
	Apoclion	A. clavatus
		A. dolini
		A. antennatus
	Anoixis	A. complanus
Cerophytidae	Necromera	N. admiranda
Nemonychidae	Brenthorrhinus	B. longidigitus
	Probelus	P. sinicus
	Microprobelus	M. liuae
	Brenthorrhinoides	B. latipecteris
		B. angustipecteris
		B. magnoculi
Ithyceridae	Abrocar	A. brachyorhinos
		A. macilentus
	Cretonanophyes	C. zherikhini

			C. punctatus
		Leptocar	L. polychaetus
	Mordellidae	Mirimordella	M. gracilicruralis
		Cretanaspis	C. lushangfenensis
		Liaoximordella	L. hongi
		Bellimordella	B. capitulifera
			B. longispina
			B. robusta
	Staphylinidae	Glabrimycetoporus	G. amoenus
		Sinoxytelus	S. euglypheus
			S. breviventer
			S. longisetus
		Oxyporus	O. yixianus
		Mesostaphylinus	M. elongatus
			M. yixianus
			M. antiquus
		Cretoprosopus	C. problematicus
		Paleothius	P. gracilis
		Thayeralinus	T. fieldi
			T. longelytratus
			T. glandulifer
			T. giganteus
			T. fraternus
		Paleowinus	P. rex
			P. fossilis
			P. ambiguus
			P. mirabilis
			P. chinensis
		Durothorax	D. creticus
		Cretoquedius	C. distinctus
			C. infractus
			C. dorsalis
	Tenebrionidae	Cretaceites	C. jinxiensis
		Karadromeus	K. xiangfanggouensis
	Scarabaeidae	Proscarabaeus	P. baissensis
	Geotrupidae	Geotrupoides	G. jiaoheense
			G. kezuoensis
		Parageotrupes	P. incanus
	Glaresidae	Glaresis	G. orthochilus
		Cretohypna	C. cristata
	Ochodaeidae	Mesochodaeus	M. daohugouensis
	Byrrhidae	Discus	D. lushangfenensis
	Hybosoridae	Fortishybosoru	F. ericeusicus
		Pulcherhybosorus	P. tridentatus
Mecoptera	Aneuretopsychidae	Jeholopsyche	J. liaoningensis
			J. completa
			J. bella

			J. maxima
	Bittacidae	*Megabittacus*	*M. beipiaoensis*
			M. colosseus
			M. spatiosus
		Sibirobittacus	*S. atalus*
	Eomeropidae	*Typhothauma*	*T. yixianensis*
	Neorthophlebiidae	*Yanorthophlebia*	*Y. hebeiensis*
	Orthophlebiidae	*Orthophlebia*	*O. liaoningensis*
		Parachorista	*P. miris*
	Mesopsychidae	*Vitimopsyche*	*V. kozlovi*
Diptera			
	Bibionidae	*Lichnoplecia*	*L. kovalevi*
	Chaoboridae	*Chironomaptera*	*C. gregaria*
			C. melanura
			C. robustus
		Mesochaoborus	*M. zhangshanyingensis*
	Chironomidae	*Guvanomyia*	*G. rohdendorfi*
		Manlayamyia	*M. dabeigouensis*
	Mesosciophilidae	*Sinosciophila*	*S. meileyingziensis*
	Mycetophilidae	*Atalosciophila*	*A. yanensis*
		Liaoxifungivora	*L. simplicis*
	Pleciofungivoridae	*Opiparifungivora*	*O. aliena*
		Pleciofungivora	*P. yangtianense*
	Protendipedidae	*Priscotendipes*	*P. mirus*
		Protendipes	*P. huabeiensis*
	Protopleciidae	*Huaxiaplecia*	*H. zhongguanensis*
		Mesoplecia	*M. xinboensis*
	Protorhyphidae	*Brachyopteryx*	*B. weichangensis*
	Simuliidae	*Mesasimulium*	*M. lahaigouense*
	Eremochaetidae	*Alleremonomus*	*A. liaoningensis*
			A. xingi
		Lepteremochaetus	*L. lithoecius*
	Nemestrinidae	*Florinemestrius*	*F. pulcherrimus*
		Protonemestrius	*P. beipiaoensis*
			P. jurassicus
	Archisargidae	*Origoasilus*	*O. pingquanensis*
		Helempis	*H. eucalla*

		Liaoropronia	*L. regia*
			L. leonina
	Serphidae	*Chengdeserphus*	*C. petidatus*
		Alloserphus	*A. saxosus*
		Gurvanotrupes	*G. exiguus*
			G. liaoningensis
			G. stolidus
		Liaoserphus	*L. perrarus*
		Ocnoserphus	*O. sculptus*
		Saucrotrupes	*S. decorosus*
		Scalprogaster	*S. fossilis*
		Steleoserphus	*S. beipiaoensis*
	Scolebythidae	*Mirabythus*	*M. lechrius*
			M. liae
	Scoliidae	*Protoscolia*	*P. imperialis*
			P. normalis
			P. sinensis
Siphonaptera	Pseudopulicidae	*Pseudopulex*	*P. magnus*
		Tyrannopsylla	*T. beipiaoensis*
	Saurophthiridae	*Saurophthirus*	*S. exquisitus*

Table 2: Taxonomic list of vertebrate fossils of the Jehol Biota.

Order or higher taxa	Family	Genus	Species
Agnatha			
Petromyzontiformes	Petromyzontidae	*Mesomyzon*	*M. mengae*
Pisces			
Elasmobranchii	Hybodontoidea	indet.	indet.
Acipenseriformes	Peipiaosteidae	*Peipiaosteus*	*P. fengningensis*
			P. pani
		Yanosteus	*Y. longidorsalis*
	Polyodontidae	*Protopsephurus*	*P. liui*
Amiiformes	Sinamiidae	*Sinamia*	*S. zdanskyi*
			S. liaoningensis
Osteoglossomorpha	Lycopteridae	*Lycoptera*	*L. davidi*
			L. fuxinensis
			L. muroii
			L. sankeyushuensis
			L. sinensis
			L. tokunagai
	Kuyangichthyidae	*Jinanichthys*	*J. longicephalus*
Teleostei	indet.	*Longdeichthys*	*L. luojiaxiaensis*
Amphibia			
Anura	indet.	*Liaobatrachus*	*L. beipiaoensis*
			(*Dalianbatrachus mengi*;
			= *L. grabaui*;
			= *Callobatrachus sanyanensis*;
			L. beipiaoensis
			(= *Mesophryne beipiaoensis*)
			L. macilentus
			(= *Yizhoubatrachus macilentus*)
			L. zhaoi
		indet.	indet.
Urodela	Cryptobranchidae	*Regalerpeton*	*R. weichangensis*
	Hynobiidae	*Liaoxitriton*	*L. zhongjiani*
		Laccotriton	*L. subsolanus*
		Sinerpeton	*S. fengshanensis*
Chelonia			
	Sinemydidae	*Ordosemys*	*O. liaoxiensis*
			(= *Manchurochelys liaoxiensis*)
		Manchurochelys	*M. manchoukuoensis*
	Eucryptodira	*Liaochelys*	*L. jianchangensis*
Choristodera			
	Simoedosauridae	*Ikechosaurus*	*I. gaoi*
			I. pijiagouensis
		Liaoxisaurus	*L. chaoyangensis*
	Hyphalosauridae	*Hyphalosaurus*	*H. lingyuanensis*
			(= *Sinohydrosaurus lingyuanensis*)
			H. baitaigouensis
	Monjurosuchidae	*Monjurosuchus*	*M. manchoukuoensis*
		Philydosaurus	*P. proseilus*
Squamata			
	indet.	*Yabeinosaurus*	*Y. tenuis*
			(?=*Jeholacerta formosa*)
	indet.	*Xianglong*	*X. zhaoi*
	indet.	*Dalinghosaurus*	*D. longidigitus*
	indet.	*Liaoningolacerta*	*L. brevirostra*, nomen dubium
	indet.	*Liushusaurus*	*L. acanthocaudata*
Pterosauria			
	?Gallodactylidae	*Feilongus*	*F. youngi*
	Anhangueridae	*Liaoningopterus*	*L. gui*
	Pteranodontidae	*Chaoyangopterus*	*C. zhangi*
			(= *Jidapterus edentus*;
			= *Eopteranodon lii*;
			= *Eoazhdarcho liaoxianesis*
			= *Shenzoupterus chaoyangens*
		Guidraco	*G. venator*
	Pterodactylidae	*Eosipterus*	*E. yangi*
		Haopterus	*H. gracilis*
		Ningchengopterus	*N. liuae*
	Tapejaridae	*Sinopterus*	*S. dongi*
			(= *Huaxiapterus jii*;
			= *Sinopterus gui*)
	Ctenochasmatidae	*Beipiaopterus*	*B. chenianus*
		Cathayopterus	*C. grabaui*

		Caudipteridae	*Caudipteryx*	*C. dongi*
				C. zoui
			Similicaudipteryx	*S. yixianensis*
		Therizinosauroidea	*Beipiaosaurus*	*B. inexpectus*
		Ornithomimosauria: Ornithomimidae	*Shenzhousaurus*	*S. orientalis*
Aves				
Jeholornithiformes	Jeholornithidae	*Jeholornis*	*J. prima* (= *Shenzhouraptor sinensis*; = *Jixiangornis orientalis*) *J. palmapenis*
Sapeornithiformes	Sapeornithidae	*Sapeornis*	*S. chaoyangensis* (*Didactylornis jii*; = *S. angustis*; = *Shenshiornis primita*)
Confuciusornithiformes	Confuciusornithidae	*Eoconfuciusornis*	*E. zhengi*
		Confuciusornis	*C. sanctus*
			C. dui
			C. suni
Confuciusornithiformes	Confuciusornithidae	*Confuciusornis*	*C. chuonzhous*
			C. feducciai
		Changchengornis	*C. hengdaoziensis*
		Jinzhouornis	*J. yixianensis*
			J. zhangjiyingia
Protopterygiformes	Protopterygidae	*Protopteryx*	*P. fengningensis*
Eoenantiornithiformes	Eoenantiornithidae	*Eoenantiornis*	*E. buhleri*
Longipterygithiformes	Longipterygithidae	*Longipteryx*	*L. chaoyangensis*
		Longirostravis	*L. hani*
		Rapaxavis	*R. pani*
		Shanweiniao	*S. cooperorum*
		Boluochia	*B. zhengi*
		Shengjingornis	*S. yangi*
Cathayornithiformes	Cathayornithidae	*Eocathayornis*	*E. walkeri*
		Cathayornis	*C. yandica* (= *Cathayornis caudatus*; = *Cathayornis aberransis*; = *Longchengornis sanyanensis*; = *Cuspirostrisornis houi*; = *Largirostrornis sexdentornis*)
		Sinornis	*S. santensis*
Liaoxiornithiformes	Liaoxiornithidae	*Liaoxiornis*	*L. delicatus* (= *Lingyuanornis parvus*)
Jibeiniaithiformes	Jibeiniaithidae	*Jibeinia*	*J. luanhera*
Enantiornithes	indet.	*Pengornis*	*P. houi*
		Vescornis	*V. hebeiensis*
		Paraprotopteryx	*P. gracilis*
		Shenqiornis	*S. mengi*
		Xiangornis	*X. shenmi*
		Bohaiornis	*B. guoi*
		Fortungulavis	*F. xiaotaizicus*
Liaoningornithiformes	Liaoningornithidae	*Liaoningornis*	*L. longidigitris*
Chaoyangornithiformes	Chaoyangornithidae	*Chaoyangia*	*C. beishanensis*
	indet.	*Songlingornis*	*S. linghensis*
Yanornithiformes	Yanornithidae	*Yanornis*	*Y. martini* (= *Aberratiodontus wui*)
Yixianornithiformes	Yixianornithidae	*Yixianornis*	*Y. grabaui*
Ornithurae Order indet.	Hongshanornithidae	*Hongshanornis*	*H. longicresta*
		Longicrusavis	*L. houi*
Order indet.		*Parahongshanornis*	*P. chaoyangensis*
Order indet.	indet.	*Archaeorhynchus*	*A. spathula*
		Maotherium	*M. sinensis*
			M. asiaticus
		Zhangheotherium	*Z. quinquecuspidens*
Metatheria	indet.	*Sinodelphys*	*S. szalayi*
Eutheria	indet.	*Eomaia*	*E. scansoria*
		Acristatherium	*A. yanensis*

Mammalia			
Triconodonta	Gobiconodontidae	*Gobiconodon*	*G. zofiae*
		Meemannodon	*M. lujiatunensis*
	Repenomamidae	*Repenomamus*	*R. robustus*
			R. giganticus
	Jeholodentidae	*Jeholodens*	*J. jenkinsi*
	Eutriconodonta indet.	*Yanoconodon*	*Y. allini*
		Liaoconodon	*L. hui*
Multituberculata	Eobaataridae	*Sinobaatar*	*S. lingyuanensis*
Symmetrodonta	Spalacotheriidae	*Akidolestes*	*A. cifellii*
		Maotherium	*M. sinensis*
			M. asiaticus
		Zhangheotherium	*Z. quinquecuspidens*
Metatheria	indet.	*Sinodelphys*	*S. szalayi*
Eutheria	indet.	*Eomaia*	*E. scansoria*
		Acristatherium	*A. yanensis*
Mammalia	indet.	*Juchilestes*	*J. liaoningensis*

Chapter 7

The Santana Formation

David M. Martill[1] and Paulo M. Brito[2]

1 School of Earth and Environmental Sciences, University of Portsmouth, UK.
2 Universidade Estadual Rio de Janeiro, R.J., Brazil.

Abstract

The mid-Cretaceous Santana Formation of northeastern Brazil is one of the most famous fossil conservation Lagerstätte. Lithologically the formation is highly heterogenous, and includes a wide range of siliciclastics, thin limestones, and palaeontologically important early diagenetic concretions rich in three dimensionally preserved vertebrates, some with preserved soft tissues. Many aspects of the Formation's palaeoenvironments remain controversial, but there is a consensus that deposition occurred in a large, fault bounded basin with tenuous links to both the south and central Atlantic Oceans. Salinity and water depth was probably variable, but for the concretion-bearing Romualdo Member, water depths were below storm wave base allowing for the preservation of fine laminae. Bottom waters were probably anoxic, but salinity may have fluctuated considerably, perhaps resulting in mass mortality events. The vertebrate fauna of the Santana Formation is dominated by abundant actinopterygian fishes, and rarer coelacanths and elasmobranchs. Tetrapods are represented by rare crocodyliforms, turtles, very rare theropod dinosaurs but surprisingly frequent occurrences of pterosaurs. Pterosaur diversity is high with tapejarids, thalassodromeids, tupuxuarids and ornithocheirids all represented by multiple species.

Invertebrates are dominated by ostracods, and rarer small bivalves and gastropods. Plant remains occur also. Soft tissue preservation of vertebrates occurs frequently in actinopterygian and elasmobranch fishes, but has also been reported for pterosaur wing membranes and muscle tissue in theropods

Introduction

The Lower Cretaceous Santana Formation of northeast Brazil is both a fossil conservation and concentration Lagerstätte (Martill, 1997). It is palaeontologically famous for the exceptional quality of preservation, abundance and high diversity of its vertebrate fossils, but the exceptional preservation of soft tissues was first noted in ostracods (Bate, 1971) and copepods found serendipitously during the acid preparation of fossil fishes (Cressey and Patterson, 1973). The Santana Formation has exceptional preservation of both original biominerals (bone and teeth) and fossilized soft tissues (Maisey, 1991; Martill, 2001). Its fossil assemblage is typified by fossil vertebrates enclosed in carbonate concretions that are usually articulated, three-dimensional and often have preserved soft tissues and stomach contents (Martill, 1988, 2001; Maisey, 1991; Wilby and Martill, 1992). Exposures of the Santana Formation are found mainly on the flanks of the Chapada do Araripe, an elevated tableland in northeast Brazil's fertile Val do Cariri in the otherwise semi-arid Caatinga at the boundaries of the states of Pernambuco, Ceará and Piauí (Fig 1). Here, while placing the formation in its geological and geographic setting, we examine both the biodiversity and evolutionary implications, and the mechanisms leading to enhanced fossil abundance and high-fidelity preservation. Although not a terrestrial deposit, it is certainly not a normal marine stenohaline deposit either, and its fossil assemblage includes a mix of aquatic vertebrates (fish, turtles), aerial forms (pterosaurs) and fully terrestrial forms (rare dinosaurs and crocodyliforms), as well as plants of arid habitats (Crane and Maisey, 1991).

The first published accounts of fossils from the formation date back to the early nineteenth century when the German naturalists Johann Baptist von Spix

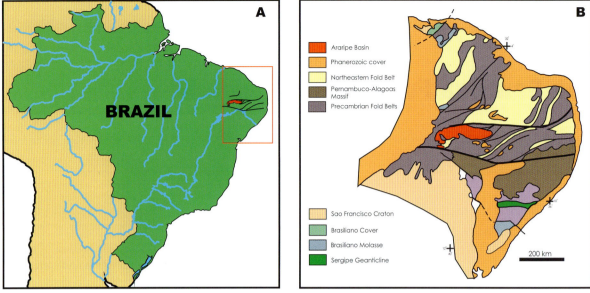
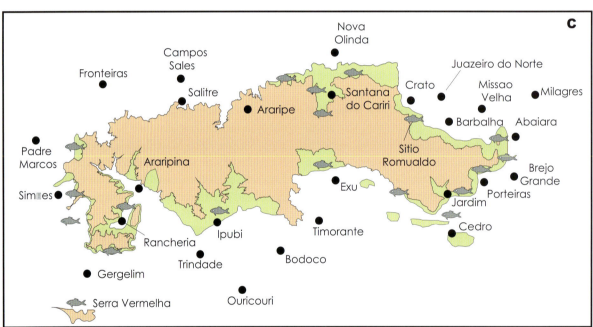

Figure 1 Locality maps. **A**, location of the Araripe Basin in north-eastern Brazil. **B**, main structural elements of north-eastern Brazil and the relationship of the Araripe Basin to the main basement megashears. **C**, the Chapada do Araripe (buff) and the main outcrops of the Araripe Group (pale yellow) with main fossiliferous localities of the Santana Formation indicated by the fish symbol. Many of the main towns are mentioned in the text.

(1781–1826) and Carl Friedrich Philipp von Martius (1794–1868) figured one of the now famous fossil fishes presented to them by the Governor of Piauí in their *Reise in Brasilien* (1823–1831) (Fig 2). Earlier collections of fossil fishes had been made and shipped to Europe (Lisbon), but no publications had arisen from this material until recently (Antunes, 2005). Also in this early period, Scottish botanist and explorer George Gardner (1809–1849) collected extensively from the Santana Formation in 1837 or 1838 and published a brief account of the geology of the area where the Santana fossils occur (Gardner, 1841). Gardner documented his travels in considerable detail in his posthumously published *Travels in the Interior of Brazil* (Gardner, 1846), and from this it has been possible to locate some of his original collecting localities (Martill,

Figure 2 First scientists to describe and illustrate fossils from the Chapada do Araripe, Johann Baptist Ritter von Spix (**A**) and Carl Friedrich Philipp von Martius (**B**) obtained a fossil of the common fish cf *Rhacolepis* sp. (**C**) from the Governor of Piauí in 1819, but they did not undertake any formal study of the geology or palaeontology of the Chapada do Araripe.

2007). Gardner had a lasting influence on later workers as he developed the idea of a 'layer-cake' stratigraphy for the Chapada do Araripe and (with Louis Agassiz's help) determined a Cretaceous age for the sequence. Gardner's fossils were shipped to the United Kingdom where some came to the attention of the ichthyologist

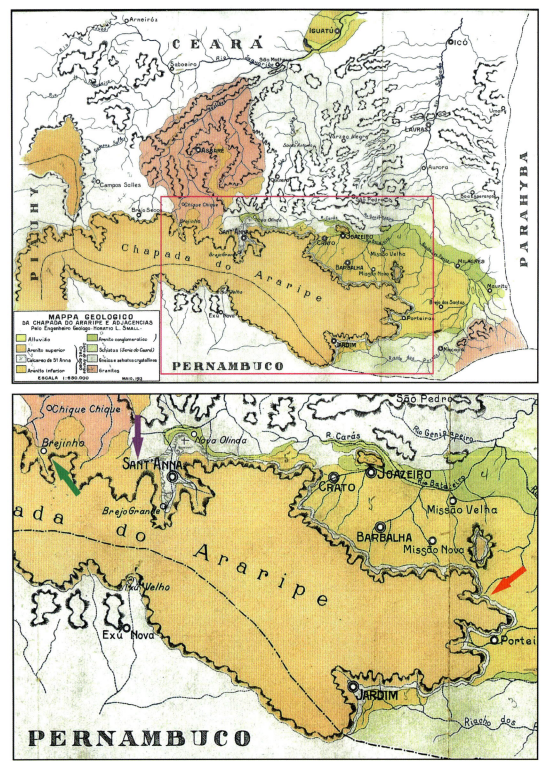

Figure 3 First geological map of the Val do Cariri, produced by Horatio Small in 1913. Top, Small's original map showing the area mapped (much of the unit mapped as granite is Proterozoic gneiss with some granitic intrusions). Below, detail of the outcrop of the Calcareo de S Anna (red arrow). The outcrop is overlapped by the Arenito Superior (Exu Formation of today) at the purple arrow. On the ground, this is where the Crato Formation Limestone is overlapped. The Santana Formation fossil Lagerstätte is overlapped at the green arrow near Brejinho. Thus Small's Calcareo de S Anna corresponds to today's concept of the Crato Formation.

Louis Agassiz (1807–1873), who began his scientific career working on the present-day Brazilian fishes collected by Spix and Martius during their expedition a few years earlier. Agassiz then turned his attention to Brazilian fossil fishes, becoming the first scientist to describe and name Santana Formation fossils (Agassiz, 1841), although at this time the formation had yet to be named.

There are few if any accounts of fossils from the Santana Formation for the latter half of the nineteenth century, and the strata of the Araripe Basin went largely ignored until Jordan and Branner (1908) produced an extensive account of the fossil fishes. Small (1913) described the stratigraphic succession of the Araripe Basin and coined the name Calcareo de Sant'Ana for a heterolithic sequence that includes the strata now called Santana Formation. Much of the history of the nomenclature of the stratigraphy is covered by Maisey (1991) and Martill (2007), but it may be worth noticing that Small (1913) published a geological map with a limestone layer mapped around the flanks of the Chapado do Araripe that corresponds with the outcrop of the Crato Formation (Fig 3). Thus, it might be fair to say that the Santana Formation *sensu stricto* should only be applied to the easily mapped Crato Formation underneath, rather than the fish concretion beds discussed here.

Following the pioneering palaeontological works of Agassiz and Jordan and Branner, there was again a long quiescence. Occasional papers describing new taxa appeared from time to time, with notable examples by Llewellyn Ivor Price (1905–1980) who described the first tetrapods from the Santana Formation. These included the crocodyliform *Araripesuchus* (Price, 1959) and the first pterosaur (Price, 1971). During the 1970s an increase in the commercial trade of the fossil fishes of the Santana Formation and the utilization of laminated limestones from the underlying Crato Formation prompted a massive increase in the palaeontological study of these two world-class deposits. There is a certain irony in that the commercial supply of fossils sparked the scientific interest, with some scientists now seeking an end to the commercial fossil trade (Langer *et al.*, 2012).

The 1980s saw a great increase in scientific interest on the Araripe Basin, with descriptions of many new discoveries, and reappraisals of much of the material described by earlier workers. Much of the renewed interest in the palaeoichthyology is attributable to Sylvie Wenz at the Muséum National d'Histoire Naturelle in Paris, followed by John Maisey at the American Museum of Natural History. But attention was now also turning to the palaeoenvironment, palaeoecology and the genesis of the Santana Formation. An early study by Mabesoone and Tinoco (1973) tried to explain the formation of the carbonate concretions around fossil fish as the adhering of sand grains to adipocere-coated fishes rolling around on a sandy shore, a scenario rebutted more than a decade later by Martill (1988, 1997) who instead invoked the sinking into soft substrates and very early diagenesis brought about by bacterial activity as the main mechanisms controlling preservation. The first conference dedicated to the Araripe Basin and its palaeontology was held in Crato in 1990. Entitled *Simpósio Sobre a Bacia do Araripe e Bacias Interiores do Nordeste*, this conference has now become an important component of the geological and palaeontological calendar in Brazil. Since then, there have been numerous studies on the Santana Formation biota, with considerable emphasis placed on the biogeographic importance of the assemblage and how it relates to the opening of the South Atlantic Ocean (e.g. Maisey, 2000).

More recently the results of a number of controlled excavations of the Romualdo Member have been presented, undertaken with the aim of determining the abundance of fossils and their stratigraphic and geographic distribution within the basin (Fara *et al.*, 2005; Saraiva *et al.*, 2007; Villa Nova *et al.*, 2011), and a considerable number of studies are ongoing, both in Brazil and on collected material housed in museums around the world. Carvalho and Santos (2005) have reviewed the history of palaeontological research in the Araripe Basin, and additional accounts can be found in Maisey (1991) and Martill (2007).

Geological context of the Santana Formation
Stratigraphic framework

The Santana Formation is one of several lithological units forming the diverse sedimentary fill of the Araripe Basin. Situated between the Pernambuco, Aurora and Patos megafractures in the Borborema Proterozoic basement (Fig 1B), the Araripe Basin developed during the Middle to Late Jurassic when opening of the South Atlantic Ocean reactivated a series of largely east–west trending shear zones developed during the Neoproterozoic closure of the

Brazilide Ocean. This ancient closure, involving the São Francisco/Congo, Rio de la Plata and Amazonian cratons of South America and central Africa and the cratons of West Africa, had a lasting effect on the geological evolution of the northeast of Brazil (Dalziel, 2012; Stump, 1987). Most of the sedimentary history of the Araripe Basin and its evolution is directly related to the opening of the South Atlantic Ocean and the break up of Gondwana (Fig 4).

The Araripe basin is largely bordered by a series of sinistral shear faults that controlled sediment thickness during much of the infill history, and generated local synsedimentary folding and faulting. There is some evidence that sedimentation spilled over the basin margins during high sea stands in the ?mid-Cretaceous (Arai and Coimbra, 1990, 2000). For example, in those places where concretion beds of the Santana Formation can be mapped adjacent to the faulted margins of the basin, there is no lateral lithological change to shallow water facies (e.g. between Tatajuba and Brejhino in southern Ceará). Similarly, there are many places where the uppermost unit, the Exu Formation, can be traced across the basin boundary faults, especially near Salitre and Araripe.

The oldest sediments in the Araripe Basin are a series of immature sandstones and conglomerates, usually named the Cariri Formation (the name Mauriti Formation has been used by some authors but Cariri has date priority), that have been dated as Siluro-Devonian based on palynomorphs (Brito and Quadros, 1995). Carvalho (2000) claimed a Cretaceous age for the Cariri Formation on the basis of dinosaur footprints, but it is more likely that the locality near Milagres where they occur is incorrectly mapped, and belongs instead to the Missao Velha Formation (see below). The Cariri Formation is extensively faulted and forms part of a topographic unconformity on which various Mesozoic strata rest. This unconformity can be easily studied around Nova Olinda in southern Ceará where it is progressively overlapped by the Missao Velha, Rio Batatiera, Crato, and probably Ipubi formations (Fig 5).

Of the Mesozoic strata, the oldest are a series of conglomerates and sandstones of fluvial origin, generally lacking diagnostic fossils. Berthou (1994) claimed that they may be as old as Triassic, but there is no substantial evidence for this. These immature fluvial strata are overlain by somewhat more mature fluvial sandstones and mudstones deposited in a meander belt system and have been referred to a lower Brejo Santos Formation and an upper Missao Velha Formation (Cavalcanti and Vianna, 1992). The upper unit yields the elasmobranchs *Planohybodus* and *Parvodus*, the coelacanth *Mawsonia*, the lungfish Neoceratodus and the actinopterygians *Vinctifer* sp. and *Lepidotes* sp. (Brito et al., 1994; Cupello et al., 2012). Both *Mawsonia* and *Vinctifer* indicate a Cretaceous age for at least the Missao Velha Formation, and Cupello et al. (2012) suggested a pre-Aptian age. These fluvio-lacustrine strata are overlain be a series of lacustrine, deltaic, saline lagoonal and marine lagoonal deposits that are richly fossiliferous at certain horizons. They include the Rio Batateira Formation that passes upwards in the Crato Formation, which includes the famous Nova Olinda Member Konservat-Lagerstätte (Martill et al., 2007), with diverse insects and an early angiosperm flora.

The Crato Formation is overlain by the extensive Ipubi Formation evaporites and is largely unfossiliferous, except for some intervening shales between evaporite beds. The Ipubi Formation generally comprises two thick seams of evaporites, mainly gypsum, with some anhydrite, with a depocentre (thickest occurrences) near Ipubi in Pernambuco. There are few published accounts on the stratigraphy of this unit. It is very thin in the eastern part of the basin (near Porteiras two gypsum beds separated by a black shale

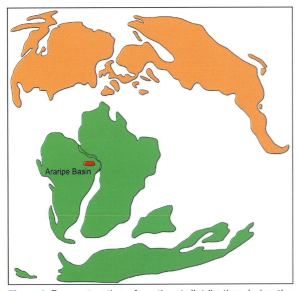

Figure 4 Reconstruction of continent distribution during the middle of the Cretaceous with the location of the Araripe Basin outlined in red. Simplified from Scotese (2001).

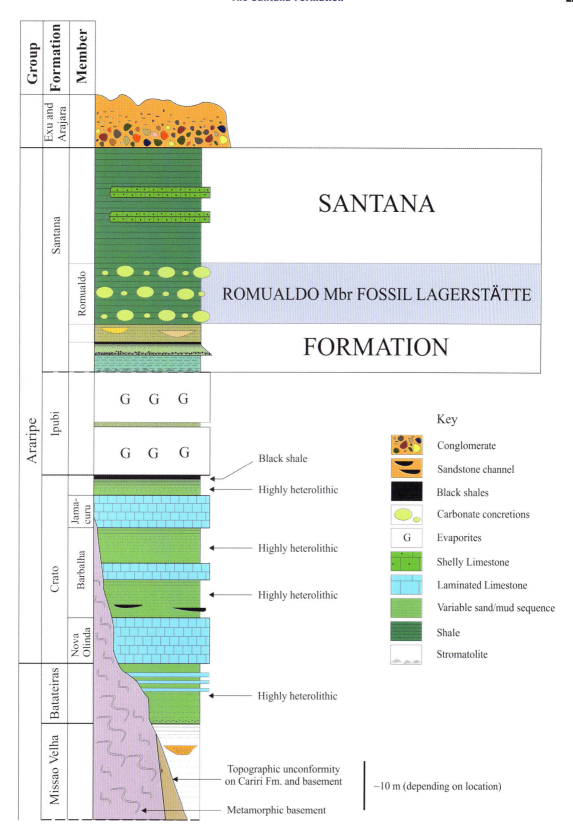

Figure 5 Composite generalized stratigraphic log for the Araripe Basin sedimentary fill and its relationship to the basement. Thicknesses may vary considerably over the basin, as may the lithology.

are no more than 2m thick) but thickens considerably around Santana do Cariri and also in the Araripina-Ipubi-Rancharia area. The unit can also be traced to the Serra Vermelha in Piauhí, and it is possible that these strata are correlates of the Cretaceous evaporites of Codo in Maranhão, some 500km to the west (Rossetti *et al.*, 2000). There is a prominent disconformity between the Ipubi Formation evaporites and the overlying Santana Formation (Fig 6). This disconformity was first identified by da Silva (1986a, 1986b, 1988)

Figure 6 Lateral lithological variation in the lower part of the Santana Formation. **A**, variegated red and green fluvio-lacustrine strata resting directly on Ipubi Formation evaporites at Holcim's Mina Sao Severino gypsum mine 7.5km SW of Ipubi, Pernambuco. The base of the Romualdo Member concretion Lagerstätte is seen in the top of the section. **B**, similar sequence to above, but with intervening laminated lacustrine limestones between the Ipubi Formation and the red and green fluvio-lacustrine strata. Seen at M.P.G, just 2km southwest of Ipubi, Pernambuco. **C**, thick sequence of deltaic sands and lacustrine mudstones at Mina Pedra Branco, 6km southwest of Nova Olinda, Ceará.

who argued for a karstic topographic unconformity, but this was a misinterpretation of modern subterranean karstic development close to the outcrop of the evaporites with subsequent collapse into solution hollows by overlying strata. In fresh exposures of this boundary in the large gypsum quarries around Ipubi and Araripina the disconformity is line sharp and largely concordant, or gently undulating. In places, a thin bone bed overlain by stromatolitic limestone is developed on the eroded surface of the gypsum, and ventifacts may also occur, as at Mina Pedra Branca 3.5km southwest of Nova Olinda.

Formerly the two formations (Crato and Ipubi) below this disconformity, of unknown duration, were included within the Santana Formation. Beurlen (1962) considered the Crato limestone unit a distinct formation, whereas da Silva (1986a, 1986b, 1988) united the Crato and Ipubi units in a single Araripina Formation. Martill and Wilby (1993) considered two units sufficiently distinct and treated the Crato and Ipubi units as distinct formations. The result of this was to leave the Santana Formation and the Romualdo Member one and the same thing. To resolve this, while maintaining the frequently used term Romualdo Member, Martill and Wilby (1993) limited the scope of the Romualdo Member to include only the fossil-bearing concretion part of the sequence. Subsequently, some authors have accepted this approach, whereas Vila Nova *et al.* (2011) elevated the Romualdo Member (*sensu* Martill and Wilby) to formation status, thus eliminating the Santana Formation. Here we follow Martill and Wilby (1993) as we feel their pragmatic approach reflects both the stratigraphic dynamics of the basin, while retaining names and definitions that have become entrenched in the literature. Thus, Santana Formation here is taken to be all strata lying above the Ipubi Formation evaporites and below the Exu Formation sandstone that forms the summit of the Chapada do Araripe and some adjacent mesas (Fig 5). The suite of strata from the base of the Rio da Batateira Formation to the top of the Santana Formation is widely referred to as the Araripe Group.

The Santana Formation

The areal extent of the Santana Formation is considerable. Its easternmost outcrop is on the flanks of the Chapada do Araripe in Ceará, where it is at least 10–12m thick, and the fossiliferous Romualdo Member is more than 6m thick (Heimhofer *et al.*, 2008). To the east, the land surface falls away beyond the chapada, and it is conceivable that the formation continued many kilometres beyond its present outcrop eastwards (see below). There are no outcrops of the Santana Formation beyond the flanks of the Chapada do Araripe toward the north, and in places Neoproterozoic strata are present at similar elevations to the Santana Formation outcrop, suggesting the northern limits of deposition may have been close to the present-day outcrop. That said, a peneplained surface on the Neoproterozoic rocks of the Borborema massif possibly represents a former Cretaceous erosion surface over which the Santana Formation (and other Araripe Group formations) may have transgressed. If so, such deposits were removed prior to deposition of the Exu Formation, which in places (i.e. near Araripe) sits on the peneplained surface with no strata of the Santana Formation interposed. The Santana Formation can be traced almost continuously along the flanks of the Chapada do Araripe in its eastern part, but is overlapped by the Exu Formation at Brejhino and does not re-emerge until the western flank of the Chapada in Piauí where it is at least 20m thick (Heimhofer *et al.*, 2008).

The topographic and outcrop situation at the western end of the chapada mirrors that of the eastern end, where the outcrop of the Santana Formation can be projected well beyond the present-day margin of the chapada. Evidence that the outcrop was considerably more extensive is found in several Mesozoic outliers forming mesas beyond the Chapada do Araripe. One of the largest of these, the Serra Vermelha, has good exposures of the Crato, Ipubi and Santana formations on its northern flank, but not its southern flank. The large Pernambuco lineament (also called the Zona de Cisalhamento) passes underneath the Serra Vermelha, but only the Exu Formation occurs on the southern side, with the Crato, Ipubi and Santana formations downthrown against this fault prior to Exu Formation deposition. It is unlikely that the ancient shoreline was adjacent to the fault because, once again, there is no significant facies change seen in any of these strata as they approach the fault zone. Similarly, the Romualdo Member facies at both the eastern and western ends is remarkably similar with only minor differences in thickness and sedimentary characteristics.

On the southern flanks of the Chapada do Araripe the outcrop of the Santana Formation is discontinuous, largely due to large fault-bounded blocks of basement projecting up into the Exu Formation, which, most likely, formed islands and peninsulas in the water body forming the Santana Formation. Excellent examples are seen at Trindade and near Araripina, both in Pernambuco. Beyond the Chapada do Araripe southwards are outliers of the older Crato Formation limestone, but the Ipubi and Santana formations are not present, presumably having been lost to post-Cretaceous erosion. Thus, while the present outcrop pattern of the Santana Formation indicates deposition in a water body approximately 196km east to west and 86km north to south, and thus with an area (estimated as an oval) of about 9850km^2, it was in reality much larger than this, although its shape was probably complex, given the topographic variability of the basement rocks on which it was deposited. Projecting the outcrop of the Santana Formation from the flanks of the chapada to the faulted margins of the Araripe Basin in the north and south gives a trapezoid with similar N–S and E–W distances, and an estimated area of 12,550km^2. Projecting the eastern and western outcrops to the nearest basement outcrop of similar altitude (Santana crops out at about 800m at the eastern end of the chapada and about 600m at the western end) suggests an even larger water body.

There are no geological barriers to a continuous outcrop extending eastwards as far as the South Atlantic Ocean, while westwards the nearest terrain of similar altitude occurs beyond the Amazon Basin in eastern Peru. It is thus plausible that the Araripe Basin section is a relic of a much more expansive system that is preserved merely because it is down-faulted between two east–west shear zones, and thus escaped much of the pre-Exu Formation erosion (Arai, 2000). Globally, sea levels were high at the beginning in the late Early and early Late Cretaceous (Aptian, Albian, Cenomanian and Turonian high stands) and may have transgressed much of South America. Remarkably similar deposits to the Romualdo Member, of possibly coeval age, and with comparable ichthyofaunas occur in Colombia (Weeks, 1956; Schultze and Stöhr, 1996) and Venezuela (Moody and Maisey, 1994). A very similar fish assemblage also occurs in the possibly Aptian Tepexi de Rodriguez Formation near Puebla, Mexico, although here the sedimentary facies of plattenkalk-like limestones is markedly different (Kashiyama et al., 2004).

Sedimentology

Sedimentologically the Santana Formation is highly variable both vertically and laterally, although much of this variability is found in the lower parts of the sequence (Fig 6). Here a fluvio-lacustrine facies dominates, but passes upwards into more laterally extensive mudstones with concretions (Romualdo Member), followed by green-grey silty mudstones with thin lumachelle limestones and lacking concretions before being truncated by a basin-wide disconformity (Fig 7). These latter two lithological units, the Romualdo Member and its overlying strata, are less variable across the basin with a similar general sequence being detected for more than 195km from Sobradinho in the east to Simões in the west. Typically the concretion-bearing horizon of the Romualdo Member passes up into greenish-grey shales and clays with thin beds of gastropod limestone dominated by *Paraglauconia* sp. In some places thin limestones with the echinoid *Bothryopneustes araripensis* attest to shallow marine influences (Manso and Hessel, 2007). Many of the gypsum quarries in the west display excellent sections through the lower Santana Formation, and often include the lower part of the Romualdo Member, but it is rare to find sections that pass up into the clays with gastropods. A notable exception is the Mina Lagoa dos Gregorios near Rancharia, Pernambuco, but here the sequence is steeply dipping and faulted (Fig 8).

Considerable sedimentological data can be derived from Romualdo Member concretions, which display a variety of fabrics reflecting extremely early diagenetic formation through to post-compaction formation at some horizons. The surrounding shales are usually dark grey or grey-green, perhaps silty (micaceous) and are often finely laminated. Fossils do occur within the shales themselves (e.g. at Mina dos Gregorios, near Rancharia in Pernambuco) but their relative rarity (except for ostracods) may be apparent, being due to concretions forming around most fossils. Some thin, horizontally extensive limestone beds are probably very large flat elliptical concretions, some of which envelop mass mortality fish horizons (Fig 9; Martill et al., 2008). There have been a limited number of geochemical appraisals of the Santana Formation in recent years. Berthou et al. (1990) analysed the clay mineralogy of the sequence and found it dominated by smectites in the lower parts of the sequence, becoming mixed smectite-illite dominant in the Romualdo Member and above. Analysis of oxygen and sulphur

Figure 7 Exposures of the Romualdo Member fossil Lagerstätte and higher beds of the Santana Formation. **A**, disconformity at the top of Santana Formation. Author (DMM) is standing on the disconformity between the Santana Formation and the overlying basal member of the Exu Formation. **B**, silty shales and clays of the upper Santana Formation with thin gastropod (*Paraglauconia* sp.) limestone beds as exposed in a stream section at Estiva, near Araporanga, Ceará. **C**, passage beds from the Romualdo Lagerstätte into the shales with gastropods. There is often a bed of large concretions (seen bottom left of image) at the top of the Romualdo Member. This exposure is at Sobradinho just a few metres downstream from A above. **D**, the concretion-bearing part of the Romualdo Member with beds of thin, laminated limestones and concretions. This exposure is in the abandoned gypsum mine at Mina Lagoa de Dentro, near Araripina, Pernambuco.

isotopes has been performed on the Ipubi Formation evaporites, indicating a strong marine influence (Berthou and Pierre, 1990), whereas carbon isotopes on the concretions from the Romualdo Member show a range of values becoming increasingly more negative away from the centre of the concretion (Heimhofer *et al.*, 2010). Organic carbon analyses were first reported by Baudin *et al.* (1990) who analysed the sequence from the Rio Bateiras Formation to the Santana Formation. A more thorough analysis was performed by Heimhofer *et al.* (2008, 2010) who also analysed the organic carbon through the Araripe Group succession in general (Heimhofer *et al.*, 2008) and in detail across the Romualdo Member concretions (Heimhofer *et al.*, 2010). Their results indicated high abundances of acyclic isoprenoids, steranes and isorenieratane and its derivatives, indicating the presence of anoxic bottom waters within the photic zone. Organic material in the Romualdo Member and other Araripe Group strata is thermally immature TAI <2, reflecting the low burial depth of these sediments (Baudin and Berthou, 1996).

Figure 8 Full sequence of Santana Formation exposed in tectonically disturbed region near Rancharia, Pernambuco. At least some of the folding here postdates Exu Formation deposition, but it is more intense (steeper-dipping) in the Santana Formation, suggesting some pre-Exu Formation disturbance occurred. Fossil fishes are as common in the shales as in the concretions at this locality.

Figure 9 Concretion split along a single lamina to reveal a surface entirely covered with small fishes: each black dot is the eye of a complete individual. This specimen comes from the Romualdo Member at Estiva, near Araporanga.

Sediment consistency

The sea floor of the Romualdo lagoon was periodically soupy, allowing the intrusion of carcasses descending from the surface waters into a liquefied substrate (Martill, 1997). Evidence for this is provided

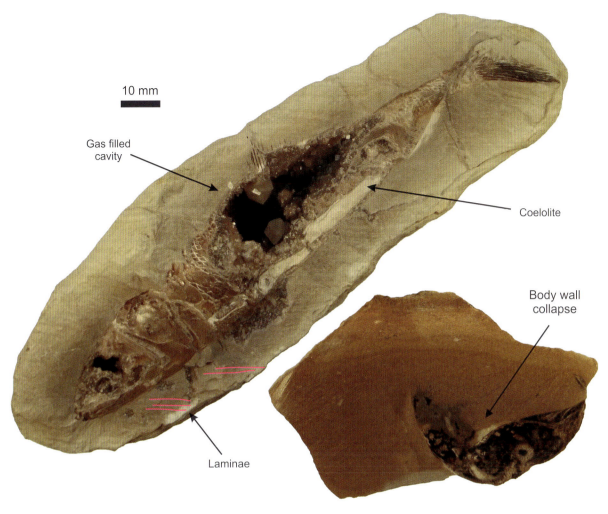

Figure 10 Section through a 3-D preserved example of *Rhacolepis* sp. OUP 96/2 from the Santana Formation at Mina Pedra Branca, Ceará. The fine-scale laminations of the concretion are cut at an angle of approximately 40 degrees. Parts of the body cavity have been filled with sediment that has ingressed through the mouth and gill openings, and perhaps through a rupture of the body wall during decay.

by three-dimensionally preserved, fully articulated fishes lying at high angles to the bedding and showing cross-cutting relationships with sedimentary laminae (Fig 10). In addition, some taphonomic features, such as rupture of the body wall, show that liquefied sediment was able to flow into bloated body cavities and through oral and gill openings. This highly fluid substrate allowed both perfect carcasses as well as some partly decomposed carcasses to enter the sediment and escape large scavengers. The introduction of significant quantities of nutrients in the form of an entire carcass caused enhanced in-sediment bacterial activity, particularly sulphate reduction and methanogenesis and consequent production of concretionary carbonate at a very early stage of diagenesis (Martill, 1988).

Fossil taphonomy and preservation

It is not surprising that several papers have addressed the taphonomy and preservation of the Santana Formation fossils. In an early study, Mabesoone and Tinoco (1973) proposed a novel method for the formation of the fish-bearing concretions of the Romualdo Member, involving the formation of adipocere coatings forming around the fish carcass during decay, followed by the adhering of sediment particles while the dead fished was rolled around on the shore, but presented no convincing evidence to support their hypothesis. Martill (1988, 1997) proposed a model in which dead fish sank to the bottom and into the sediment. These introductions of a large volume of nutrients to the bacterial community within the sediment promoted rapid production of the mineral phases involved in both soft-tissue preservation (calcium phosphates) and concretion formation (calcium carbonate and lesser amounts of iron sulphides). Later diagenesis saw the generation of late calcite cements, barite and perhaps celestine that formed in the gas- and fluid-generated cavities within the fish and within septarian cracks.

Preservation of original biominerals

Original biominerals of Romualdo Member fossils are restricted to the calcite and bone and dental apatite. Originally siliceous fossils have not been reported (but silicified fossils do occur) and aragonitic shells such as those of gastropods are only represented by internal moulds or have been transformed into calcite. By far the commonest fossils are the calcitic valves of ostracods, which can occur in such profusion as to form ostracod limestones, and many of the fish-bearing concretions are 'ostracodite', though not usually those where the fish are three-dimensionally preserved.

Thin-section petrography through ostracodite concretions reveals original microstructure of the shells with little or no recrystallizaton, but all internal voids are filled with fringing cements and euhedral calcites, and occasionally barite. Bones are preserved in original biominerals and show excellent preservation of original histology (Fig 11). In unweathered concretions bone usually appears dark (chocolate) brown, and thin sections through this bone reveals excellent preservation of osteocyte lacunae and canaliculi, while SEM analysis of the bone may reveal details of the original collagen fabric.

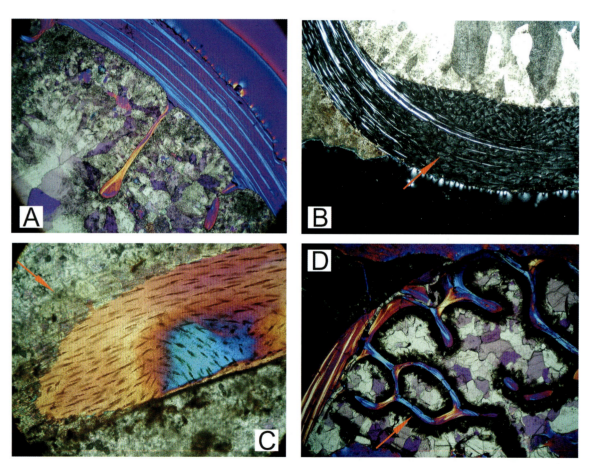

Figure 11 Thin sections through pterosaur bones from Santana Formation concretions. **A**, wall of long bone revealing laminated cortical bone and single trabeculum (cross-strut). **B**, cortical bone with secondary thickening of internal surface. **C**, detail of single trabeculum with only lightly mineralized cartilaginous growing margin. There is a high density of osteocyte lacunae and, clearly seen in the light blue zone, large numbers of short canaliculae. **D**, cross-section through a long bone showing ordered trabeculae forming a polygonal internal framework, presumably imparting strength near a joint or point of stress. All photographs taken under polarized light with crossed Nichols. A, C and D taken with quartz wedge inserted to enhance contrast.

Preservation of soft tissues

Soft tissues occur in two broad styles of preservation in the Romualdo Member concretions. In many of the fishes, but particularly examples of *Notelops*, *Rhacolepis* and *Brannerion* among the teleosts and the elasmobranchs *Iansan* and *Tribodus*, soft tissues are preserved in the calcium phosphate mineral francolite (Martill, 1988, 1994, 1998; Wilby, 1993). This mineral also preserves soft tissues in arthropods including decapod shrimps and ostracods (Fig 12; Bate, 1971, 1972, 1973; Smith, 1999, 2000) and has been reported in pterosaurs (Martill and Unwin, 1989; Martill *et al.*, 1990) and theropod dinosaurs (Kellner, 1996a). In addition to preservation in calcium phosphate, some presumed organic substances preserve soft tissues in the eyeballs of fishes (Fig 9) and the wing membranes of pterosaurs (Campos *et al.*, 1984), but this material has not been chemically analysed. It appears as a black, amorphous and soft material, or as a black layer often with some visible, but vague, fibrous structure (Fig 13).

By far the best preservation of soft tissue is found in the francolitic material where resolution of detail

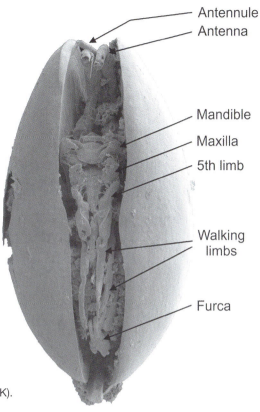

Figure 12 Scanning electron micrograph of Santana Formation ostracod *Pattersoncypris* (=*Harbinia*) *micropapillosa* Bate, 1972. Complete animal with all appendages preserved in calcium phosphate. Such well-preserved material is often found as part of the insoluble residue after acetic/formic acid preparation of fishes preserved in concretions. Image courtesy of Dr Giles Miller (NHMUK).

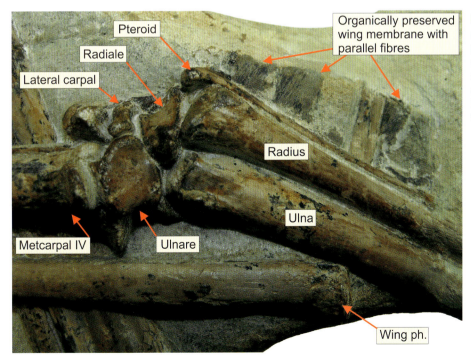

Figure 13 Portion of pterosaur wing preserved as black organic residue with some preservation of aktinofibrils (arrowed). Undescribed new taxon.

may be at the macromolecular level in muscle and connective tissue (Fig 14; Martill, 1990), and at the organelle level in some tissues where cells are preserved (Martill, 1998). Entire organs may be preserved, including the gill apparatus (Fig 15), stomach and other parts of the intestinal tract.

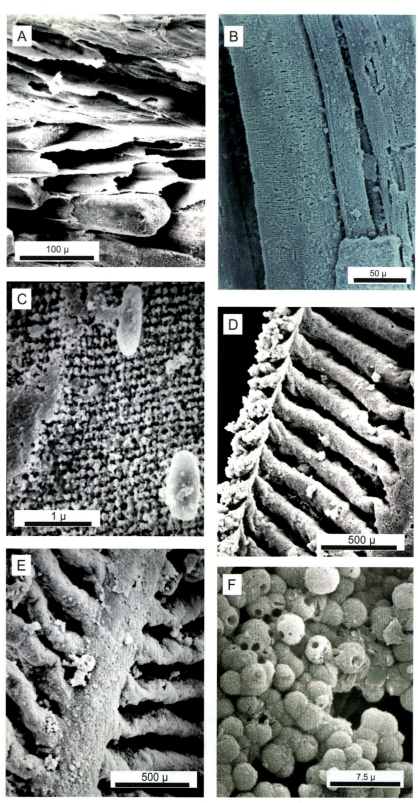

Figure 14 Scanning electron micrographs of phosphatized soft tissues from Santana Formation fishes. **A**, muscle fibres with sarcolemmic membranes preserved around individual fibres. **B**, striated muscle fibres exhibiting cross-banding. **C**, surface of striated muscle fibre with two possible mitochondria or nuclei. **D**, portion of gill filament showing open space between individual secondary lamellae. Such spacing suggests extremely early fossilization, as lamellae collapse shortly after death. **E**, gill filament with secondary lamellae and ?afferent blood vessel. **F**, possible lithified coccoid bacteria. Often such structures are found associated with preserved soft tissue. They may represent bacteria that were responsible for partial decay of the fish during a post-mortem floating phase at the surface.

Figure 15 Soft tissues preserved in Santana Formation fishes. **A**, near complete example of *Notelops brama* with extensive patches of phosphatized muscle tissues, especially in the caudal peduncle. **B**, skull and anterior body of *Notelops brama* after light acid digestion. Gill filaments can be seen behind the operculum and muscle fibres, **C**, are seen in the rib cage.

In some cases soft tissues appear to be preserved by a metasomatic replacement of original biopolymers, perhaps analogous in living systems to the osteogenesis of collagen. In such circumstances, the preserved soft tissues show well-organized crystallites of cryptocrystalline francolite located in protein-rich parts of muscle fibres and in connective tissues, sometimes reflecting the ultrastructure of the original material. Some preserved tissues show only the gross anatomy, and are usually composed of clusters of spheroidal crystal aggregates. These spheroids are usually in the size range of 1–3 micrometres, and composed of randomly oriented crystallites with diameters of just a few tens of Ångströms (Wilby, 1993). Soft tissues may also be preserved as thin crusts of francolite that have overgrown a tissue surface or cell wall/membrane, with an underlying void generated by decomposition of the original tissue/cell (Martill, 1998).

Martill and Harper (1990) examined the timing of soft-tissue preservation, undertaking a series of actualistic experiments in which they tried to relate fabrics found in fossil gills to those developed during decomposition of recently killed fishes. They found that many of the fabrics found in fossilized fish gills from the Santana Formation resembled either near perfect structures, or fabrics (artefacts) developed after just a few hours of decay. From these experiments, they concluded that phosphatization occurred (or at least began) almost as soon as the dead fish entered that part of the water column or sediment pile where fossilization was occurring. In some cases, this time period may have been as little as a few hours for the initiation of phosphatization.

Biodiversity of the Santana Formation

It is important to remember that the Santana Formation is a heterolithic sequence and that different parts of the succession possess distinctive fossil assemblages, either due to taphonomic effects of sediment/water chemistry on preservation, or because differences in sediment type represent significant differences in palaeoenvironment and original biotic composition. Many authors have chosen to ignore this aspect, and have considered the biota as a whole, while others have even included the biota from the Crato Formation as a part of the assemblage (e.g. Maisey, 1991). Here we consider the composition of the concretion-bearing Romualdo Member as distinct from that of the somewhat higher strata of the Santana Formation with thin limestone beds mentioned above and from that of the fluvio-deltaic facies of the lower Santana Formation.

The biota of the concretion beds

By far the most abundant faunal component of the Santana Formation concretion beds are ostracods (Fig 12). These are mainly mostly smooth-shelled forms and were variously described as *Hourqia* or *Pattersoncypris* but are now referred to the Chinese genus *Harbinia* (Carmo *et al.*, 2008). In some places they form ostracod limestones, and occur in countless numbers as articulated mature instars as well as isolated valves. Phosphatized examples associated with fish skeletons in the concretions are preserved with their appendages and gonads intact (Bate, 1971, 1972, 1973; Smith, 2000), and apparently males possessed giant spermatozoans, as revealed by holotomography (Matzke-Karasz *et al.*, 2009). Ostracod eggs may also be preserved (Smith, 1999).

Other invertebrates occur more rarely, and are often only encountered in unusual circumstances or local geographical regions. The parasitic copepod *Kabatarina pattersoni* Cressey and Boxshall, 1989 (Fig 16; holotype now lost) was discovered on the gills of the ichthyodectid fish *Cladocyclus gardneri*, and Maisey and Carvalho (1995) and Wilby and Martill (1992) recorded sergestid decapods and brachyuran crab larvae in the stomach of the teleost fish *Rhacolepis*. Decapod shrimps are known from a few of the concretion horizons at Jardim and near Simões, but these have yet to be studied.

Molluscs are rather rare in the concretions, possibly as a result of early dissolution of aragonite. Small (1–2mm) gastropods occur occasionally as pyritized internal moulds in acid-preparation residues, whereas larger, but much rarer examples occur as internal sediment fills. The relative abundance of fishes with durophagous dentitions (such as the pycnodonts *Iemanja palma*

Figure 16 Reconstruction of the female parasitic copepod *Kabatarina pattersoni* Cressey and Boxshall, 1989, in ventral view found adhering to the gills of the fish *Cladocyclus gardneri* Agassiz, 1841. Such exceptional fossil occurrences demonstrate the importance of conservation Lagerstätten for a wide range of palaeobiological and evolutionary studies. Scale bar equals 100 micrometres. Image courtesy of the Micropaleontology Project, New York.

and *Neoproscinetes penalvai* [Fig 17] and the elasmobranchs *Tribodus limae* and *Iansan beurleni*) suggests that perhaps more 'crunchy' food was available than the preserved record indicates. Benthic foraminifera occur at some levels with *Quinqueloculina* sp. occurring rarely in the Romualdo concretions (Berthou *et al.*, 1990).

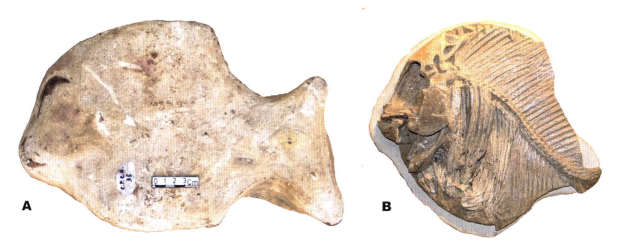

Figure 17 Two concretions, one entire (**A**) and unprepared, the other (**B**) split to reveal inside the pycnodont fish *Neoproscinetes penalvai* (Silva Santos, 1970). Concretion outline in the Santana Formation often mirrors closely the enclosed fossil, and in the case of fishes with a distinctive outline, they can be identified without recourse to the geological hammer. Pycnodonts possessed durophagous dentitions ideal for feeding on shelled invertebrates and were probably benthic feeders.

By contrast, the vertebrate assemblage is remarkably diverse with a variety of elasmobranch, actinopterygian and actinistian fishes, turtles, crocodyliforms, pterosaurs (see below) and dinosaurs. Brito and Yabumoto (2011) most recently reviewed the fish assemblage, taking into account several recent discoveries of new taxa, but there has not been a review of the tetrapod assemblage in recent years. The account of Santana Formation tetrapods in Maisey (1991) is the most comprehensive, but there have been many discoveries of new pterosaurs, dinosaurs and turtles since that time.

Fishes

Most palaeontological attention on the Santana Formation has focused on the fishes with numerous studies of their diversity, biogeography and preservation (Brito and Yabumoto, 2011). The earliest detailed studies date back to Louis Agassiz (1833–44, 1841, 1844) who identified the specimens collected by George Gardner. Jordan and Branner (1908) undertook a larger and more comprehensive study, with many subsequent authors describing new genera and species as and when they arose (e.g. Silva Santos, 1945, 1958, 1971, 1983 and many other works; Dunkle, 1940; Brito and Ferreira, 1989; Wenz, 1989; Wenz and Brito, 1992; Yabumoto, 2002; Brito and Gallo, 2003; Mayrinck *et al.*, 2009). Aside from the descriptions of new taxa, there has also been considerable interest in their preservation (e.g. Martill, 1988, 1998, 1990), palaeoecology (Mabesoone and Tinoco, 1973; Wilby and Martill 1992; Maisey 1994; Martill *et al.*, 2008), palaeobiogeography (Maisey, 1993b) and palaeohistology (Brito and Meunier, 2000; Brito *et al.*, 2000, 2010), aspects of which are considered in the brief systematic review of the ichthyofauna below.

Elasmobranchs

Elasmobranchs are a scarce component of the Santana fish assemblage, with only two known forms. The rajiform skate *Iansan beurleni* (Silva Santos, 1968) is a small form with a length of around 50cm (Fig 18). Originally placed in the genus *Rhinobatus*, it was re-evaluated by Brito and Séret (1996) who recognized it as a distinct genus. *Tribodus limae* is a hybodont shark that reached a length of a little more than 1m. Both taxa have 'under-slung' mouths and a durophagous dentition suitable for processing benthic shelled invertebrates. Their presence in the assemblage clearly attests to periods when bottom waters of the Santana lagoon were well oxygenated, but their rarity might suggest that oxygenation events were perhaps punctuated by longer periods of dysoxia or even anoxia. *Tribodus* has been recorded from Europe and North Africa and is suspected of having been tolerant of a considerable range of salinities (Vullo and Neraudeau, 2008).

Actinopterygians

Actinopterygian fishes form the most diverse group of vertebrates in the Santana Formation, with some 21

Figure 18 The rhinobatoid ray *Iansan beurleni* (Silva Santos, 1968) is one of just two elasmobranchs in the Santana Formation. It may have been tolerant of a range of salinities, and was a feeder on small shelled benthos.

genera recognized (Table 1). The assemblage includes a wide range of morphs (Figs 17, 19), sizes and ecologies from surface living planktivores (*Vinctifer*), forms taking small prey but of relatively large size (e.g. *Cladocyclus*), large (possibly apex) predators (*Calamopleurus*) to benthic feeders of invertebrates (pycnodonts and lepidotids). In one of the first studies of its kind for a Cretaceous fossil fish assemblage, Maisey (1994) analysed the trophic structure of the fish community in detail. Many of the fish genera from

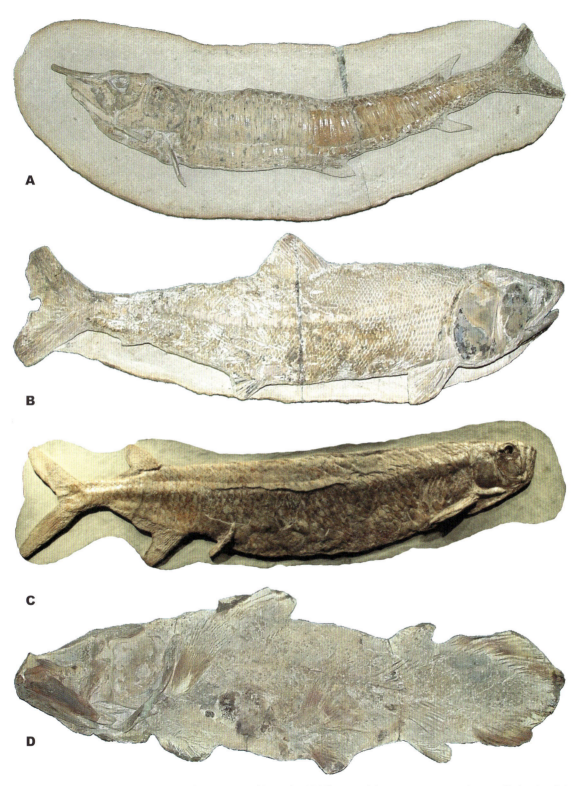

Figure 19 Fishes in concretions. **A**, *Vinctifer comptoni* (Agassiz, 1841), one of the commonest and most distinctive fishes from the Santana Formation. It is easily recognized by its shiny, elongate flank scales and extended rostrum. It has been used as an index fossil for the Cretaceous in South America. **B**, *Paraelops cearensis* Silva Santos, 1968, a large albuloid with extremely small teeth coating the jaws. **C**, *Cladocyclus gardneri* Agassiz, 1841. Another very common Santana fish, reaching a length of over 1 m. **D**, the coelacanth *Axelrodichthys araripensis* Maisey, 1986.

the Santana Formation have been found elsewhere in Brazil or South America, and a few are known also from Africa, notably the Cenomanian deposits of the Kem Kem in Morocco (e.g. *Mawsonia*, *Calamopleurus*; Cavin *et al.*, 2001).

Actinistia

Coelacanths are not uncommon in the Santana Formation, with two latimeroid genera, *Axelrodichthys* (Fig 19D) and *Mawsonia*, described. The most abundant appears to be *Axelrodichthys araripenis*, a moderately sized form attaining a length of around 1.5 m. This contrasts markedly with the much rarer *Mawsonia brasiliensis*, which may have reached several metres in length and was the largest fish in the Santana Formation (Maisey, 1991). *Mawsonia* from the Santana Formation appears to be a distinct species from *Mawsonia gigas* of Bahia and Africa (Yabumoto, 2002). *Axelrodichthys* appears to be edentulous, whereas *Mawsonia brasiliensis* has a pair of toothplates on the dentary, but with extremely small teeth, contrasting with the extant *Latimeria*, which has robust, well spaced conical teeth in both jaws.

Turtles

The Santana Formation is particularly significant with regard to fossil turtles, firstly, in that their remains are frequently well preserved (Fig 20; Oliveira, 2007), and secondly, the assemblage contains a mix of usually freshwater forms and the earliest of the modern marine turtles (Meylan and Gaffney, 1991; Gaffney and Meylan, 1991; Meylan, 1996; Hirayama, 1998). Six genera have been named comprising five species of pleurodires and a single cryptodire. The cryptodiran *Santanachelys gaffneyi* Hirayama, 1998 is regarded as the oldest known sea turtle and possessed the large salt glands necessary for life in the marine realm. However, it differed from all extant sea turtles in that the digits of the forelimb were separate rather than enclosed in a single, paddle-like structure. The pleurodiran chelonians include the bothremyidid *Cearachelys placidoi* Gaffney *et al.*, 2001, *Araripemys barretoi* Price, 1973, *Brasilemys josai* Broin, 2000, *Euraxemys essweini* Gaffney, Tong and Meylan, 2006, and *Caririemys violetae* Oliveira and Kellner, 2007. *Araripemys* is close to the base of Pleurodira and shares features with both pelomedusids and chelids (Meylan and Gaffney, 1991), whereas *Brasilemys* is probably the oldest pre-podocnemidid turtle known so far. *Cearachelys placidoi* Gaffney *et al.*, 2001 is only known from fragments and is possibly synonymous with *Euraxemys essweini*, while *Caririemys violetae* is also thought to be closely related to *Euraxemys*.

Crocodyliformes

Crocodyliform remains are extremely rare in the Santana Formation, but of those specimens known, most are exquisitely preserved skeletons of juveniles to possible adults of more than 2m length. Two genera are known: *Araripesuchus*, a notosuchian first described by Price (1959), is known from a single species in the Araripe Basin. *Araripesuchus gomesii* is a short-snouted form probably not reaching more than 1m in length and was probably terrestrial. The other crocodyliform from the Santana Formation was first reported from the Santana Formation by Kellner (1987) who proposed a new genus and species *Caririsuchus camposi*. Buffetaut

Figure 20 The pelomedusid turtle *Araripemys barretoi* Price, 1973. **A**, carapace seen in dorsal view and **B**, shell in ventral aspect.

(1991) regarded *Carirsuchus* as a subjective junior synonym of *Itasuchus*, first reported from the Upper Cretaceous Marília Formation, Bauru Group of Minas Gerais (Price, 1955), but Kellner and Campos (1999) retained *Carirsuchus* without commenting on Buffetaut's synonymy. *Carirsuchus* (if that is the valid name) was around about 1.5m long and probably aquatic. The holotype of *Carirsuchus camposi* comprises a few scraps of a snout that appear to have been salvaged from a near complete skeleton whose whereabouts are unknown (Buffetaut, 1991).

Dinosauria

Despite being a quasi-marine deposit, the concretions of the Romualdo Member have yielded the remains of several dinosaurs, including juveniles and adults, some complete, fully articulated skeletons, and others represented only by parts of skeletons and isolated bones. By far the best known example is the spinosaurid *Irritator challengeri* Martill *et al.*, 1996, represented by a nearly complete skull, missing only the anterior end of the snout (Sues *et al.*, 2002). Interestingly, the tip of a snout of a spinosaurid named *Angaturama limae* with teeth identical to those of *Irritator challengeri*, and of a similar size, was described later the same year (Kellner and Campos, 1996) and is almost certainly synonymous with the latter. A series of sacral vertebrae referred to a spinosaurid were described by Bittencourt and Kellner (2004) and may represent the first postcranial remains of *Irritator*. Small coelurosaurian dinosaurs have also been reported, including a possible compsognathid *Mirischia asymmetrica* (Naish *et al.*, 2004) and the possible maniraptoriform *Santanaraptor placidus* Kellner, 1999. *Mirischia* is known from a single specimen comprising a pelvis, sacral vertebrae, femur and proximal end of the tibia (Martill *et al.*, 2000), and an undescribed second juvenile example that is nearly complete (Fig 21). *Santanaraptor* is known from an associated partial pelvis, hind limbs and vertebrae with extensive soft tissues, including muscle fibres (Kellner, 1996c).

Frey and Martill (1995) described a series of theropod vertebrae as having affinities with Oviraptorosauria, although other authors have questioned this interpretation (e.g. Makovicky and Sues, 1998; Kellner, 1999; but see Naish *et al.*, 2004), and it

Figure 21 A near complete, almost perfect example of a baby theropod dinosaur. This specimen has yet to be studied, but it appears to represent a juvenile of *Mirischia asymmetrica*

is by no means certain that oviraptorosaurs were present in South America at all (Agnolin and Martinelli, 2007). In addition to the described material there are accounts of other dinosaur specimens, including a new partial coelurosaur (Naish et al., in review). Of particular significance is the overall composition of the Santana dinosaur assemblage, which is dominated by theropods. Naish et al. (2004) considered this apparent bias, noting that it was premature to draw too many conclusions from what is a very small data set. Nevertheless, the dinosaur assemblage from the Santana Formation is comparable with those from the Upper Jurassic Rögling, Solnhofen and Mörnsheim formations of Bavaria (Germany), which also contain an exclusively theropod assemblage (*Archaeopteryx*, *Compsognathus*, *Juravenator*, *Sciurumimus*; Göhlich et al., 2006; Wellnhofer, 2009; Rauhut et al., 2012). The lack of herbivorous sauropodomorphs and ornithischians may reflect poorly vegetated, arid hinterlands surrounding the Santana lagoon.

The dinosaur assemblage from the Santana Formation should be considered semi-allochthonous in that the specimens most likely drifted from shorelines (conceivably small examples might have been dropped by large predatory pterosaurs) where they may have scavenged on dead fish or fed directly on invertebrates (crustaceans) and other small animals.

An ancient ecosystem

The great abundance and diversity of fossils from the Santana Formation fossils has promoted attempts to reconstruct this ancient ecosystem. Some of these analyses have included the biota of the underlying Crato Formation, leading to a composite assemblage that may fall a long way short of a real community. The marked differences between these two Lagerstätten are a consequence of different palaeoenvironmental and sedimentological settings, which would have had a considerable effect on the nature of the palaeobiotas existing at the time. The Crato Lagerstätte contrasts with that from the Romualdo Member in its high abundance of insects, arachnids and plants but a very low diversity of fishes (Brito, 2008). It is true that there are similarities at higher taxonomic levels, but the differences more than outweigh arguments for a composite Romualdo/Crato ecosystem.

The palaeoecology of the fish assemblage from the Romualdo Member as a whole has been examined in detail by Maisey (1994) who examined the predator–prey relationships of the fishes, taking into account morphology, comparisons with extant relatives, and fossil fish stomach contents. Notably, pelagic fish species are more abundant that benthic forms, which is consistent with the findings of Heimhofer et al. (2008) that bottom conditions were often anoxic. However, benthic feeding did occur, and many forms with durophagous dentitions (the elasmobranchs *Tribodus* and *Iansan*, *Lepidotes*, and the pycnodonts *Neoproscinetes* and *Iamanja*) were most likely feeding at lower levels in the water column, but these may have been around topographic highs that projected above the oxic/anoxic boundary. However, ordinary benthos (gastropods, bivalves, larger decapods) is surprisingly rare (but can be abundant locally) in the Romualdo Member. Decapods are common at a few localities, most notably Simões (Martill and Wilby, personal observations) and Jardim (specimens in Jardim Museum) but benthic molluscs are absent, probably due to a combination of factors. The soupy nature of the sediment inhibited burrowers and crawlers, requiring purchase on firm ground. There may have been post-burial dissolution of aragonite, which would strongly bias perceptions of original composition of the benthos. The superabundance of ostracods at many horizons within the Romualdo Member also attests to oxygenated bottom waters. It is thus highly likely that bottom-water anoxia was seasonal or at least intermittent. Possibly anoxic water reached high levels and may have been the cause of fish mass-mortality events (Martill et al., 2008).

Water depth

There is no reliable sedimentological evidence for accurate water depth determination for the Romualdo Member. Below the concretion beds of the Romualdo Member are fluvio-lacustrine sediments in the west of the basin and deltaic facies in the north and east showing that the starting position is sea level or nearly so. The boundary between the Romualdo Member and the underlying strata forms a sharp line in the field, but is not apparently unconformable, although the disparity between the underlying strata over a wide area might hint at this. Strata above the Romualdo Formation are greenish-grey clays with shallow water echinoderms (*Bothriopneustes*) and cassiopid gastropods

(*Paraglauconia* sp.). With shallow water sequences both above and below, it might be expected that the strata of the Romualdo Member were not laid down in deep water. The organic geochemistry indicates that anoxia developed within the photic zone by photosynthetic Chlorobiaceae (the brown strain of the Green Sulphur bacteria) (Heimhofer *et al.*, 2008). However, in the Black Sea and some Norwegian fjords, Chlorobiaceae occur at the interface between anoxic, hydrogen-sulphide-rich waters and the oxygenated zone, a chemocline that commonly lies within the water column, rather than on the sea floor. Thus, photic-zone euxinia cannot be used to gauge water depth with precision. Also of little help are the abundant fishes. The coelacanths *Mawsonia* and *Axelrodichthys* are rather different from the extant coelacanth *Latimeria* (Forey, 1991), which inhabits water depths from 55 to 150 m between night and day, and therefore may not be reliable depth indicators for the Romualdo Member. The rhinobatid ray *Iansan* was a benthic form; present-day guitar fishes (Rhinobatidae) range from water depths of 1 to 112m. The ubiquitous cypridid ostracods today range from water depths of a few centimetres to well over 100m. Thus, little of the biota is of value for determining the water depth during deposition of the Romualdo Member. It seems that neither the age nor the water depth can be determined with any precision.

Palaeosalinity

It is beyond question that some of the upper part of the Santana Formation was deposited in marine waters. However, the fossil assemblage from the Romualdo Member concretions contains a variety of forms ranging from aerial (pterosaurs), terrestrial (theropod dinosaurs) to freshwater or euryhaline (the gars *Obaichthys decoratus* and *Dentilepisosteus laevis*, the hybodont *Tribodus*), normal marine (*Cladocyclus*, *Araripichthys*) and possibly hypersaline (*Tharrias* sp.), and it cannot be ruled out that some taxa may have been anadromous seasonally. Among the invertebrates, typical marine Cretaceous forms such as ammonites, belemnites, brachiopods, bryozoans or corals have never been recorded from the Romualdo Member. Decapod crustaceans do occur, and although today they are predominantly marine, many forms also occur in freshwater. Similarly, the parasitic copepod *Kabatrina* (Fig 16) is regarded as a dichelestiid siphonostomatan by Cressey and Boxshall (1989). Siphonostomatans are well documented from freshwater (Piasecki *et al.*, 2004), but the family Dichelesthiidae is mainly marine, with just a few forms parasitizing anadromous fishes (Wilson, 1922). This may suggest that the ichthyodectiform fish *Cladocyclus* was anadromous, entering freshwater for spawning (or parasite removal?). The turtle *Santanachelys gaffneyi* is the oldest known sea turtle. Belonging to the extinct family Protostegidae, it is by no means certain that this turtle was a fully stenohaline form, but it possessed large interorbital foramina that indicate large lacrimal salt glands surrounding the eyes (Hirayama, 1998). If not fully marine, *Santanachelys* was certainly salt tolerant. However, it is exceedingly rare, and it might be expected that if the Romualdo water body was of normal marine salinity, this turtle might occur more frequently, as might other marine fossils. Similarly, if it were a freshwater body, it might be expected to encounter the gars *Obaichthys* and *Dentilepisosteus* more commonly. The Romualdo water body was an unusual inland sea, and its diverse, but restricted fossil assemblage is probably a result of differences in salinity, water depth, nutrient input, and more or less anything else that affected the physical nature of the water and the sea floor. Whatever the sum total of these factors, they led to exceptional preservation of (at least some) of the biota. Some levels of the Romualdo Member are so rich in fossils that, when examining some bedding planes, it is tempting to consider that perhaps all bones that reached the sea floor were preserved.

Significance of the Santana Formation Fossil Lagerstätte: a case study using the Pterosauria

Although the Santana Formation is not strictly a terrestrial deposit, it nevertheless yields a large number of non-marine tetrapods preserved in such an excellent condition that it provides important insights into the nature and evolution of Early Cretaceous Gondwanan terrestrial faunas. However, it is also fair to say that this significance is only just being appreciated, despite the Santana Formation being among the best-known Mesozoic Lagerstätten of Gondwana. Its significance can be considered from three perspectives. Firstly, the quality of preservation of the biota provides opportunities to examine the morphology, biomechanics and perhaps physiology and behaviour of individual components of the biota and to evaluate their autecology. While this can be achieved in theory for many deposits, the Santana Formation is so rich in fossils, that specimens may be 'sacrificed' to gain an understanding of palaeohistology, internal structure, stomach contents, etc. for many taxa. Secondly, the Santana Formation Lagerstätte has 'captured' a significant sample of the biota of the Araripe Basin, facilitating attempts to construct the ancient ecosystem and trace some of its complexity from predation to parasitism, with all the caveats that go with such interpretations using fossil data (Maisey, 1994). Thirdly, the entire assemblage can be studied in an evolutionary context, both to examine the complexity of the evolving ecosystem and to place its individual components in an evolutionary framework for specific clades: all of which can be placed in a biogeographic framework, specifically with regard to the opening of the South and Central Atlantic basins during the mid-Cretaceous (Maisey, 1993b, 1994, 2000). Much emphasis has been placed on the significance of the ichthyofauna of the Santana Formation, which has been reviewed extensively by Maisey (1990, 1993, 1994, 2000) and Brito and Yabumoto (2011). Here we discuss the evolutionary significance of the Santana Formation with special reference to the Pterosauria.

The first pterosaur remains from the Santana Formation concretions were reported by Price (1971) who described some associated wing bones as *Araripesaurus castilhoi*. Most subsequent authors have considered this taxon indeterminate, as it lacks diagnostic features that, in pterosaurs, are mainly found in the skull (Unwin, 2006). This original material, which was figured by Wellnhofer (1991b), comprises bones of a single wing, loosely articulated, and including most of the carpus and metacarpus. Vila Nova *et al.* (2010) regarded *Araripesaurus* as an ornithocheirid, but its humerus, which might be diagnostic for Ornithocheiroidea, is not known. Since that initial discovery (pterosaur remains were often discarded by commercial fossil diggers in the 1970s and 1980s as they did not look like fish and could not easily be sold on the tourist markets of Ipanema and Copacabana) scientific interest in their remains began to increase, and spectacular specimens (Figs 22, 23) began finding their way into museums in Europe and

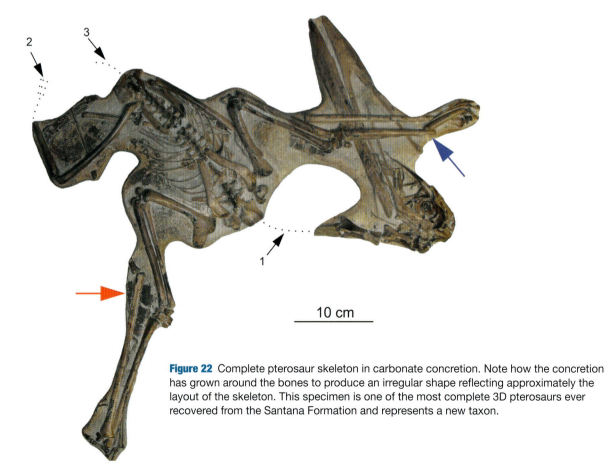

Figure 22 Complete pterosaur skeleton in carbonate concretion. Note how the concretion has grown around the bones to produce an irregular shape reflecting approximately the layout of the skeleton. This specimen is one of the most complete 3D pterosaurs ever recovered from the Santana Formation and represents a new taxon.

Figure 23 An acid-prepared three-dimensional skull of Anhanguera (NHMUK R11978). Such specimens allow a complete understanding of the structure of the pterosaur skull and, when CT scanned, the internal anatomy of the braincase (see, for example, http://digimorph.org/specimens/Anhanguera_santanae/).

North America (Wellnhofer, 1977, 1985, 1987a, 1988, 1991a,b; Wellnhofer and Kellner, 1991). The demand for pterosaur fossils escalated, but so did the number of specimens coming out of the ground. Today, many excellent, nearly complete skeletons and three-dimensionally preserved skulls of pterosaurs from the Santana Formation are displayed in museums in Germany, Japan, The Netherlands, Portugal, Switzerland, the United Kingdom and the United States (Eck et al., 2011; Elgin et al., 2011; Kellner and Tomida, 2000; Martill, 2011; Veldmeijer, 2002, 2003; Veldmeijer and Hense, 2004; Wellnhofer, 1977, 1985, 1987a, 1991a). These specimens have formed the basis for a wide range of studies including phylogenetic analyses (Kellner, 2003; Unwin, 2003; Martill and Naish, 2006), biomechanics (Unwin et al., 1996; Wilkinson et al., 2006; Humphries et al., 2007), systematics (Kellner and Tomida, 2000; Veldmeijer, 2003; Martill, 2011) and palaeohistology (Sayão, 2003).

Santana Formation pterosaurs: taxic diversity

In recent times the pterosaurs of the Santana Formation have overshadowed the fish in terms of palaeontological attention. To date some 13 genera have been reported from the Romualdo Member concretions (Table 2), although many of these names are likely subjective junior synonyms of the others: for example, specimens of *Anhanguera* are probably juveniles of *Coloborhynchus* whereas *Cearadactylus* may be a subjective junior synonym of *Coloborhychus* (Unwin, pers. comm.). It is difficult to establish just how many pterosaur taxa are present in the Santana Formation concretions, as pterosaurian taxonomy is in a flux and many of the listed taxa are based on material that is difficult to diagnose. Here we follow the more robust phylogenetic analysis by Lü et al. (2010) rather than that by Kellner (2003) in our appraisal. The Santana Formation concretions have yielded members of the major clades Ornithocheiridae, ?Ctenochasmatoidea, Tapejaridae and Thalassodromidae (Lü et al., 2010). It is possible that Ctenochamatidae may be represented in the Santana assemblage on the basis of a partial rostrum named *Unwindia* by Martill (2011), although Unwin (pers. comm.) and Witton (pers. comm.) both suggest this taxon is more likely a lonchodectid, while a previous claim by Unwin (2003) for ctenochamatoids in this assemblage was based on a misinterpretation of a 'faked' specimen (Vila Nova et al., 2010). Of these, Ornithocheiridae and Ctenochasmatoidea (or Lonchodectidae) have teeth, whereas Tapejaridae and Thalassodromidae are edentulous. Although some Santana pterosaurs have been based on clearly inadequate material (e.g. *Tupuxuara longicristatus*; Witton, 2009), much as was the case with the pterosaurs from the similar-aged Cambridge Greensand (Cenomanian with derived Albian assemblage [Unwin, 2001]) of England by Harry Govier Seeley in the 1880s and 1890s, many species from the Santana Formation are now known from nearly complete skeletons in an excellent state of preservation (Elgin et al., 2011; Kellner and Tomida, 2000; Veldmeijer, 2003) and from multiple specimens.

There have been two recent reviews of the ornithocheirids from the Santana Formation, with both coming to slightly different conclusions about the validity of the taxa considered. Kellner and Tomida (2000) reviewed all of the named material while describing a new species of *Anhanguera piscator*, whereas Veldmeijer (2003), in his description of *Coloborhynchus spielbergi*, reviewed other ornithocheirids from the Santana Formation, especially those in museums in Europe. Some Santana pterosaurs are notable for their size. For example, the skull of *Thalassodromeus sethi* is 1.42m long (Kellner and Campos, 2002), and Dalla Vecchia and Ligabue (1993) described a single distal phalanx from a pterosaur estimated to have had a wingspan of between 8.2 and 9.3m. Some small pterosaurs have been reported, including a probable juvenile tapejarid described by Eck *et al.* (2011) (Fig 24C).

By far the most diverse pterosaurs are those assigned to the Ornithocheiridae. These have been referred to the genera *Anhanguera*, *Araripesaurus*, *Barbosania*, *Brasileodactylus*, *Coloborhynchus* and *Tropeoganthus*. Of these, *Brasileodactylus* was erected on a fragmentary anterior end of a mandible and is best considered a *nomen dubium*, whereas *Araripesaurus* was erected on a partial wing skeleton and is also probably indeterminate. *Barbosania* is most likely an example of *Anhanguera*, which, as mentioned above, may be a juvenile morph or sexual dimorph of *Coloborhynchus*. The single rostral fragment named *Unwindia* by Martill (2011) is significant if it belongs to Lonchodectidae, in that it would be the first record of this clade from South America.

The edentulous pterosaurs are easily distinguished on the basis of their overall skull shape, bearing elaborate head crests and having a distinctive mandible. *Tupuxuara* has an elongate triangular skull in lateral view, with narrow, spear-like lower jaws, while its sister-taxon *Thalassodromeus* has narrow, slightly upturned mandibular symphysis with a blade-like occlusal surface. Kellner and Campos (2002) interpreted *Thalassodromeus* as a skim feeder by, but this claim did not stand up to scrutiny (Humphries *et al.*, 2007), although quite how this pterosaur did feed remains to be determined. A third edentulous pterosaur, *Tapejara* is typified by a high crest above the rostrum, a long narrow crest behind the cranium and a ventral crest on the mandibular symphysis. Related taxa from the Crato Formation show that the bony crest supported an even larger soft tissue structure, perhaps analogous to the keratin crest of some bird bills (e.g. hornbills) (Unwin and Martill, 2007). Recently, Pinheiro and Schultz (2012) described a distinctive partial palate of a pterosaur from the Romualdo Member that may represent a new taxon of Azhdarchoidea, but it is too incomplete to be diagnosable.

Preservation of Santana Formation pterosaurs

The taphonomic and diagenetic features displayed by Santana pterosaurs are diverse, but there has been no in-depth study of their taphonomy. Kellner (1994) made some passing comments on the taphonomy of the Santana Formation pterosaurs, but presented no new analysis. His comment that pterosaurs occur as isolated elements is a notion that is almost certainly an artefact of the way the pterosaur remains are collected. Specimens do seem to occur in the concretions as isolated bones, but this observation is rarely made in the field. Concretions with associated remains in a disarticulated condition, as fully articulated portions of skeletons, and as near complete articulated skeletons occur more frequently. Pterosaur bones frequently extend beyond the boundaries of the concretions, suggesting that they also occur within the surrounding shales, probably in a crushed condition, but commercial fossil collectors discard such remains. It is common to find the bones in three-dimensional preservation in concretions (Fig 24A, B) but highly compacted remains also occur, suggesting that concretion formation at some horizons postdates an initial compaction phase. Where bones occur as three-dimensional remains the internal void spaces may be filled with diagenetic minerals such as calcite, but it is not unusual to find bones hollow with little or no mineral infill. Thin-section petrography of the bones reveals beautiful preservation of the original bone histology (Sayão, 2003).

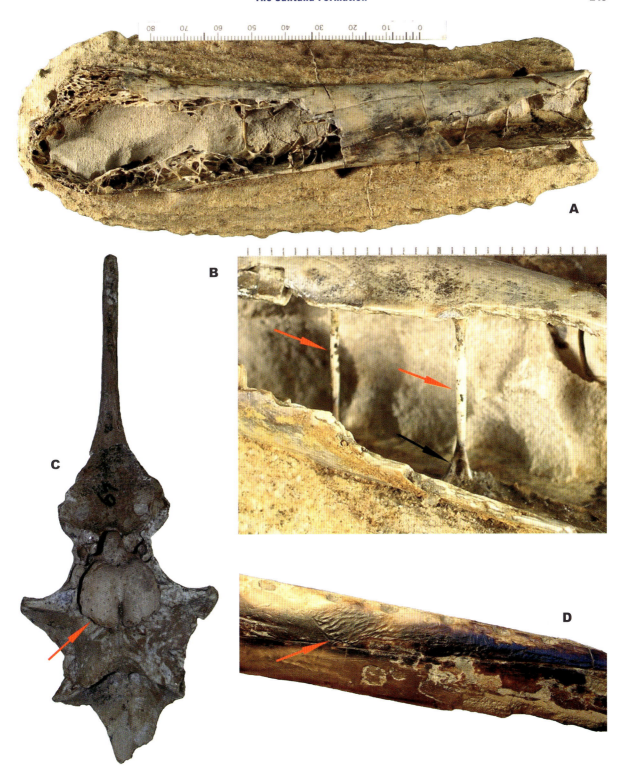

Figure 24 A selection of pterosaur bones from the Santana Formation with the internal architecture preserved. **A**, proximal portion of a first wing phalanx of an indeterminate pterosaur. The internal trabeculae are restricted to the margins, allowing for a large central lumen, probably occupied by an air sac. **B**, details of cross-supporting structures in the diaphysis of the same bone. **C**, internal view of braincase of *Tapejara wellnhoferi* SMNK PAL 1137. **D**, external surface of bone with highly detailed muscle attachment site.

In rare instances Santana pterosaurs are found with soft tissues associated with the skeletons. Campos et al. (1984) first reported preserved wing membrane where carbonaceous material appears to preserve a fibrous membrane. Martill and Unwin (1989) reported phosphatized wing membrane in a pterosaur that displayed preservation of several distinct tissue types, leading to the realization that pterosaur wing membranes were complex structures involved not only in flight but probably in temperature regulation (Martill and Unwin, 1989; Frey et al., 2003). Martill and Unwin's interpretation was rejected by Kellner (1996b) who argued instead that the specimen represented integument of the body. However, the presence of stiffening fibres like those in the 'Zittel' wing of a Late Jurassic rhamphorhynchid pterosaur would be unusual in body integument, and has not been reported for any other pterosaurs (Wellnhofer, 1987b).

Occurrence and abundance of Santana Formation pterosaurs

Pterosaurs are the most abundant tetrapods in the Santana Formation, occurring frequently and over a wide area of the outcrop. According to local fossil diggers we have met, pterosaurs have been found around the Serra da Maozinha in the eastern Chapada, in the vicinity of Santana do Cariri in the north central region, and one of us (DMM) has seen specimens in the field near Simões in the western chapada. Therefore they do not seem to be restricted to any particular part of the Lagerstätte's outcrop. Estimating their abundance is difficult, but local fossil diggers have commented that pterosaur bones are commonly picked up in the field brash of the Romualdo Member outcrop, and especially so in the western Chapada. Recently, Eck et al. (2011) described a concretion from the Santana Formation containing the remains of two partially articulated examples of *Tapejara wellnhoferi*. This suggests that either there was an intimate relationship between these two individuals (one is smaller than the other – perhaps sexual dimorphs) or they were killed in a mass mortality event (such events are known for the Santana fishes [Martill et al., 2008])

Pterosaur locomotion studies

There have been numerous studies on the locomotion of pterosaurs, but most have been hampered to some degree by the nature of most pterosaur specimens: they are either displayed on slabs of limestone in a flat and often crushed condition and cannot be removed, or they are highly fragmentary and exceedingly delicate. In both cases it becomes nearly impossible to determine how one bone articulated with another, and consequently, many analyses have been undertaken on misinterpretations of skeletal relationships (e.g. Wilkinson et al., 2006). Therefore, the three-dimensionally preserved, articulated pterosaurs from the Santana Formation have provided new insights into the functions of the pterosaur skeleton, especially the locomotion of these animals (Wellnhofer, 1988; Bennett, 1990, 2007; Wilkinson et al., 2006; Humphries et al., 2007; Costa et al., 2010).

Flying and walking

Although flight was established for pterosaurs as early as 1801 by Cuvier (earlier studies considered them to have been marine rather than volant reptiles or even mammals), many details of the precise flight mechanisms, distribution of the wing membranes, function of the membranes, and mobility at the joints of the wing skeleton remained to be determined. Over the course of the next few decades, pterosaur specimens from the Upper Jurassic Solnhofen Formation of Germany provided details of the wing membrane distribution, and later also details of its fine surface structure (Zittel, 1882), but most aspects of the wing biomechanics remained unknown. Martill and Unwin (1989) provided details of the internal structure of the wing membrane. They found a multi-layered structure rich in possibly stiffening fibres, previously termed aktinofibrils by Wellnhofer (1987b).

Arguments about terrestrial locomotion in pterosaurs have also long persisted. Some workers considered pterosaurs bipedal (e.g. Padian, 1983), whereas others regarded them as obligate quadrupeds (e.g. Unwin, 1988; Mazin, 2003). The discovery of abundant undoubted pterosaur trackways in France (Mazin, 2003) went a long way toward solving this problem, but it was the discovery of three-dimensionally preserved pterosaur pelves in the Santana Formation that permitted detailed analysis of the orientation of the acetabulum, enabling determination of the stance and range of movement of the hind limb (Bennett, 1990).

Of particular concern for evaluating pterosaur flight dynamics is the origin, position and function of the pteroid, an element that projects from the wrist of the pterosaur wing and is assumed to be involved in flight control, perhaps linked with a propatagium (a membrane extending from the leading edge of the wing to the body at the base of the neck). It was unclear whether this element was an ossified cartilage (a sesamoid like the human patella) or a true bone, perhaps a highly modified carpal. Isolated specimens of pteroids examined in thin section revealed the element to be bone rather than ossified cartilage (Unwin et al., 1996), whereas articulated examples clearly showed that the pteroid did not articulate with the lateral carpal as previously thought (Bennett, 2007). In addition, the availability of three-dimensionally preserved fossils allowed artificial wings to be constructed and tested in wind tunnel environments, resulting in new interpretations for the pteroid (Wilkinson et al., 2006).

Discussion

There can be no gainsaying the importance of the Santana Formation fossil Lagerstätte for the study of mid-Cretaceous pterosaurs. Although the assemblage now appears somewhat less diverse than that of the older (Barremian) Yixian Formation of northeastern China (~7 vs ~17 genera, respectively), the quality of preservation is superior in many ways. The uncrushed, three-dimensional nature of much of the Brazilian material, and the opportunities for completely extracting bones from the matrix, allow manipulation of joints not possible for any other pterosaur material. The exceptional preservation of soft tissues in three dimensions, with mineral replacement at the cellular level, allows unique insights into the histology and physiology of these flying archosaurs. The abundance of specimens in different growth stages provides opportunities for testing hypotheses of sexual dimorphism for pterosaur headcrests between taxa and for examining aspects of their growth. Some of these studies will be dependent on more robust systematic analyses determining the validity of the Santana taxa, and it is to be hoped that legislation restricting fossil collecting will not diminish the number of specimens finding their way into legitimate collections.

Acknowledgements

Dino Frey, Darren Naish, Helmut Tishlinger, David Unwin and Mark Witton are thanked for discussions on things pterosaurian and Santana. Mark Witton is also thanked for allowing us to use his pterosaur artwork. Over the years fieldwork of DMM has been supported by DNPM staff in Crato, especially Betimar Filgueiras and Arthur Andrade. Help in the field has also come from Uli Heimhoffer, Robert Loveridge, John Nudds, Paul Selden, Lorna Steel, Jane Evans and several undergraduates from the University of Portsmouth. Thanks to Burkhard Pohl and the Dinosaur Centre Wyoming for allowing me access to the baby theropod illustrated in Figure 8 and allowing me to include a photograph of it.

References

Agassiz, L. 1841. On the fossil fishes found by Mr. Gardner in the Province of Ceará, in the North of Brazil. Edinburgh New Philosophical Journal 30:82–84.

Agassiz, L. 1833–1844. Recherches sur les poissons fossiles. 5 volumes and atlas. Imprimerie Petitpierre, Neuchâtel.

Agassiz, L. 1844. Sur quelques poissons fossiles du Brésil. Compte Rendus Hebdomadaires des Séances de l'Académie des Sciences, Paris 18:1007–1015.

Agnolin, F. L., and A. G. Martinelli. 2007. Did oviraptorosaurs (Dinosauria; Theropoda) inhabit Argentina? Cretaceous Research 28:785–790.

Antunes, M. T., A. C. Balbino, and I. Freitas. 2005. Early (18th century) discovery of Cretaceous fishes from Chapada do Araripe, Ceará, Brazile specimens kept at the 'Academia das Ciencias de Lisboa' Museum. Comptes Rendus Palevol 4:375–384.

Arai, M. 2000. Chapadas: relict of mid-Cretaceous interior seas in Brazil. Revista Brasileira de Geociências 30:436–438.

Arai, M., and J. C. Coimbra. 1990. Análise paleoecológica do registro das primeiras ingressões marinhas na Formação Santana (Cretáceo Inferior da Chapada do Araripe). Pp. 225–239 in D. A. Campos, M. S. S. Viana, P. M. Brito, and G. Beurlen (eds), Atas do I Simpósio sobre a Bacia do Araripe e bacias interiores do nordeste, Crato, Anais 1.

Bate, R. H. 1971. Phosphatized ostracods from the Cretaceous of Brazil. Nature 230:397–398.

Bate, R. H. 1972. Phosphatized ostracods with appendages from the Lower Cretaceous of Brazil. Palaeontology 15:379–393.

Bate, R. H. 1973. On *Pattersoncypris micropapillosa* Bate. Pp. 101–108 in The Stereo-Atlas of Ostracod Shells, Volume 1. The Micropalaeontological Society.

Baudin, F., and P.-Y. Berthou. 1996. Depositional environments of the organic matter of Aptian-Albian sediments from the Araripe Basin (NE Brazil). Bulletin des Centres de Recherches Exploration-Production Elf-Aquitaine 20:213–227.

Baudin, F., P.-Y. Berthou, J. P. Herbin, and D. A. Campos. 1990. Matière organique et sedimentation argileuse dans le Crétacé du basin d'Araripe. Comparaison avec les données de Crétacé d'autres bassins Brésiliens. Pp. 83–94 in D. A. Campos, M. S. S. Viana, P. M. Brito, and G. Beurlen (eds), Atas do Simpósio Sobre a Bacia do Araripe e Bacias Interiores do Nordeste, Crato, 14–16 de Junho de 1990.

Bennett, S. C. 1990. A pterodactyloid pterosaur pelvis from the Santana Formation of Brazil; implications for terrestrial locomotion. Journal of Vertebrate Paleontology 10:80–85.

Bennett, S. C. 2007. Articulation and function of the pteroid bone in pterosaurs. Journal of Vertebrate Paleontology 27:881–891.

Berthou, P.-Y. 1994. Critical analysis of the main publications about the stratigraphical framework of the Palaeozoic and Mesozoic sedimentary deposits in the Araripe Basin (northeastern Brazil). Boletim do 3rd Simpósio Sobre o Cretáceo do Brasil 123–126.

Berthou, P.-Y. and C. Pierre. 1990. Analyse isotopique du soufre et de l'oxygène des quelques gypses des basins du N. E. du Brésil. Implications pour le paléoenvironnement. Pp. 95–98 in D. A. Campos, M. S. S. Viana, P. M. Brito, and G. Beurlen (eds), Atas do Simpósio Sobre a Bacia do Araripe e Bacias Interiores do Nordeste, Crato, 14–16 de Junho de 1990.

Berthou, P.-Y., M. S. L. T. Teles, and D. A. Campos. 1990. Sedimentation argileuse Cretacée dans le basin d'Araripe et quelques bassins annexes (N. E. du Brésil). Pp. 143–162 in D. A. Campos, M. S. S. Viana, P. M. Brito, and G. Beurlen (eds), Atas do Simpósio Sobre a Bacia do Araripe e Bacias Interiores do Nordeste, Crato, 14–16 de Junho de 1990.

Beurlen, K. 1962. Geologia da Chapada do Araripe. Anais da Academia Brasileira de Ciências 34:365–370.

Bittencourt, J. S. and A. W. A. Kellner. 2004. On a sequence of sacrocaudal theropod dinosaur vertebrae from the Lower Cretaceous Santana Formation, northeastern Brazil. Arquivos do Museum Nacional, Rio de Janeiro 62:309–320.

Brito, I. M., and L. P. Quadros. 1995. Retrabalhamento do Devoniano no Cretáceo Inferior da Bacia do Araripe. Anais da Academia Brasileira de Ciências 67:493–496.

Brito, P. M. 1997. Révision des Aspidorhynchidae (Pisces-Actinopterygia) du Mésozoïque: ostéologie et relations phylogénétique. Geodiversitas 19:681–772.

Brito, P. M. 2000. A new halecomorph with two dorsal fins, *Placidichthys bidorsalis* n. g., n. sp. (Actinopterygii: Halecomorphi) from the Lower Cretaceous of the Araripe Basin, northeast Brazil. Comptes Rendus de l'Académie des Sciences Paris, Sciences de la Terre et des planètes 331:749–754.

Brito, P. M. 2008. The Crato Formation fish fauna. Pp. 429–443 in D. M. Martill, G. Bechly and R. F. Loveridge (eds), The Crato Fossil Beds of Brazil: Window into An Ancient World. Cambridge University Press, Cambridge.

Brito, P. M. and P. L. N. Ferreira. 1989. The first hybodont shark, *Tribodus limae* n.g., n. sp., from the Lower Cretaceous of Chapada do Araripe (N-E Brazil). Anais da Academia Brasileira de Ciências, 61:53–57.

Brito, P. M., and V. Gallo. 2003. A new species of *Lepidotes* (Neopterygii: Semionotiformes: Semionotidae) from the Santana Formation, Lower Cretaceous of northeastern Brazil. Journal of Vertebrate Palaeontology 23:47–53.

Brito, P. M., and F. J. Meunier. 2000. The morphology and histology of the scales of aspidorhynchids (Actinopterygii, Halecostomi). Geobios 33:105–111.

Brito, P. M., and B. Seret. 1996. The new genus *Iansan* (Chondrichthyes, Rhinobatoidea) from the Early Cretaceous of Brazil and its phylogenetic relationships. Pp. 47–62 in G. Arratia and G. Viohl (eds), Mesozoic Fishes: Systematics and Paleoecology. Verlag Dr. Friedrich Pfeil, Munich.

Brito, P. M. and Y. Yabumoto. 2011. An updated review of the fish faunas from the Crato and Santana formations in Brazil, a close relationship to the Tethys fauna. Bulletin of the Kitakyushu Museum of Natural History and Human History, Series A, 9:107–136.

Brito, P. M., F. J. Meunier, and M. Gayet. 2000. The morphology and histology of the scales of the Cretaceous gar *Obaichthys* (Actinopterygii, Lepisosteidae): phylogenetic implications. Comptes Rendus de l'Académie des Sciences Paris, Sciences de la Terre et des planètes 331:823–829.

Brito, P. M., R. J. Bertini, D. M. Martill, and L. O. Salles. 1994. Vertebrate fauna from the Missao Velha Formation (Lower Cretaceous, N. E. Brazil). Boletim do 3rd Simpósio Sobre o Cretáceo do Brasil, Rio Claro, Sao Paulo:139–140.

Brito, P. M., F. J. Meunier, G. Clément, and D. Geffard-Kuriyama. 2010. The histological structure of the calcified lung of the fossil coelacanth *Axelrodichthys araripensis* (Actinistia: Mawsoniidae). Palaeontology 53:1281–1290.

Broin, L. de. 2000. The oldest pre-podocnemidid turtle (Chelonii, Pleurodira), from the early Cretaceous, Ceará state, Brasil, and its environment. Treballs del Museu de Geologia de Barcelona 9:43–95.

Buisonjé, P. H. de. 1980. *Santanadactylus brasilensis* nov. gen. nov. sp., a long-necked, large pterosaurier from the Aptian of Brazil, Part I & II. Proceedings, Koninklijke Akademie der Wetenschappen, B 83:145–172.

Buisonjé, P. H. de. 1981. *Santanadactylus brasilensis*: Skelet-reconstructie van een vliegend reptiel met zes meter vlucht. Geia 14:37–48.

Buisonjé, P. H. de. 1993. Provisional evaluation of a large, rather flat concretion from the Santana Formation, Chapada do Araripe, Brazil, containing the complete skull and the lower jaw, as well as more than twenty post-cranial skeletal remains of a large pterosaur: *Tropeognathus* cf. *robustus* (Wellnhofer 1987), examined on Thursday Eighth January 1993 at the National Museum of Natural History in Leiden. Internal Report, National Museum of Natural History, Leiden.

Buffetaut, E. 1991. *Itasuchus* Price, 1955. Pp. 348–350 in J. G. Maisey (ed), Santana Fossils: An Illustrated Atlas. T.F.H. Publications, Neptune City, NJ.

Buffetaut, E., D. M. Martill, and F. Escuillié. 2004. Pterosaurs as part of a spinosaur diet. Nature 429:33.

Campos, D. A., and A. W. A. Kellner. 1985a. Um novo exemplar de *Anhanguera blittersdorffi* (Reptilia, Pterosauria) da Formação Santana, Cretáceo Inferior do nordeste do Brasil. Resumos, 9° Congresso Brasileiro de Paleontologia:13.

Campos, D. A., and A. W. A. Kellner. 1985b. Panorama of the flying reptiles study in Brazil and South America. Anais da Academia Brasileira de Ciências 57:453–466.

Campos, D. A., and S. Wenz. 1982. Première découverte de Coelacanthes dans le Crétacé inférieur de la Chapada do Araripe (Brésil). Comptes Rendus de l'Académie des Sciences, Série II 294:1151–1154.

Campos, D. A., G. Ligabue, and P. Taquet. 1984. Wing membrane and wing supporting fibres of a flying reptile from the Lower Cretaceous of the Chapada do Araripe Aptian, Ceará State, Brazil. Pp. 37–40 in W.-E. Reif and F. Westphal (eds), Third Symposium on Mesozoic Terrestrial Ecosystems: Short Papers. ATTEMPTO Verlag, Tübingen.

Carmo, D. A., R. Whatley, J. V. Queiroz Neto, and J. C. Coimbra. 2008. On the validity of two Lower Cretaceous non-marine ostracode genera: biostratigraphic and paleogeographic implications. Journal of Paleontology 82:790–799.

Carvalho, I. de S. 2000. Geological environments of dinosaur footprints in the intracratonic basins of northeast Brazil during the Early Cretaceous opening of the South Atlantic. Cretaceous Research 21:255–267.

Carvalho, M. S. S., and M. E. C. M. Santos. 2005. Histórico das pesquisas paleontológicas na Bacia do Araripe, Nordeste do Brasil. Anuário do Instituto de Geociências, UFRJ 28:15–34.

Castro Leal, M. E. de, and P. M. Brito. 2004. The ichthyodectiform *Cladocyclus gardneri* (Actinopterygii: Teleostei) from the Crato and Santana Formations, Lower Cretaceous of Araripe Basin, North-Eastern Brazil. Annales de Paléontologie 90:103–113.

Cavalcanti, V. M. M., and M. S. S. Viana. 1992. Revisaõ estratigráfica da Formação Missão Velha, Bacia do Araripe, Nordeste do Brasil. Anais da Academia Brasileira de Ciências 64:155–168.

Cavin, L., L. Boudad, S. Duffaud, L. Kabiri, J. Le Loeuff, I. Rouget, and H. Tong. 2001. L'évolution paléoenvironnementale des faunes de poissons du Crétacé supérieur du bassin du Tafilalt et des régions avoisinantes (Sud-Est du Maroc): implications paléobiogéographiques. Comptes Rendus l'Académie de Sciences, Paris, Sciences de la Terre et des planètes 333:677–683.

Cope, E. D. 1886. A contribution to the vertebrate paleontology of Brazil. Proceedings of the American Philosophical Society 23:3–4.

Costa, F. R., O. Rocha-Barbosa, and A. W. A. Kellner. 2010. Stance biomechanics of *Anhanguera piscator* Kellner & Tomida, 2000 (Pterosauria, Pterodactyloidea) using tridimentional virtual animation. Acta Geoscientica Sinica 31:15–16.

Crane, P., and J. G. Maisey. 1991. Fossil plants. Pp. 414–419 in J. G. Maisey (ed), Santana Fossils: An Illustrated Atlas. T.F.H. Publications, Neptune City, NJ.

Cressey, R., and C. Patterson. 1973. Fossil parasitic copepods from a Lower Cretaceous fish. Science 180:1283–1285.

Cressey, R., and G. Boxshall. 1989. *Kabatarina pattersoni*, a fossil parasitic copepod (Dichelesthiidae) from a Lower Cretaceous fish. Micropalaeontology 35:150–167.

Cupelo, C. D., D. Bermudez-Rochas, D. M. Martill, and P. M. Brito. 2012. The hybodontiform sharks (Chondrichthyes: Elasmobranchii) from the Missão Velha Formation, Lower Cretaceous of Araripe Basin, North-East Brazil. Comptes Rendus Palevol 11:41–47.

Dalla Vecchia, F. M. 1993. *Cearadactylus*? *ligabuei* nov.sp., a new early Cretaceous (Aptian) pterosaur from Chapada do Araripe (northeastern Brazil). Bollettino della Società Paleontologica Italiana 31:401–409.

Dalla Vecchia, F. M., and G. Ligabue. 1993. On the presence of a giant pterosaur in the Lower Cretaceous (Aptian) of Chapada do Araripe (northeastern Brazil). Bollettino della Società Paleontologica Italiana 32:131–136.

Dalziel, I. W. D. 2012. Neoproterozoic-Paleozoic geography and tectonics: review, hypothesis, environmental speculation. Geological Society of America Bulletin 109: 16–42.

da Silva, M. A. M. 1986a. Lower Cretaceous unconformity truncating evaporite-carbonate sequences, Araripe Basin, northeastern Brazil. Revista Brasileira de Geociências 16:306–310.

da Silva, M. A. M. 1986b. Lower Cretaceous sedimentary sequences in the Araripe Basin. Revista Brasileira de Geociências 16:311–319.

da Silva, M. A. M. 1988. Evaporitos do Cretaceo da Bacia do Araripe: ambientes de deposiçao e historia diagenetica. Boletim de Geociências da Petrobrás 2:53–63.

Dunkle, D. H. 1940. The cranial osteology of *Notelops brama* (Agassiz), an elopid fish of the Cretaceous of Brazil. Lloydia 3:157–190.

Eck, K., R. Elgin, and E. Frey. 2011. On the osteology of *Tapejara wellnhoferi* Kellner 1989 and the first occurrence of a multi-specimen assemblage from the Santana Formation, Araripe Basin NE-Brazil. Swiss Journal of Palaeontology 130:277–296.

Elgin, R. A., and E. Frey. 2011. A new ornithocheirid, *Barbosania gracilirostris* gen. et sp. nov. (Pterosauria, Pterodactyloidea) from the Santana Formation (Cretaceous) of NE Brazil. Swiss Journal of Palaeontology 130:259–275.

Fara, E., A. A. F. Saraiva, D. A. Campos, J. K. R. Moreira, D. C. Siebra, and A. W. A. Kellner. 2005. Controlled excavations in the Romualdo Member of the Santana Formation (Early Cretaceous, Araripe Basin, northeastern Brazil): stratigraphic, palaeoenvironmental and palaeoecological implications. Palaeogeography, Palaeoclimatology, Palaeoecology 218:145–160.

Fastnacht, M. 2001. First record of *Coloborhynchus* (Pterosauria) from the Santana Formation (Lower Cretaceous) of the Chapada do Araripe, Brazil. Paläontologische Zeitschrift 75:23–36.

Figueiredo, F. J., and V. Gallo. 2004. A new teleost fish from the Early Cretaceous of northeastern Brazil. Boletim do Museu Nacional, Rio de Janeiro, Nova Série, Geologia 73:1–23.

Filleul, A., and J. G. Maisey. 2004. Redescription of *Santanichthys diasii* (Otophysi, Characiformes) from the Albian of the Santana Formation and comments on its implications for otophysan relationships. American Museum Novitiates 3455:1–21.

Forey, P. L. 1991. *Latimeria chalumnae* and its pedigree. Environmental Biology of Fishes 32:75–91.

Forey, P. L., and L. Grande. 1998. An African twin to the Brazilian *Calamopleurus* (Actinopterygii: Amiidae). Zoological Journal of the Linnean Society of London 123: 179–195.

Frey, E., and D. M. Martill. 1995. A possible oviraptorsaurid theropod from the Santana Formation (Lower Cretaceous, ?Albian) of Brazil. Neues Jahrbuch für Geologie und Paläontologie Monatshefte 1995:397–412.

Frey, E., and D. M. Martill. 1998. Late ontogenetic fusion of the processus tendinosis extensoris in Cretaceous pterosaurs from Brazil. Neues Jahrbuch für Geologie und Paläontologie Monatshefte 1998:587–549.

Frey, E., H. Tischlinger, M.-C. Buchy, and D. M. Martill. 2003. New specimens of Pterosauria (Reptilia) with soft parts with implications for pterosaurian anatomy and locomotion. Pp. 233–266 in E. Buffetaut and J.-M. Mazin (eds), Evolution and Palaeobiology of Pterosaurs. Geological Society, London, Special Publications 217.

Gaffney, E. S., and P. A. Meylan. 1991. Primitive pelomedusid turtle. Pp. 335–339 in J. G. Maisey (ed), Santana Fossils, an

Illustrated Atlas. T.F.H. Publications, Neptune City, New Jersey.

Gaffney, E. S., D. A. Campos, and R. Hirayama. 2001. *Cearachelys*, a new side-necked turtle (Pelomedusoides: Bothremydidae) from the Early Cretaceous of Brazil. American Museum Novitates 3319:1–20.

Gaffney, E. S., H. Tong, and P. A. Meylan. 2006. Evolution of the side-necked turtles: the families Bothremydidae, Euraxemydidae, and Araripemydidae. Bulletin of the American Museum of Natural History 300:1–698.

Gallo, V., F. J. Figueiredoa, and S. A. Azevedo. 2009. *Santanasalmo elegans* gen. et sp. nov., a basal euteleostean fish from the Lower Cretaceous of the Araripe Basin, northeastern Brazil. Cretaceous Research 30:1357–1366.

Gardner, G. 1841. Geological notes made during a journey from the coast into the interior of the Province of Ceará, embracing an account of a deposit of fossil fishes. Edinburgh New Philosophical Journal 30:75–82.

Gardner, G. 1843. On the existence of an immense deposit of chalk in the northern province of Brazil. Proceedings of the Glasgow Philosophical Society 1:146–153.

Gardner, G. 1846. Travels in the Interior of Brazil, Principally through the Northern Provinces and the Gold and Diamond Districts during the Years 1836–1841. Reeve, Benham and Reeve, London, 603 pp.

Göhlich, U. B., and L. M. Chiappe. 2006. A new carnivorous dinosaur from the Late Jurassic Solnhofen archipelago. Nature 440:329–332.

Grande, L., and W. E. Bemis. 1998. A comprehensive phylogenetic study of amiid fishes (Amiidae) based on comparative skeletal anatomy. An empirical search for interconnected patterns of natural history. Society of Vertebrate Paleontology Memoir 4:1–690.

Heimhofer, U., S. P. Hesselbo, R. D. Pancost, D.M. Martill, P. A. Hochuli, and J. V. P. Guzzo. 2008. Evidence for photic-zone euxinia in the Early Albian Santana Formation (Araripe Basin, NE Brazil). Terra Nova 20:347–354.

Heimhofer, U., L. Schwark, D. Ariztegui, D. M. Martill, and A. Immenhauser. 2010. Geochemistry of fossiliferous carbonate concretions from the Cretaceous Santana Formation – assessing the role of microbial processes. Geophysical Research Abstracts 12: EGU2010–9390, 2010, 2 pp.

Hirayama, R. 1998. Oldest known sea turtle. Nature 392:705–708.

Humphries, S., R. H. C. Bonser, M. P. Witton, and D. M. Martill. 2007. Did pterosaurs feed by skimming? Physical modelling and anatomical evaluation of an unusual feeding method. PLoS Biology 5(8):e204. doi: 10.1371/journal.pbio.0050204

Jordan, D. S. 1923. Peixes Cretáceos do Ceará e Piauhy. Monografia Serviço Geológico Mineralógico do Brasil, Rio de Janeiro:31–97.

Jordan, D. S., and J. C. Branner. 1908. The Cretaceous fishes of Ceará, Brazil. Smithsonian Miscellaneous Collections 25:1–29.

Kashiyamaa, Y., D. E. Fastovsky, S. Rutherford, J. King, and M. Montellano. 2004. Genesis of a locality of exceptional fossil preservation: paleoenvironments of Tepexi de Rodríguez (mid-Cretaceous, Puebla, Mexico). Cretaceous Research 25:153–177.

Kellner, A. W. A. 1984. Ocorrência de uma mandíbula de Pterosauria (*Brasileodactylus araripensis*, nov. gen., nov. sp.) na Formação Santana, Cretáceo da Chapada do Araripe, Ceará–Brasil. Anais 33 Congresso Brasileiro de Geologia 2:578–590.

Kellner, A. W. A. 1987. Ocorrência de um novo crocodiliano no Cretáceo inferior da bacia do Araripe, Nordeste do Brasil. Anais da Academia Brasileira de Ciências 59:219–232.

Kellner, A. W. A. 1989. A new edentate pterosaur of the Lower Cretaceous from the Araripe Basin, Northeast Brazil. Anais da Academia Brasileira de Ciências 61:439–446.

Kellner, A. W. A. 1994. Remarks on pterosaur taphonomy and palaeoecology. Acta Geologica Leopoldensia 39:175–189.

Kellner, A. W. A. 1996a. Description of the braincase of two Early Cretaceous pterosaurs (Pterodactyloidea) from Brazil. American Museum Novitates 3175:1–34.

Kellner, A. W. A. 1996b. Reinterpretation of a remarkably well preserved pterosaur soft tissue from the Early Cretaceous of Brazil. Journal of Vertebrate Paleontology 16: 718–722.

Kellner, A. W. A. 1996c. Fossilized theropod soft tissues. Nature 379:32.

Kellner, A. W. A. 1996d. Remarks on Brazilian dinosaurs. Memoirs of the Queensland Museum 39:611–626.

Kellner, A. W. A. 1999. Short note on a new dinosaur (Theropoda, Coelurosauria) from the Santana Formation (Romualdo Member, Albian), northeastern Brazil. Boletim do Museu Nacional, Rio de Janeiro, Nova Serie 49:1–8.

Kellner, A. W. A. 2002. Membro Romualdo da Formação Santana, Chapada do Araripe, CE. Pp. 121–130 in C. Schobbenhaus, D. A. Campos, E. T. Queiroz, M. Winge, and M. Berbert-Born (eds), Sítios Geológicos e Paleontológicos do Brasil 6.

Kellner, A. W. A. 2003. Pterosaur phylogeny and comments on the evolutionary history of the group. Pp. 105–137 in E. Buffetaut and J.-M. Mazin (eds), Evolution and Palaeobiology of Pterosaurs. Geological Society, London, Special Publications 217.

Kellner, A. W. A., and D. A. Campos. 1988. Sobre um nova pterossauro com crista sagital da Bacia do Araripe, Cretáceo Inferior do Nordeste do Brasil. Anais da Academia Brasileira de Ciências 60:459–469.

Kellner, A. W. A., and D. A. Campos. 1990. Preliminary description of an unusual pterosaur skull of the Lower Cretaceous from the Araripe Basin. Pp. 401–405 in D. A. Campos, M. S. S. Viana, P. M. Brito, and G. Beurlen (eds), Atas do simpósio sobre a Bacia do Araripe e Bacias Interiores do Nordeste, Crato, 14-16 de Junho de 1990. Crato.

Kellner, A. W. A., and D. A. Campos. 1994. A new species of *Tupuxuara* (Pterosauria, Tapejaridae) from the Early Cretaceous of Brazil. Anais da Academia Brasileira de Ciências 66, 467–473.

Kellner, A. W. A., and D. A. Campos. 1996. First Early Cretaceous theropod dinosaur from Brazil with comments on Spinosauridae. Neues Jahrbuch für Geologie und Paläontologie, Abhandlungen 199:151–166.

Kellner, A. W. A., and D. A. Campos. 1999. Vertebrate paleontology in Brazil: a review. Episodes 22:238–251.

Kellner, A. W. A. and D. A. Campos. 2002. The function of the cranial crest and jaws of a unique pterosaur from the Early Cretaceous of Brazil. Science 297:389–392.

Kellner, A. W. A., and Y. Tomida. 2000. Description of a new species of Anhangueridae (Pterodactyloidea) with comments on the pterosaur fauna from the Santana Formation (Aptian–Albian), northeastern Brazil. National Science Museum Monographs 17:1–135.

Langer, M. C., R. Iannuzzi, A. A. S. da Rosa, R. P. Ghilardi, C. S. Scherer, and V. G. Pitana. 2012. A reply to Martill - The bearable heaviness of liability. Geoscientist Online Special 12 March http://www.geolsoc.org.uk/en/Geoscientist/March%202012.

Leonardi, G., and G. Borgomanero. 1985. *Cearadactylus atrox* nov. gen., nov. sp.: novo pterosauria (Pterodactyloidea) da Chapada do Araripe, Ceará, Brasil. DNPM, Coletânea de Trabalhos Paleontológicos. Série Geologia 27:75–80.

Leonardi, G., and G. Borgomanero. 1987. The skeleton of a pair of wings of a pterosaur (Pterodactyloidea, ?Ornithocheiridae, cf. *Santanadactylus*) from the Santana Formation of the Araripe Plateau, Ceará, Brasil. Anais, Congresso Brasileiro de Paleontologia, Rio de Janeiro 1:123–129.

Lü, J., D. M. Unwin, X. Jin, Y. Liu, and Q. Ji. 2010. Evidence for modular evolution in a long-tailed pterosaur with a pterodactyloid skull. Proceedings of the Royal Society B 277:383–389.

Mabesoone, J. M., and I. M. Tinoco. 1973. Palaeoecology of the Aptian Santana Formation (northeastern Brazil). Palaeogeography, Palaeoclimatology, Palaeoecology 14:97–118.

Maisey, J. G. 1986. Coelacanths from the Lower Cretaceous of Brazil. American Museum Novitates 2866:1–30.

Maisey, J. G. (ed). 1991. Santana Fossils: An Illustrated Atlas. T.F.H. Publications, Neptune City, NJ, 459 pp.

Maisey, J. G. 1993a. A new clupeomorph fish from the Santana Formation (Albian) of NE Brazil. American Museum Novitates 3076:1–15.

Maisey, J. G. 1993b. Tectonics, the Santana Lagerstätten, and the implications for Late Gondwanan biogeography. Pp. 435–454 in P. Goldblatt (ed), Biological Relationships between Africa and South America. Yale University Press, New Haven, CT.

Maisey, J. G. 1994. Predator-prey relationships and trophic level reconstruction in a fossil fish community. Environmental Biology of Fishes 40:1–22.

Maisey, J. G. 2000. Continental break up and the distribution of fishes of Western Gondwana during the Early Cretaceous. Cretaceous Research 21:281–314.

Maisey, J. G., and M. G. P. Carvalho. 1995. First records of fossil sergestid decapods and fossil brachyuran crab larvae (Arthropoda, Crustacea), with remarks on some supposed palaemonid fossils, from the Santana Formation (Aptian-Albian, NE Brazil). American Museum Novitates 3132:1–17.

Maisey, J. G., and J. M. Moody. 2001. A review of the problematic extinct teleost fish *Araripichthys*, with a description of a new species from the Lower Cretaceous of Venezuela. American Museum Novitates 3324:1–27.

Makovicky, P. J., and H.-D. Sues. 1998. Anatomy and phylogenetic relationships of the theropod dinosaur *Microvenator celer* from the Lower Cretaceous of Montana. American Museum Novitates 3240:1–27.

Manso, C. L. C., and M. H. Hessel. 2007. Revisão sistemática de *Pygidiolampas araripensis* (Beurlen, 1966), (Echinodermata Cassiduloida) da Bacia do Araripe, nordeste do Brasil. São Paulo UNESP Geosciências 26:271–277.

Martill, D. M. 1988. Preservation of fish in the Cretaceous Santana Formation of Brazil. Palaeontology 31:1–18.

Martill, D. M. 1990. Macromolecular resolution of fossilized muscle tissues from an elopomorph fish. Nature 346:171–172.

Martill, D. M. 1990. The significance of the Santana Biota. Pp. 253–264 in D. A. Campos, M. S. S. Viana, P. M. Brito, and G. Beurlen (eds), Atas do simpósio sobre a Bacia do Araripe e Bacias Interiores do Nordeste, Crato, 14–16 de Junho de 1990. Crato.

Martill, D. M. 1994. La fossilisation instantanée. La Recherche 269:996–1002.

Martill, D. M. 1997. Fish oblique to bedding in early diagenetic concretions from the Cretaceous Santana Formation of Brazil – implications for substrate consistency. Palaeontology 41:1011–1026.

Martill, D. M. 1998. Fidelity of fossilization: the Santana Formation of Brazil. Pp. 55–74 in S. K. Donovan (ed), The Fidelity of the Fossil Record. Bellhaven Press, London.

Martill, D. M. 2001. The Santana Formation. Pp. 351–356 in D. E. G. Briggs and P. R. Crowther (eds), Palaeobiology II. Blackwell Science, Oxford.

Martill, D. M. 2007. The age of the Cretaceous Santana Formation fossil Konservat Lagerstätte of north-east Brazil: a historical review and an appraisal of the biochronostratigraphic utility of its palaeobiota. Cretaceous Research 28: 895–920.

Martill, D. M. 2011. A new pterodactyloid pterosaur from the Santana Formation (Cretaceous) of Brazil. Cretaceous Research 32:236–243.

Martill, D. M., and E. Harper. 1990. An application of critical point drying to the comparison of modern and fossilized soft tissues of fishes. Palaeontology 33:423–428.

Martill, D. M., and D. Naish. 2006. Cranial crest development in the azhdarchoid pterosaur *Tupuxuara*, with a review of the genus and tapejarid monophyly. Palaeontology 49:925–941.

Martill, D. M., and D. M. Unwin. 1989. Exceptionally well preserved pterosaur wing membrane from the Cretaceous of Brazil. Nature 340:138–140.

Martill, D. M., and P. R. Wilby. 1993. Stratigraphy. Pp. 20–50 in D. M. Martill (ed), Fossils of the Santana and Crato Formations, Brazil. The Palaeontological Association Field Guides to Fossils 5.

Martill, D. M., G. Bechly, and R. F. Loveridge (eds). 2007. The Crato Fossil Beds of Brazil: Window into an Ancient World. Cambridge University Press, Cambridge, 625 pp.

Martill, D. M., P. M. Brito, and J. Washington-Evans. 2008. Mass mortality of fishes in the Santana Formation (Lower Cretaceous, ?Albian) of northeast Brazil. Cretaceous Research 29:649–658.

Martill, D. M., D. M. Unwin, and P. Wilby. 1990. Stripes on a pterosaur wing – Reply. Nature 346:116–116.

Martill, D. M., E. Frey, H.-D. Sues, and A. R. I. Cruickshank. 2000. Skeletal remains of a small theropod dinosaur with associated soft structures from the Lower Cretaceous Santana Formation of northeastern Brazil. Canadian Journal of Earth Sciences 37: 891–900.

Martill, D. M., A. R. I. Cruickshank, E. Frey, P. G. Small, and M. Clarke. 1996. A new crested maniraptoran dinosaur from the Santana Formation (Lower Cretaceous) of Brazil. Journal of the Geological Society, London 153:5–8.

Martins Neto, R. G. 1986. *Pricesaurus megalodon* nov. gen. nov. sp. (Pterosauria, Pterodactyloidea), Cretáceo Inferior, Chapada do Araripe (NE-Brasil). Ciência e Cultura 38:756–757.

Matzke-Karasz, R., R. J. Smith, R. Symonova, C. G. Miller, and P. Tafforeau. 2009. Sexual intercourse involving giant sperm in Cretaceous ostracode. Science 324:1535.

Mayrinck, D., P. M. Brito, and O. Otero. 2009. A new albuliform (Teleostei: Elopomorpha) from the Lower Cretaceous Santana Formation, Araripe Basin, northeastern Brazil. Cretaceous Research 31:227–236.

Mazin, J.-M. 2003. Ichnological evidence for quadrupedal locomotion in pterodactyloid pterosaurs: trackways from the Late Jurassic of Crayssac (southwestern France). Pp. 283–296 in E. Buffetaut and J.-M. Mazin (eds), Evolution and Palaeobiology of Pterosaurs. Geological Society, London, Special Publications 217.

Meylan, P. A. 1996. Skeletal morphology and relationships of the Early Cretaceous side-necked turtle, *Araripemys barretoi* (Testudines: Pelomedusoides: Araripemydidae), from the Santana Formation of Brazil. Journal of Vertebrate Paleontology 16:20–33.

Meylan, P., and E. S. Gaffney. 1991. *Araripemeys* Price, 1973. Pp. 326–334 in J. G. Maisey (ed), Santana Fossils: An Illustrated Atlas. T.F.H. Publications, Neptune City, NJ.

Moody, J. M., and J. G. Maisey. 1994. New Cretaceous marine vertebrate assemblage from north-western Venezuela and their significance. Journal of Vertebrate Paleontology 14:1–8.

Naish, D., D. M. Martill, and E. Frey. 2004. Ecology, systematics and biogeographical relationships of dinosaurs, including a new theropod, from the Santana Formation (?Albian, Early Cretaceous) of Brazil. Historical Biology 16:57–70.

Oliveira, G. R. 2007. Aspectos tafonômicos de Testudines da Formação Santana (Cretáceo Inferior), Bacia do Araripe, Nordeste do Brasil. Anuário do Instituto de Geociências – UFRJ 30:8–93.

Oliveira, G. R., and A. W. A. Kellner. 2007. A new sidenecked turtle (Pleurodira, Pelomedusoides) from the Santana Formation (Early Cretaceous), Araripe Basin, northeastern Brazil. Zootaxa 1425:53–61.

Padian, K. 1983. A functional analysis of flying and walking in pterosaurs. Paleobiology 9: 218–239.

Piasecki, W., A. E. Goodwin, J. C. Eiras, and B. F. Nowak. 2004. Importance of Copepoda in freshwater aquaculture. Zoological Studies 43:193–205.

Pinheiro, F. L., and C. L. Schultz. 2012. An unusual pterosaur specimen (Pterodactyloidea, ?Azhdarchoidea) from the Early Cretaceous Romualdo Formation of Brazil, and the evolution of the pterodactyloid palate. PLoS One 7(11): e50088. doi:10.1371/journal.pone.0050088

Price, L. I. 1955. Novos crocodilídeos dos arenitos da Série Baurú, Cretáceo do Estado de Minas Gerais. Anais da Academia Brasileira de Ciencias 27:487–498.

Price, L. I. 1959. Sobre um crocodilídeo notossuquio do Cretácico Brasileiro. Boletim, Divisão de Geologia e Mineralogia, Rio de Janeiro 118:1–55.

Price, L. I. 1971. A presenca de Pterosauria no Cretáceo Inferior da Chapada do Araripe, Brasil. Anais da Academia Brasileira de Ciências 43(Suppl.):452–461.

Price, L. I. 1973. Quelônio Amphychelydia no Cretáceo inferior do nord-este do Brasil. Revista Brasileira de Geociencias 3:84–96.

Rauhut, O. W. M., C. Foth, H. Tischlinger, and M. A. Norell. 2012. Exceptionally preserved juvenile megalosauroid theropod dinosaur with filamentous integument from the Late Jurassic of Germany. Proceedings of the National Academy of Sciences of the United States of America 109:11746–11751.

Rossetti, D. F., A. M. Góes, J. S. Paz, and M. Macambira. 2000. Sequential analysis of the Aptian deposits from the São Luís and Grajaú basins, Maranhão State (Brazil) and its implication for unraveling the origin of evaporites. Revista Brasileira de Geociências 30:466–469.

Saraiva, A. A. F., M. Hessel, N. C. Guerra, and E. Fara. 2007. Concreções calcárias da Formação Santana, Bacia do Araripe: uma proposta de classificação. Estudos Geológicos 17:40–57.

Sayão, J. M. 2003. Histovariability in bones of two pterodactyloid pterosaurs from the Santana Formation, Araripe Basin, Brazil: preliminary results. Pp. 335–342 in E. Buffetaut and J.-M. Mazin (eds), Evolution and Palaeobiology of Pterosaurs. Geological Society, London, Special Publications 217.

Schultze, H.-P., and D. Stöhr. 1996. *Vinctifer* (Pisces, Aspidorhynchidae) aus der Unterkreide (oberes Aptium) von Kolumbien. Neues Jahrbuch für Geologie und Paläontologie Abhandlungen 199:395–415.

Scotese, C. R. 2001. Atlas of Earth History. PALEOMAP Project, The University of Texas at Arlington.

Silva Santos, R. 1945. Revalidação de *Aspidorhynchus comptoni* Agassiz, do Cretáceo do Ceará, Brasil. D.G.M./D.N.P.M., Notas Preliminares, Estudos 42:1–7.

Silva Santos, R. 1950. *Anaedopogon*, *Chiromystus* e *Ennelichthys*, como sinônimos de *Cladocyclus*, família Chirocentridae. Anais da Academia Brasileira de Ciências 22: 123–134,

Silva Santos, R. 1958. *Leptolepis diasii*, novo peixe fóssil da Serra do Araripe, Brasil. Notas Preliminares e Estudos, Divisão de Geologia e Mineralogia, Departamento Nacional de Produção Mineral 108:1–15.

Silva Santos, R. 1968. A paleoictiofauna da formação Santana-Euselachii. Anais da Academia Brasileira de Ciências 40:491–497.

Silva Santos, R. 1971. Nouveau genre et espéce d'Elopidae du Bassin Sédimentaire de la Chapada do Araripe. Anais da Academia Brasileira Ciências 43:439–442.

Silva Santos, R. 1983. *Araripichthys castilhoi* novo gênero et espécie de acantopterigio da Formação Santana, Chapada do Araripe, Brasil. VIII Congresso Brasileira Paleontologia, Rio de Janeiro, Resumos Comminicados:27.

Small, H. 1913. Geologia e suprimento de agua subterranea no Ceará e parte do Piaui. Inspetoria Obras Contra Secas, Serie Geologia 25:1–180.

Smith, R. J. 1999. Possible fossil ostracod (Crustacea) eggs from the Cretaceous of Brazil. Journal of Micropalaeontology 18:81–87.

Smith, R. J. 2000. Morphology and ontogeny of Cretaceous

ostracods with preserved appendages from Brazil. Palaeontology 43:63–98.
Spix, J. B. von, and C. F. P. von Martius. 1828-1831. Reise in Brasilien: auf Befehl Sr. Majestät Maximilian Joseph I., Königs von Baiern, in den Jahren 1817 bis 1820 gemacht und beschrieben. M. Lindauer, Munich.
Stump, E. 1987. Construction of the Pacific margin of Gondwana during the Pannotios cycle. Pp. 77–87 in G. D. McKenzie (ed), Gondwana Six: Stratigraphy, Sedimentology, and Paleontology. American Geophysical Union, Geophysical Monograph 41.
Sues, H.-D., E. Frey, D. M. Martill, and D. M. Scott. 2002. *Irritator challengeri*, a spinosaurid (Dinosauria: Theropoda) from the Lower Cretaceous of Brazil. Journal of Vertebrate Paleontology 22:535–547.
Torrington, G. B. 1849. Death of George Gardner, Esq. Superintendent of the Botanical Garden, Ceylon. Hooker's Journal of Botany and Kew Garden Miscellany 1:154–156.
Unwin, D. M. 1988. New remains of the pterosaur *Dimorphodon* (Pterosauria: Rhamphorhynchoidea) and the terrestrial ability of early pterosaurs. Modern Geology 13:57–68.
Unwin, D. M. 2001. An overview of the pterosaur assemblage from the Cambridge Greensand (Cretaceous) of eastern England. Mitteilungen aus dem Museum für Naturkunde Berlin, Geowissenschaftliche Reihe 4:189–221.
Unwin, D. M. 2002. On the systematic relationships of *Cearadactylus atrox* Leonardi and Borgamanero, 1985, an enigmatic Lower Cretaceous pterodactyloid pterosaur from the Santana Formation of Brazil. Mitteilungen aus dem Museum für Naturkunde, Berlin, Geowissenschaftliche Reihe 5:237–261.
Unwin, D. M. 2003. On the phylogeny and evolutionary history of pterosaurs. Pp. 139–190 in E. Buffetaut and J.-M. Mazin (eds), Evolution and Palaeobiology of Pterosaurs. Geological Society, London, Special Publications 217.
Unwin, D. M. 2006. The Pterosaurs from Deep Time. Pearson Press, New York.
Unwin, D. M., and D. M. Martill. 2007. Pterosaurs of the Crato Formation. Pp. 475–524 in D. M. Martill, G. Bechly, and R. F. Loveridge (eds), The Crato Fossil Beds of Brazil: Window into an Ancient World. Cambridge Univeristy Press, Cambridge.
Unwin, D. M., E. Frey, D. M. Martill, J. B. Clarke, and J. Riess. 1996. On the nature of the pteroid in pterosaurs. Proceedings of the Royal Society of London B 277:1121–1127.
Veldmeijer, A. J. 2002. Pterosaurs from the Lower Cretaceous of Brazil in the Stuttgart Collection. Stuttgarter Beiträge zur Naturkunde B 327:1–27.
Veldmeijer, A. J. 2003. Description of *Coloborhynchus spielbergi* sp. nov. (Pterodactyloidea) from the Albian (Lower Cretaceous) of Brazil. Scripta Geologica 125:35–139.
Veldmeijer, A. J., and A. M. Hense. 2004. Supplement to: Pterosaurs from the Lower Cretaceous of Brazil in the Stuttgart collection, in: Stuttgarter Beiträge zur Naturkunde, Serie B (Geologie und Paläontologie) 2002, 327, 1–27. PalArch, Vertebrate Paleontology 1:14–21.
Veldmeijer, A. J., M. Signore, and E. Bucci. 2007. Predator-prey interaction of Brazilian Cretaceous toothed pterosaurs: a case example. Pp. 295–308 in A. M. T. Elewa (ed), Predation in Organisms: A Distinct Phenomenon. Springer, Berlin and Heidlberg.
Vila Nova, B. C., and J. M. Sayão. 2012. On wing disparity and morphological variation of the Santana Group pterosaurs. Historical Biology 24:567–574.
Vila Nova, B. C., A. W. A. Kellner, and J. M. Sayão. 2010. Short note on the phylogenetic position of *Cearadactylus atrox*, and comments regarding its relationships to other pterosaurs. (Flugsaurier 2010, Third International Symposium on Pterosaurs, 5–10 August 2010, Beijing, China.) Acta Geological Sinica 31 (Supplement 1):73–75.
Vila Nova, B. C., A. A. F. Saraiva, J. K. R. Moreira, and J. M. Sayão. 2011. Controlled excavations in the Romualdo Formation Lagerstätte (Araripe Basin, Brazil) and pterosaur diversity: remarks based on new findings. Palaios 26:173–179.
Vullo, R., and D. Neraudeau. 2008. When the 'primitive' shark *Tribodus* (Hybodontiformes) meets the 'modern' ray *Pseudohypolophus* (Rajiformes): the unique co-occurrence of these two durophagous Cretaceous selachians in Charentes (SW France). Acta Geologica Polonica 58:249–255.
Weeks, L. G. 1956. Origin of carbonate concretions in shales, Magdalena Valley, Colombia. Geological Society of America Bulletin 68:95–102.
Wellnhofer, P. 1977. *Araripedactylus dehmi* nov. gen., nov. sp., ein neuer Flugsaurier aus der Unterkreide von Brasilien. Mitteilungen der Bayerischen Staatssammlung für Paläontologie und historische Geologie 17:157–167.
Wellnhofer, P. 1985. Neue Pterosaurier aus der Santana–Formation (Apt.) der Chapada do Araripe, Brasilien. Palaeontographica A 187:105–182.
Wellnhofer, P. 1987a. New crested pterosaurs from the Lower Cretaceous of Brazil. Mitteilungen der Bayerischen Staatssammlung für Paläontologie und historische Geologie 27:175–186.
Wellnhofer, P. 1987b. Die Flughaut von *Pterodactylus* (Reptilia, Pterosauria) am Beispiel des Wiener Exemplares von *Pterodactylus kochi* (Wagner). Annalen des Naturhistorischen Museums Wien 88:149–162.
Wellnhofer, P. 1988. Terrestrial locomotion in pterosaurs. Historical Biology 1:3–16.
Wellnhofer, P. 1991a. Weitere Pterosaurierfunde aus der Santana-Formation (Apt) der Chapada do Araripe, Brasilien. Palaeontographica A 215:43–101.
Wellnhofer, P. 1991b. The Illustrated Encyclopedia of Pterosaurs. Crescent Books, New York, 192 pp.
Wellnhofer, P. 2009. *Archaeopteryx*: The Icon of Evolution. Verlag Dr. Friedrich Pfeil, Munich, 208 pp.
Wellnhofer, P., and A. W. A. Kellner. 1991. The skull of *Tapejara wellnhoferi* Kellner (Reptilia, Pterosauria) from the Lower Cretaceous Santana Formation of the Araripe Basin, northeastern Brazil. Mitteilungen der Bayerischen Staatssammlung für Paläontologie und historische Geologie 31:89–106.
Wellnhofer, P., E. Buffetaut, and G. Gigase. 1983. A pterosaurian notarium from the Lower Cretaceous of Brazil. Paläontologische Zeitschrift 57:147–157.
Wenz, S. 1989. *Iemanja palma* n. g., n. sp., Gyrodontidae nouveau (Pisces, Actinopterygii) du Crétacé inférieur de la Chapada do Araripe (N-E du Brésil). Comptes Rendus de l'Académie des Sciences, Série II 308:975–980.
Wenz, S., and P. M. Brito. 1990. L'ichthyofaune des nodules

fossilifères de la Chapada do Araripe (NE du Brésil). Pp. 309–328 in D. A. Campos, M. S. S. Viana, P. M. Brito, and G. Beurlen (eds), Atas do simpósio sobre a Bacia do Araripe e Bacias Interiores do Nordeste, Crato, 14-16 de Junho de 1990. Crato.

Wenz, S., and P. M. Brito. 1992. Première découverté de Lepisosteidae (Pisces, Actinopterygii) dans le Crétacé inférieur de la Chapada do Araripe (N-E du Braisil). Conséquences sur la phylogénie des Ginglymodi. Comptes Rendus de l'Académie des Sciences, Série II 314:1519–1525.

Wenz, S., and P. M. Brito. 1996. New data about the lepisosteids and semionotids from the Early Cretaceous of Chapada do Araripe (NE Brazil): phylogenetic implications. Pp. 153–165 in G. Arratia and G. Viohl (eds), Mesozoic Fishes – Systematics and Paleoecology. Verlag Dr. Friedrich Pfeil, Munich.

Wilby, P. R. 1993. The Mechanisms and Timing of Mineralization of Fossil Phosphatized Soft Tissues. PhD dissertation. The Open University, Milton Keynes.

Wilby, P. R., and D. M. Martill. 1992. Fossil fish stomachs: a microenvironment for exceptional preservation. Historical Biology 6:25–36.

Wilkinson, M. T., D. M. Unwin, and C. P. Ellington. 2006. High lift function of the pteroid bone and forewing of pterosaurs. Proceedings of the Royal Society of London B 273:119–126.

Wilson, C. B. 1922. North American parasitic copepods belonging to the family Dichelesthiidae. Proceedings of the U.S. National Museum 60:1–100.

Witton, M. P. 2009. A new species of *Tupuxuara* (Thalassodromidae, Azhdarchoidea) from the Lower Cretaceous Santana Formation of Brazil, with a note on the nomenclature of Thalassodromidae. Cretaceous Research 30:1293–1300.

Woodward, A. S. 1887. On the fossil teleostean genus *Rhacolepis*, Agass. Proceedings of the Zoological Society of London 1887:535–542.

Yabumoto, Y. 2002. A new coelacanth from the Early Cretaceous of Brazil (Sarcopterygii, Actinistia). Palaeontological Research 6:343–350.

Zittel, K. A. von. 1882. Über Flugsaurier aus dem lithographischen Schiefer Bayerns. Palaeontographica 29:47–80.

Tables

Table 1: Fishes of the Romualdo Member, Santana Formation

Taxon	Author	Date	Abundance
Sarcopterygii Coelacanthiformes			
Axelrodichthys araripensis	Maisey	1986	F
Mawsonia brasiliensis	Yabumoto	2008	R
Actinopterygii Holostei Ginglymodi Semionotiformes Semionotidae			
Araripelepidotes temnurus	(Agassiz)	1841	F
Lepidotes wenzae	Brito and Gallo	2003	VR
Lepisosteiformes Obaichthyidae			
Obaichthys decoratus	Wenz and Brito,	1992	VR
Dentilepisosteus laevis	(Wenz and Brito)	1992	VR
Halecomorphi Amiiformes Amiidae			
Calamopleurus cylindricus	Agassiz	1841	C
Ionoscopiformes Ophiopsidae			
Placidichthys bidorsalis	Brito	2000	VR
Oshuniidae			
Oshunia brevis	Wenz and Kellner	1986	R
Pycnodontiformes Pycnodontidae			
Neoproscinetes penalvai	(Santos)	1968	F
Iemanja palma	Wenz,	1989	VR
Teleosteomorpha Aspidorhynchiformes Aspidorhynchidae			
Vinctifer comptoni	(Agassiz)	1841	VC
Teleostei Ichthyodectiformes Cladocyclidae			
Cladocyclus gardneri	Agassiz	1841	VC
Crossognathiformes Pachyrhizodontoidei Notelopidae			
Notelops brama	Agassiz	1841	C
Pachyrhizodontidae			
Rhacolepis buccalis	Agassiz	1841	VC
Elopomorpha Albuliformes			
Branneriun latum	(Agassiz)	1841	?
Brannerion vestitum	(Jordan and Branner)	1908	VC
Paraelops cearensis	Santos	1971	C
Bullichthys santanensis	Mayrinck, Brito Otero	2010	?

Clupeocephala Ostarioclupeomorpha Gonorynchiformes Chanidae *Tharrhias araripis*	(Jordan and Branner)	1908	VC
Clupeocephala order and family *incertae sedis* *Beurlenichthys ouricuriensis*	Figueiredo and Gallo	2004	F
Clupeomorpha *incertae sedis* *Santanaclupea silvasantosi*	Maisey	1993	F
Otophysi incertae sedis *Santanichthys diasii*	(Santos)	1958	F
Teleostei *insertae sides* Araripichthyidae *Araripichthys castilhoi*	Santos	1985	R
Chondrichthyes Hybodontiformes Hybodontidae *Tribodus limae*	Brito and Ferreira,	1989	R
Batoidea Rajiformes family *incertae sedis* *Iansan beurleni*	(Santos)	1968	U

Table 2: Pterosaur taxa considered valid for the Santana Formation fossil conservation Lagerstätte. Artwork with kind permission of Dr Mark Witton.

Ornithocheiridae
Anhanguera santanae
Anhanguera blittersdorfi
Coloborhynhcus robustus
Ornithocheirus mesembrinus

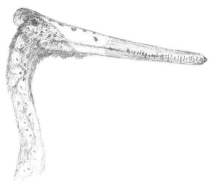

Lonchodectidae
Unwindia trigonus

Thalassodromidae
Thalssodromeus sethi
Tupuxuara deliridamus
Tupuxuara leonardi

Tapejaridae
Tapejara wellnhoferi

Table 3: Taxa certainly recorded from the Santana Formation concretion fossil conservation Lagerstätte. Taxa from the Santana Formation above or below the Romualdo concretion horizon are omitted.

Mollusca
Gastropoda
Unidentified protoconchs

Arthropoda
Copepoda
Kabatarina pattersoni Cressey and Boxshall 1989
Ostracoda
Pattersoncypris (=Harbinia) micropapillosa (Bate, 1972)
Crustacea
Paleomattea deliciosa Masiey and Carvalho, 1995
Unamed brachyuran crab larvae

Vertebrata
Fishes (see Table 1 for taxonomic details)
Tribodus limae
Iansan beurleni
Araripelepidotes temnurus
Lepidotes wenzae
Obaichthys decoratus
Dentilepisosteus laevis
Calamopleurus cylindricus
Placidichthys bidorsalis,
Oshunia brevis
Neoproscinetes penalvai
Iemanja palma
Vinctifer comptoni
Cladocyclus gardneri
Notelops brama
Rhacolepis buccalis
Branneriun latum
Brannerion vestitum
Paraelops cearensis
Bullichthys santanensis
Tharrhias araripis
Beurlenichthys ouricuriensis
Santanaclupea silvasantosi
Santanichthys diasii
Araripichthys castilhoi

Tetrapoda
Testudines
Pleurodira
Pelomedusidae
Araripemys barretoi Price, 1973
Bothremyididae
Cearachelys placidoi Gaffney et al., 2001
Cryptodira Gray, 1825
Eucryptodira Gaffney, 1975
Chelonioidea Agassiz, 1857
Protostegidae Cope, 1873
Santanachelys gaffneyi Hirayama, 1998
Pleurodira Cope 1864
Brasilemydidae Lapparent de Broin, 2000
Brasilemys josai Lapparent de Broin, 2000
Crocodylomorpha
Notosuchia
Araripesuchus gomesii Price, 1959
Neosuchia
Itasuchus camposi (Kellner, 1987)
Pterosauria (see Table 2)
Dinosauria
Theropoda
Santanaraptor placidus Kellner, 1999
Irritator challengeri Martill et al., 1996
Mirischia asymmetrica Naish, Martill and Frey, 2004

Chapter 8

The Messel Pit Fossil Site

Stephan F. K. Schaal

Senckenberg Forschungsinstitut, Frankfurt, Germany.

Abstract

Messel Pit Fossil Site in the state of Hesse, Germany is a UNESCO World Heritage Site. The exposed strata are notable for the preservation a remarkable diverse flora and fauna from 48 million years ago in the Eocene. The unique depositional conditions in a volcanic crater lake have led to unparalleled preservation of remains of extraordinary large numbers of fossils representing animals and plants living in a large lake and in its surrounds. They include complete insects with colour patterns preserved and numerous mammals and birds complete with fossilized fur and feathers. Together they indicate a warm climate and document a tropical rainforest setting for the Messel community. The isolation of Europe as an archipelago during the early Eocene led to the development of endemic European animal species living alongside groups that immigrated from elsewhere. Excavations continue to this day thereby constantly enhancing our understanding of mid-Eocene life in Europe.

Discovery and history of exploration of the Messel Pit Fossil Site

The UNESCO World Heritage Site Messel Pit Fossil Site is situated 9km NE of Darmstadt and about 35km SE of Frankfurt am Main in the state of Hesse, Germany (Fig 1) (Schaal and Rabenstein, 2012). In the nineteenth century geologists identified a suite of sedimentary rocks discovered here as an 'oil shale'. (There had been some earlier mining at the location for iron ore and lignite.) From 1884 (when the mining for the oil shale began) until 1971, these rocks yielded more than one million metric tons of crude oil for industrial purposes. During this time, the Messel pit, a 60 m deep open-cast mine, was created. It now has a length of 1km and a width of 700 m. During the late nineteenth and early twentieth centuries, the first scientific studies of fossil plants, insects and vertebrates from the oil shale were published. During the 1960s, private fossil collectors started excavations in the pit and remained very active there until the 1980s.

In 1973 termination of open-cast mining led to the proposal to fill the pit with waste, which resulted in a long period of public protests and legal actions. It took nearly 20 years to resolve this issue until the state government of Hesse decided to purchase the Messel pit in 1991. Since 1992 the scientific management of the pit has been entrusted to the Forschungsinstitut Senckenberg (Frankfurt am Main) where a department explicitly dedicated to scientific research on the Messel Pit Fossil Site was established (Schaal and Schneider, 1995). Regular excavations in the Messel pit by the Forschungsinstitut Senckenberg and the Hessisches Landesmuseum Darmstadt commenced in 1975. Both institutions have continued their highly successful excavation efforts, leading to about 1400 scientific and popular publications to date (Schaal et al., 2004). Avocational collectors and professional palaeontologists have recovered a wealth of animal and plant fossils, often in an extraordinary state of preservation, and prepared them using a resin-transfer method. The results of these efforts testify to the extraordinary importance of the Messel Pit Fossil Site as a fossil Konservat-Lagerstätte (Schaal, 2012). Specimens of insects, fish, amphibians, reptiles, birds and mammals preserve complete body silhouettes, with traces of hair and feathers, as well as gastrointestinal contents. Some skeletons of early horse-like perissodactyls contain foetal remains. This remarkable type of preservation opens an unparallelled window into the history of life and of the Earth during the Eocene. Thus, the Messel Pit Fossil Site was designated a UNESCO World Heritage site in 1995. Temporary and permanent exhibitions with Messel fossils were created, and a visitor centre was constructed near the locality in

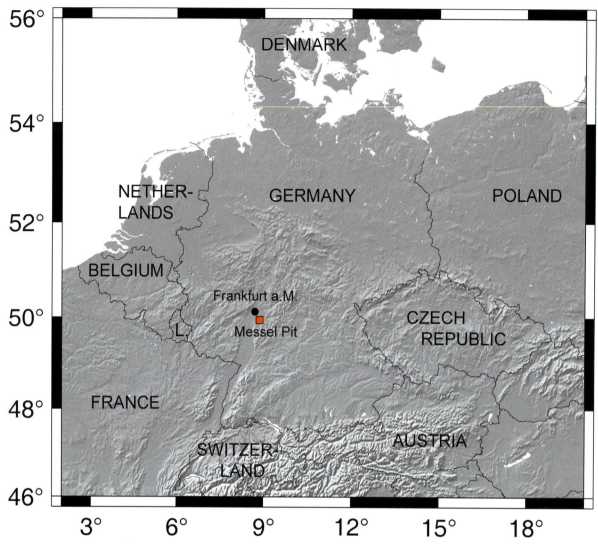

Figure 1 Position of the Messel Pit Fossil Site on a physiographic map of Central Europe.

2010 to serve as the interface between research and the public (Mangel, 2011).

The amount of information on the Messel Pit Fossil Site and its fossil content is so vast that this chapter can only provide a brief review augmented by some particularly pertinent references. Schaal and Ziegler (1988, 1992) have edited a major survey of the geology, fauna and flora and provided a comprehensive bibliography to the extensive primary literature. Schaal et al. (2004) provided an updated list of references, but many additional papers have been published since that time.

Geological context

During the Eocene Europe was an archipelago, separated by seas from neighbouring lands, and was situated further south than today. The Messel Pit Fossil Site was located at nearly 40°N palaeolatitude, equivalent to the present-day locations of southern Italy and Spain (Meulenkamp and Sissingh, 2003). Extensive tectonic activities took place in the Messel region, and molten rock rose through tectonic faults and came into contact with groundwater (Fig 2). This led to phreatomagmatic explosions that created a diatreme, which was 200–300m deep and more than 1km wide (Pirrung et al., 2001; Lorenz and Kurzlaukis, 2007). This crater was situated in a granite or granodiorite and was encircled by a ring wall of volcanic debris and ash probably up to 50m high. Groundwater and rainwater accumulated in the crater, and a maar lake formed. The Messel oil shale represents the deposits of this maar lake (Schulz et al., 2002). The lacustrine laminites from Messel were dated astronomically by

Figure 2 Diagrams illustrating the formation of a diatreme and maar lake through phreatomagmatic explosion.

orbitally-controlled changes in vegetation and fluctuations of the pollen rain, respectively, in combination with a revised $^{40}Ar/^{39}Ar$ age of a basalt fragment from a lapilli tuff section below the first lacustrine sediments. The Messel oil shale has an age of 48Ma and includes the Ypresian/Lutetian boundary (Lenz et al., 2014). Messel fossils therefore belong to MP11 (Mammal Palaeogene, zone 11). Sedimentation of the Messel Formation lasted for up to one million years. The base of the sequence is formed by rock fragmented by the volcanic eruption, followed by volcaniclastics, and the middle Messel Formation with layers of sandstone, siltstone, and claystone, and especially an algal laminated deposit containing a high percentage of organic compounds known as the Messel oil shale (Felder and Harms, 2004; Fig 3). The latter does not actually contain oil but rather kerogen, which formed by the decomposition of green algae (Goth, 1990). This form of hydrocarbon was exploited for nearly 80 years (Schaal and Schneider, 1995). The upper part of the Messel Formation consisted of clay with intercalations of sand and lignite (Pirrung, 1998).

After the aggradation of the lake to the present day the European continent (Suhr, 2006), as part of the Eurasian plate, was pushed northwards to its present location by the African plate. The Messel Pit Fossil Site is now located at UTM coordinates 32U 5530000 32482700 (WGS84) (geographic latitude and longitude coordinates: 49°55'0.012"N 8°45'14.004"E). Later, global climates cooled considerably. Sea levels dropped, and the European archipelago of Eocene times became the contiguous European part of the Eurasian tectonic plate. The Rhine graben formed much later, and the Messel maar is situated now on an uplift of the Odenwald, east of the Rhine (Buness et al., 2005). Until mining commenced at Messel in 1884, most of the maar sediments were protected from erosion by the surrounding basement rocks of the Odenwald horst. Today some 100 maar craters with lakes are known worldwide; notable examples occur in the Eifel region of Germany (Fig 4) and Xiaolongwan in northeastern China (Figs 5–6) (Büchel, 1988). Some (like Xiaolongwan) still preserve the original ring wall.

Sedimentological aspects and consequences for the preparation of fossils

The laminae in the Messel oil shale, each only about 0.1mm thick, were formed by seasonal blooms of the tiny green alga *Tetraedron minimum*, which sank to the

Figure 3 — Four sections of core from the Messel drilling project. Outer left (between 23 and 23.40m depth): finely laminated dark shale with yellowish bands of siderite. Second from left (between 140.55 and 140.95m depth): sandstone, siltstone and individual laminae of dark shale, finely bedded to finely laminated, with ripple marks. Second from right (between 362.55 and 362.95m depth): lapilli tuff with many xenoliths from Rotliegend deposits and granodiorite as well as granodiorite debris. Outer right (between 395 and 395.40m depth): breccia of the diatreme, probably composed of amphibolite with granodiorite filling in fractures, in places stained red.

Figure 4 View of the Weinfelder Maar (Totenmaar) in the Eifel region of Germany.

bottom of the lake after dying off. Lenz et al. (2011) calculated an average sedimentation rate of 0.14mm per year for the still-exposed deposits of oil shale. The lamination of the sediment caused by algal blooms and small amounts of clay permits splitting of slabs of rock using knives during excavation. The clay material was deposited constantly but in small quantities, and may have its source on the ring wall of the crater lake. The

Figure 5 View of Lake Xiaolongwan on the northwestern margin of the Changbai mountain chain in northeastern China. Note part of the ring wall and early accumulation of silt and organic debris in the lake.

Figure 6 Ring wall of Lake Xiaolongwan, with volcaniclastic sediments and larger volcanic bombs.

most abundant minerals in the oil shale are smectites, clay minerals that can absorb a large amount of water. Today the oil shale still contains up to 40% of water.

Thousands of 0.1–3mm thick, yellowish-brown layers of siderite can be seen in the black sedimentary rock (Fig 3). They were probably formed by lithified microbial mats that grew in the organic-rich mud on the lake floor and concentrated iron ions (Felder and Harms, 2004). Some fossils were enclosed in early diagenetic siderite concretions. Sometimes dispersed pyrite and pyrite layers are present. Inclusions of phosphate minerals such as messelite and montgomeryite are irregularly distributed but concentrated in a 2cm thick marker horizon known from the centre of the maar lake. Occasional turbidity currents and debris flows formed sand and gravel layers. The upper part of the Messel Formation, which was completely removed by mining, consisted of clay with intercalations of sand and lignite (Matthess, 1966).

The oil shale can be split along its bedding planes using knives. The fossils, which are often split along the sagittal plane during this process, are saved as part and counterpart plates. The fresh, gleaming black-brown rock still contains much water and, rapidly drying out after exposure to air, turns grey in colour and disintegrates. This poses a problem for specimen recovery, especially during warm summer weather. To prevent destruction of the rock and fossils, vertebrate specimens must be quickly transferred from oil shale to a different material in a costly and difficult procedure. Part and counterpart are transferred during preparation onto two plates (Fig 7). Epoxy resins are used for this purpose, which explains the yellow or amber colour of the final preparations. This technique makes it possible to conserve traces of soft parts like skin, feathers and hair and any

Figure 7 Skeleton of *Messelobunodon schaeferi* prepared as part and counterpart by resin-transfer method. Note dark outlines of soft tissues. *Messelobunodon* attained a total length of approximatley 70 cm (size of a dachshund).

gastrointestinal contents. Only robust vertebrate fossils like crocodylian, large mammal, and turtle bones can be conserved by treatment with cyanoacrylate glues.

Insect and plant remains are usually stored permanently in glycerine to prevent fungal growth on the specimens. Fossils of these organisms are only prepared using the transfer method if the needs of scientific study mandate this procedure. In the photographs showing fossils of insects and plants in this chapter, the original oil shale is still visible.

Taphonomical aspects

The configuration of maar lakes, with a diameter of 1km or less and a depth of several hundred metres, promotes conditions for creating a fossil *Konservat-Lagerstätte*, especially in warm climates such as those prevalent in Europe during the Eocene. The oxygenated top layer of the lake waters was a few metres deep and suitable as a habitat for fishes, amphibians, insects and plants. Warm water with temperatures up to 30°C has a lower density than, and thus does not mix with, cooler bottom water as long as there is no mixing due to seasonal changes. Because of this stable stratification of the water body, the lake must have been meromictic and had a chemocline. The lack of circulation, as well as low oxygen levels in the deeper layers of the lake, created anaerobic conditions at the bottom and in the sediment. Animal carcasses and plant remains from the surrounding areas introduced into the lake experienced rapid burial along with the lacustrine flora and fauna. Anaerobic bacteria consumed most of the organic material. When they became lithified the bacteria replicated soft body outlines (Wuttke, 1983) (Figs 8–9), and sets of melanosomes have preserved details of feathers although the beta keratin has degraded (Vinther *et al.*, 2010). All insects are flattened by postmortem compression and may be deformed. Some have retained the structural colours of their exoskeleton, which lend them iridescent metallic colours. Clay and organic substances formed the mud of the lake, and subsequently claystone. The vertebrate remains were flattened by compaction of the overlying sediments, during which process bones of articulated skeletons overlying each other were usually deformed and/or crushed.

During excavations plant, insect and vertebrate fossils are still found on a daily basis, but they are not numerous relative to the considerable time span represented by the sedimentary deposits. During each day of excavation, field crews process up to one metre thickness of oil shale deposits, which represents about 10,000 years of sedimentation.

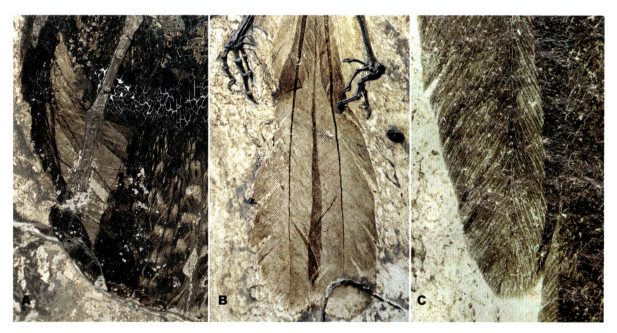

Figure 8 Three examples of bird plumage from the Messel oil shale. Image height: A 6 cm, B 8 cm, C 1 cm.

Figure 9 Skeleton of *Macrocranion tupaiodon* (find made in 2012) with body outline and traces of hair, still embedded in oil shale before resin-transfer preparation. Detail shows close-up of hair traces just behind the head. Length of the specimen approximately 25 cm.

Biodiversity

Plants

Plants are the most common macrofossils at the Messel Pit Fossil Site. Thus, after 35 years of excavations, the site has become world-famous for its Eocene plant fossils. Leaves, flowers, fruits and seeds are still found daily during excavations (Figs 10–13). The diverse flora provides evidence for an equable, warm and humid palaeoclimate with some seasonality. Common features of the foliage, such as large leaf size and development of drip tips, reflect moist conditions (Schaarschmidt, 1992).

The plants recovered from the oil shale represent pteridophytes, conifers and angiosperms (Wilde, 2004). Remains of more than 100 families indicate a high diversity of flowering plants in the mid-Eocene flora of Europe and make up the majority of plant fossils. It is significant that all the families found in Messel occur in warm and moist tropical or subtropical settings today (Storch and Schaarschmidt, 1992). A monographic study of the fruit and seeds from the Messel site demonstrated that the vegetation formed a multi-layered canopy forest (Collinson *et al.*, 2012). This comprised tall trees such as walnut

Figure 10 Slightly opened male flower of a palm. Length: 8mm.

Figure 11 Fruits of a moonseed, *Martinmuellera tuberculata* (Menispermaceae). Diameter of each fruit approximately 5.5 cm.

Figure 12 Leaf of a wine grape (Vitaceae). Length of leaf stalk 4 cm.

Figure 13 Compound leaf of a walnut (Juglandaceae). The leaflet on the left is approximately 10 cm long.

(Fig 13), trees and shrubs at an intermediate level, and a forest floor covered with herbaceous plants. The flora included a high proportion of lianas and palms. Marshy settings bordered the lake in some places but the forest extended apparently right up to the lake in other places. The lake flora was dominated by green algae. No remains of free-floating or submerged plants have been found, but some aquatic plants with floating leaves are known. Leaves of arums are common. These plants lived in shallow water or in marshy areas, together with ferns and sedges. Wood fragments are rare (Wilde, 2005).

The setting of the maar lake and its vicinity changed over the course of time as the ring wall of the crater was eroded. The steep shores of the young, deep maar lake gave way to a more open area of shallow water. The slow disappearance of the ring wall facilitated access to the vicinity of the lake for many species and also altered the influx of water into the lake. After several hundred thousand years the lake was filled by silt and forest came to cover the area.

Insects

Several hundred insect species (Figs 14–19) have been identified among some 16,000 insect fossils recovered at the Messel Pit Fossil Site by Senckenberg teams since 1975. Most of them have yet to be formally described.

Surprisingly, aquatic insects living in shallow and open water are rare. However, some insight into their diversity is provided by research on the coprolites of small fish (Richter and Wedmann, 2005). This source has yielded fragments such as mandibles of larvae of Chaoboridae and pieces of filter apparatus of larvae of Culicidae. These insects must have been present in vast numbers and likely formed the base of the food web in the top layer of lake Messel. The characteristic larval cases of caddisflies are common. Also recorded are larvae of mayflies (Ephemeroptera), dragonflies, damselflies and their larvae (Odonata), stone flies (Plecoptera), scavenger beetles (Hydrophilidae), water striders (Gerridae), backswimmers (Notonectidae) and water pennies (Psephenidae).

Eighteen orders of insects have been identified to date. Most belong to the terrestrial fauna (Wedmann, 2005). Fossils of compact insects such as beetles (Coleoptera) are frequently found and comprise 60% of the insect finds. Some insects, especially jewel beetles (Buprestidae), still retain their structural colours with iridescent metallic hues (Figs 14–16). Some of the buprestid genera recorded from Messel became extinct in Europe but others still survive, mainly in Central and South America. Coleopteran families found at Messel include scarabs (Scarabaeidea), stag beetles (Lucanidae), leaf beetles (Chrysomelidae) (Fig 14), ground beetles (Carabidae), and longhorn and darkling beetles (Cerambycidae and Tenebrionidae) (Figs 15–16).

Other insects include cockroaches (Blattaria), flies (Diptera), true bugs (Heteroptera), cicadas and hoppers (Homoptera), bees and ants (Hymenoptera), termites (Isoptera), butterflies (Lepidoptera), stick insects (Phasmatodea), grasshoppers (Saltatoria) (Wedmann and Yeates, 2008; Wedmann et al., 2009). One noteworthy record is of parasitic bat flies (Nycteribiidae), which feed on the blood of bats (Wedmann and Habersetzer, unpublished data). Fossils of *Formicium giganteum*, the largest known ant of all time, are found during every field season. The females of this species attained a wingspan of up to 15cm and a body length of about 6cm. The ants apparently fell into the lake during mating flights. Weaver ants are also present; their extant relatives live in the canopy of tropical forests by weaving living leaves together (Dlussky et al., 2008). The leaf insect *Eophyllium messelensis* (Phasmatodea) from the Messel oil shale has the same body form as extant leaf insects and represents the geologically oldest example of leaf mimicry among insects to date (Wedmann et al., 2007; Fig 17).

Fossils of wasps (Vespidae) and bees (Apidae; Fig 18) are rare but they may have lived in large colonies. Like their extant relatives, many of the bees had pollen-gathering structures on the hind legs (Wappler and Engel, 2003). The cicadas (Cicadidae), attaining lengths ranging from 2 to 7cm, show a size range comparable to that found today in the tropics and subtropics. Mosquitoes, horseflies, snipe flies, hover flies, robber flies and soldier flies represent a diversity of families of Diptera. On the other hand, dipteran fossils are rare at Messel, with only about 125 specimens found to date (Wedmann, pers. comm.). This fact does not imply that they were rare during the Eocene or in the area of lake Messel. Their remains probably did not reach the bottom of the lake, which is the condition for becoming fossilized. Other rare insects include cockroaches, earwigs, stick insects, lacewings and scorpion flies. The first find of an almost completely preserved moth referable to the family Zygaenidae (Fig 19), made in 2012 after 37 years of excavation effort, underscores

Figure 14 Leaf beetle (Chrysomelidae) preserving brilliant structural colours. Length of the specimen is 10 mm.

Figure 15 Ground beetle *Ceropria messelense* (Tenebrionidae) preserving brilliant structural colours. Length of the specimen 20 mm.

Figure 16 Another example of *Ceropria messelense* associated with small flower. Length of the specimen 20 mm.

Figure 17 Leaf insect *Eophyllium messelensis*. Length of the specimen 63 mm.

Figure 18 Unidentified bee. Length of the specimen 7 mm.

Figure 19 Zygaenid moth preserving some of its coloration. Wing length 22 mm.

Fishes

As inhabitants of the lake, the fishes are the most common fossils at the Messel Pit Fossil Site and are found daily during the excavations. The amiid *Cyclurus* and the lepisosteid *Atractosteus* are the most common taxa (first reported by Andreae, 1893), whereas *Thaumaturus* and perciforms like *Amphiperca*, *Rhenanoperca* and *Palaeoperca* are rarer. The composition and diversity of fish fauna varies, indicating fluctuations in water quality in the top water layers of the Messel lake (Micklich, 2002, 2012). The short-snouted gar *Masillosteus* and a single specimen of the eel *Anguilla* are rarities among the Messel fish fossils. Remains of non-predatory fish have not yet been found.

Amphibians

It is noteworthy that amphibian fossils are uncommon at the Messel Pit Fossil Site. Only six species have been discovered thus far. Only a single skeleton of a salamander, *Chelotriton robustus*, has been found. Frogs are much more common, in particular *Eopelobates*, which has a robust skull. Less common are aquatic frogs of the family Palaeobatrachidae. It is possible that they could colonize the lake only when water quality was excellent (Wuttke et al., 2012). This would explain the fact that tadpoles are only rarely found and that another group of aquatic frogs, the clawed frogs, is also uncommon.

Reptiles

Three major groups of reptiles are known from the Messel Pit Fossil Site: Squamata (snakes and lizards), Testudines (turtles), and Crocodylia (crocodylians). The largest snake is *Palaeopython fischeri*, a boid that reached a length of 2m (Schaal, 2004). Whereas only isolated vertebrae of snakes are found at most fossil sites, a number of complete ophidian skeletons have been discovered at Messel. They provide valuable information concerning the length, morphology and relationships of these Eocene snakes. Only half as long as *Palaeopython* are the two species of *Messelophis*, which have been assigned to the dwarf boas (Schaal and Baszio, 2004). There are additional, as yet undescribed, snake species. Gastrointestinal contents of snakes are rarely found because the strong digestive enzymes of these reptiles usually rapidly break down ingested prey. But some snake fossils, for instance of the family Boidae, preserve gut contents, including, in one example, a complete specimen of the lizard *Geiseltaliellus* (Fig 20) and a juvenile *Diplocynodon* in another. As the crocodylian was almost as large as the snake, this act of predation may have resulted in the snake's death. A spectacular find was a regurgitate of a snake (probably *Palaeopython*), which consists of the skeletal remains of the creodont mammal *Lesmesodon edingeri* compacted into a 30-cm long 'sausage' (Morlo et al., 2012; Fig 21).

Only seven species of lizards, most of them terrestrial, have been described from Messel to date, but this figure represents at most half of their actual diversity (Smith, pers. comm.). They belong to Iguania, Anguimorpha and Scincomorpha. Only geckos (Gekkota) have not yet been recognized from Messel. A particularly impressive species is the large-headed arboreal lizard *Ornatocephalus*, which attains a length of up to 90cm, 70cm of which represent the prehensile tail. Its relationships are problematic, but it was tentatively referred to Scincoidea (Weber, 2004), a clade that today includes skinks and night lizards. A variety of forms are similar to extant wall lizards and their relatives (Lacertidae), the dominant extant lizard group in Europe. One lacertid burrowed in leaf litter and shows certain similarities to worm-lizards (Amphisbaenia; Müller et al., 2011). Armoured terrestrial species include the nearly limbless *Ophisauriscus* and the stout-bodied *Placosauriops*.

The most common lizard at Messel is *Geiseltaliellus*, which belongs to the predominantly New World Iguanidae, and in certain features resembles basiliscs (Smith, 2009). Iguanid squamates probably crossed from North America into Europe during the earliest Eocene warm interval. The tail is missing in more than half the specimens of *Geiseltaliellus* from Messel, suggesting that it was shed intentionally. Unexpectedly, autotomy occurred *between* vertebrae, rather than within individual vertebrae as in other lizards. The evolution of this novel tail-loss mechanism suggests an unusually high predation risk for *Geiseltaliellus* and an arboreal mode of life (Smith and Wuttke, 2012). Its previously mentioned occurrence in the gut content of a snake further implies that the snake was arboreal. *Geiseltaliellus* thus emerges as an important taxon for reconstructing reptilian ecology around the Messel lake.

The only larger predatory lizard reported from

Figure 20 Skeleton of snake with gut content comprising a lizard (insert **A**) that preserves coloured insect remains in its own gut content (insert **B**). Skull length 3 cm.

Figure 21 Regurgitate of a snake (presumably *Palaeopython*) comprising skeletal remains of the creodont *Lesmesodon edingeri*. Inserts **A–C** illustrate details of the dentition of this mammal. Length of the specimen 27.5 cm.

Messel is an animal with powerful limbs related to the Palaeogene *Necrosaurus*. Additional predatory forms will be added to this tally in the coming years.

The Messel oil shale preserves a much more diverse squamate fauna than is present in Europe today, including a variety of terrestrial and arboreal species.

Before mining commenced at Messel, the first crocodylian specimens were found at the top of the site and described by Ludwig (1877). To date excavations have yielded dozens of complete skeletons pertaining to eight species. The most common form is the up to 2m long alligatoroid *Diplocynodon*. *Diplocynodon* and the closely related *Baryphracta*, which attained a length of only 80cm, are both also known from very young individuals. The number of such finds indicates that these alligatoroids reproduced in and around the lake. The other six crocodylian species probably visited the lake occasionally or seasonally, because fossils of these forms are rare. The largest crocodyloid found is almost 4m long, referred to *Asiatosuchus*, and may have resembled the Nile crocodile in its mode of life. It was probably the apex predator in the ecosystem and would have preyed even on the larger mammals. The alligatoroid *Allognathosuchus*, for which a terrestrial mode of life has been inferred, grew to a maximum of 1.5m. It is characterized by stout, crushing teeth in the back of the jaws. *Boverisuchus* ('*Pristichampsus*') had hoof-like claws and was probably a more terrestrially adapted predator. Instead of the conical teeth, as in all extant crocodilians, it had mediolaterally compressed tooth crowns with serrated cutting edges. Of special interest is the presumably land-dwelling *Bergisuchus*, estimated to have reached a length of up to 1.5m. It may be related to certain Palaeogene crocodile-like reptiles from South America and elsewhere, and raises interesting palaeobiogeographical questions. The presence of the family Tomistomidae is documented only by a single jaw fragment. Gastrointestinal contents, commonly found in the mammals and squamates, are rarely found in Messel crocodylians. However, many skeletons do contain clusters of stones in the stomach (gastroliths), which may have contributed to negative buoyancy and aided in the mechanical breakdown of food. Rare food remains found in the stomach and in coprolites mostly represent fish. One specimen of *Diplocynodon* contains remains of a turtle in its stomach.

Pairs of the fossil carettochelyid turtle *Allaeochelys crassesculpta* (Fig 22) provide a dramatic insight into the environment 47 million years ago (Joyce *et al.*, 2012; Schaal and Joyce, 2012). They represent the first known fossil reptiles preserved in the act of mating. *Allaeochelys* had a permeable skin like the extant members of this group of turtles, which allowed them to breathe and stay under water for long periods of time. This adaptation could become lethal if the reptiles came in contact with toxic water. Most likely, the mating animals drifted calmly in the lake and sank from well-oxygenated surface water into noxious deep water where they perished. Nine mating pairs, among a total of about 200 turtle fossils, have been recovered at Messel to date. The females can be distinguished from males by their longer tail and the development of plastral kinesis. *Allaeochelys crassesculpta* was the most common turtle in the Messel lake. It had flipper-like limbs for swimming. Altogether four genera of turtles are known from Messel (Joyce *et al.*, 2012). The trionychid was a typical freshwater form and the largest turtle at Messel. Its extant relatives inhabit freshwater settings in Africa, Asia and North America. The habitat of the trionychid from Messel was limited to the nearshore environment. Another turtle belongs to the Geoemydidae, Asian pond turtles that prefer wetlands, marshes and the water's edge. A rare turtle at Messel is the pleurodire *Neochelys*. Extant relatives of this genus are known from Madagascar.

Many reptiles are good indicators for climate. Warmth is one of the most important prerequisites for crocodylians, large snakes, and particularly turtles and lizards. They prefer warm regions and are most diverse where the average temperature is at least 20°C. The diversity of fossil reptiles would thus indicate warm temperatures for Messel 47 million years ago.

Birds

The bird fauna of the Messel Pit Fossil Site shows a great diversity of genera and species. Hundreds of fossil finds represent more than 60 species belonging to 30 families, most of them small forms. They are predominantly land birds, similar to the ratites (Struthioniformes),

Figure 22 Pair of the carettochelyid *Allaeochelys crassesculpta* that perished during copulation. Length of the carapace on the right is 17.5 cm.

game birds (Galliformes), potoos (Caprimulgiformes), swifts (Apodiformes; Fig 23), hoopoes (Upupiformes) as well as many groups belonging to extinct lineages and a number of species yet to be identified. All described species recovered from the Messel oil shale either belong to extinct groups or are stem members of extant bird lineages (Mayr, 2006, 2009).

An especially noteworthy find was the skeleton of a tiny apodiform bird with short, broad wings and a long tail. Detailed examination revealed that this bird is an early relative of hummingbirds. This fossil represents but one of many examples of Neotropic bird groups present in the Palaeogene of Europe (Mayr, 2009, 2011).

Some large bird species rarely found at Messel indicate connections with North and South America. A ratite of the family Palaeotididae has its present-day relatives living on the Southern Hemisphere. Another example is *Gastornis*. This large flightless bird is probably related to the anseriforms and grew to a height of 2m. Fossils of *Gastornis* have also been found in France and in North America. Together with some of the other vertebrates from Messel, this bird indicates extensive dispersal of animals between North America and Europe during the Palaeogene (Mayr, 2000, 2006, 2009).

Some bird fossils preserve stomach contents, which in some cases reveal remarkable detail. Perhaps the most striking example is the occurrence of fossilized cytoplasm with moulds of storage organelles in plant parenchyma found in the stomach contents of a large bird of uncertain affinities (Mayr and Richter, 2011).

The preservation of the bird fossils from Messel is extraordinary; many skeletons have feathers showing many details and even colour patterns (Figs 23–24). While it was initially assumed that the excellent feather preservation was due to lithification of feather-decomposing bacteria, it has recently been shown that these feathers consist of fossilized layers of melanosomes, while the beta keratin has been degraded. This was the first evidence of preservation of a colour-producing nanostructure in a fossil feather (Vinther *et al.*, 2010). This

Figure 23 Skeleton of apodiform *Scaniacypselus szarskii* with completely preserved plumage. Wing length is 8 cm.

Figure 24 Skeleton of messelornithid ('Messel rail') *Messelornis* with traces of body outline and plumage (see insert for magnified detail). Length of skull 4 cm.

peculiar mode of preservation and the large number of fossils recovered by the excavation teams each year offer much potential for the determination of colours in a variety of bird species.

Fossils of typical water birds are surprisingly rare, and there is no record of ducks, which are known from Eocene strata elsewhere (Mayr, 2009). The long-legged *Juncitarsus*, which may be related to flamingos and grebes, was the only form adapted to life in the water. A lakeshore inhabitant, the 'Messel rail' *Messelornis cristata*, is known from hundreds of skeletons and was probably the most common bird in the shoreline settings (Fig 24).

Mammals

The mammalian finds had made the Messel Pit Fossil Site world-famous by the end of the twentieth century, especially the early horses. Along with one of the bats from Messel, the latter were even depicted on two postal stamps issued in Germany in 1978. To date, 45 species of mammals have been documented from Messel and provide a detailed picture of mammalian diversity during the middle Eocene (Morlo *et al.*, 2004). At the beginning of the Palaeogene, mammals rapidly diversified worldwide. Many of these early groups became extinct during the Eocene and only some of the evolutionary lineages survived until the present day (Storch *et al.*, 2005). The species recovered in Messel belong to the orders Marsupialia, Edentata, Insectivora, Pholidota, Proeutheria, Lipotyphla, Chiroptera, Primates, Rodentia, Creodonta, Carnivora, Condylarthra, Artiodactyla, and Perissodactyla (see Schaal and Ziegler 1988, 1992).

The origins of many of the mammalian groups at Messel can be traced to Africa, Asia, and North and South America. Short periods during which land bridges connected Europe to other continents made the dispersal of mammals and other animals possible (Storch, 1990, Brikiatis, 2014). The pangolins are related to Asian species, and rodents such as *Ailuravus macrurus*, the largest rodent found at Messel, show relationships to North American species. The controversial anteater-like *Eurotamandua joresi* has been interpreted by some authors as implying a connection to South America, perhaps via Africa.

Marsupials live today mainly in Australia and on neighbouring islands, but also in North and South America. The marsupials from Messel belong to the family Didelphidae, which includes the extant opossum in North America. They include *Peradectes* with a length of 25 cm. The prehensile tail of this genus suggests arboreal habits. The other genus from Messel, *Amphiperatherium*, was probably a ground-dwelling form (Koenigswald and Storch, 1992).

A connection to North America can be inferred from the presence of stem carnivorans such as Miacidae and insectivorans, the latter of which are first known from the Late Cretaceous of North America. The best-known insectivoran is the amphilemurid erinaceomorph, *Macrocranion tupaiodon*, which was an omnivorous form. Two species of *Macrocranion* are known, a larger and a much smaller one. The preserved body outlines around some skeletons show large ears, which could be evidence of nocturnal activity, and a woolly pelt. The hind legs of the larger species, *M. tupaiodon* (Fig 9) are long and powerful, and it was a ground-dwelling runner. Two dozen specimens of this form have been recovered to date. Investigation of its gut contents shows that fish, insects, and fragments of leaves comprised this mammal's diet (Maier, 1979).

Another amphilemurid, *Pholidocercus hassiacus* (Fig 25), is characterized by scale-like plates encasing the tail like a tube. It attained a length of 20 cm. A horny plate probably covered the forehead, as distinct pits and vascular grooves cover the dorsal surfaces of the

Figure 25 Skeleton of erinaceomorph *Pholidocercus hassiacus*. Note the tube of bony plates around the tail and the bristle-like hair along the back. Length of head and trunk 19 cm.

Figure 26 Skeleton of leptictid *Leptictidium nasutum*. Note very long tail (45cm long, with overall length of animal 75cm), short forelegs and long, robust hind legs.

nasals and frontals. In its mode of life *Pholidocercus* may have been similar to the extant hairy hedgehog (Koenigswald and Storch, 1983).

An unusual mammal known only from the Eocene of Europe is the leptictid *Leptictidium nasutum* (Fig 26), nearly one metre long and noteworthy for its specialized feeding habits and locomotion (Maier *et al.*, 1986). This animal had a very long tail, long hind legs and very short forelegs. Whether it jumped or (less likely) ran on its hind legs has been the subject of an extensive debate. The skull shows muscular depressions in front of the orbits, suggesting that *Leptictidium* had a highly mobile snout. Three species of this genus are known from Messel, and their gut contents comprise bone fragments of small reptiles and mammals and remains of insects. *Leptictidium* clearly hunted small, agile prey (Koenigswald *et al.*, 1992). The pantolestid *Buxolestes* was an otter-like form that preyed on fish. It attained a body length of about 45cm and had a strong tail with a length of 35cm.

The Messel Pit Fossil Site is the most important locality in the world for early equoid perissodactyls (Franzen, 2007). Some teeth of these animals were first recovered during mining operations as early as 1910, but it was the innovation in preparation methods and the beginning of regular excavations that made it possible to recover more than 60 specimens. Four species have been identified: *Hallensia matthesi*, *Propalaeotherium hassiacum*, *Propalaeotherium voigti* and *Eurohippus messelensis*. *P. hassiacum*, reaching the size of a German shepherd, and *E. messelensis*, attaining the size of a fox terrier, are the most common species. They have low-crowned molars, four digits on each forefoot and three digits on each hind foot. Skeletons of several pregnant mares have been discovered (Fig 27).

Other perissodactyls comprise the tapir-like *Hyrachyus* and *Lophiodon*. They are the largest mammals in the Messel community. *Hyrachyus* is an early rhinocerotoid also known from the early Eocene of North America. *Lophiodon* is widely known from the Eocene of Europe and reached nearly the size of a modern horse. The specimen of *Lophiodon* found at Messel represents a very young animal.

The artiodactyls include *Aumelasia*, *Eurodexis*, *Masillabune* and *Messelobunodon* and represent some of the earliest known even-toed ungulates (Franzen and Richter, 1992). The best-known form is the dichobunid *Messelobunodon schaeferi* (Fig 7), which attained the size

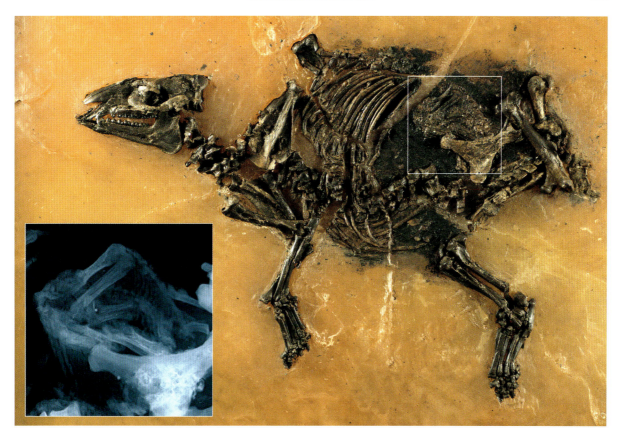

Figure 27 Skeleton of a pregnant mare of equoid *Propalaeotherium parvulum* with foetal remains (see also x-ray image in inset). *P. parvulum* was about the size of a fox terrier.

of a dachshund and was a fast runner. Its gut contents comprise fungi and seeds.

The paroxyclaenid *Kopidodon macrognathus* (Fig 28) was an arboreal omnivorous mammal. The structure of its elbow and ankle suggest that it was an agile form. The fingers and toes ended in strong, laterally compressed claws and facilitated locomotion in the trees. The tail was long and bushy. More than a dozen specimens of this mammal were recovered, most of them with a length of about 1m.

The most publicized recent discovery was a skeleton of the early primate *Darwinius masillae*. It is 50cm long including the tail and represents a juvenile female. This conclusion is justified by the stage of development of the dentition (evident from radiographs), and the fact that this specimen lacks baculum. *Darwinius* was an agile, nail-bearing, arboreal animal and belongs to the Adapoidea (Franzen *et al.*, 2009). The fossil preserves the body outline and contents of the digestive tract, which consist of leaves and fruits (Franzen and Wilde, 2003). The investigation of the teeth using a virtual 3D model of the dentition will allow further studies about functional morphology, ecology and the development of teeth in this early primate (Engels *et al.*, 2012). Known from partial skeletons, another adapoid from Messel, *Europolemur*, resembles the North American *Notharctus* in its body proportions and skeletal structure.

Bats are usually rare in fossil sites but hundreds of skeletons have been recovered during 37 years of excavation in an approximately 30m thick section of oil shale of the middle Messel Formation. As bats are excellent flyers mishaps during hunting are not a likely cause for this high mortality rate. Furthermore, most of the recovered bat fossils represent presumably healthy adults. A possible cause for the preservation of so many bat fossils is carbon dioxide, which may have been released deep in the maar crater and accumulated on the water surface during still air (Habersetzer *et al.*, 1992). Bats flying into these gas clouds while hunting for insects or approaching the water surface to drink would have been asphyxiated and sunk to the bottom of the lake. Bats can be found in many layers of the oil shale, which shows that the high number of finds was not due to a single mass kill.

Figure 28 Skeleton of paroxyclaenid *Kopidodon macrognathus* with body silhouette. Note the long, bushy tail. Overall length 1.1 m.

Figure 29 Skeleton of bat *Palaeochiropteryx tupaiodon* with gut content (**A**), which includes scales of moths (inset). The length of the forearm is approximately 37 mm and the SEM magnification is 1700X.

Bats are represented in Messel by eight species. The genera *Archaeonycteris*, *Hassianycteris*, *Palaeochiropteryx* (Fig 29) and *Tachypteron* can be allocated to three ecological niches based on wing structure, inner ear morphology and gut contents – close to the ground, in mid-canopy and above the tops of the trees. Thus, the Messel bats represent the oldest known bat community (Habersetzer et al., 1994). Most of the specimens belong to the species that hunted close to the ground. The rare *Tachypteron*, known from only two specimens, can definitely be assigned to the extant family Emballonuridae (Storch et al., 2002). Detailed examination of the inner ear in several skulls of Messel bats using 3D microtomography allowed inferences concerning the evolution and quality of echolocation in these animals (Habersetzer, 2004; Habersetzer et al., 2008). Three finds of female bats were embedded in the lake sediments with their two babies still attached to their abdomens (Habersetzer, pers. comm.).

Significance of the Messel site as a window into the history of Life

Following the extinction of non-avian dinosaurs about 66 million years ago, mammals rapidly diversified around the world. Some aspects of this evolutionary success story are documented at the Messel Pit Fossil Site, where this remarkable fossil Lagerstätte formed some 48 million years ago. The peculiar depositional conditions in a volcanic crater lake led to unmatched preservation of remains of animals and plants living in the lake as well as in its surroundings. Anaerobic conditions at the lake bottom prevented significant decay, accounting for the extraordinary fossilization of even soft tissues.

The diverse crocodylians and great diversity of plants indicate a warm climate and document a rainforest setting for the Messel community. The isolation of Europe as an archipelago during the early Eocene led to the development of endemic European animal species along with groups that immigrated from elsewhere.

The biodiversity preserved in the Messel oil shale has opened a unique 'window' on the world of the Eocene 48 million years go. The Forschungsinstitut Senckenberg has conducted field and laboratory research on the Messel Pit Fossil Site since 1975 and will continue to do so as long as possible. The deposits continue to yield amazing new finds including a number over the past year. These will be published in future and serve to demonstrate that it is important to continue research efforts as they are offering many new insights into mid-Eocene life in Europe.

References

Andreae, A. 1893. Vorläufige Mitteilung über die Ganoiden (*Lepisosteus* und *Amia*) des Mainzer Beckens. Verhandlungen des Naturhistorisch-Medizinischen Vereins zu Heidelberg. Neue Folge 5:7–15.

Buness, H., M. Felder, G. Gabriel, and F.-J. Harms. 2005. Explosives Tropenparadies – Geologie und Geophysik im Zeitraffer. UNESCO-Welterbe 21/05, 13(151):6–11. Vernissage, Heidelberg.

Büchel, G. 1988. Geophysik der Eifel-Maare 2: Geomagnetische Erkundung von Trockenmaaren im Vulkanfeld der Westeifel. Mainzer geowissenschaftliche Mitteilungen, 17:357–376.

Brikiatis, L. 2014. The De Geer, Thulean and Beringia routes: key concepts for understanding early Cenozoic biogeography. Journal of Biogeography, 41: 1036–1054.

Collinson, M. E., S. R. Manchester, and V. Wilde. 2012. Fossil fruits and seeds of the Middle Eocene Messel biota, Germany. Abhandlungen der Senckenberg Gesellschaft für Naturforschung 570:1–251.

Dlussky, G. M., T. Wappler, and S. Wedmann. 2008. New middle Eocene formicid species from Germany and the evolution of weaver ants. Acta Palaeontologica Polonica 53:615–626.

Engels, S., J. Habersetzer, and O. Kullmer. 2012. Gebiss von Ida (*Darwinius masillae*) in 3D virtuell rekonstruiert. Senckenberg – Natur, Forschung, Museum 142: 372–375.

Felder, M., and F.-J. Harms. 2004. Lithologie und genetische Interpretation der vulkano-sedimentären Ablagerungen aus der Grube Messel anhand der Forschungsbohrung Messel 2001 und weiterer Bohrungen. Courier Forschungsinstitut Senckenberg 252:151–203.

Franzen, J. L. 2007. Eozäne Equoidea (Mammalia, Perissodactyla) aus der Grube Messel bei Darmstadt (Deutschland). Funde der Jahre 1969–2000. Schweizerische Paläontologische Abhandlungen 127:1–245.

Franzen, J. L., and G. Richter. 1992. Primitive even-toed ungulates: loners in the undergrowth. Pp. 249–256 in S. Schaal and W. Ziegler (eds), Messel – An Insight into the History of Life and of the Earth. Clarendon Press, Oxford.

Franzen, J. L., and V. Wilde. 2003. First gut content of a fossil primate. Journal of Human Evolution 44:373–378.

Franzen, J. L., P. D. Gingerich, J. Habersetzer, J. H. Hurum, W. von Koenigswald, and B. H. Smith. 2009. Complete primate skeleton from the Middle Eocene of Messel in Germany: morphology and paleobiology. PLoS ONE 4(5): e5723. doi:10.1371/journal.pone.0005723.

Goth, K. 1990. Der Messeler Ölschiefer – ein Algenlaminit. Courier Forschungsinstitut Senckenberg 131:1–143.

Gruber, G., and N. Micklich (eds). 2007. Messel – Schätze der Urzeit. Hessisches Landesmuseum, Darmstadt, 159 pp.

Habersetzer, J. 2004. Röntgenverfahren zur Untersuchung Messeler Fossilien. Courier Forschungsinstitut Senckenberg 252: 211–218.

Habersetzer, J., G. Richter, and G. Storch. 1992. Bats: already highly specialized insect predators. Pp. 179–190 in S. Schaal and W. Ziegler (eds), Messel – An Insight into the History of Life and of the Earth. Clarendon Press, Oxford.

Habersetzer, J., G. Richter, and G. Storch. 1994. Paleoecology of early middle Eocene bats from Messel, FRG. Aspects of flight, feeding and echolocation. Historical Biology 8:235–260.

Habersetzer, J., N. Simmons, K. Seymour, G. F. Gunnell, and E. Schlosser-Sturm. 2008. Die Evolution des Fluges und der Echoortung: Fledermäuse. Biologie unserer Zeit 38:246–254.

Joyce, W. G., N. Micklich, S. F. K. Schaal, and T. M. Scheyer. 2012. Caught in the act: the first record of copulating fossil vertebrates. Biology Letters 8:846–848.

Koenigswald, W. von, and G. Storch. 1992. The marsupials: inconspicious opossums. Pp. 153–158 in S. Schaal and W. Ziegler (eds), Messel – An Insight into the History of Life and of the Earth. Clarendon Press, Oxford.

Koenigswald, W. von, and G. Storch. 1983. *Pholidocercus hassiacus*, ein Amphilemuride aus dem Eozän der Grube Messel bei Darmstadt (Mammalia, Lipotyphla). Senckenbergiana lethaea 64: 447–495.

Lenz, O. K., V. Wilde, and W. Riegel. 2011. Short-term fluctuations in vegetation and phytoplankton during the Middle Eocene greenhouse climate: a 640-kyr record from the Messel oil shale (Germany). International Journal of Earth Sciences 100:1851–1874.

Lenz, O. K., V. Wilde, D. F. Mertz and W. Riegel. 2014. New palynology-based astronomical and revised $^{40}Ar/^{39}Ar$ ages for the Eocene maar lake of Messel (Germany). International Journal of Earth Sciences. 104:873–889.

Lorenz, V., and S. Kurzlaukis. 2007. Root zone processes in the phreatomagmatic pipe emplacement model and consequences for evolution of maar-diatreme volcanoes. Journal of Volcanology and Geothermal Research 159:4–32.

Ludwig, R. 1877. Fossile Crocodiliden aus der Tertiärformation des Mainzer Beckens. Palaeontographica, Supplement 3:1–52.

Maier, W. 1979. *Macrocranion tupaiodon*, an adapisoricid(?) insectivore from the Eocene of Grube Messel (western Germany). Paläontologische Zeitschrift 53:38–62.

Maier, W., G. Richter, and G. Storch. 1986. *Leptictidium nasutum* – ein archaisches Säugetier aus Messel mit aussergewöhnlichen biologischen Anpassungen. Natur und Museum 116:1–19.

Mangel, G. 2011. Faszination Grube Messel. Zu Besuch in einer Welt vor 47 Millionen Jahren. Kleine Senckenberg-Reihe 52:1–160.

Matthess, G. 1966. Zur Geologie des Ölschiefervorkommens von Messel bei Darmstadt. Abhandlungen des Hessischen Landesamtes für Bodenforschung 51:1–87.

Mayr, G. 2000. Die Vögel der Grube Messel – ein Einblick in die Vogelwelt Mitteleuropas vor 49 Millionen Jahren. Natur und Museum 130:365–378.

Mayr, G. 2006. Fine feathered fossils of the Eocene – the birdlife of Messel. UNESCO-World Heritage 13/21:38–43. Vernissage. Heidelberg.

Mayr, G. 2009. *Paleogene Fossil Birds*. Springer, Berlin.

Mayr, G. 2011. Two-phase extinction of "Southern Hemispheric" birds in the Cenozoic of Europe and the origin of the Neotropic avifauna. Palaeobiodiversity and Palaeoenvironments 91:325–333.

Mayr, G., and G. Richter. 2011. Exceptionally preserved plant parenchyma in the digestive tract indicates a herbivorous diet in the Middle Eocene bird *Strigogyps sapea* (Ameghinornithidae). Paläontologische Zeitschrift 85:303–307.

Mertz, D. F., and P. R. Renne. 2005. A numerical age for the Messel fossil deposit (UNESCO World Heritage Site) derived from $^{40}Ar/^{39}Ar$ dating on a basaltic rock fragment. Courier Forschungsinstitut Senckenberg 255:67–75.

Meulenkamp, J. E., and W. Sissingh. 2003. Tertiary palaeogeography and tectonostratigraphic evolution of the northern and southern Peri-Tethys platforms and the intermediate domains of the African-Eurasian convergent plate boundary zone. Palaeogeography, Palaeoclimatology, Palaeoecology 196:206–228.

Micklich, N. 2002. Die Ichthyofauna des eozänen Messelsees – Besonderheiten und paläoökologische Implikationen. Schriftenreihe der Deutschen Geologischen Gesellschaft 21:240–241.

Micklich, N. 2012. Peculiarities of the Messel fish fauna and their palaeoecological implications: a case study. Palaeobiodiversity and Palaeoenvironments 92:585–629.

Morlo, M., G. F. Gunnell, and K. T. Smith. 2012. Mammalian carnivores from Messel and a comparison of non-volant predator guilds from the middle Eocene of Europe and North America. Pp. 120–121 in T. Lehmann, and S. K. F. Schaal (eds), The World at the Time of Messel: Puzzles in Palaeobiology, Palaeoenvironment, and the History of Early Primates. Senckenberg Gesellschaft für Naturforschung, Frankfurt am Main.

Morlo, M., S. Schaal, G. Mayr, and C. Seiffert. 2004. An annotated taxonomic list of the Middle Eocene (MP11) Vertebrata of Messel. Courier Forschungsinstitut Senckenberg 252:95–108.

Müller, J., C. A. Hipsley, J. J. Head, N. Kardjilov, A. Hilger, M. Wuttke, and R. R. Reisz. 2011. Eocene lizard from Germany reveals amphisbaenian origins. Nature 473:364–367.

Pirrung, M., G. Büchel, and W. Jacoby. 2001. The Tertiary volcanic basins of Eckfeld, Enspel and Messel (Germany). Zeitschrift der Deutschen Geologischen Gesellschaft 152:27–59.

Pirrung, M. 1998. Zur Entstehung isolierter alttertiärer Seesedimente in zentraleuropäischen Vulkanfeldern. Mainzer Naturwissenschaftliche Abhandlungen, Beiheft, 20: 1–117.

Richter, G., and S. Wedmann. 2005. Ecology of the Eocene Lake Messel revealed by analysis of small fish coprolites and sediments from a drilling core. Palaeogeography, Palaeoclimatology, Palaeoecology 223:147–161.

Schaal, S. 2004. *Palaeopython fischeri* n. sp. (Serpentes: Boidae), eine Riesenschlange aus dem Eozän (MP11) von Messel. Courier Forschungsinstitut Senckenberg 252:35–45.

Schaal, S. F. K. 2012. Messel Pit fossil site – the legacy of the environment and life of Eocene. Pp. 225–236 in J. A. Talent (ed), Earth and Life: Global Biodiversity, Extinction Intervals and Biogeographic Perturbations Through Time. Springer, Dordrecht.

Schaal, S., and S. Baszio. 2004. *Messelophis ermannorum* n. sp., eine neue Zwergboa (Serpentes: Boidae: Tropidopheinae) aus dem Mittel-Eozän von Messel. Courier Forschungsinstitut Senckenberg 252:67–77.

Schaal, S. F. K., and W. Joyce. 2012. Vereint, vergiftet, versteinert. Senckenberg – Natur, Forschung, Museum 142:368–369.

Schaal, S. F. K., and R. Rabenstein. 2012. Der Tagebau in Messel in Linien und Zahlen. Senckenberg – Natur, Forschung, Museum 142:376–377.

Schaal, S., and U. Schneider. 1995. Chronik der Grube Messel 1965-1995. Pp. 195–276 in S. Schaal and U. Schneider (eds), Chronik der Grube Messel. Kempkes, Gladenbach.

Schaal, S., and W. Ziegler (eds). 1988. Messel – Ein Schaufenster in die Geschichte der Erde und des Lebens. (Senckenberg-Buch 64). Verlag Waldemar Kramer, Frankfurt am Main, 315 pp.

Schaal, S., and W. Ziegler (eds). 1992. Messel – An Insight into the History of Life and of the Earth. Clarendon Press, Oxford, 322 pp. [English translation of Schaal and Ziegler (1988).]

Schaal, S., S. Baszio, and J. Habersetzer. 2005. Differenzierung von Schlangenarten anhand qualitativer und quantitativer Merkmale sowie konventioneller Streckenmaße und Indizes. Courier Forschungsinstitut Senckenberg 255:133–169.

Schaal, S., E. E. Brahm, J. Habersetzer, A. Hebs, M. Müller, and E. Schlosser-Sturm. 2004. Literaturübersicht und Schriftenverzeichnis zur wissenschaftlichen Erforschung der Fossilienfundstätte Messel. Courier Forschungsinstitut Senckenberg 252:243–245.

Schaarschmidt, F. 1992. The vegetation: fossil plants as witnesses of a warm climate. Pp. 27–52 in S. Schaal and W. Ziegler (eds), Messel – An Insight into the History of Life and of the Earth. Clarendon Press, Oxford.

Schulz, R., F.-J. Harms, and M. Felder. 2002. Die Forschungsbohrung Messel 2001: Ein Beitrag zur Entschlüsselung der Genese einer Ölschieferlagerstätte. Zeitschrift für angewandte Geologie 4:9–17.

Smith, K. T. 2009. Eocene lizards of the clade *Geiseltaliellus* from Messel and Geiseltal, and the early radiation of Iguanidae (Reptilia: Squamata). Bulletin of the Peabody Museum of Natural History, Yale University 50:219–306.

Smith, K. T., and M. Wuttke. 2012. From tree to shining sea: taphonomy of the arboreal lizard *Geiseltaliellus maarius* from Messel, Germany. Palaeobiodiversity and Palaeoenvironments 92:45–65.

Storch, G. 1990. The Eocene mammalian fauna from Messel – a paleobiogeographical jigsaw puzzle. Pp. 23–32 in G. Peters and R. Hutterer (eds), Vertebrates in the Tropics. Museum Alexander Koenig, Bonn.

Storch, G. 1992. The mammals of island Europe. Scientific American 226(2):64–69.

Storch, G., and F. Schaarschmidt. 1992. The Messel fauna and flora: a biogeographical puzzle. Pp. 291–297 in S. Schaal and W. Ziegler (eds), Messel – An Insight into the History of Life and of the Earth. Clarendon Press, Oxford.

Storch, G., B. Sigé, and J. Habersetzer. 2002. *Tachypteron franzeni* n.gen., n.sp., earliest emballonurid bat from the Middle Eocene of Messel (Mammalia, Chiroptera). Paläontologische Zeitschrift 76:189–199.

Storch, G., J. Habersetzer, T. Martin, M. Morlo, and J. L. Franzen. 2005. Die "Stars" im Ölschiefer – Die Säugetiere. UNESCO-Welterbe 21/05, 13(151):44–59. Vernissage, Heidelberg.

Suhr, P., Goth, K., Lorenz, V. & Suhr, S. (2006) Long lasting subsidence and deformation in and above maar-diatreme volcanoes – a never ending story. Zeitschrift der deutschen Geowissenschaften, 157: 491–511.

Vinther, J., D. E. G. Briggs, J. B. Clarke, G. Mayr, and R. O. Prum. 2010. Structural coloration in a fossil feather. Biology Letters 6:128–131.

Wappler, T., and M. S. Engel. 2003. The middle Eocene bee faunas of Eckfeld and Messel, Germany (Hymenoptera, Apoidea). Journal of Paleontology 77:908–921.

Weber, S. 2004. *Ornatocephalus metzleri* gen. et spec. nov. (Lacertilia, Scincoidea) – Taxonomy and paleobiology of a basal scincoid lizard from the Messel Formation (Middle Eocene: basal Lutetian, Geiseltalium), Germany. Abhandlungen der Senckenbergischen Naturforschenden Gesellschaft 561:1–159.

Wedmann, S. 2005. Annotated taxon-list of the invertebrate animals from the Eocene fossil site Grube Messel near Darmstadt, Germany. Courier Forschungsinstitut Senckenberg 255:103–110.

Wedmann, S., S. Bradler, and J. Rust. 2007. The first fossil leaf insect: 47 million years of specialized cryptic morphology and behavior. Proceedings of the National Academy of Sciences of the United States of America 104:565–569.

Wedmann, S., and D. K. Yeates. 2008. Eocene records of bee flies (Insecta, Diptera, Bombyliidae, *Comptosia*): their paleobiogeographic implications and remarks on the evolutionary history of bombyliids. Palaeontology 51:231–240.

Wedmann S., T. Wappler, and M. S. Engel. 2009. Direct and indirect fossil records of megachilid bees from the Paleogene of Central Europe (Hymenoptera: Megachilidae). Naturwissenschaften 96:703–712.

Wilde, V. 2004. Aktuelle Übersicht zur Flora aus dem mitteleozänen "Ölschiefer" der Grube Messel bei Darmstadt (Hessen, Deutschland). Courier Forschungsinstitut Senckenberg 252:109–114.

Wilde, V. 2005. Es grünte so grün ... – Eine artenreiche Flora im "paratropischen" Klima. UNESCO-Welterbe 21/05, 13(151):14–19. Vernissage, Heidelberg.

Wilde, V., O. K. Lenz, and W. Riegel. 2012. Die Dynamik des Treibhausklimas vor 48 Millionen Jahren. Senckenberg – Natur, Forschung, Museum 142:358–367.

Wuttke, M. 1983. Weichteil-Erhaltung durch lithifizierte Mikroorganismen bei mittel-eozänen Vertebraten aus den Ölschiefern der Grube Messel bei Darmstadt. Senckenbergiana lethaea 64:509–527.

Wuttke, M., T. Prikryl, V. Y. Ratnikov, Z. Dvorák, and Z. Roček. 2012. Generic diversity and distributional dynamics of the Palaeobatrachidae (Amphibia: Anura). Palaeobiodiversity and Palaeoenvironments 92:367–395.

Chapter 9

Extraordinary Lagerstätten in Amber, with particular reference to the Cretaceous of Burma

David A. Grimaldi[1] and Andrew J. Ross[2]

1 Division of Invertebrate Zoology, American Museum of Natural History, New York, USA.
2 Department of Natural Sciences, National Museums Scotland, Edinburgh, UK.

Abstract

Deposits of fossilized tree resins, or amber, occur throughout the world and range in age from Carboniferous to Holocene, produced by myriads of vascular plants. Small organisms encapsulated within amber are commonly preserved with fidelity at cellular and even subcellular scales, which in turn profoundly improves phylogenetic interpretation of inclusions. Burmese amber, from Kachin Province in northern Myanmar, is approximately 100Ma and is the most productive source of Cretaceous amber. It also contains the most diverse taxonomic assemblage in amber from the Cretaceous Period, with 591 named arthropods in 342 families and 54 orders. It also preserves protists, myxomycetes, fungi, vascular and non-vascular plants (including flowers), worms from several phyla, gastropods, and small vertebrates. There is a greater diversity of lizards in Burmese amber than even in large Tertiary amber deposits (i.e., Dominican and Baltic ambers), as well as small frogs, and various remains of Eumaniraptora, especially feathers (most probably from birds). Examples are presented that illustrate the phylogenetic significance of select taxa in Burmese amber, the evolution of major adaptive features such as sociality in termites and ants, insect phytophagy, as well as biogeography.

Introduction

Amber is one of the few natural substances that is highly prized both aesthetically and scientifically. It has been collected from the shores of the Baltic Sea and carved into amulets and charms for at least 13 millennia (Beck, 1986; Grimaldi, 1996; Ross, 2010), making amber one of the original precious substances. The scientific study of amber essentially began with Sendelio (1742), but not seriously until the nineteenth century in Prussia, when monographs on Baltic amber (Eocene) sumptuously illustrated, in hand-tinted copper-plate etchings and lithographs, the myriad plant and animal inclusions (Berendt, 1845–56, etc.). The Baltic region in fact has the largest deposits of amber in the world. As optical microscopy was perfected, the exquisite, microscopic fidelity of preservation in amber allowed detailed study of the inclusions; today, Baltic amber is also the richest palaeofauna of insects among all deposits in the 425 million-year fossil record of terrestrial arthropods. Global scientific interest in amber waned, however, through much of the twentieth century with the development of genetics and molecular biology, but was gradually resurrected with the resurgence of organismal biology, the landmark publication by Langenheim (1969), and a frenzy of field discoveries made over the past 50 years. Public interest was piqued in 1993 with the release of the film *Jurassic Park*, and has remained high ever since.

The past half-century has seen an order-of-magnitude more discoveries regarding amber compared to the previous two centuries. Major deposits of highly fossiliferous Miocene amber from the Dominican Republic and Mexico were discovered in the 1960s and 1970s (Sanderson and Farr, 1960; Durham and Hurd, 1957), which are now heavily marketed commercially. Initial discoveries of large outcrops of amber from the Cretaceous were made in western Canada (Carpenter *et al.*, 1937; McAlpine and Martin, 1969; these and other deposits subsequently studied by Pike, 1995; McKellar and Engel, 2012), then Lebanon (Schlee and Dietrich, 1970; subsequently studied by Azar [e.g., Azar *et al.*, 2010]), and from northern Siberia (Zherikhin and Sukatsheva, 1973; subsequently reviewed by Eskov, 2002). A prolific outcrop from New Jersey was discovered in 1992 (Grimaldi *et al.*, 2000); then various ones from western France (Perrichot *et al.*, 2007; Nel *et al.*, 2010), and throughout Spain (Alonso *et al.*, 2000; Peñalver and Delclòs, 2010). Very large outcrops of early Eocene amber from the Cambay Formation of western India have recently been discovered and are being scientifically exploited (Rust *et al.*, 2010). An outcrop of abundant small droplets of amber has even been found from the Late Triassic (Carnian) of Italy, which contains tiny arthropod inclusions (Schmidt *et al.*, 2012). Reports in the 1990s of DNA from inclusions certainly fuelled popular and scientific interest in amber, even though it appears that these reports were premature and based on sequences that were probably modern contaminants (Austin *et al.*, 1997). At the very least, the cellular and ultrastructural (subcellular) preservation of cuticular details and tissues in amber insects and plants is unparalleled for most modes of fossilization. Reviews of world amber deposits have been provided by Langenheim (1969); Martinez-Delclòs *et al.* (2004) and in the book edited by Penney (2010). See Figure 1 for the ages of all the arthropod-bearing ambers in the world, and see Appendix 1 for more information on these localities. Appendix 2 summarizes all the arthropod orders and the ambers in which they have been recorded.

Despite the mounting intensity of scientific research on ambers, particularly with the development of analytic techniques in molecular analysis, micro-CT scanning and other digital imaging techniques, the romance and appeal of amber appears to only be intensifying. The deposit that has attracted global attention over the past decade involves amber from the mid-Cretaceous of northern Burma or Myanmar, also called Burmese amber or Burmite, the subject of this article. Previous review articles (Zherikhin and Ross, 2000; Grimaldi *et al.*, 2002; Ross *et al.*, 2010) have focused on the history of use of Burmese amber or reviews, principally, of its arthropod inclusions, and new discoveries are being made on nearly a monthly basis. There have been over 400 scientific papers since 2000, principally on the organismal inclusions in Burmese amber.

General Aspects of Amber
Molecular composition

Amber is a highly polymerized, cross-linked form of natural hydrocarbon (terpenoid-based) plant resins, to be distinguished from water-soluble saps and gums (Langenheim, 2003). The terms 'resinite' and 'fossil resin' have also been used, but increasingly the term 'amber' is being applied to any form of fossilized resin. At what point in time does resin become amber? Since resins are easily dated using ^{14}C techniques, it was proposed that amber pertains to those resins that are greater than 40,000 years old (the maximum reliable age of ^{14}C dating); all younger resins that are naturally buried are either 'subfossil resins' or 'copal' (Anderson, 1997). Fossils are generally regarded as any natural remains that have been buried for a reasonably long time; they do not have to be mineralized replicas of the remains. Amber and subfossil resins retain the complex mixture of organic compounds distinct to the species of plant that originally produced them, even though some of the compounds cross-link with time.

Modern resins and amber are readily 'fingerprinted' using analytical techniques such as FTIR (Fourier Transform Infrared Spectroscopy) (e.g., Beck, 1986; Peñalver *et al.*, 2007), NMR (Nuclear Magnetic Resonance) (e.g., Lambert *et al.*, 1996), and Py-GC/MS (Pyrolyis-Gas Chromatography and Mass Spectroscopy) (e.g., Bray and Anderson, 2009). A classification of ambers and other resins was presented by Anderson (e.g., Anderson *et al.*, 1992; Bray and Anderson, 2008), who proposed three major classes based on backbone molecular composition: Class I, II, and III, with subgroups a, b, and c for some of these. Burmese amber, for example, is Class Ib. Cadinene-based resins, such as ambers from Borneo and Sumatra, modern dammar resins, and Eocene amber from India and Arkansas USA, are Class II.

Botanical Origins

While Anderson's classification has good general utility, there are some pitfalls. One is that it is not a natural classification in the phylogenetic sense, since plants can metabolically produce similar or identical compounds via independent pathways, and because the classification is strictly based on molecular features, not on botanical sources. In other words, molecular homologies are not considered. Indeed, it is now known that extraordinary convergence in the molecular composition of amber occurs, the prime example being Class Ic resins, which include Dominican and Mexican amber (both Miocene), resin produced by living *Hymenaea* trees (a legume), and Carboniferous amber from Illinois USA that was produced, presumably, by tree ferns (Bray and Anderson, 2009). The approach of giving botanical attribution to an amber deposit based entirely on molecular composition is likewise problematic. In papers by Poinar (e.g., Poinar et al., 2007) most of the Cretaceous deposits are said to be derived from araucarian conifers, including Burmese amber, despite macrofossil evidence that contradicts this (e.g., New Jersey and Canadian ambers, which are from Cupressaceae, based on macrofossil and molecular evidence [Grimaldi *et al.*, 2000; McKellar and Wolfe, 2010]). Moreover, organic geochemists who examine molecular composition do not analyse their results phylogenetically, so the presence and absence of compounds such as α-pinene or agathic acid are only putatively diagnostic for higher taxa. Ideally, the botanical source of an amber deposit must be based on amber found in situ within unambiguously identified macrofossil remains of the plant, such as wood, leaves, and/or cones (e.g., Cambay amber [Rust *et al.*, 2010]; Alabama amber [Knight *et al.*, 2010], Alaskan amber [Sunderlin *et al.*, 2011], and the New Jersey and Canadian deposits mentioned above). Unfortunately, Burmese amber strata and the plant fossils within it have not been carefully studied, so the proposed araucarian origin of this amber is circumstantial and needs to be confirmed. In fact, inclusions of leafy shoots of the dawn redwood genus *Metasequoia* (Cupressaceae) (Fig 4E) are not uncommon in Burmese amber, so it is quite possible that Burmese amber was formed by one or more species of this genus.

Deposition

It is a common misconception that amber is found in only certain rich outcrops, such as the Baltic coast, Dominican Republic, or the deposits in northern Burma. Trace quantities of amber occur throughout most terrestrial deposits, from Carboniferous coals to Pleistocene peats; like most fossils, amber requires particular conditions for its concentration, burial, and preservation. Typically, these are sediments of fluvial and lacustrine origin or (as seems to be the case for Burmese amber) near-shore estuary and lagoonal basins, always ones rich in coals, lignites, or other vegetative remains, with clay/mudstones and sand. Generally, the compact, anoxic sediments of ancient wet sediments preserve amber from the oxidation that will destroy it over a period of years to centuries. It is exceptional for amber to be deposited in palaeosols, such as the Triassic amber from the Italian Alps (Schmidt *et al.*, 2012).

Preservation of inclusions

The scientific, as well as aesthetic, hallmark of amber is the preservation of its inclusions. The external cuticular details of delicate arthropods in amber have been observed for centuries with light microscopy. Electron microscopy (scanning and transmission) has revealed that the preservation of internal organs and soft tissues can be exquisite, including the preservation of individual cells and their organelles, such as cell membranes, nuclei, mitochondria, and chloroplasts (Mierzejewski, 1976; Henwood, 1992; Grimaldi *et al.*, 1994; Kohring, 1998; Wier *et al.*, 2002). Ambers preserve unequally, apparently uncorrelated with age but rather depending on the molecular composition of the fresh resin and whether this allowed quick embalming of internal tissues. Insect inclusions in Baltic amber, for example, commonly are covered with a milky froth, a product of internal decomposition, and good preservation of internal tissues is highly inconsistent. Perhaps the most consistent and finest preservation is in Dominican amber (e.g., Grimaldi *et al.*, 1994). Preservation in Burmese amber is variable; some inclusions are very well preserved (including even rare colour patterning), whereas others are poor. With the development of synchrotron microtomography and (micro- and nano-) CT scanning, with micrometre-level resolution, examining gross internal preservation does not necessarily require destructive sampling of an inclusion. Pohl *et al.* (2010), for example, examined the flight muscles, gut, and even the brain of the tiny parasitoidal strepsipteran insect *Mengea* in Baltic amber. These techniques require a contrast or differential in the density between the amber matrix and covering (e.g., cuticle) and internal tissues of the inclusion. Despite what appears to be excellent external preservation, the senior author has found that inclusions in Burmese amber are actually quite difficult to image with CT scanning, compared to ones in Baltic and Dominican amber. This may be due to the hardness of Burmese amber or interaction of some components of the original resin with the cuticle.

Given the subcellular scale of preservation in amber, the natural and most common question regards molecular preservation of organismal inclusions in amber. Early reports of DNA sequences from insects preserved in amber (DeSalle et al., 1992; Cano et al., 1993) were not repeatable (Austin et al., 1997), so the consensus is that DNA is not preserved in amber (reviewed in Grimaldi and Engel, 2005b). A recent attempt to extract DNA from bees in copal also failed (Penney et al., 2013). Interestingly, even macromolecules as seemingly durable as chitin (a polysaccharide sheet laced with proteins) are not preserved in amber or copal (Stankiewicz et al., 1998), although there are suggestions that this varies with analytic method (e.g., Raman spectroscopy) and the amber source (e.g., Baltic vs. Mexican amber) (Edwards et al., 2007).

Even if macromolecular preservation within amber is minimal, the anatomical preservation in amber clearly has the highest and most consistent fidelity of any mode of fossilization that is millions of years old. This has profound implications for interpreting the phylogenetic relationships of myriad arthropods and other life forms preserved in amber. Modelling has shown that character sampling greatly affects the resolution of phylogenetic analyses (Wiens, 2003), so fine preservation of fossils facilitates phylogenetic interpretation, which in turn facilitates the understanding of general aspects of evolution like extinctions, biogeography, and the origins of major adaptive features. Of course, amber preserves only smaller organisms (generally less than a few centimetres in length), but since this size range includes 90% of Recent terrestrial eukaryotes – from protists to small vertebrates – amber is, indeed, a unique window to the evolutionary past.

Historical and geological context of Burmese amber

Burmese amber has been exploited for at least two millennia, principally for use as carvings in China (Fig 3E), where most of the material is still exported. It is hard; it polishes to a fine lustre, and even facets well (Fig 3E); pieces can be very large (Fig 3B); and the amber ranges from transparent to opaque, and in colour from light yellow to ruby red. Zherikhin and Ross (2000) and Ross et al. (2010) summarized the history of use of Burmese amber; Cruikshank and Ko (2003) is the primary source on the geology of Burmese amber.

Outcrops of the fossiliferous Cretaceous amber from Myanmar are in the Hukawng Valley of Kachin State, northern Myanmar. They are very localized, originating from a relatively tiny area of Cretaceous exposure $c.12km^2$ at 26°15'N, 96°34'E, called Noijebum (Fig 2). These outcrops are entirely surrounded by extensive Quaternary and Tertiary exposures, and some amber was reworked into younger Eocene deposits, which were also exploited historically (Zherikhin and Ross, 2000). Amber was recently recorded from the Magway Region, Central Myanmar, but as yet no arthropod inclusions have been reported (Sun et al., 2015).

The first observations of Burmese amber outcrops by western geologists were published in the mid-nineteenth century; Chhibber (1934) provided the most comprehensive account. These deposits are by far the most prolific source of Cretaceous amber: according to the Geological Survey of India, between 1898 and 1940 some 83 tonnes of Burmese amber were mined. During this time the Natural History Museum, London (NHMUK) acquired the only scientific collection of Burmese amber, donated by R. C. J. Swinhoe between 1919 and 1921. That collection consisted of about 117 pieces containing 1200 arthropods (incredibly, one piece contained 458 arthropods). Mining of Burmese amber has not changed much from the early twentieth century; it is still excavated from deep open-pit mines by hand, only now pumps are used to remove water (Fig 3A). Mining was minimal after 1940, resuming c.1999 (Levinson, 2001; Grimaldi et al., 2002). Since 2002 the American Museum of Natural History in New York has developed a collection of approximately 3500 fossil inclusions in this amber (Grimaldi et al., 2002; unpublished). Burmese amber is now actively marketed worldwide, and several other institutions have developed collections, such as the Staatliches Museum für Naturkunde Stuttgart, Germany, National Museums Scotland in Edinburgh, Scotland (examples figured by Ross and Sheridan, 2013) and Nanjing Institute of Geology and Palaeontology, China. Many of the most prized specimens are acquired by private collectors, who have more funds than museums.

The age of the amber was originally thought to be Miocene (Noetling, 1893), but was presciently estimated to be Cretaceous in age by the entomological polymath T. D. A. Cockerell (1917). Cockerell described 42 species of arthropods in the NHMUK collection of Burmese amber. Rasnitsyn (1996) later confirmed Cockerell's estimate, based on finding, in Burmese amber, several families of insects known only from the Cretaceous. Grimaldi *et al.* (2002) estimated a more specific time period of Cenomanian to Turonian, based on the stratigraphic distributions of various Cretaceous insect families. Cruikshank and Ko (2003) reported an age of late Albian (105–100Ma) based on the ammonite *Mortoniceras* from the outcrops; this age corresponds closely to the early Cenomanian age at (99Ma) estimated by Shi *et al.* (2012), based on U-Pb radiometric dating of zircons in the amber sediments. However, Burmese amber pieces are almost always rounded, with a surface that appears to have been smoothed by abrasive sediments, probably having been tumbled by water currents. This is consistent with pholadid bivalve (shipworm) borings in the surfaces of most pieces (Fig 1D). So clearly the amber was transported into a brackish-marine environment and the surface was hard enough to be bored prior to burial (Smith and Ross, in press).

Palaeobiology
1 Summary of overall diversity

Of the six known major deposits of Cretaceous amber, Burmese amber preserves by far the greatest diversity of taxa. As expected, the great bulk of species in all ambers consist of terrestrial arthropods, such as mites, spiders, myriapods, isopods, and insects (Figs 5–15). As of December 2016 there were 591 species of arthropods in 54 orders and 342 families recorded in Burmese amber. In 2010 there were 216 families in 36 orders known (Ross *et al.*, 2010), which can be directly compared to the numbers of arthropod orders and families in the five other major Cretaceous outcrops as follows: Canadian amber with 126 families in 14 orders (McKellar and Wolfe, 2010); Raritan (New Jersey) amber with 60 families in 13 orders (Grimaldi and Nascimbene, 2010); Charentes (France) amber with 82 families in 29 orders (Perrichot *et al.*, 2010); Spanish amber with 83 families in 22 orders (Peñalver and Delclòs, 2010); and Lebanese amber with 76 families in 18 orders (Azar *et al.*, 2010). Thus, Burmese amber contains two to nearly three times the number of arthropod orders as some other Cretaceous deposits (e.g., Canada, Lebanon, New Jersey), and three to four times the number of families. Rarefaction curves are needed, but it appears that most of the species in Burmese amber are represented by just one or a few specimens. If excavations continue at the present pace it would not be surprising if over 1000 species of arthropods were found in Burmese amber. This deposit may end up being the most diverse Cretaceous deposit for terrestrial arthropods, including the vast outcrops of lithified remains like the Crato Formation of Brazil and Yixian Formation of China.

Among the extinct families present, about half are only known from Burmese amber, whereas others are also known from other ambers, or even as compression fossils in rock. For the Hymenoptera these include Falsiformicidae, Galloromatidae, Maimetshidae, Serphitidae, Spathiopterygidae, and Stigmaphronidae. With the exception of Falsiformicidae, these families are known only in Cretaceous amber from Siberia, France, Lebanon, Spain, New Jersey, and/or Canada. Among Diptera, the family Chimeromyiidae is known in Cretaceous amber from Lebanon and Spain, as well as Myanmar; Archizelmiridae (also placed as a subfamily of Sciaridae) is known from Early Cretaceous compressions of Eurasia and in amber from Lebanon, Myanmar, and New Jersey. Tethepomyiidae is a distinctive family of tiny, reduced flies known in amber from Myanmar, New Jersey, and Spain. The family Zhangsolvidae was originally described as compression fossils from China, but is also known from fossils from Brazil and in Burmese and Spanish ambers (Arillo *et al.*, 2015).

Burmese amber also contains diverse plants (Fig 4), fungi, protists, vertebrates (Fig 15), and non-arthropod invertebrates (Fig 5). There are non-vascular plants such as bryophytes (mosses and liverworts: Figs 4B, C), ferns (Fig 4D), leafy shoots of conifers (*Metasequoia* being the most common) (Fig 4E), and the leaves and flowers of various angiosperms (Figs 4F–H). In lieu of detailed study it is difficult to say if the ferns were epiphytes, but clearly the relatively abundant liverworts were, indicating that the Burmese amber forest was relatively wet. Angiosperm leaves with long, slender 'drip tips' (Fig 4H) also reflect an everwet palaeoenvironment. The forest was probably dominated by the coniferous tree that produced the amber, and the forest had a relatively open canopy that could support an understory of angiosperm shrubs, herbs, and vines. The only other Cretaceous amber deposit that contains flowers is from New Jersey (Grimaldi et al., 2000). Burmese amber also contains the oldest definitive slime moulds (Myxomycophyta: Amoebozoa) (Fig 4A). Another indication of a tropical environment is the presence of the velvet worm phylum Onychophora (Grimaldi et al., 2002, unpublished), a group of approximately 180 species that today inhabits leaf litter in tropical forests around the world and in wet temperate forests of South America and New Zealand. The only other worms in this amber are in the phylum Nematoda, with fossils that are free-living and parasitic (e.g., Fig 5A). Terrestrial gastropods (Figs 5B, C) are relatively abundant and diverse in Burmese amber, indicating that the forest was not fully coniferous; land snails today do not thrive well in conifer forests since the acidic soil interferes with calcium metabolism.

Burmese amber is revealing a high diversity of groups that, until recently, were thought to have appeared and diversified relatively late. One such group is the ants (Hymenoptera: Formicidae). For many years the primitive ant *Sphecomyrma freyi* from New Jersey amber was the only known ant in Cretaceous amber. However, Burmese amber, which is older, has yielded a high diversity of ants with 20 species described so far. Another group is the woodlice (Crustacea: Oniscidea), which, due to its poor fossil record, was also thought to have diversified relatively late. The first known woodlouse in Burmese amber, *Myanmariscus deboiseae* (Fig 5D) belongs to an advanced clade (Broly et al., 2015). Together with other newly discovered specimens, this demonstrates that the woodlouse fauna at that time was diverse.

Burmese amber preserves the highest diversity of vertebrates of all amber deposits, even the huge deposits of Eocene Baltic amber or Miocene amber from the Dominican Republic. As expected, vertebrate remains in amber are very rare; these consist mostly of small lizards (Squamata) (Figs 15A, B), feathers (Figs 15C, D), and extraordinarily rare small frogs (Anura) (Fig 15E). Mammalian hair occurs in Dominican and Baltic amber but is unknown from any Cretaceous amber, including Burmese amber. Even though arthropod inclusions in Burmese amber are very challenging to observe with CT scanning, bones of the vertebrate inclusions are resolved extremely well with this technique, allowing complete osteological study. The lizards in Burmese amber include basal Squamata (not assignable to an extant family), geckos (Gekkota), Lacertoidea (Lacertidae and/or Teiidae), Agamidae (Acrodonta), and a probable stem-group chameleon (Daza et al., 2016). By comparison, the lizards in Baltic amber are just Lacertidae; those in Dominican amber are *Sphaerodactylus* (Gekkonidae) and *Anolis*. There are diverse bird feathers in Burmese amber, but not identified to any major group (if at all identifiable). Incredibly, there are even the bony remains of several birds (Fig 16). The two known frogs in Burmese amber are in private collections in China (Fig 15E); their taxonomic identity has not been reported.

2 Phylogenetic significance

Typical of the microscopic preservation within amber, the phylogenetic relationships of various arthropod taxa in Burmese amber can be very well clarified. Out of six examples discussed below, five are insects that are only several millimetres in body length; accurately discerning their relationships required preservation and resolution of microscopic structures in the order of several microns.

Diets and long proboscides

Two examples regard insects that convergently evolved a long, narrow, rigid proboscis, 'long-tongued' taxa whose mouthparts are developed from several pairs of appendages and individual central ones. One of these groups involves scorpionflies (order Mecoptera) in the Pseudopolycentropodidae, a Mesozoic family of four genera and 15 species from the Middle Triassic to mid-Cretaceous (Grimaldi et al., 2005; Ren et al., 2009). The genus *Parapolycentropus* in Burmese amber is the latest known occurrence of the family (Grimaldi et al., 2005) (Fig 13A); most of the fossils are lithified taxa from the Jurassic and Cretaceous of central and eastern Asia (Ren et al., 2009). *Parapolycentropus* is highly unusual: besides the long, styletiform proboscis, its hind wings are almost completely lost, the thoracic sclerites are very similar to those of flies (order Diptera), and the antenna is modified into a fine, whip-like structure. The genus looks superficially like a mosquito. Proboscides of pseudopolycentropodids are similar in proportions to those of slender, long pollen tubes in the Bennettitales and Gnetales, and this led Ren et al. (2009) to propose that these scorpionflies were specialized pollinators of certain Mesozoic gymnosperms. Microscopic structure, however, of the mouthparts and pretarsal claws of *Parapolycentropus* in Burmese amber indicate that this genus was probably insectivorous (as are some living scorpionflies) (Grimaldi and Johnston, 2014), not anthophilous (or hematophagous, as suggested by Ren et al. [2009]). The family has been considered an extinct stem group of extant Mecoptera, but excellent preservation of the complex male genitalia of *Parapolycentropus* indicate a close relationship with the living family Boreidae (Grimaldi and Johnston, 2014). Boreidae, or 'snow fleas', are a small family of apterous and brachypterous scorpionflies (all completely flightless), which occur in boreal regions of the Northern Hemisphere.

Another 'long-tongued' group in Burmese amber involves flies in the Cretaceous family Zhangsolvidae. These are stout-bodied flies approximately one centimetre in body length, with a proboscis about half the body length (Fig 14E); several species have long, flagellate antennae twice the length of the body (Fig 14E), which is unique within the brachyceran Diptera. Zhangsolvidae was first discovered based on lithified remains from the Early Cretaceous of China (Zhang et al., 1993), then from the Early Cretaceous Crato Formation of Brazil (Mazzarolo and Amorim, 2000). Completely unexpectedly, three species were discovered in Cretaceous amber: two in late Albian Spanish amber and one species in Burmese amber (Arillo *et al.*, 2015). Based on the Crato specimens, it was hypothesized that Zhangsolvidae (referred to as Cratomyiidae, a synonym) were in the Stratiomyomorpha, based largely on the best-preserved character system, the wing venation. This infraorder of flies contains two small living families of flies (the Xylomyidae and Pantophthalmidae, the latter including the largest known flies), and the large family of soldier flies, the Stratiomyidae. Stratiomyids and xylomyids are usually boldly patterned in yellow and black, many mimicking wasps and bees that forage on the same flowers. Species in amber allowed observation not just of wing venation but of critical microscopic details such as structure of the palps, tibial spurs, antennal structure, etc., and this confirmed placement in the Stratiomyomorpha. Even more recently, it was found that adhering to the body of one of the Spanish amber specimens is a clump of *Exesipollenites* pollen, produced by Mesozoic plants probably in the Bennettitales (Peñalver *et al.*, 2015). This fly visited gymnosperms, providing direct evidence that one cannot necessarily infer that Cretaceous insects specialized for pollination visited angiosperms (contra, e.g., Ren, 1998).

Transitional Taxa

Examples of the phylogenetic significance of certain arthropods in Burmese amber involve two families in the Paraneoptera, a group that includes the sucking insects (aphids, scale insects, true bugs), thrips (Thysanoptera), as well as the true lice (Phthiraptera) and 'bark' lice (Psocodea). True lice live their entire life cycle in the pelage of their bird and mammal hosts, which is why these insects have only a few fossil records (for a remarkable one from Messel, see Wappler *et al.*, 2004). However, there is keen interest in the times of divergence of lice, since these insects, more than any others, appear to co-speciate with their hosts (e.g., Reed *et al.*, 2004; Paterson *et al.*, 2000). The free-living sister group to the true, ectoparasitic lice is the booklouse family Liposcelididae, relationships that were first presented based on morphology and behaviour (Lyal, 1985), then confirmed by various molecular studies (e.g., Misof *et al.*, 2014). The nominal genus of the Liposcelididae, *Liposcelis*, is the common booklouse, small (1.0–1.5 mm) pale, flat, wingless insects commonly found on damp paper, and which consume various animal products such as pinned insects, hair, glue, shed skin, etc. They are often found in animal nests but are not ectoparasitic. Burmese amber contains the oldest Liposcelididae, *Cretoscelis burmitica*. This relationship was disputed by Mockford *et al.* (2013) on the basis that *Cretoscelis* does not possess all of the characters of the family. It lacks, for example, Pearman's organ (a delicate stidulatory structure at the base of the hind legs) and possesses the Rs vein in the forewing (rather than having lost this) – microscopic features preservable only in amber. This information reveals that *Cretoscelis* is a *stem-group* liposcelidid, and provides important data for fossil calibration of divergence-time analyses of lice.

Another paraneopteran in Burmese amber is the most recent occurrence of the hemipteran family Protopsyllidiidae (Fig 9A), otherwise known entirely from lithified specimens from the Permian to Late Jurassic (Grimaldi, 2003). Most of the fossil species are known only as disarticulated wings, and are quite small (*c*.1–3mm), so the phylogenetic position of the family was ambiguous, but thought to be basal within Sternorrhyncha (which includes aphids, whiteflies, mealybugs, scales, and plant lice). Fine preservation of *Postopsyllidium*, a very bristly, jumping insect in Burmese and New Jersey ambers, confirms that this family is an extinct sister-group to the Sternorrhyncha (Grimaldi, 2003).

Oldest records

Lastly, Burmese amber preserves species that are the oldest representatives of a small, enigmatic order of insects, the Strepsiptera (twisted-winged parasites) (Fig 13F). These are minute, generally 2mm or less in length: a well-defined monophyletic group, but whose relationship to other orders has been ambiguous and highly controversial. These are truly bizarre animals; all but the most basal of the 600 world species have females that are neotenic, larviform endoparasitoids of other insects; males have one pair of fan-like wings and large raspberry-like eyes (Fig 11F), dispersing for mating. The oldest known member of the order occurs in Burmese amber, *Cretostylops engeli* (Grimaldi et al., 2005). This species is basal, but interestingly, *Protoxenos* in Baltic amber (less than half the age of *Cretostylops*) is actually the most basal species of the order; it was probably a relict even in the Eocene.

Another order of minute insects, the Zoraptera, contains about 40 species worldwide; hypotheses on relationships of the order have varied widely, from their being close relatives of termites (Isoptera), earwigs (Dermaptera), webspinners (Embiodea), to barklice (Psocoptera), and other groups. The only fossils are two species in Dominican amber, four species in Burmese amber (Engel and Grimaldi, 2002) (Figs 8D, 8E), one species in Jordanian amber (Engel, 2008), and they have been recorded from Ethiopian amber (Schmidt et al., 2010). They have not been found in Baltic amber. Interestingly, with the exception of *Xenozorotypus*, three of the four Burmese amber zorapterans are very similar to other (extant and Miocene) species of *Zorotypus* (Engel and Grimaldi, 2002), indicating that the group is at least Jurassic in age.

The Burmese amber zorapterans and strepsipteran have too many autapomorphies of their extant crown-groups in order to resolve relationship issues, but the fossils do reveal that both groups probably extend well into the Jurassic. Also, Burmese fossils of both orders provide information on the sequence of character evolution, which would never have been traced based on just the living species.

3 Evolution of major adaptive features
Sociality

Advanced sociality, or eusociality, is essentially an arthropod phenomenon. With the exception of two species of African naked mole rats, all other animals that live in colonies of closely related individuals from several generations and that have castes are arthropods, mostly insects (reviewed in Grimaldi and Engel, 2005b). The two largest groups of social animals are the ants (family Formicidae, order Hymenoptera) and termites (infraorder Isoptera, order Blattodea). All species of the approximately 13,000 extant species of ants and 3100 species of termites are eusocial, which is not particularly impressive species diversity in arthropod terms, but these groups have a combined ecological impact that is absolutely profound. For example, it has been estimated that 94 out of every 100 arthropods in some tropical forests are ants (Davidson et al., 2003), and that 15% of the animal biomass in lowland, Amazonian rain forest is composed of ants (Fittkau and Klinge, 1973). Termites are one of the main consumers of lignocellulose in tropical and subtropical regions, and in dry tropical forests they consume 40–100% of the dead wood, and 20% of grass biomass in savannas (Dettling, 1988). Both groups, moreover, process prodigious quantities of soil and are essential in humification. Near Lake Malawi there is a geological formation comprising 44 million cubic metres that spans 10,000 to 100,000 ybp, formed entirely from the eroded mounds of the living termite *Macrotermes falciger* (Crossley, 1986).

Driving this ecological supremacy is eusociality. A division of labour amongst nestmates allows great efficiency in tasks, and the recruitment and mobilization of large numbers of workers and soldiers in foraging and defence allows a colony to easily outcompete solitary species and overwhelm enemies. Thus, understanding the origins of insect sociality is not a trivial consideration. For both ants and termites, Burmese amber has been the single most important source of fossils.

Isoptera are now known to be myopic or blind, wood-eating, social roaches (Inward et al., 2007); their closest living relatives are the colonial wood roaches, *Cryptocercus* (reviewed by Grimaldi and Engel, 2005b). True, eusocial termites are hypothesized to have originated in the Late Jurassic (Engel *et al.*, 2009), but until recently the Cretaceous fossil record for the group has been based almost entirely on winged (alate) reproductive individuals (e.g., Engel *et al.*, 2007). The oldest workers and soldier termites were preserved in Miocene Dominican amber (reviewed in Krishna *et al.*, 2013). Recently, the first definitive worker termites from the Cretaceous, and the first Cretaceous soldiers, have been discovered in Burmese amber (Engel *et al.*, 2016) (Figs 8F, G). One of the species is, in fact, one of the largest termite soldiers known, extant or extinct (only soldiers of the extant genus *Syntermes* are larger). At this point there are at least 10 species of termites in 9 genera known from Burmese amber; the next most diverse Cretaceous deposit is from the Crato Formation of Brazil, with six species.

Ants have attracted enormous scientific focus because of their social behavior and ecological significance (e.g., Hölldobler and Wilson, 1990). The origin of ants has been hypothesized to have occurred anywhere from the Late Jurassic to Early Cretaceous (*c*.160 to 120 mya), although the former of these is implausible and the latter of these estimates seems most reasonable. Until the recent study of scores of ant specimens in Burmese amber (e.g., Barden and Grimaldi, 2012, 2013, 2014, 2015), sociality in Cretaceous ants has been ambiguous. Indeed, there are 20 species of ants in Burmese amber known thus far, fully half of all known Cretaceous species, and this wealth of specimens has entirely revised earlier concepts of Cretaceous ants as extremely rare and very generalized. There is abundant, direct evidence that early ants had alate and de-alate females (queens, who shed their wings after a nuptial flight), wingless females (workers), and winged males: they had castes. There is also direct behavioural evidence for ant sociality preserved in Burmese amber: several pieces have conspecific aggregations of workers, reflective of group foraging and recruitment, and one piece even has a pair of heterospecific ants fighting (Fig 10E). Interspecific territorial aggression and fighting in aculeate (stinging) wasps is known only in ants, which are, again, entirely eusocial.

With the exception of a few species of undescribed formicine ants among Burmese ants, all other species of ants and all termites in Burmese amber are stem groups to Recent and Tertiary crown-groups. Though phylogenetically basal, mid-Cretaceous ants and termites were already eusocial.

Phytophagy

The spectacular diversity of insects is attributed in part to their association with plants, specifically phytophagy or herbivory. Some 40% of living insect species feed directly on plants, especially angiosperms (Engel and Grimaldi, 2005B). Indeed, in comparisons of sister groups of phytophagous insect taxa, one feeding on conifers or lower vascular plants and the other feeding on angiosperms, the latter are always far more diverse (Mitter *et al.*, 1988). This has led to the widespread view that 'colonization' of angiosperm hosts facilitates adaptive radiation of phytophagous insects (e.g., Farrell, 1998), and may explain the enormous radiations of groups like the Phytophaga (*c*.120,000 species that include leaf beetles, long-horned beetles, and the largest family of insects, weevils), and the Lepidoptera (*c*.160,000 species, the largest lineage of plant-feeding animals). There are various phytophagous insect groups preserved in Burmese amber: Orthoptera (crickets, katydids, grasshoppers), Hemiptera (Auchenorrhyncha, Sternorrhyncha), thrips (Thysanoptera), various moths, weevils, etc. Eventually, thorough phylogenetic study of these fossils will be very revealing.

One group that has shown particular promise for studying the apparent co-diversification with angiosperms is the Coccoidea, or scale insects. This is a group of approximately 8000 Recent species of generally small to minute insects (5mm to <1mm long). Female scale insects are very modified (wings lost, eyes lost or highly reduced; appendages reduced or even lost); they are pedomorphic and are very sedentary to entirely sessile. Males are minute, winged, ephemeral, and do not feed; they disperse to mate. Though their minute size makes study of male coccoids in amber very challenging (it requires meticulous trimming, polishing and microscopy), the abundance and diversity of these 'aerial plankton' insects in ambers affords unique study of the evolution of a major group of phytophagous insects through the Cretaceous and Tertiary. Thus far, there are 15 species and eight families of Coccoidea described from Burmese amber – the highest Cretaceous diversity and more than in New Jersey amber (10 species in six families), where coccoids are exceptionally abundant. Phylogenetic study of the coccoids in Cretaceous and Tertiary ambers reveals that some groups thought to be recently evolved, like mealybugs (Pseudococcidae) actually appeared in the Cretaceous, and that more basal lineages of Coccoidea extend probably into the early Mesozoic, well before the angiosperm radiations (Vea and Grimaldi, 2015a, b). These findings lend support to theories that the diversification of plant-feeding insects does not follow strictly along the lines of host-plant divergence, but instead is probably related to host-plant use making environments more coarse-grained for these insects, and thus facilitating allopatric divergence of populations (Janz, 2011; Futuyma and Agrawal, 2009; Vea and Grimaldi, 2015b).

4 Biogeography

Fossils provide unique insight into understanding modern distributions; for understanding the history of distributions in geological time they are indispensible. Fossils provide unique, direct data on the ages of taxa, and so can support or refute geological explanations for distributions, but perhaps the most powerful application of fossils is the potential to reveal dramatic changes in distributions over geological time, which we discuss below with examples in Burmese amber. Again, the detailed preservation in amber allows unambiguous phylogenetic interpretation, which in turn allows rigorous biogeographic interpretation.

Repeated patterns of modern distributions can be very compelling, an excellent and classic example being a southern temperate distribution, which is a highly disjunct area that usually involves Australia and New Zealand, Chile and Argentina, and/or southern Africa. It is also called an austral (Darlington, 1965), southern Gondwanan, or transantarctic (Brundin, 1967) distribution, involving myriad lineages of plants and animals (reviewed for insects by Grimaldi and Engel, 2005b). An austral distribution has commonly been interpreted as a relict of Gondwanan drift (e.g, Raven and Axelrod, 1974), but fossils have increasingly proven that Gondwanan taxa often had Laurasian histories.

For Burmese amber, two examples will illustrate the formerly widespread nature of presently austral groups, one example involving spiders of the family Archaeidae and the other involving liverworts of the genus *Gackstroemia* (Lepidolaenaceae). Archaeidae are called 'pelican spiders', because the cephalothorax in living species is extended into a giraffe-like neck, with long chelicerae or 'jaws' hanging downward, appearing in profile like a pelican at rest. The 54 extant species are specialized predators of other spiders, occurring in Madagascar, South Africa, and Australia. There are 18 named species of archaeid fossils, oddly all from the Northern Hemisphere, ranging in age from Jurassic to Eocene; one genus is in Burmese amber, *Burmesarchaea* (Fig 6A). Even though the living species appear to form a monophyletic group and the fossils form a paraphyletic grade basal to this group (Wood *et al.*, 2012), clearly the evolutionary history of Archaeidae extended far beyond the Austral Region. Interpreting the evolutionary history of the family based just on living species is misleading. *Gackstroemia* liverworts are similar, with three species in Australasia, six species in southern South America, and a species in Burmese amber, *Gackstroemia cretacea* (Heinrichs *et al.*, 2014). Fossil data continues to support the emerging view, summarized by Sanmartin and Ronquist (2004), that common austral distributions involve complex patterns of extinction.

A dramatic example of geographic extinction in Burmese amber involves another scorpionfly (order Mecoptera), of the family Meropeidae (Fig 13B). This family is sometimes called 'earwig flies', because the male terminalia are developed into large forceps. There are three living species of Meropeidae, with a distribution about as disjunct as possible for vegetated land: *Merope tuber* in eastern North America, *Austromerope brasiliensis* in southeastern Brazil, and *Austromerope poultoni* in southwestern Australia (reviewed in Grimaldi and Engel, 2013). A small relative of these occurs in Burmese amber, *Burmomerope eureka*, which possesses the distinctive forcipate male genitalia. There is also a lithified (but diagnostic) fossil of the family, *Boreomerope antiquaa*, from the Middle Jurassic of Siberia. Clearly, the highly disjunct distribution of the three living species of meropeid scorpionflies is due to extinction, but the geographical distribution of the family was even more widespread than suggested by the living species.

Conclusions

There is a wealth of future work required on Burmese amber. First and foremost there is a need for comprehensive fieldwork on the amber outcrops. Cruikshank and Ko (2003) is the only modern study of the outcrops, and it is based on only a few microfossils and one ammonite specimen. Direct and thorough study is required to prospect for other index fossils and sample for diverse other microfossils, particularly Foraminifera and pollen. It is to be hoped that on-site prospecting would also uncover wood, and perhaps other macrofossil plant remains like cones that contain amber, which are essential for unambiguously determining the botanical source of the amber.

At present, Burmese amber is actively marketed globally. A few museums have modest funds to procure specimens, but the vast proportion of Burmese amber is sold into China as decorative carvings. In fact, the second known specimen of the oldest strepsipteran (Fig 13F) was found in a carved piece similar to that shown in Figure 3D. The amount of scientifically valuable specimens lost as jewellery and charms must be enormous. Fortunately amber, and Burmese amber in particular, has an avid and devoted community of private collectors who are enamoured with the fossil inclusions; the private collectors are in total much more effective in securing Burmese amber fossils than any museum can be. Many of the private collectors assist museums with procuring or loaning specimens, which is citizen science at its best.

It is essential that descriptive taxonomy on species in Burmese amber be done in a comprehensive, comparative context that is based on related fossil and living species, and taxonomists must know the literature on other diverse deposits from the late Mesozoic. There also needs to be detailed re-study of Cockerell's types of Burmese amber arthropods in the Natural History Museum, London, since the original descriptions are very superficial. At this point, palaeobiogeographic affinities of the Burmese palaeobiota appear to be primarily with taxa from the Late Jurassic and Early Cretaceous of Eurasia and Asia, like Karatau, Baissa, and the Yixian Formation, but this needs quantification. Phylogenetics provides the necessary framework for evolutionary interpretation of fossils. Other evidence indicates significant provincialism of the Burmese palaeobiota. Was the Burmese amber forest, for example, geographically isolated? Many questions remain.

Acknowledgments

Both authors are indebted to Scott Anderson and Sieghard Ellenburger, who have provided specimens and materials for study. The senior author is also indebted particularly to James Zigras, but also to Federico Berloecher, Scott Davies, Keith Luzzi, Ru Smith, and others who have also provided specimens. The years of work by Paul Nascimbene in the amber lab at the AMNH have been a crucial part of research and collection development at the museum. Michael S. Engel, Alexander Schmidt, Jeff Cumming, Enrique Penalver and others are thanked for years of collaboration on the systematics of various inclusions in Burmese and other ambers. Robert G. Goelet, trustee and chairman emeritus of the AMNH, generously supported fieldwork, lab work, and specimen acquisition. Further funding was provided to D. G. by the US National Science Foundation and the Institute for Museum and Library Services. The junior author would also like to thank Claire Mellish at the Natural History Museum, London for access to the collections, to Peter York and Bill Crighton for their expert photography of inclusions from the NHMUK and National Museums Scotland collections respectively, to Mr Fan Yong of the Fushun Amber Institute for kindly donating the specimen in Figure 1e, to David Nicholson (Natural History Museum, London) for tracking down obscure references and supplying data from his PhD thesis, to Monja Knoll for translating German text, and to Rosina Buckland (NMS) and Tomoko Snyder for translating Japanese text. He would also like to thank Dany Azar (Lebanese University, Beirut), Enrique Peñalver (Instituto Geológico y Minero de España, Madrid), Vincent Perrichot (Université de Rennes), Alexandr Rasnitsyn (Palaeontological Insititute, Moscow)and to Yasuyuki Shirota (Ikadogen Apple Institute) for supplying data on amber inclusions for Appendix 2.

References

Alonso, J., A. Arillo, E. Barrón, J. C. Corral, J. Grimalt, et al. 2000. A new fossil resin with biological inclusions in Lower Cretaceous deposits from Álava (northern Spain, Basque-Cantabrian Basin). Journal of Paleontology 74:158–178.

Anderson, K. B. 1997. The nature and fate of natural resins in the geosphere – VII. A radiocarbon (^{14}C) age-scale for description of immature natural resins: an invitation to scientific debate. Organic Geochemistry 25:251–253.

Anderson, K. B., E. E. Winans, and R. E. Botto. 1992. The nature and fate of natural resins in the geosphere – II. Identification, classification, and nomenclature of resinites. Organic Geochemistry 18:829–841.

Antoine, P.-O., D. De Franceschi, J. J. Flynn, A. Nel, P. Baby, M. Benammi, Y. Calderón, N. Espurt, A. Goswami, and R. Salas-Gismondi. 2006. Amber from western Amazonia reveals neotropical diversity during the middle Miocene.

Proceedings of the National Academy of Sciences of the United States of America 103:13595–13600.

Arillo, A., D. A. Grimaldi, E. Peñalver, R. Perez-de la Fuente, X. Delclòs, J. Criscione, P. Barden, and M. Riccio. 2015. Long-proboscid brachyceran flies in Cretaceous amber (Diptera: Brachycera: Stratiomyomorpha: Zhangsolvidae). Systematic Entomology 40:242–267.

Austin, J. J., A. J. Ross, A. B. Smith, R. A. Fortey, and R. H. Thomas. 1997. Problems of reproducibility – does geologically ancient DNA survive in amber-preserved insects? Proceedings of the Royal Society of London B 264:467–474.

Azar, D., and A. Nel. 2013. A new beaded lacewing from a new Lower Cretaceous amber outcrop in Lebanon (Neuroptera: Berothidae). Pp. 111–130 in D. Azar, M. S. Engel, E. Jarzembowsky, L. Krogman, A. Nel and J. Santiago-Bley (eds), Insect Evolution in an Amberiferous and Stone Alphabet. Brill, Leiden.

Azar, D., R. Geze, and F. Acra. 2010. Lebanese amber. Pp. 271–298 in Penny, D. (ed.), Biodiversity of Fossils in Amber from the Major World Deposits. Siri Scientific Press, Manchester.

Barden, P., and D. Grimaldi, 2012. Rediscovery of the bizarre Cretaceous ant *Haidomyrmex* (Hymenoptera: Formicidae), including two new species. American Museum Novitates 3755:1–16.

Barden, P., and D. A. Grimaldi. 2013. A new genus of highly specialized ants in Cretaceous Burmese amber (Hymenoptera: Formicidae). Zootaxa 3681:405–412.

Barden, P. and Grimaldi, D. 2014. A diverse ant fauna from the mid-Cretaceous of Myanmar (Hymenoptera: Formicidae). PLoS ONE 9(4):e93627. doi:10.1371/journal.pone.0093627.

Barden, P. and D. Grimaldi. 2016. Advanced sociality and adaptive radiation in Cretaceous stem-group ants. Current Biology 26:515–521.

Beck, C. 1986. Spectroscopic investigations of amber. Applied Spectroscopy Reviews 22:55–110.

Berendt, C. G. (ed.) 1845–1856. Die im Bernstein befindlichen organischen Reste der Vorwelt. Vol. I–II. Nicolai, Berlin.

Blagoderov, V., and D. Grimaldi. 2004. Fossil Sciaroidea (Diptera) in Cretaceous ambers, exclusive of Cecidomyiidae, Sciaridae, and Keroplatidae. American Museum Novitates 3433: 1–76.

Borkent, A. 1997. Upper and Lower Cretaceous biting midges (Ceratopogonidae: Diptera) from Hungarian and Austrian amber and the Koonwarra Fossil Bed of Australia. Stuttgarter Beiträge zur Naturkunde, B 249:1–10.

Bray, P. S. and K. B. Anderson. 2008. The nature and fate of natural resins in the geosphere XIII: A probable pinaceous resin from the Early Cretaceous (Barremian), Isle of Wight. Geochemical Transactions 9:3–13.

Bray, P. S., and K. B. Anderson. 2009. Identification of Carboniferous (320 million-year-old) class 1c amber. Science 326:132–134.

Broly, P., S. Maillet, and A. J. Ross. 2015. The first terrestrial isopod (Crustacea: Isopoda: Oniscidea) from Cretaceous Burmese amber of Myanmar. Cretaceous Research 55:220–228.

Brundin, L. 1967. Insects and the problem of austral disjunctive distribution. Annual Review of Entomology 12:149–168.

Cano, R. J., H. N. Poinar, N. J. Pieniazek, A. Acra, and G. O. Poinar, Jr. 1993. Amplification and sequencing of DNA from a 120–135 million-year-old weevil. Nature 363:536–538.

Carpenter, F. M., J. W. Folsom, E. O. Essig, A. C. Kinsey, C. T. Brues, M. W. Boesel, and H. E. Ewing. 1937. Insects and arachnids from Canadian amber. University of Toronto Studies, Geology 40:7–62.

Chhibber, H. L. 1934. The Mineral Resources of Burma. Macmillian & Co., London.

Clauer, N., J. M. Huggett, and S. Hillier. 2005. How reliable is the K-Ar glauconite chronometer? A case study of Eocene sediments from the Isle of Wight. Clay Minerals 40:167–176.

Cockerell, T. D. A. 1917. Arthropods in Burmese amber. Psyche 24:40–45.

Cognato, A. I. and D. A. Grimaldi. 2009. 100 million years of morphological stasis in bark beetles (Coleoptera: Curculionidae: Scolytinae). Systematic Entomology 34:93–100.

Crossley, R. 1986. Sedimentation by termites in the Malawi Rift Valley. Pp 191–199 in L. E. Frostik *et al.* (eds), Sedimentation in the Africa Rifts. Geological Society of London Special Publication 25.

Cruikshank, R. D., and K. Ko. 2003. Geology of an amber locality in the Hukawng Valley, northern Myanmar. Journal of Asian Earth Sciences 21:441–455.

Darlington, P. J. 1965. Biogeography of the Southern End of the World: Distribution and History of Far-Southern Life and Land, with an Assessment of Continental Drift. Harvard University Press, Cambridge, Massachusetts.

Davidson, D. W., S. C. Cook, R. R. Snelling, and T. H. Chua. 2003. Explaining the abundance of ants in lowland tropical rainforest canopies. Science 300:969–972.

Daza, J. D., E. L. Stanley, P. Wagner, A. M. Bauer, and D. A. Grimaldi. 2016. Mid-Cretaceous amber fossils shed light on the past diversity of tropical lizards. Science Advances 2:e1501080. doi:10.1126/sciadv.1501080

DePalma, R., F. Cichocki, M. Dierick, and R. Feeney. 2010. Preliminary notes on the first recorded amber insects from the Hell Creek Formation. Journal of Paleontological Sciences 2:1–7.

DeSalle, R., J. Gatesy, W. Wheeler, and D. Grimaldi. 1992. DNA sequences from a fossil termite in Oligo-Miocene amber and their phylogenetic implications. Science 257:1933–1936.

Dettling, D. K. 1988. Grasslands and savannas: regulation of energy flow and nutrient cycling by herbivores. Pp 131–148 in L. R. Pomeroy and J. A. Alberts (eds), Concepts of Ecosystem Ecology: A Comparative Review. Springer Verlag, Berlin.

Dunlop, J. A. 2010. Bitterfeld amber. Pp. 57–68 in D. Penney (ed), Biodiversity of Fossils in Amber from the Major World Deposits. Siri Scientific Press, Manchester.

Dunlop, J. A., and D. Penney. 2012. Fossil Arachnids. Siri Scientific Press, Manchester, 192 pp.

Durham, J. W. 1956. Insect-bearing amber in Indonesia and the Philippine islands. Pan-Pacific Entomologist 32:51–53.

Durham, H. W., and P. D. Hurd. 1957. Fossiliferous amber of Chiapas, Mexico. Bulletin of the Geological Society of America 68:18–24.

Edwards, H. G. M., D. W. Farwell, and S. E. Jorge Villar. 2007.

Raman microspectroscopic studies of amber resins with insect inclusions. Spectrochimica Acta A: Molecular and Biomolecular Spectroscopy 68:1089–1095.

Engel, M. S. 2008. A new apterous *Zorotypus* in Miocene amber from the Dominican Republic (Zoraptera: Zorotypidae). Acta Entomologica Slovenica 16:127–136.

Engel, M. S., and D. Grimaldi. 2002. The first Mesozoic Zoraptera (Insecta). AmericanMuseum Novitates 3362:1–20.

Engel, M. S., D. Grimaldi, and K. Krishna. 2007. Primitive termites from the Early Cretaceous of Asia. Stuttgarter Beiträge zur Naturkunde, B371:1–32.

Engel, M. S., D. Grimaldi, and K. Krishna. 2009. Termites: their phylogeny and rise to ecological dominance. American Museum Novitates 3650:1–27.

Engel, M. S., P. Barden, M. Riccio, and D. Grimaldi. 2016. Morphologically specialized termite castes and advanced sociality in the Early Cretaceous. Current Biology 26:522–530.

Eskov, K. Y. 2002. Fossil resins. Pp. 444–446 in A. P. Rasnitsyn and D. L. J. Quicke (eds), History of Insects. Kluwer Academic Publishers.

Farrell, B. D. 1998. "Inordinate fondness" explained: why are there so many beetles? Science 281:555–559.

Fittkau, E. J. and H. Klinge. 1973. On biomass and trophic structure of the central Amazonian rain forest ecosystem. Biotropica 5:2–14.

Folger, D. W., J. C. Hathaway, R. A. Christopher, P. C. Valentine, and C. W. Poag. 1978. Stratigraphic test well, Nantucket Island, Massachusetts, U. S. Geological Survey Circular 773: 1–28.

Futuyma, D. J. and A. A. Agrawal. 2009. Macroevolution and the biological diversity of plants and herbivores. Proceedings of the National Academy of Sciences of the United States of America USA. 106:18054–18061.

Ghiurca, V. 1990. New considerations on Romanian amber. Prace Muzeum Ziemi 41:158.

Goldsmith, E. 1879. On amber containing fossil insects. Proceedings of the Academy of Natural Sciences of Philadelphia 31:207–208.

Greenwood, D. R., A. J. Vadala, and J. G. Douglas. 2000. Victorian Paleogene and Neogene macrofloras: a conspectus. Proceedings of the Royal Society of Victoria 112: 65–92.

Grimaldi, D. 1996. Amber: Window to the Past. Abrams/American Museum of Natural History, New York, 216 pp.

Grimaldi, D. 2003. First amber fossils of the extinct family Protopsyllidiidae, and their phylogenetic significance among Hemiptera. Insect Systematics and Evolution 34:329–344.

Grimaldi, D. A., and M. S. Engel. 2005a. Fossil Liposcelididae and the lice ages (Insecta: Psocodea). Proceedings of the Royal Society of London B 273:625–633.

Grimaldi, D. A., and M. S. Engel. 2005b. Evolution of the Insects. Cambridge University Press, Cambridge, 755 pp.

Grimaldi, D. A., and M. S. Engel. 2013. The relict scorpionfly family Meropeidae (Mecoptera) in Cretacous amber. Journal of the Kansas Entomological Society 86:253–263.

Grimaldi, D. A., and M. A. Johnston. 2014. The long-tongued Cretaceous scorpionfly *Parapolycentropus* Grimaldi and Rasnitsyn (Mecoptera: Pseudopolycentropodidae): new data and interpretations. American Museum Novitates 3793:1-23.

Grimaldi, D. A., and P. C. Nascimbene. 2010. Raritan (New Jersey) amber. Pp. 167–191 in D. Penney (ed), Biodiversity of Fossils in Amber from the Major World Deposits. Siri Scientific Press, Manchester.

Grimaldi, D. A., M. S. Engel, and P. C. Nascimbene. 2002. Fossiliferous Cretaceous amber from Myanmar (Burma): its rediscovery, biotic diversity, and paleontological significance. American Museum Novitates 3361:1–72.

Grimaldi, D. A., J. Kathirithamby, and V. Schawaroch. 2005. Strepsiptera and triungula in Cretaceous amber. Insect Systematics and Evolution 36:1–20.

Grimaldi, D. A., A. Shedrinsky, and T. P. Wampler. 2000. A remarkable deposit of fossiliferous amber from the Upper Cretaceous (Turonian) of New Jersey. Pp. 1–76 in D. A. Grimaldi (ed), Studies on Fossils in Amber, with Particular Reference to the Cretaceous of New Jersey. Backhuys Publishers, Leiden510 pp.

Grimaldi, D. A., E. Bonwich, M. Delannoy, and S. Doberstein. 1994. Electron microscopic studies of mummified tissues in amber fossils. American Museum Novitates 3097:1–31.

Grimaldi, D. A., Z. Junfeng, A. Rasnitsyn, and N. Fraser. 2005. Revision of the bizarre Mesozoic scorpionflies in the Pseudopolycentropodidae (Mecopteroidea). Insect Systematics and Evolution 36:443–458.

Grimaldi, D. A., J. A. Lillegravan, T. W. Wampler, D. Bookwalter. and A. Shedrinsky. 2000. Amber from Upper Cretaceous through Paleocene strata of the Hanna Basin, Wyoming, with evidence for source and taphonomy of fossil resins. Rocky Mountain Geology 35: 163–204.

Hand, S., M. Archer, D. Bickel, P. Creaser, M. Dettmann, H. Godthelp, A. Jones, B. Norris, and D. Wicks. 2010. Australian Cape York amber. Pp. 69–79 in D. Penney (ed), Biodiversity of Fossils in Amber from the Major World Deposits. Siri Scientific Press, Manchester.

Heinrichs, J., A. Schaefer-Verwimp, K. Feldberg, and A. R. Schmidt. 2014. The extant liverwort *Gackstroemia* (Lepidolaenacae, Porellales) in Cretaceous amber from Myanmar. Review of Palaeobotany and Palynology 203:48–52.

Heinrichs, J., M. E. Reiner-Drehwald, K. Feldberg, D. A. Grimaldi, P. C. Nascimbene, M. von Konrat, and A. R. Schmidt. 2011. *Kaolakia borealis* nov. gen. et sp. (Porellales, Jungermanniopsida): a leafy liverwort from the Cretaceous of Alaska. Review of Palaeobotany and Palynology 165:235–240.

Henwood, A. 1992. Exceptional preservation of dipteran flight muscle and the taphonomy of insects in amber. Palaios 7:203–212.

Hillmer, G., W. Weitschat, and N. Vávra. 1992. Bernstein aus dem Miozän von Borneo. Naturwissenschaftliche Rundschau 45:72–74.

Hillmer, G., W. Weitschat, and P. C. Voight. 1992. Bernstein im Regenwald von Borneo. Fossilien 6:336–340.

Hölldobler, B. and E. O. Wilson. 1990. The Ants. Belknap Press of Harvard University Press, Cambridge, Massachusetts, 732pp.

Inward, D., G. Beccaloni, and P. Eggleton. 2007. Death of an order: a comprehensive molecular phylogenetic study confirms that termites are eusocial cockroaches. Biology Letters 3:331–335.

Janz, N. 2011. Ehrlich and Raven revisited: mechanisms underlying codiversification of plants and enemies. Annual Review of Ecology, Evolution, and Systematics 42:71–89.

Jarzembowski, E. A. 1999. British amber: a little-known resource. Estudios del Museo de Ciencias Naturales de Alava 14(Núm. Esp. 2):133–140.

Jarzembowski, E. A., D. Azar, and A. Nel. 2008. A new chironomid (Insecta: Diptera) from Wealden amber (Lower Cretaceous) of the Isle of Wight (UK). Geologica Acta 6:285–291.

Kaddumi, H. F. 2005. Amber of Jordan. The Oldest Prehistoric Insects in Fossilized Resin. 2nd Edition. Eternal River Museum of Natural History, Jordan, 224 pp.

Knight, T., P. S. Bingham, D. A. Grimaldi, K. Anderson, R. D. Lewis, and C. E. Savrda. 2010. A new Upper Cretaceous (Santonian) amber deposit from the Eutaw Formation of eastern Alabama, USA. Cretaceous Research 31:85–93.

Kohring, R. 1998. REM-Untersuchungen an harzkonservierten Arthropoden. Entomologia Generalis 23:95–106.

Kohring, R., and T. Schlüter. 1989. Historische und paläontologische Bestandsaufnahme des Simetits, eines fossilen Harzes mutmasslich mio/pliozänen Alters aus Sizilien. Documenta Naturae 56:33–58.

Koteja, J. and G. O. Poinar, Jr. 2001. A new family, genus, and species of scale insect (Hemiptera: Coccinea: Kukaspididae, new family) from Cretaceous Alaskan amber. Proceedings of the Entomological Society of Washington 103:356–363.

Kraemer, M. M. S. 2010. Mexican amber. Pp. 42–56 in D. Penney (ed), Biodiversity of Fossils in Amber from the Major World Deposits. Siri Scientific Press, Manchester.

Krishna, K., D. Grimaldi, M. S. Engel, and V. Krishna. 2013. Treatise on the Isoptera of the World. Bulletin of the American Museum of Natural History 377:1–2704.

Lambert, J. B., S. C. Johnson, and G. O. Poinar, Jr. 1996. Nuclear magnetic resonance characterization of Cretaceous amber. Archaeometry 38:325–335.

Langenheim, J. H. 1969. Amber: a botanical inquiry. Science 136:1157–1169.

Langenheim, J. H. 2003. Plant Resins: Chemistry, Evolution, Ecology, Ethnobotany. Timber Press, Portland, Oregon, 586 pp.

Levinson, A. A. 2001. Amber ("burmite") from Myanmar: production resumes. Gems and Gemology 37:142–143.

Lyal, C. H. C. 1985. Phylogeny and classification of the Psocodea, with particular reference to the lice (Psocodea: Phthiraptera). Systematic Entomology 10:145–165.

Martinez-Delclòs, X., D. E. G Briggs, and E. Peñalver. 2004. Taphonomy of insects in carbonates and amber. Palaeogeography, Palaeoclimatology, Palaeoecology 203:19–64.

Mazzarolo, L. A. and D. S. Amorim. 2000. *Cratomyia macrorrhyncha*, a Lower Cretaceous brachyceran fossil from the Santana Formation, Brazil, representing a new species, genus and family of the Stratiomyomorpha (Diptera). Insect Systematics and Evolution 31:91–102.

McAlpine, J. F., and J. E. H. Martin. 1969. Canadian amber – a paleontological treasure-chest. Canadian Entomologist 101:819–838.

McKellar, R. C., and M. S. Engel. 2012. Hymenoptera in Canadian Cretaceous amber (Insecta). Cretaceous Research 35:258–279.

McKellar, R. C., and A. P. Wolfe. 2010. Canadian amber. Pp. 149–166 in D. Penney (ed). Biodiversity of Fossils in Amber from the Major World Deposits. Siri Scientific Press, Manchester.

Mierzejewski, P. 1976. On application of scanning electron microscopy to the study of organic inclusions from the Baltic amber. Annals of the Geological Society of Poland 46:291–295.

Misof, B., S. Liu, K. Meusemann and 98 other co-authors. 2014. Phylogenomics resolves the timing and pattern of insect evolution. Science 346:763–767.

Mitter, C., B. D. Farrell, and B. M. Wiegmann. 1988. The phylogenetic study of adaptive zones: has phytophagy promoted insect diversification? American Naturalist 132:107–128.

Mockford, E. L., C. Lienhard, and K. Yoshizawa. 2013. Revised classification of Psocoptera from Cretaceous ambers: a reassessment of published information. Journal of the Faculty of Agriculture of Hokkaido University, Series Entomology, New Series 69:1–26.

Mustoe, G. E. 1985. Eocene amber from the Pacific Coast of North America. Geological Society of America Bulletin 96:1530–1536.

Nel, A., and N. Brasero. 2010. Oise amber. Pp. 137–148 in: D. Penney (ed), Biodiversity of Fossils in Amber from the Major World Deposits. Siri Scientific Press, Manchester

Nel, A., R. A. DePalma, and M. S. Engel. 2010. A possible hemiphlebiid damselfly in Late Cretaceous amber from South Dakota (Odonata: Zygoptera). Transactions of the Kansas Academy of Science, 113: 231–234.

Néraudeau, D., V. Perrichot, J. -C. Colin, V. Girard, B. Gomez, F. Guillocheau, E. Masure, D. Peyrot, F. Tostain, B. Videt, and R. Vullo. 2008. A new amber deposit from the Cretaceous (uppermost Albian–lowermost Cenomanian of southwestern France). Cretaceous Research 29:925–929.

Noetling, F. 1893. On the occurrence of burmite, a new fossil resin from Upper Burma. Records of the Geological Survey of India 26:31–40.

Paterson, A. M., G. P. Wallis, L. J. Wallis, and R. D. Gray. 2000. Seabird and louse coevolution: complex histories revealed by 12S rRNA sequences and reconciliation analyses. Systematic Biology 49:383–399.

Peñalver, E., and X. Delclòs. 2010. Spanish amber. Pp. 236–270 in D. Penney (ed), Biodiversity of Fossils in Amber from the Major World Deposits. Siri Scientific Press, Manchester.

Peñalver, E., E. Álvarez-Fernández, P. Arias, X. Delclòs, and R. Ontañón. 2007. Local amber in a Paleolithic context in Cantabrian Spain: the case of La Garma A. Journal of Archaeological Science 34:843–849.

Peñalver, E., A. Arillo, R. Perez-de la Fuente, M. Riccio, X. Delclòs, E. Barrón, and D. A. Grimaldi. 2015. Long-proboscid flies as pollinators of Mesozoic gymnosperms. Current Biology 25:1–7.

Penney, D., C. Wadsworth, G. Fox, S. L. Kennedy, R. F. Preziosi, and T. A. Brown. 2013. Absence of ancient DNA in sub-fossil insect inclusions preserved in 'Anthropocene' Colombian copal. PLoS ONE, 8(9):e73150. doi:10.1371/journal.pone.0073150

Perkovsky, E. E., V. Y. Zosimovich, and A. P. Vlaskin. 2010. Rovno amber. Pp. 116–136 in D. Penney (ed), Biodiversity of Fossil in Amber from the Major World Deposits. Siri Scientific Press, Manchester.

Perrichot, V., D. Néradeau, and P. Tafforeau. 2010. Charentese amber. Pp. 192–207 in Penny, D. (ed), Biodiversity of Fossils in Amber from the Major World Deposits. Siri Scientific Press, Manchester.

Perrichot, V., D. Néraudeau, A. Nel, and G. de Ploëg. 2007. A reassessment of the Cretaceous amber deposits from France and their palaeontological significance. African Invertebrates 48:213–227.

Pike, E. M. 1995. Amber Taphonomy and the Grassy Lake, Alberta, Amber Fauna. PhD dissertation. University of Calgary, Calgary, Alberta.

Pohl, H., B. Wipfler, D. Grimaldi, F. Beckman, and R. G. Beutel. 2010. Reconstructing the anatomy of the 42 million-year-old fossil *Mengea tertiaria* (Insecta, Strepsiptera). Naturwissenschaften 97: 855–859.

Poinar, G. O., Jr. 1992. Life in Amber. Stanford University Press, Stanford, California, 350 pp.

Poinar, G. O., Jr. 2008. [Book Review.] Amber of Jordan. Third Edition. By Hani Faig Kaddumi. The Eternal River Museum, Jordan. 2007. 298 pp. U.S. Proceedings of the Entomological Society of Washington 110:1251–1252.

Poinar, G. O., Jr., B. Archibald, and A. Brown. 1999. New amber deposit provides evidence of early Paleocene extinctions, paleoclimates, and past distributions. Canadian Entomologist 131:171–177.

Poinar, G. O., Jr., J. B. Lambert, and Y. Wu. 2007. Araucarian source of fossiliferous Burmese amber: spectroscopic and anatomical evidence. Journal of the Botanical Research Institute of Texas 1:449–455.

Prokop, J. and A. Nel. 2005. New scuttle flies from the early Paleogene amber of eastern Moravia of the Czech Republic (Diptera: Phoridae). Studia Dipterologica 12:13–22.

Rasnitsyn, A. P. 1996. Burmese amber at the Natural History Museum [London]. Inclusion 23:19–21.

Raven, P. H. and D. Axelrod. 1974. Angiosperm biogeography and past continental movements. Annals of the Missouri Botanical Garden 61:539–673.

Reed, D. L., V. S. Smith, S. L. Hammond, A. R. Rogers, and D. H. Clayton. 2004. Genetic analysis of lice supports direct contact between modern and archaic humans. PLoS Biology 2(11): e340. doi:10.1371/journal.pbio.0020340

Ren, D. 1998. Flower-associated Brachycera flies as fossil evidence for Jurassic angiosperm origins. Science 280:85–88.

Ren, D., C. C. Labandeira, J. A. Santiago-Blay, A. Rasnitsyn, C. Shih, A. Bashkuev, M. A. V. Logan, C. L. Hotton, and D. Dilcher. 2009. A probable pollination mode before angiosperms: Eurasian, long-proboscid scorpionflies. Science 326:840–847.

Ritzkowski, S. 1997. K-Ar-Altersbestimmungen der bernsteinführenden Sedimente des Samlandes (Paläogen), Bezirk Kaliningrad. Metalla 66:19–23.

Ross, A. J. 2010. Amber: The Natural Time Capsule. Second Edition. Natural History Museum, London, 112 pp.

Ross, A. J. 2017. Burmese (Myanmar) amber taxa, on-line checklist v.2017.1. 67pp. http://www.nms.ac.uk/collections-research/collections-departments/natural-sciences/palaeobiology/

Ross, A. and A. Sheridan. 2013. Amazing Amber. NMS Enterprises Limited – Publishing, Edinburgh, 64 pp.

Ross, A. J., C. Mellish, P. York, and B. Crighton. 2010. Burmese amber. Pp. 208–235 in D. Penney (ed), Biodiversity of Fossils in Amber from the Major World Deposits. Siri Scientific Press, Manchester.

Rust, J., H. Singh, R. S. Rana, T. McCann, L. Singh, K. Anderson, N. Sarkar, P. C. Nascimbene, F. Gerdes, J. C. Thomas, M. Solórzano-Kraemer, C. J. Williams, M. S. Engel, A. Sahni, and D. Grimaldi. 2010. Biogeographic and evolutionary implications of a diverse paleobiota in amber from the Early Eocene of India. Proceedings of the National Academy of Sciences of the United States of America 107:18360–18365.

Sanderson, M. W. and T. H. Farr. 1960. Amber with insect and plant inclusions from the Dominican Republic. Science 131: 1313.

Sanmartin, I. and F. Ronquist. 2004. Southern hemisphere biogeography inferred by even-based models: plant vs. animal patterns. Systematic Biology 53:216–243.

Saunders, B., R. H. Mapes, F. M. Carpenter, W. C. Elsik. 1974. Fossiliferous amber from the Eocene (Claiborne) of the Gulf Coastal Plain. Geological Society of America Bulletin 85:979–984.

Schlee, D. 1990. Das Bernstein-Kabinett. Stuttgarter Beiträge zur Naturkunde, C 28:1–100.

Schlee, D. and H.-G. Dietrich. 1970. Insektenführender Bernstein aus der Unterkreide des Libanon. Neues Jahrbuch für Geologie und Paläontologie Monatshefte 1970:40–50.

Schmidt, A., S. R. Jancke, E. Ragazzi, G. Roghi, E. E. Lindquist, P. C. Nascimbene, K. Schmidt, T. Wappler, and D. A. Grimaldi. 2012. Arthropods in amber from the Triassic Period. Proceedings of the National Academy of Sciences of the United States of America 109:14796–14801.

Schmidt, A. R., V. Perrichot, M. Svojtka, K. B. Anderson, K. H. Belete, R. Bussert, H. Dörfelt, S. Jancke, B. Mohr, E. Mohrmann, P. C. Nascimbene, A. Nel, P. Nel, E. Ragazzi, G. Roghi, E. E. Saupe, K. Schmidt, H. Schneider, P. A. Selden, and N. Vávra. 2010. Cretaceous African life captured in amber. Proceedings of the National Academy of Sciences of the United States of America 107:7329–7334.

Sendelio, N. 1742. Historia succinorum corpora aliena involventium et naturae opere pictorum et caelatorum ex Augustorum I et II cimeliis Dresdae conditis aeri insculptorum. Verlag Gleditsch, Leipzig.

Shi, G., D. A. Grimaldi, G. E. Harlow, J. Wang, J. Wang, M. Yang, W. Lei, Q. Li, and X. Li. 2012. Age constraint on Burmese amber based on U-Pb dating of zircons. Cretaceous Research 37:155–163.

Skalski, A. W., and A. Veggiani. 1990. Fossil resin in Sicily and the Northern Apennines: Geology and organic content. Prace Museum Ziemi 41:37–49.

Smith, R. D. A. and Ross, A. J. (in press) Amberground pholadid bivalve borings and inclusions in Burmese amber: Implications for proximity of resin-producing forests to brackish-marine waters, and the age of the amber. Earth and Environmental Science Transactions of the Royal Society of Edinburgh.

Soom, M. 1984. Bernstein vom Nordrand der Schweizer Alpen. Pp. 15–20 in Bernstein-Neuigkeiten. Stuttgarter Beiträge zur Naturkunde, C18.

Standke, G. 2008. Bitterfelder Bernstein gleich Baltischer Bernstein? – Eine geologische Raum-Zeit-Betrachtung und genetische Schlussfolgerungen. Pp. 11–33 in: R. Wimmer, G. Krumbiegel, S. Schmiedel and J. Rascher (eds), Bitterfelder Bernstein versus Baltischer Bernstein – Hypothesen, Fakten, Fragen.. Exkursionsführer und Veröffentlichungen der Deutschen Gesellschaft für Geowissenschaften 236.

Stankiewicz, B.A., H. N. Poinar, D. E. G. Briggs, R. P. Evershed, and G. O. Poinar, Jr. 1998. Chemical preservation of plants and insects in natural resins. Proceeding of the Royal Society of London B 265:641–647.

Sun, T. T., A. Kleismanta, T. T. Nyunt, Z. Minrui, M. Krishnaswamy, and L. H. Ying. 2015. Burmese amber from Hti Lin. 34th International Gemmological Conference, Vilnius:26–29.

Sunderlin, D., G. Loope, N. E. Parker, and C. J. Williams. 2011. Paleoclimatic and paleoecological implications of a Paleocene-Eocene fossil leaf assemblage, Chickaloon Formation, Alaska. Palaios 26:335–345.

Szwedo, J., B. Wang, and H. Zhang. 2013. The Eocene Fushun amber – known and unknown. The International Amber Researcher Symposium, Abstracts:33–36.

Tschudy, R. H. 1975. *Normapolles* pollen from the Mississippi Embayment. U. S. Geological Survey Professional Paper 865:1–42.

Vea, I. M., and D. A. Grimaldi. 2015a. Diverse new scale insects (Hemiptera: Coccoidea) in amber from the Cretaceous and Eocene, with a phylogenetic framework for fossil Coccoidea. American Museum Novitates 3823:1–80.

Vea, I. M., and D. A. Grimaldi. 2015b. Putting scales into evolutionary time: the divergence of major lineages of scale insects (Hemiptera) predates the radiation of modern angiosperm hosts. Scientific Reports 6:23487. doi:10.1038/srep23487.

Wade, B. 1926. The fauna of the Ripley Formation on Coon Creek, Tennessee. Geological Survey of Tennessee, Professional Paper 137:1–272.

Wang, B., J. Rust, M. S. Engel, J. Szwedo, S. Dutta, A. Nel, Y. Fan, M. Meng, G. Shi, E. A. Jarzembowski, T. Wappler, F. Stebner, Y. Fang, L. Mao, D. Zheng, H. Zhang. 2014. A diverse paleobiota in Early Eocene Fushun Amber from China. Current Biology 24:1606–1610.

Wappler, T., V. S. Smith, and R. C. Dalgleish. 2004. Scratching an ancient itch: an Eocene bird louse fossil. Proceedings of the Royal Society of London B 271:5255–5258.

Wiens, J. J. 2003. Incomplete taxa, incomplete characters and phylogenetic accuracy: is there a missing data problem? Journal of Vertebrate Paleontology 23:297–310.

Wier, A., M. Dolan, D. Grimaldi, R. Guerreo, J. Wagensberg, and L. Margulis. 2002. Spirochete and protist symbionts of a termite (*Mastotermes electrodominicus*) in Miocene amber. Proceedings of the National Academy of Sciences of the United States of America 99: 1410–1413.

Weitschat, W., and W. Wichard. 2010. Baltic amber. Pp. 80–115 in D. Penney (ed), Biodiversity of Fossils in Amber from the Major World Deposits. Siri Scientific Press, Manchester.

Wood, H. M., C. E. Griswold, and R. G. Gillespie. 2012. Phylogenetic placement of pelican spiders (Archaeidae, Araneae), with insight into evolution of the "neck" and predatory behaviours of the superfamily Palpimanoidea. Cladistics 28:598–626.

Zhang. J. F., S. Zhang, and L. Y. Li. 1993. Mesozoic gadflies (Insecta: Diptera). Acta Palaeontologica Sinica 26:595–603.

Zherikhin, V. V. and A. J. Ross. 2000. A review of the history, geology and age of Burmese (burmite). Bulletin of the Natural History Museum (London), Geology Series 56:3–10

Zherikhin, V. V. and I. D. Sukacheva. 1973. On the Cretaceous insectiferous "ambers" (retinites) in North Siberia. Pp. 3–48 in E. P. Narchuk (ed), Problems of Insect Paleontology. Lectures on the XXIV Annual Readings in memory of N. A. Kholodkovsky, 1–2 April 1971. Nauka, St. Petersburg.

Figures

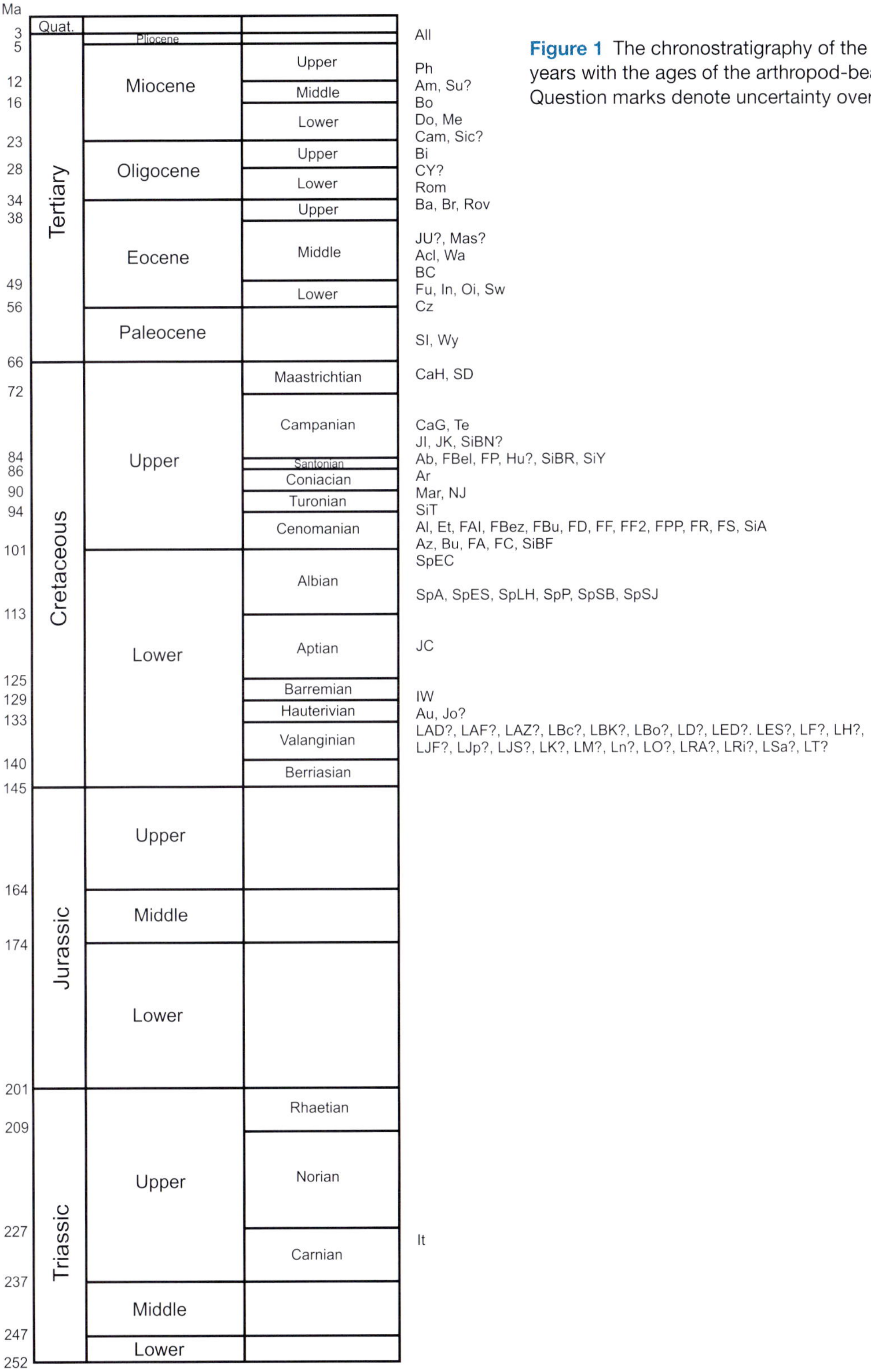

Ma	Era	Period	Epoch	Stage	Ambers
3, 5		Quat.	Pliocene		All
12, 16	Tertiary		Miocene	Upper	Ph
				Middle	Am, Su?
					Bo
23				Lower	Do, Me
					Cam, Sic?
28			Oligocene	Upper	Bi
					CY?
34				Lower	Rom
38			Eocene	Upper	Ba, Br, Rov
					JU?, Mas?
				Middle	Acl, Wa
					BC
49				Lower	Fu, In, Oi, Sw
56					Cz
66			Paleocene		SI, Wy
72	Cretaceous	Upper		Maastrichtian	CaH, SD
				Campanian	CaG, Te
84					JI, JK, SiBN?
86				Santonian	Ab, FBel, FP, Hu?, SiBR, SiY
90				Coniacian	Ar
94				Turonian	Mar, NJ
					SiT
101				Cenomanian	Al, Et, FAI, FBez, FBu, FD, FF, FF2, FPP, FR, FS, SiA
					Az, Bu, FA, FC, SiBF
					SpEC
113		Lower		Albian	SpA, SpES, SpLH, SpP, SpSB, SpSJ
				Aptian	JC
125				Barremian	IW
129				Hauterivian	Au, Jo?
133				Valanginian	LAD?, LAF?, LAZ?, LBc?, LBK?, LBo?, LD?, LED?. LES?, LF?, LH?, LJF?, LJp?, LJS?, LK?, LM?, Ln?, LO?, LRA?, LRi?, LSa?, LT?
140				Berriasian	
145	Jurassic	Upper			
164		Middle			
174		Lower			
201	Triassic	Upper		Rhaetian	
209				Norian	
227				Carnian	It
237		Middle			
247		Lower			
252					

Figure 1 The chronostratigraphy of the past 250 million years with the ages of the arthropod-bearing ambers. Question marks denote uncertainty over the exact age.

Figure 2 Geological map of northern Myanmar, showing the location of the mid-Cretaceous Burmese amber outcrops.

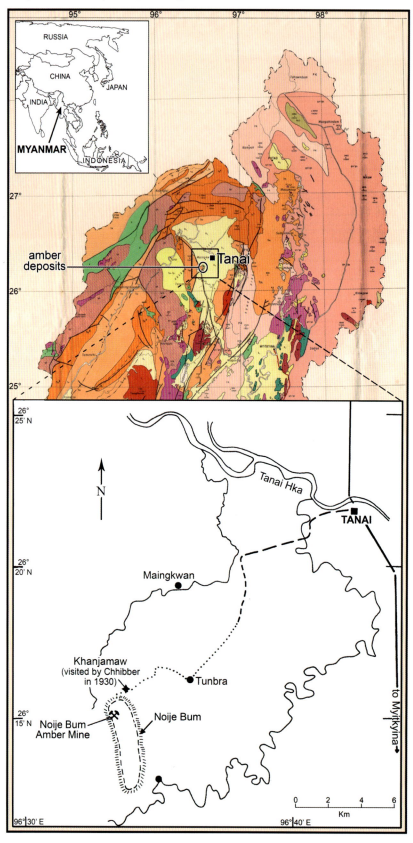

Figure 3 **A** Excavations of Burmese amber at one of many open-pit mines in the vicinity of Noijebum, Kachin, northern Myanmar. Photo by Federico Berloecher, c.2009. **B** The largest piece of Cretaceous amber, from the same vicinity. The piece was excavated in 2014 and sold to a buyer in China. **C** Recently excavated rough Burmese amber being sorted for polishing and screening of inclusions. **D** Rounded piece of Burmese amber with conical pholadid bivalve crypts (NMS G.2015.21.1). **E** Faceted and carved pieces; the central piece (32mm height) is carved into a bunch of fruits, each fruit framing an insect inclusion (pieces in AMNH).

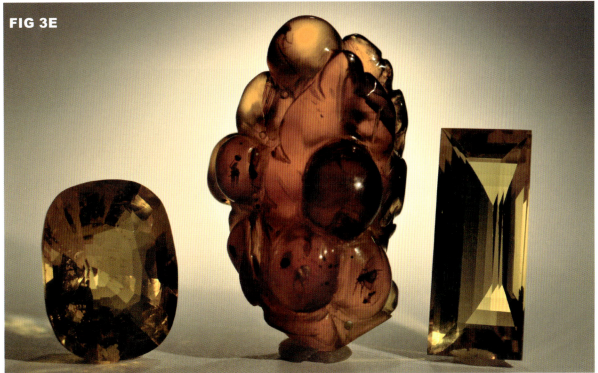

Figure 4 Fungi, myxomycetes and assorted plants in Burmese amber. **A** Fruiting body of the only Cretaceous slime mould (Myxomycophyta) (JZC Bu266). **B** Liverwort, holotype of *Frullania cretacea* (AMNH Bu011). **C** Moss (Bryophyta), holotype of *Calymperites burmensis* (AMNH Bu-ASJ2). **D** Portion of fern frond (JZC Bu1463) (depth of piece is 20mm). **E** Leafy shoot of *Metasequoia* (JZC Bu1831) (length 24mm). **F** Unusual angiosperm inflorescence covered in trichomes (JZC Bu418) (length 15mm). **G** Trimerous flower (JZC Bu1449). **H** Angiosperm leaf with long drip tip (JZC BuF3c) (length 48mm).

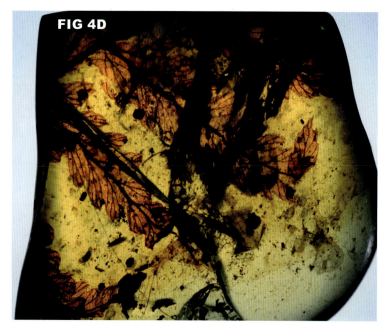

Figure 4

FIG 4E

FIG 4G

FIG 4F

FIG 4H

Figure 5 Various invertebrates in Burmese amber. **A** Two mermithid nematodes emerging from the body of their midge host (Diptera: Chirononomidae) (AMNH Bu320). **B** Small, ribbed snail (Gastropoda) (JZC Bu1668). **C** Snail (AMNH Bu099). **D** Holotype of the woodlouse (Isopoda) *Myanmariscus deboiseae* (NMS G.2010.20.42).

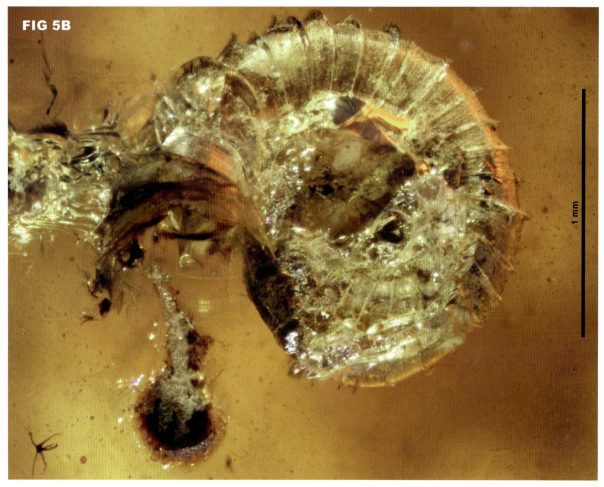

Figure 5

FIG 5C

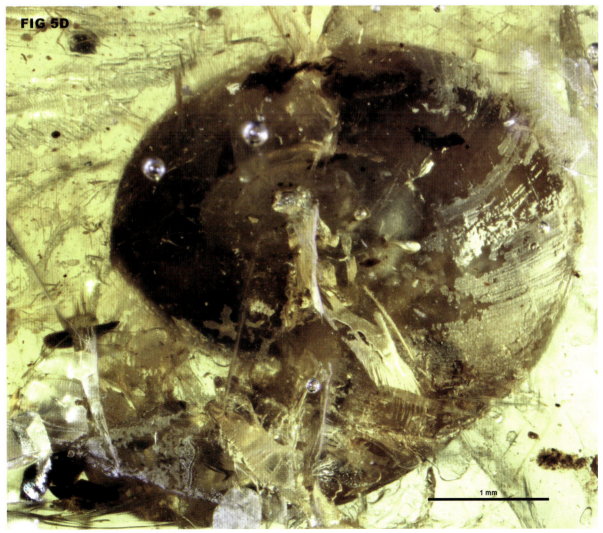

FIG 5D

1 mm

Figure 6 Arachnida (**A–G**), Myriapoda (**H**), basal (wingless) Hexapoda (**I, J**). **A** Holotype of the spider *Burmesarchaea grimaldii*, family Archaeidae (AMNH Bu256). **B** Holotype of the spider *Burlagonegops eskovi* (AMNH Bu1353). **C** Scorpion of the family Buthidae (AMNH Bu337). **D** Pseudoscorpion, family Chernetidae (AMNH Bu230). **E** Schizomidan (JZC Bu63). **F** Photo and rendering of the holotype of the tailless whip spider (Amblypygida), *Kronocharon prendinii* (JZC Bu150). The specimen includes an adult female with three of her nymphs. **G** Tick (Ixodida), JZC Bu061. **H** Millipede (Diplopoda), AMNH. **I** Diplura (family Campodeidae), with a springtail (Collembola: Sminthuridae) between its antennae (JZC). **J** Silverfish (Zygentoma), AMNH.

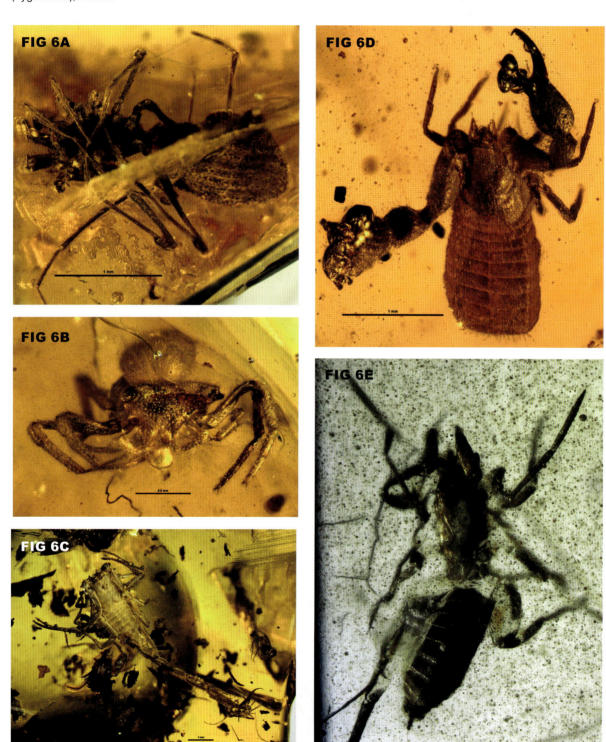

Figure 6

FIG 6F

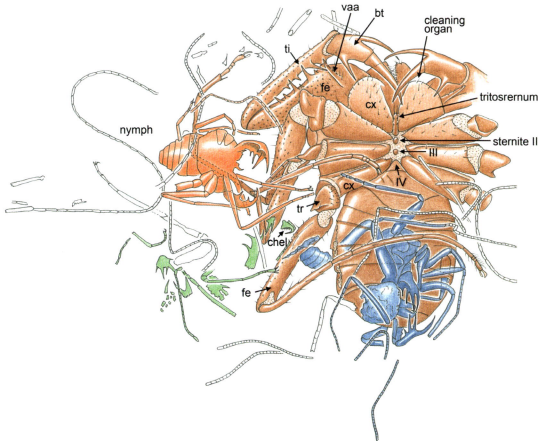

Figure 6

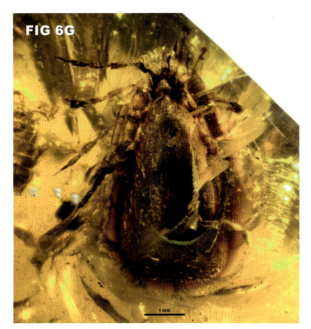

FIG 6G

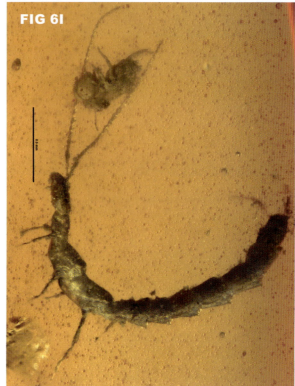

FIG 6I

FIG 6H

FIG 6J

Figure 7 **A** Small male mayfly (Ephemeroptera: Baetidae) (AMNH Bu203). The long filaments from the tip of the abdomen are the cerci. **B** A virtually complete, small damselfly (Odonata: Zygoptera) (JZC Bu314). **C** Anterior half of a very unusual damselfly with flag-shaped antennae and forelegs (JZC).

Figure 8 Assorted polyneopterous insects in Burmese amber. **A** Stonefly (order Plecoptera) (JZC Bu1442). **B** Pygmy mole cricket, family Tridactylidae (order Orthoptera) (NMS G.2010.20.7). **C** Male webspinner, order Embioidea (JZC Bu312). **D** Holotype of Zorotypus nascimbenei (Zorotypidae: order Zoraptera) (AMNH Bu341). **E** Wingless zorapteran (NMS G.2010.20.9) **F** Earwig (order Dermaptera), with its short forewing outstretched and the underlying, fan-like hind wing unfolded (JZC). **G** Worker termite (JZC Bu235). **H** Soldier termite (JZC Bu183). Burmese amber contains the first definitive non-winged termite casts from the Cretaceous, indicating advanced sociality.

Figure 8

FIG 8E

FIG 8G

FIG 8F

FIG 8H

Figure 9 Hemipteran insects in Burmese amber. **A–C**: Sternorrhyncha. **D–H**: Heteroptera. **A** Holotype of *Postopsyllidium rebeccae* (AMNH Bu137), which is the latest occurrence of the extinct family Protopsyllidiidae. **B** Holotype of the scale insect *Burmorthezia insolita* (AMNH Bu1095), family Ortheziidae. The light lobes attached to the body are bundles of secreted waxy filaments. **C** Winged aphid, AMNH Bu561. **D** Predatory water bug, family Naucoridae (JZC Bu274). **E** Predatory shore bug, family Ochteridae (JZC Bu303). **F** Predatory shore bug, family Gelastocoridae (JZC Bu1647). **G** Holotype of the water measurer bug, *Carinametria burmensis*, family Hydrometridae (AMNH Bu1098). **H** Lace bug, family Tingidae (JZC Bu005).

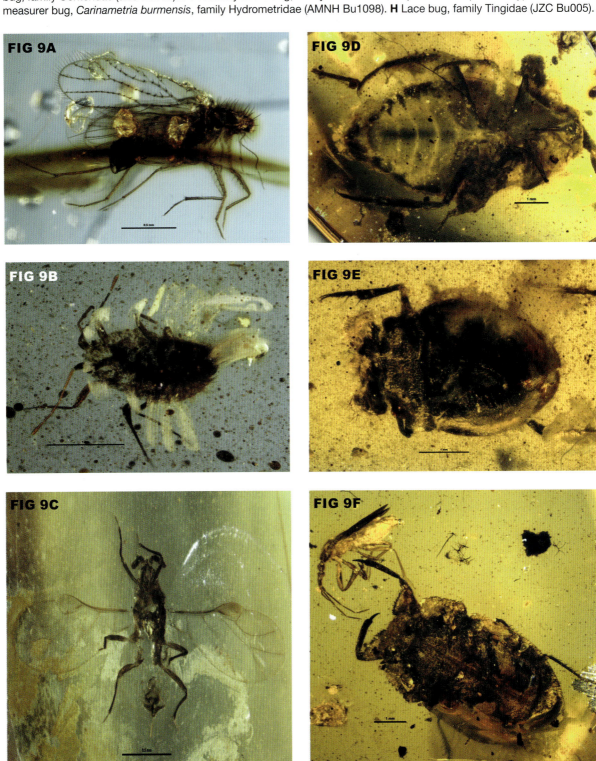

Figure 9

Figure 10 Assorted wasps (order Hymenoptera) in Burmese amber. **A** Holotype of *Zoropelecinus zigrasi*, family Pelecinidae (JZC Bu229). The long, jointed abdomen was probably used to probe soil for its insect hosts. **B** Specimen of the family Megalyridae (JZC). Note that the original metallic-green iridescence of the body and red pigment of the eyes are preserved. **C** A wasp of the family Vespidae, which includes modern paperwasps and mud-daubers (JZC Bu1839). **D** Primitive worker ant (family Formicidae), *Gerontoformica gracilis* (JZC Bu324). Of the 12 species in the genus *Gerontoformica*, 11 are known only in Burmese amber. **E** Two species of *Gerontoformica* workers fighting (JC Bu1646). Highly aggressive territoriality towards other species occurs in eusocial insects. **F** Renderings of two species of the bizarre ant genus *Haidomyrmex*, which has sickle-like trap jaws (from Barden and Grimaldi, 2012).

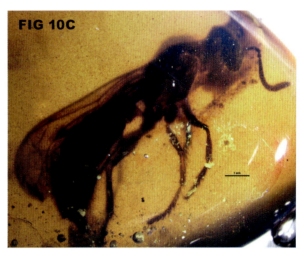

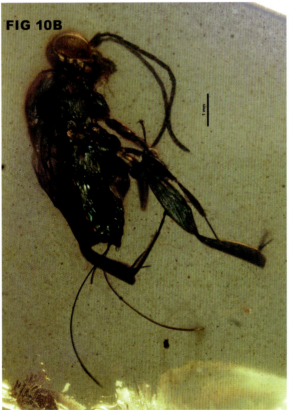

Figure 10

FIG 10E

FIG 10F

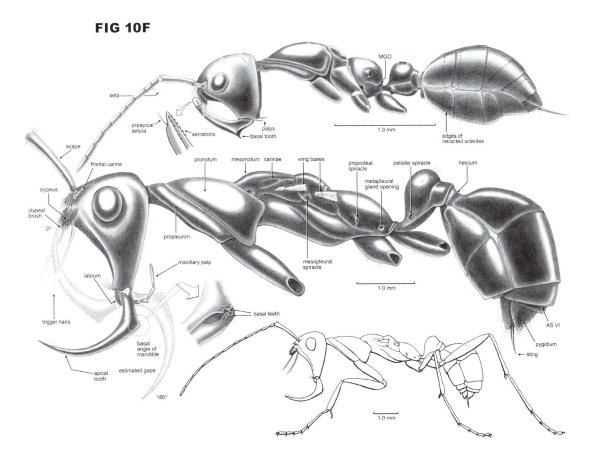

Figure 11 Assorted adult and larval specimens of predatory Neuroptera (lacewings, owlflies, and allies). **A** Snakefly (order Raphidioptera) (JZC Bu244). **B** Berothidae (JZC). **C** Psychopsidae adult, which resembles a small moth (JZC Bu260). **D** Psychopsidae larva (AMNH Bu197). **E** Larva of probably the small relict family Nevrorthidae (JZC). **F** Larva of an owlfly (family Ascalaphidae), which has camouflaged itself with sand grains placed onto its back (JZC Bu031).

Figure 12 Representative beetles (Coleoptera). **A** Rove beetle, family Staphylinidae (AMNH Bu255). **B** Holotype of the prostomid beetle, *Vetuprostomis consimilis* (AMNH Bu1422). **C** Family Ptilodactylidae, distinctive for its highly branched antennae (AMNH Bu016). **D** Flower beetle, family Anthicidae (AMNH Bu1367). **E** Tumbling flower beetle, family Mordellidae (AMNH Bu367). **F** Specimen in the diverse cucujoid-family group (AMNH Bu1370). **G** Exuvium (shed skin) of a scarab larva; head capsule is to the lower right, abdomen is at left. **H** Renderings of the holotype of an early bark weevil (Curculionidae: Scolytinae), *Microborus inertus* (AMNH Bu1607) (from Cognato and Grimaldi, 2009).

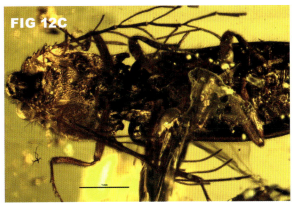

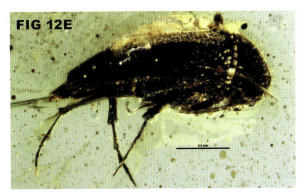

FIG 12H

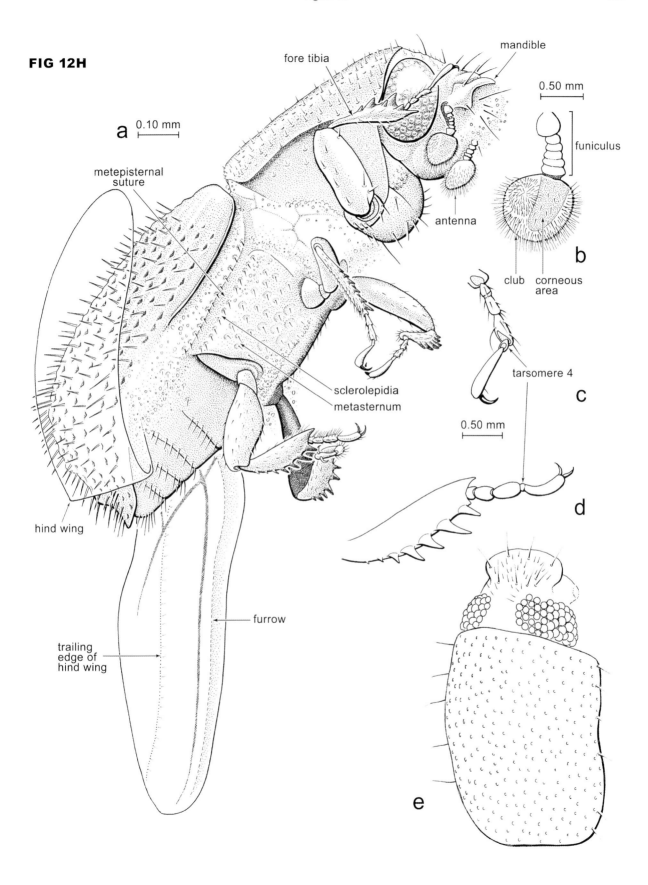

Figure 13 Exemplar Holometabola in Burmese amber. **A** The bizarre, two-winged, 'long-tongued' scorpionfly (order Mecoptera), *Parapolycentropus*, which superficially resembles a mosquito (JZC Bu082). **B** Drawing of the holotype of *Burmomerope eureka* (JZC Bu084), which belongs to the highly relict scorpionfly family Meropeidae. **C** Small moth (order Lepidoptera) of the basal family Micropterigidae (AMNH Bu701). **D** Small moth (AMNH Bu203). **E** Large moth (AMNH Bu-SD1). **F** Specimen of a twisted-wing parasite (order Strepsiptera), *Cretostylops engeli* (AMNH Bu-FB31). The relationships of Strepsiptera are arguably the most enigmatic of all insect orders. **G** Caddisfly, *Palerasnitsynus* sp. (NMS G.2010.20.35).

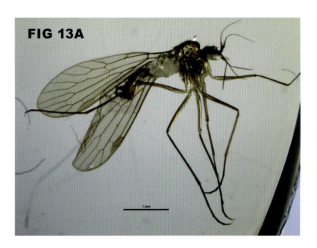

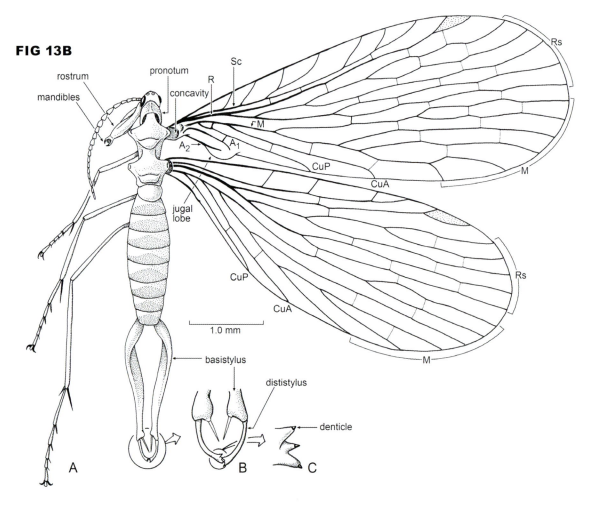

Figure 13

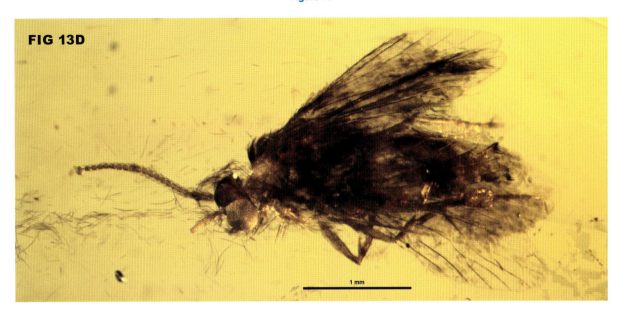

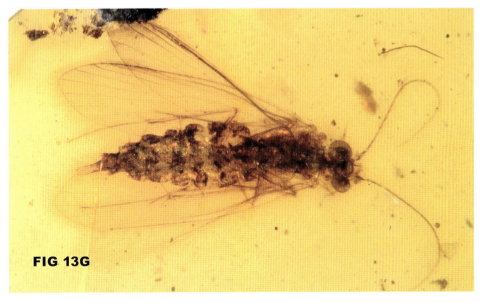

Figure 14 Assorted flies (order Diptera) in Burmese amber. **A** Midge of the family Blephariceridae (AMNH). **B** Holotype of the sciaroid midge *Burmazelmira* asiatica (AMNH Bu178). **C** Oldest mosquito, *Burmaculex antiquus* (JZC Bu213). **D** Mothfly, family Psychodidae (AMNH BuSE2). **E** The genus *Alavesia* (AMNH BuSE11)(Empidoidea), known also from the Cretaceous of Spain and Botswana. One living species exists, in South Africa. **F** Whole piece with, and rendered reconstruction of, the zhangsolvid fly *Linguatormyia teletacta*. The fly used its long proboscis for pollination; function of the bizarre, whip-like antennae is unknown. **G** Early robber fly (family Asilidae), holotype of *Burmapogon bruckschi* (AMNH BuKB1). **H** Primitive cyclorrhaphan fly, family Platypezidae (AMNH BuSE8).

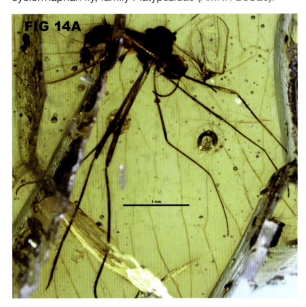

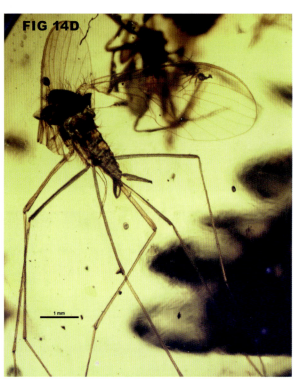

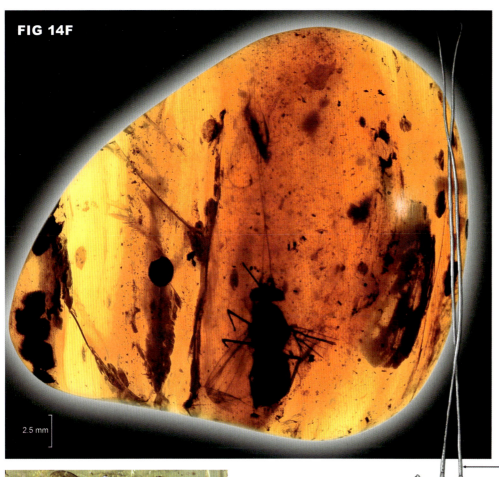
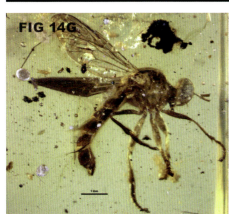

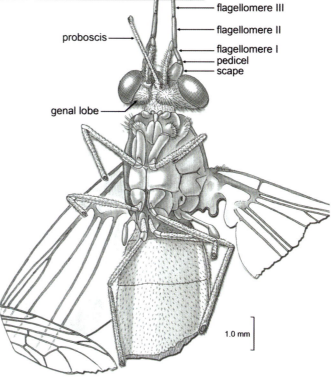

Figure 15 Assorted small vertebrates and their remains in Burmese amber. **A** Small lizard adjacent to a millipede (JZC Bu154). **B** Small lizard (JZC Bu267). Burmese amber contains the greatest diversity of lizards among all deposits of amber. **C** Bird feather (JZC Bu105). **D** Detail of another bird feather (JZC BuF8). The dark spots in each barbule are clumps of small, circular melanosomes. This feather was black. **E** Small frog, in a private collection in China.

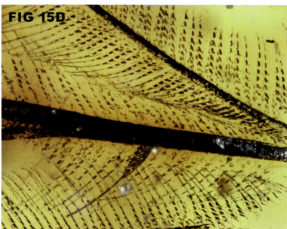

Fig. 16 Avialean remains in Burmese amber (JZC). **A**, **B**: Views of the specimen imaged with light photomicrography (**A**) and CT X-Ray scanning (**B**), to the same scale. Note how the feather shafts are nearly parallel, originating close to an appendage that is apparently the wing limb. **C**, **D**: Magnified views of feather barbs. **E**, **F**: Enlarged views of the apparent wing limb, showing bony portions; webby portions are probably decaying soft tissue. **G**, **H**: Bony remains of two toes, shown in full lateral (**G**) and oblique views. **I**: Remains of claw tips and segment of a toe. CT imaging by Henry Towbin, AMNH MIF.

Appendix 1 – Amber localities that have yielded arthropod inclusions

The following is a list of all the amber localities that have yielded arthropod inclusions, with a brief description of where the amber occurs, its age and key references.

Africa

Ethiopian amber (Et). This amber occurs near Alem Ketema in the northwestern plateau of Ethiopia. It was found within the Debre Libanos Sandstone and was considered to be Late Cenomanian in age (Schmidt *et al.*, 2010) although its age is being reviewed.

Asia

Armenian amber (Ar). This amber occurs at Shavarshavan, Armenia and is considered to be Coniacian in age (Eskov, 2002).

Azerbaijanian amber (Az). This amber occurs at Yukhary Agdzhakend, Azerbaijan and is considered to be Early Cenomanian in age (Eskov, 2002).

Bornean amber (Bo). Amber occurs in the Sarawak and Sabah regions of Malaysia and Kalimantan region of Indonesia. As yet only the first two have yielded arthropod inclusions. Most of the Sarawak amber came from the disused Merit-Pila coal mine, considered to be Lower–Middle Miocene in age (Hillmer, Weitschat and Vávra, 1992; Hillmer, Weitschat and Voight, 1992), whereas the Sabah amber is Middle Miocene (John Noad pers. comm.; Ross, 2010).

Burmese amber (Bu). Burmese amber occurs in the Hukawng Valley in northern Myanmar and has a very diverse fauna. It was originally considered to be Tertiary in age and then Albian. Recent dating of volcanic zircons in the amber deposit yielded an early Cenomanian age (Shi *et al.*, 2012); however, this is a minimum age as it may have taken a while for the amber to be deposited in the marine environment. The presence of pholadid bivalve borings indicates that some of the pieces were hard before being buried, so an Albian age for this amber is possible. Some pieces were later reworked into younger Tertiary deposits (Ross *et al.*, 2010).

Chinese amber. Although amber is known from a number of localities in China, only amber from Fushun (Fu) in Liaoning Province has yielded arthropod inclusions. Fushun amber is found in coal beds from the Guchengzi Formation and is Lower Eocene (Ypresian) in age (Szwedo *et al.*, 2013; Wang *et al.*, 2014).

Indian amber (In). This amber occurs at Gujarat in western India. It occurs in lignitic muddy sediments known as the Cambay Shale and is Lower Eocene (Ypresian) in age (Rust *et al.*, 2010).

Japanese amber. Amber is found in many localities in Japan (see Schlee, 1990), however, most have not yielded insects. Four localities on Honshu Island have yielded insects, but the insects have hardly been studied. Amber from Kuji, Iwate Prefecture (JK) has yielded the most insects and is considered to be Campanian in age. Amber of a similar age is also known from Iwaki, Fukushima Prefecture (JI). Amber of Aptian age is known from Choushi, Chiba Prefecture (JC) (Shirota pers. comm.), and amber of Paleogene age is known from Ube, Yamaguchi Prefecture (JU) (Schlee, 1990).

Jordanian amber (Jo). This amber occurs in the Zarqa River Basin, north of Amman, Jordan. It is found in the Kurnub Sandstone Formation, which is considered to be Valanginian to Barremian in age (Kaddumi, 2005; Poinar, 2008).

Lebanese amber. There are many amber localities in Lebanon. Some of them have yielded arthropod inclusions, and they are among the earliest known arthropods in amber (Azar *et al.*, 2010). Lebanon is divided into regions known as Governorates. North Lebanon has five arthropod-bearing amber localities: Bcharreh (LBc), Beqaa Kafra (LBK), El-Dabsheh (LED), Nabeh El-Sukkar Sadd Brissa (LN) and Tannourine (LT). Mount Lebanon has eight localities: Ain Dara 1 (LAD), Ain Zhalta (LAZ), Daychouniyyeh (LD), Falougha (LF), Mdeyrij-Hammana (LH), Kfar Selouan (LK), Ouata El-Jaouz (LO) and Sarhmoul (LSa). South Lebanon has six localities: Jezzine Fall (LJF), Jouar Es-Souss (=Jezzine of traditional usage, LJS), Jezzine Fall prime (LJp), Maknouniyeh (LM), Roum-Azour-Homsiyeh (LRA) and Rihan (Azar & Nel, 2013, LRi). Beqaa has three localities: Aita El-Foukhar (LAF),

Bouarij (LBo) and Esh-Sheaybeh (LES). The exact ages of the arthropod-bearing amber localities are not known but they are considered to be Berriasian to Barremian in age, and do not range into the Aptian as previously reported (Azar pers. comm.).

Philippino amber (Ph). A single insect has been recorded in this amber from the west bank of Sapa Tubigbinukot, Bondoc Peninsula, Tayabas Province on Luzon Island in the Philippines. It occurs in the Canguinsa Formation and is considered to be Upper Miocene in age (Durham, 1956).

Sakhalinian amber (Sl). This amber occurs at the Naiba River and is washed up on the Okhotsk Sea beach, near Starodubskoye, Dolinsk District, on the east side of Sakhalin Island. It occurs in coals of the Lower Due Formation and is considered to be Palaeocene in age (Eskov, 2002).

Siberian amber. Several fossiliferous amber localities occur on the Taimyr Peninsula in northern Siberia. The Agapa (SiA) locality is situated at Nizhnaya Agapa, on the north shore of the Agapa River in the Yst'-Enisey depression, western Taimyr. The amber comes from the Dolgan Formation and is considered to be late Cenomanian in age (Eskov, 2002). Amber has been recovered from the Bechigev Formation (SiBF) at four localities in the Khatanga River basin, eastern Taimyr, and considered to be Albian to early Cenomanian in age (Eskov, 2002). Amber occurs in cross-bedded sands at Baikura-Neru Bay (SiBN), Taimyr Lake in central Taimyr. The age of this amber is uncertain, but likely to be Upper Cretaceous (Eskov, 2002). The most abundant amber deposit occurs at Yantardakh (SiY), at the Maimecha River, eastern Taimyr. It occurs in sand of the Kheta Formation and is considered to be Santonian in age (Eskov, 2002). Another locality in the Kheta Formation is at the middle course of the Bulun River (SiBR), 18km south of post Novaya (Blagoderov and Grimaldi, 2004). Apart from the Taimyr Peninsula, amber is also found at Timmerdyakh-Khaya (SiT) at the Vilyui River, central Yakutia. It occurs in the Timmerdyakh Formation and is considered to be Cenomanian to Turonian in age (Eskov, 2002).

Sumatran amber (Su). A single insect and possible fragments have been recorded in this amber from the Goenoeng Toea area of Central Sumatra, Indonesia. The amber occurs in the Telisa, Lower Palembang and Middle Palembang formations, which are considered to be Miocene in age (Durham, 1956). This amber is now easily available, however, no further arthropod inclusions have been reported.

Australia

Allendale amber (All). A large piece of fossiliferous resin was found at Allendale, Victoria, Australia and probably came from the 'deep lead' at about 300ft in the Madam Berry West Company Shaft, Smeaton. It is considered to be Pliocene to Quaternary (Pleistocene) in age (Greenwood *et al.*, 2000).

Cape York amber (CY). This amber occurs on a beach, between Cape Grenville and Cape Weymouth, on the eastern side of the Cape York Peninsula, north Queensland. Although most is found loose, it has been found *in situ* in lignites. Its age is uncertain, though it probably formed sometime in the Tertiary (Hand *et al.*, 2010).

Europe

Austrian amber (Au). This amber occurs at Golling, Austria and is considered to be Hauterivian in age (Borkent, 1997).

Baltic amber (Ba). Baltic amber is the best known of all the inclusion-bearing ambers. It is washed up on the shores around the Baltic Sea; however, the primary mined source is the Blue Earth that outcrops on the Samland Peninsula, forming part of Kaliningrad Oblast, an enclave of Russia sandwiched between Poland and Lithuania. The age of Baltic amber is in dispute. The majority of Baltic amber inclusions come from the Blue Earth, however, Baltic amber also occurs in an older horizon called the Lower Blue Earth and a younger horizon called the Lower Gestreifer Sand. Biostratigraphical dating using microfossils puts them at Upper Eocene (Priabonian), Middle Eocene (Lutetian) and Oligocene respectively (see Standke, 2008). Ritzkowski (1997) dated the Blue Earth as Lutetian based on Glauconite dating, which has been widely cited as the age of Baltic amber. However, this method of dating can be unreliable as Glauconite can be re-worked and contaminated with mica/illite which gives an older result (see Clauer *et al.*, 2005). Although, as the amber is deposited in a marine setting, it is uncertain as to the length of time between its formation and burial, it has been suggested that the amber has been re-worked (Weitschat and Wichard, 2010). For younger deposits this is probably true, however, the amber from the Blue Earth is generally fresh (lemon yellow) and unoxidized (except for a thin crust), and has not been eroded into pebble shapes, so is unlikely to have been reworked. Therefore an Upper Eocene (Priabonian) age is most likely.

Bitterfeld (Saxonian) amber (Bi). This amber occurs at Goitzsche, Germany and was excavated from a brown coal mine. Its age is under dispute, and it is also uncertain whether it is contemporary with or identical to Baltic amber, as some species occur in both ambers. Most authors accept an Upper Oligocene (Chattian) age (Dunlop, 2010), therefore younger than Baltic amber and implying that some species lived unchanged for at least 10 million years, which is not unreasonable, given that there are some extant species in Dominican and Mexican amber.

British amber (Br). 'British amber' is usually used when referring to Baltic amber that has been washed up on the east coast of Great Britain, although other deposits also occur in the UK (Jarzembowski, 1999; Ross and Sheridan, 2013). Also see Isle of Wight amber.

Campaolo amber (Cam). This amber comes from Campaolo, east of Paderno, in the Northern Apennines, Italy. It occurs in sandstones of the Formazione di Campaolo (Skalski and Veggiani, 1990), which is Upper Oligocene (Chattian) to Lower Miocene (Aquitanian) in age.

Czech amber (Cz). This amber occurs at Študlov, eastern Moravia, Czech Republic. The amber comes from marine sediments of the Belovž Formation and is considered to be Palaeocene to Lower Eocene (Prokop and Nel, 2005).

French amber. Inclusion-bearing amber of mid-Cretaceous age is known from many localities in France (Perrichot et al., 2007). From north to south, the Pays De La Loire Region has two localities: Bezonnais (FBez) and Durtal (FD), both of Cenomanian age. A cluster of localities occur in the Poitou-Charente Region: Archingeay (FA), Aix Island (FAI), La Buzinie (FBu), Cadeuil (FC), Fouras (FF), Puy-Puy (FPP) and Les Renardières (FR), of which Archingeay has yielded the highest number of inclusions by far. These localities are Cenomanian except for Archingeay and Cadeuil, which are Albian to Cenomanian in age (Néraudeau et al., 2008; Perrichot et al., 2010). The Provence–Alpes–Côte-d'Azur Region has three localities: Belcodene (FBel), Piolenc (FP) and Salignac (FS), of which the first two are Santonian and the last one is Cenomanian in age. The Languedoc–Roussillon Region has one locality: Fourtou (FF2), of Cenomanian age. Also see Oise amber.

Hungarian amber (Hu). This amber comes from Ajka, Hungary. It occurs in a coal seam of Coniacian to Campanian age (Borkent, 1997).

Isle of Wight amber (IW). This amber occurs at Chilton Chine on the southwest coast of the Isle of Wight, southern England. It occurs in mudstones with lignite and is Barremian in age (Jarzembowski et al., 2008).

Italian amber (It). Although there are several localities that yield amber in Italy, the oldest known arthropod-bearing amber in the world comes from near Cortina in the Dolomite Alps. It occurs in the Heiligkruez Formation and is late Carnian in age (Schmidt et al., 2012). Also see Campaolo and Sicilian amber.

Oise amber (Oi). This amber occurs near Creil, at Le Quesnoy in the Oise Department of the Picardie Region of northern France. It was found in a brown sand known as the Argiles à lignite du Soissonais and is Lower Eocene (Ypresian) in age (Nel and Brasero, 2010).

Romanian amber (Rom). The main deposits of this amber are in the Lower Kliwa Sandstone in the Buzau District of Romania. It is considered to be Lower Oligocene (Rupelian) in age (Ghiurca, 1990).

Rovno amber (Rov). This amber occurs in the Rovno region, northern Ukraine. Some has been reworked into Quaternary deposits in southern Belarus. Most of it has been collected from the Mezhygorje Formation (Repelian) at Klesov, where it has been commercially mined since 1993. It has also been reported from the older Obukhov Formation (Upper Eocene). Some of the species are the same as those found in Baltic amber (Perkovsky et al., 2010).

Sicilian amber (Sic). This amber has been found at several localities around Sicily, often transported by rivers to the sea where it was washed up on beaches. The primary source is most likely to be in central and north Sicily, though possible ages vary from Oligocene to late Upper Miocene/Lower Pliocene (Kohring and Schlüter, 1989; Skalski and Veggiani, 1990).

Spanish amber. Amber is known from many localities in Spain, but only eight have yielded arthropod inclusions. Some of these were only discovered recently, so it is likely that many more arthropod records will be discovered in due course. There are three arthropod-bearing localities in the Maestrat Basin in eastern Spain: San Just (SpSJ) and Arroyo de la Pascueta (SpA) in Teruel Province, and La Hoya (SpLH) in Castellón Province. There are four localities in the Basque–Cantabrian Basin in northern Spain: Peñacerrada I & II (SpP), and Salinillas de Buradón (SpSB) in Álava Province, and El Soplao (SpES) in Cantabria. For all these localities the amber occurs in the Escucha Formation and is early Albian in age. There is one locality in the Central Asturian Depression in northern Spain: El Caleyu (SpEC) in Asturias. Here the amber occurs in the Ullaga Formation and is late Albian in age (Peñalver and Delclòs, 2010).

Swiss amber (Sw). This amber has been found at three localities in the southern geosyncline – Plaffeien, Plasselbschlund and Gurnigel. It is Lower Eocene in age. (Soom, 1984).

North America

Alabama amber (Ab). This amber comes from eastern Alabama, USA and occurs within the Eutaw Foramtion. It is considered to be Santonian in age (Knight *et al.*, 2010).

Alaskan amber (Al). This amber occurs between the Brooks Range and northern coastline of Alaska, USA, comprising the Arctic Coastal Plain and Arctic Foothills (Poinar, 1992). Koteja and Poinar (2001) considered it to be late-middle Albian in age; however, more recent isotopic dating has resulted in a Cenomanian age (Heinrichs *et al.*, 2011).

Arkansas amber (ACl). This amber comes from near Malvern, Arkansas, USA and occurs within lignitic clays of the Claiborne Group. This amber is considered to be Middle Eocene in age (Saunders *et al.*, 1974).

British Columbia amber (BC). This amber occurs in the Hat Creek coalfield, British Columbia, Canada, and is considered to be Lower to Middle Eocene (Poinar *et al.*, 1999).

Canadian amber. The primary deposit of Canadian amber is at Grassy Lake (CaG), southern Alberta, where it occurs in the upper sub-bituminous coal and overlying shales, within the Taber Coal Zone. This amber is Campanian in age. Amber derived from this locality has also been found at Cedar Lake in western Manitoba. Insects have also been reported of early Maastrictian age from the Horseshoe Canyon Formation (CaH) (McKellar and Wolfe, 2010). Also see British Columbia amber.

Dominican amber (Do). This amber is the most productive inclusion-bearing amber in the New World. It is extracted by many small underground mines from the Cordillera Septentrional (La Toca Formation) and Cordillera Oriental (Yanigua Formation) of the Dominican Republic, on the island of Hispaniola in the Caribbean (Penney, 2010). It is considered to be Lower Miocene (Burdigalian) in age.

Maryland amber (Mar). Insects were reported from a large piece of amber found near the Magothy River at Cape Sable, Maryland, USA. Its age is presumably similar to that of New Jersey amber (Turonian) (Grimaldi and Nascimbene, 2010).

Massachusetts amber (Mas). Insects and a leaf were recorded in a piece of amber from Nantucket Island, Massachusetts, USA by Goldsmith (1879), though nothing else has been recorded since. There are no pre-Quaternary deposits exposed on this island, so the amber must be re-worked from elsewhere. The age of the amber is uncertain, though believed to be Tertiary; however, given that both Tertiary and Cretaceous sediments are exposed on nearby Martha's Vineyard (see Folger *et al.*, 1978), a Cretaceous age cannot be ruled out. Interestingly, it contains succinic acid, like Baltic amber.

Mexican amber (Me). This amber occurs in several localities in Chiapas State, Mexico, however the majority has come from Simojovel de Allende, where it occurs in the La Quinta Formation, Mazantic Shale and Balumtum Sandstone (Kraemer, 2010). It is considered to be Lower Miocene (predominantly Burdigalian) in age.

New Jersey amber (NJ). This amber occurs in the Raritan Formation and, more rarely, in the Magothy Formation. Arthropods have been recorded from Kinkora in Burlington County and Cliffwood and Sayreville in Middlesex County, New Jersey, USA. The majority of inclusions came from Crossman's Clay Pit in Sayreville from within the South Amboy Fire Clay and Old Bridge Sand Members of the Raritan Formation, and are considered to be Turonian in age (Grimaldi and Nascimbene, 2010).

South Dakota amber (SD). This amber occurs at the Bone Butte site, Harding County, South Dakota, USA. It is found in the Hell Creek Formation and considered to be Maastrichtian in age (DePalma *et al.*, 2010; Nel *et al.*, 2010).

Tennessee amber (Te). A single insect has been described in this amber from Coffee Bluff, Hardin County, Tennessee, USA. It was found in the Coffee Sand (Wade, 1926) and considered to be Campanian in age (Tschudy, 1975).

Washington amber (Wa). This amber occurs near Issaquah, Washington, USA and is found in coal-bearing siltstones of the Tiger Mountain Formation, considered to be Middle Eocene in age (Mustoe, 1985). Although insects have been recorded from it, none have been identified to order level.

Wyoming amber (Wy). This amber comes from the Hanna Basin, Carbon County, Wyoming, USA and is found in carbonaceous non-marine deposits. The amber ranges from Campanian to Palaeocene in age; however, only amber from the Hanna Formation (Palaeocene) has yielded one insect inclusion (Grimaldi *et al.*, 2000).

South America

Amazonian amber (Am). This amber occurs in the Pebas Formation, on the eastern bank of the Amazon River, about 30km upstream of Iquitos, Peru. It is considered to be Middle Miocene in age (Antoine *et al.*, 2006).

Appendix 2 – Arthropod orders in amber

The following is a list of the arthropod orders and the different ambers in which they have been recorded. The traditional orders Acari and Collembola are kept for convenience; however, both have each recently been divided into three orders. The Onychophora and Tardigrada are also included although they are related to, but not part of, Arthropoda. The records were taken from the published literature with additional data supplied by Dany Azar, Enrique Peñalver, Vincent Perrichot, Alexandr Rasnitsyn and Ankoh Y. Shirota (pers. comms) for the specific Lebanese, Spanish, French, Siberian and Japanese localities respectively. † denotes extinct orders.

Arachnida
- Acari — Ab, All, Am, Az, Ba, Bi, Bu, CaG, CY, Cz, Do, Et, FA, FBu, FC, FF, FF2, FS, Fu, In, It, Jo, LAD, LAZ, LBo, LF, LH, LJF, LJS, LK, LO, LRA, LRi, LT, Me, NJ, Rov, SI, SiA, SiBF, SiBN, Sic, SpES, SpP, SpSB, SpSJ, SY
- Amblypygi — Bu, Do, Me
- Araneae — Ab, ACl, All, Ar, Az, Ba, Bi, Bo, Bu, CaG, CY, Cz, Do, Et, FA, FAI, FC, FF, Fu, In, IW, Jo, LAZ, LBo, LH, LJS, LRi, Me, NJ, Oi, Rom, Rov, SI, SiA, SiBF, SiBN, SiBR, Sic, SiT, SiY, SpEC, SpES, SpP, SpSJ
- Opiliones — Ba, Bi, Bu, Do, FBez, Fu, LBK?, Rom, Rov, Sic
- Palpigradi — Bu
- Pseudoscorpiones — Ba, Bi, Bu, CaG, CY, Do, FA, Fu, In, LH, LJS, Me, Rom, Rov, Sic, SpP
- Ricinulei — Bu
- Schizomida — Bu, Do
- Scorpiones — Ba, Bu, Do, FA, In, LBc, Me, Rov
- Solfugae — Ba, Bu, Do
- Thelyphonida — Bu

Chilopoda
- Geophilomorpha — Ba, Bu, FBu, Me, Rov, SI
- Lithobiomorpha — Ba, Bi, Rov
- Scolopendromorpha — Ba, Bu, Do
- Scutigeromorpha — Ba, Do

Diplopoda
- Chordeumatida — Ba, Rov
- Glomerida — Ba, Rov
- Glomeridesmida — Do
- Julida — Ba
- Polydesmida — Ba, Do, In, Rov
- Polyxenida — Ba, Bi, Bu, Do, FA, FS, LBo, LH, LJS, Rov, SpES, SpP
- Polyzoniida — Ba, Do

	Siphonophorida	Bu, Do
	Spirobolida	All
	Spirostreptida	Do
	Undet.	CY, FF2, LH

Pauropoda   Ba

Symphyla   Ba, Bi, Do

Crustacea
    Amphipoda   Ba, Do, Me

    Cladocera   FC
    Decapoda   Bu, Me
    Isopoda   Ba, Bi, Bu, Do, FA, Me, Rov, SpP
    Ostracoda   Ba
    Tanaidacea   FA, FBu, FF2, SpES, SpP

Onychophora   Ba, Bu, Do

Tardigrada   CaG, NJ

Entognatha
    Collembola   ACl, Am, Ba, Bi, Bu, CaG, Do, Et, FA, FBu, FP, FS, Fu, In, LAZ, LBo, LH, LJS, LS, Me, Rov, SiA, SiBN, Sic, SiY, SpES, SpP, SpSJ
    Diplura   Ba, Bu, Do

Insecta
    †Aethiocarenodea   Bu
    †Alienoptera   Bu
    Archaeognatha   Ba, Bi, Bu, Do, Me, LAZ, LH, LJS, LRi, Rov, SiY, SpP
    Blattodea   ACl, Az, Ba, BC, Bi, Bo, Bu, Do, FA, FBu, FC, Fu, In, IW, JK, Jo, LAZ, LBo, LES, LH, LJF, LJp, LJS, LK, LM, LRi, Me, Rov, SI, SiA, SiBF, SiBN, Sic, SiT, SiY, SpES, SpLH, SpP, SpSJ
    Coleoptera   ACl, All, Am, Az, Ba, BC, Bi, Bo, Bu, CaG, Cam, CY, Cz, Do, Et, FA, FBu, FC, FF, FF2, FPP, FR, FS, Fu, In, IW, JK, Jo, LAD, LAZ, LBo, LES, LF, LH, LJS, LK, LM, LN, LRi, LT, Mas, Me, NJ, Oi, Rov, SI, SiA, SiBF, SiBN, Sic, SiT, SiY, SpES, SpP, SpSJ, Su
    Dermaptera   Ba, Bi, Bu, Do, FA, Fu, LH, Me, Oi, Rov, SpP
    Diptera   ACl, Al, Am, Ar, Au, Az, Ba, BC, Bi, Bo, Bu, CaG, Cam, CY, Cz, Do, Et, FA, FAl, FBel, FBez, FBu, FC, FD, FF, FF2, FP, FR, FS, Fu, Hu, In, It, IW, JK, Jo, LAD, LAZ,

	LBc, LBK, LBo, LD, LED, LES, LF, LH, LJF, LJp, LJS, LK, LM, LN, LO, LRA, LRi, LS, LT, Mas, Me, NJ, Oi, Ph, Rom, Rov, SD, SI, SiA, SiBF, SiBN, SiBR, Sic, SiT, SiY, SpA, SpEC, SpES, SpLH, SpP, SpSJ, Sw
Embioptera	Ba, Bu, Do, Fu, Me, Rov, Sic
Ephemeroptera	Ba, Bi, Bu, Do, Fu, Jo, LH, LO, LJS, Me, NJ, Rov, SiA, SiBN, SiBR, SiT, SiY
Grylloblattodea	Bu
Hemiptera	ACl, Al, Am, Az, Ba, BC, Bi, Bo, Bu, CaG, Cz, Do, FA, FBu, FC, FF, FF2, FPP, FS, Fu, In, JK, Jo, LAD, LAZ, LBc, LBo, LD, LED, LES, LH, LJF, LJS, LK, LN, LRA, LRi, LT, Mar, Me, NJ, Oi, Rom, Rov, SI, SiA, SiBF, SiBN, SiBR, Sic, SiY, SpA, SpES, SpP, SpSB, SpSJ
Hymenoptera	Ab, ACl, Al, All, Am, Au, Az, Ba, BC, Bi, Bo, Br, Bu, CaG, Cam, CY, Cz, Do, Et, FA, FAI, FBez, FBu, FC, FF, FF2, FP, FR, FS, Fu, In, IW, JC, JK, Jo, JU, LAD, LAF, LAZ, LBc, LBK, LBo, LED, LH, LJF, LJS, LM, LRA, LRi, LT, Mas, Me, NJ, Oi, Rom, Rov, SI, SiA, SiBF, SiBN, SiBR, Sic, SiY, SpA, SpEC, SpES, SpLH, SpP, SpSJ, ST, Sw
Isoptera	Ba, Bi, Bo, Bu, Cam, CY, Do, FA, FS, In, Jo, LBc, LH, LJS, Me, NJ, Oi, Rov, Sic, SiY, SpES, SpP, SpSJ
Lepidoptera	Ba, Bi, Bu, CY, Do, Et, FA, FAI, FC, FD, FF, Fu, Jo, LAZ, LBo, LH, LJF, LJS, LK, Me, Rom, Rov, SI, Sic, SiY, SpP
Mantodea	Ba, Bu, Do, FA, In, LBc, LRi, NJ, Rov, SiY, SpES, SpSJ
Mantophasmatodea	Ba, Bi
Mecoptera	Ba, Bi, Bu, Me, Rov, SpES
Megaloptera	Ba, Bu, Do, Oi, SiY
Neuroptera	Ba, BC, Bi, Br, Bu, CaG, CY, Cz, Do, FA, FF2, Fu, In, LAZ, LBo, LH, LJS, LRi, LS, Me, NJ, Oi, Rov, SiA, SiY, SpES, SpP, SpSJ
Odonata	Ba, Bu, Do, FA, Jo, LAD, LAZ, Me, Oi, SD
Orthoptera	Am, Ba, Bi, Bu, Cz, Do, FA, FBu, Fu, In, JK?, LBo, LH, LJS, LRi, Me, Oi, Rov, Sic, SpP, SpSJ
†Permopsocida	Bu
Phasmatodea	Ba, Bu, Do, Me, Oi
Plecoptera	Ba, Bi, Bu, Do, Rov, SiY
Psocoptera	Am, Ar, Az, Ba, BC, Bi, Bu, CaG, Cz, Do, Et, FA, FC, FF2, Fu, In, JK, Jo, LAZ, LBc, LBo, LF, LH, LJS, LK, LM, LO, LRi, Me, Oi, Rov, SI, SiA, SiBF, SiBN, SiBR, Sic, SiT, SiY, SpES, SpP, SpSJ
Raphidioptera	Ba, Bu, CaG, FA, LJS, NJ, Rov, SpES, SpP
Siphonaptera	Ba, Do
Strepsiptera	Ba, Bu, Ch, Do, FA, FC, Rov, SpP
Thysanoptera	Az, Ba, BC, Bi, Bo, Bu, CaG, Cam, CY, Do, Et, FA, FC, Fu, In, Jo, LBo, LH, LJF, LJS, LM, LN, LRi, LT, Me, NJ,

	Oi, Rov, Sl, SiA, SiBF, SiBN, SiBR, Sic, SiT, SiY, SpES, SpP, SpSJ, Wy
Trichoptera	ACl, Am, Ba, Bi, Bu, CaG, Cam, Do, FAl, LH, LJS, Me, NJ, Rov, Sl, SiA, SiBF, SiBN, SiBR, Sic, SiY, SpP, Te
Zoraptera	Bu, Do, Et, Jo
Zygentoma	Ba, Bi, Bu, CaG, Do, FC, Jo, Rov, Sic, SiY

Appendix 3 – Burmese amber arthropod families

*extinct

ARACHNIDA
Acari
    Anystidae
    Argasidae
    Bdellidae
    Cheyletidae
    Erythraeidae
    Eupodidae
    Ixodidae
    Opilioacaridae
    Resinacaridae
Amblypygi
Araneae
    Archaeidae
    *Burmascutidae
    Deinopidae?
    Dictynidae?
    *Eopsilodercidae
    Lagonomegopidae
    Linyphiidae?
    *Micropalpimanidae
    Myrmeciidae
    Mysmenidae?
    Nephilidae
    Oecobiidae
    Oonopidae
    *Plumorsolidae?
    *Praeterleptonetidae
    Segestriidae?
    Telemidae?
    Tetragnathidae?
    Thomisidae
    Uloboridae
Opiliones
    Sironidae
Pseudoscorpiones
    Cheiridiidae
    Chernetidae
    Garypinidae
Schizomida
Scorpiones
    Buthidae
    Chaerilidae
    *Chaerilobuthidae
    *Palaeotrilineatidae
    *Sucinolourencoidae
Solfugae

CHILOPODA
Geophilomorpha
    Geophilidae
Scolopendromorpha?

DIPLOPODA
Polyxenida
    Synxenidae
Siphonophorida
    Siphonophoridae

CRUSTACEA
Isopoda
    Styloniscidae?

ENTOGNATHA
Collembola
    Isotomidae
    Neanuridae
    *Praentomobryidae
    Sminthuridae
    Tomoceridae
Diplura
    Campodeidae

INSECTA
Archaeognatha
    Machilidae
    Meinertellidae
Zygentoma
    Lepismatidae
Ephemeroptera
    *Australiphemeridae
    Baetidae
Odonata
    *Burmaphlebiidae
    Platycnemididae
    Platystictidae
Blattodea (incl. Isoptera)
    *Archeorhinotermitidae

*Caloblattinidae
Corydiidae (=Polyphagidae)
Ectobiidae (=Blattellidae)
Hodotermitidae?
Kalotermitidae
*Manipulatoridae
*Ponopterixidae (transferred from Umenocoleidae)
Termitidae?
Dermaptera
Anisolabididae?
Diplatyidae
Labiduridae
Pygidicranidae
Embioptera
Notoligotomidae (=Burmitembiidae)
*Sorellembiidae
Mantodea
Orthoptera
*Elcanidae
Gryllidae
Mogoplistidae
Tetrigidae
Tettigonidae
Tridactylidae
Phasmatodea?
Plecoptera
Zoraptera
Zorotypidae
Hemiptera
Achilidae
*Albicoccidae
Aleyrodidae
Aphrophoridae
Aradidae
*Burmacoccidae
*Burmitaphidae
Cicadellidae
Cicadidae
Cimicidae
Cixiidae
Coccidae
Coreidae
Cydnidae
Dipsocoridae
Enicocephalidae
Fulgoridae?
Gelastocoridae
Hydrometridae
*Kozariidae
Leptopodidae
Miridae
Naucoridae
Ochteridae
Ortheziidae
*Palaeoleptidae
*Parvaverrucosidae (=Verrucosidae)
*Perforissidae
*Protopsyllidiidae
Pseudococcidae
Reduvidae
Schizopteridae
*Tajmyraphididae
Tingidae
*Weitschatidae
Xylococcidae?
Psocoptera
*Archaeatropidae
Compsocidae
Liposcelididae (=Liposcelidae)
Pachytroctidae
Trogiidae
Thysanoptera
Aeolothripidae
*Ectinothripidae (*nomen nudum*)
Thripidae
Coleoptera
Aderidae
Anobiidae
Anthicidae
Buprestidae
Cantharidae
Carabidae
Cerambycidae
Chrysomelidae
Ciidae (=Cisidae)
Cleridae
Cucujidae
Curculionidae (=Platypodidae, Scolytidae)
Dermestidae
Elateridae
Endomychidae
Eucinetidae
Eucnemidae
Histeridae
Ithyceridae (=Eccoptarthridae)
Latridiidae (=Lathridiidae)
Lepiceridae (=Haplochelidae)

    Lymexylidae
Melandryidae
Meloidae
Melyridae
Mordellidae
Nemonychidae
Nitidulidae
Oedemeridae
Prostomidae
Ptiliidae
Ptilodactylidae
Ripiphoridae
    Salpingidae
    Scarabaeidae
    Scirtidae (=Helodidae)
    Scraptiidae
    Silphidae
    Silvanidae
    Sphaeriusidae (=Microsporidae)
    Staphylinidae (=Pselaphidae, Scydmaenidae)
    Throscidae
    Trogossitidae
    Zopheridae (=Colydiidae)
Diptera
    Acroceridae
    Anisopodidae
    Apsilocephalidae
    Asilidae
    Blephariceridae
    *Cascopleciidae
    Cecidomyiidae
    Ceratopogonidae
    Chaoboridae
    *Chimeromyiidae
    Chironomidae
    Corethrellidae
    Culicidae
    Diadocidiidae
    Dolichopodidae
    Empididae
    Hilarimorphidae
    Hybotidae
    Keroplatidae
    Limoniidae
    Lygistorrhinidae
    Mycetophilidae
    Phoridae (=Sciadoceridae)
    Platypezidae
    Psychodidae (=Phlebotomidae)
    Ptychopteridae (=Eoptychopteridae)
    Rhagionidae
    Scatopsidae
    Sciaridae (=Archizelmiridae)
    Tanyderidae
    Therevidae
    Tipulidae
    Valeseguyidae
    *Zhangsolvidae
Hymenoptera
    Ampulicidae
    Bethylidae
    Braconidae
    Chalcididae
    Crabronidae
    Diapriidae
    Dryinidae
    Embolemidae
    Evaniidae
    *Falsiformicidae
    Formicidae
    *Gallorommatidae
    Gasteruptiidae (=Aulacidae)
    Ichneumonidae
    *Maimetshidae
    Megalyridae
    Megaspilidae
    *Melittosphecidae
    Mymaridae
    Mymarommatidae
    Pelecinidae
    Platygastridae
    Pompilidae
    *Praeaulacidae
    Rhopalosomatidae
    Sapygidae
    Scelionidae
    *Serphitidae
    Sierolomorphidae
    *Spathiopterygidae
    Sphecidae
    Stephanidae
    *Stigmaphronidae
    Tiphiidae
    Vespidae
Lepidoptera
    Gelechiidae
    Gracillariidae
    Micropterigidae (=Micropterygidae)

Mecoptera
    Meropeidae
    *Pseudopolycentropodidae
Megaloptera
    Sialidae
Neuroptera
    Ascalaphidae
    Berothidae (=Rachiberothidae)
    Coniopterygidae
    Dilaridae
    Mantispidae
    Nevrorthidae?
    Nymphidae
    Osmylidae
    Psychopsidae

Raphidioptera
    *Mesoraphidiidae
Strepsiptera
    *Mengeidae?
Trichoptera
    Hydroptilidae
    Leptoceridae
    Philopotamidae
    Polycentropodidae
    Psychomyiidae
Order?
*Gryllomantidae
*Lophioneuridae
    *Mantoblattidae

Index

Page numbers in **bold** denote figures.

Aberdeen University 4–5
acanthodians 54–55
Acanthodopsis 54
acipenseriforms 133, 145, 176, 177
Acrodonta 292
Actinistia *see* coelacanths
actinopterygians 54, 55–56, **56**, 70, **70**, 84, **85**, 87, 93, 220, 233–236, **233**, **235**
adapoids 281
adephagan beetles 82
Aeschnidium 177
aetosaurs 106
Agamidae 292
Agassiz, Louis 219
Aglaophyton 3, 16, 17–18, **18**, 19, 20–24, **20–23**, 28
Ailuravus macrurus 279
aïstopods 50
Alavesia **326**
algae 19, **19**, 259–260
Allaboilus gigantus **144**
Allaeochelys crassesculpta 275–276, **276**
alligatoroids 275
Allognathosuchus 275
Almatium gusevi 75, 82
Alvinia serrata 84, 87
amber *see* Burmese amber
amiids 178, 273
ammonites 291
Amnifleckia guttata **144**
amniotes 49–50, 54
amphibians
    Jehol Biota 177, 178, **179**, 181
    Messel Pit Fossil Site 273
    Yanliao Biota 135–136, **136**, 137, 145, **146**, 152, 153, 155
    see also tetrapods
Amphiesmenoptera **117**
amphilemurids **264**, 279–280, **279**
Amphiperatherium 279
Amphiperca 273
Anchiornis huxleyi 135, **136**, 148, **149**, 152, 153, 155, 157
Angaturama limae 237
angiosperms 153, 154
    in amber 291, **306**, **307**
    Jehol Biota 180, 181, 193, **193**
    Messel Pit Fossil Site 265–266, **265**, **266**
    Solite Quarry 111
    Yanliao Biota 135, 138, 141, **142**, 143
Anguilla 273
Anguimorpha 273
Anhanguera 241, **241**, 242
Anhanguera piscator 242
Anhangueridae 182
Anomozamites **142**, 143, 151
Anseriformes 277
Anthicidae **321**
Anthracodesmus 59

anthracosauroids 44–47, **46**, 52, 54
ants 59, 267, 292, 294, 295, **318–319**
anurans 178, **179**, 181
anurognathid pterosaurs **136**, 139, 147, **148**, 152, 180
Apidae 267, **271**
Apodiformes 277, **277**
aquatic mammaliaforms 135, 153
arachnids
    in amber 291, 296, **310–312**
    East Kirkton Lagerstätte **58**, 59
    Rhynie and Windyfield cherts 4, **25**, **26**, 27
    Solite Quarry 112
    Yanliao Biota 145, **145**, 151, 155
Araripe Basin *see* Santana Formation
Araripemys barretoi 236, **236**
Araripesaurus 242
Araripesaurus castilhoi 240
Araripesuchus gomesii 236
Araucaria beipiaoensis 143
Arboroharamiya jenkinsi 137, 149, **150**, 156
Archaefructus 143
Archaeidae 296
Archaeofructus 181, 193, **193**
Archaeoistiodactylus linglongtaensis 148
Archaeonycteris 283
Archaeopelecinus tebbei **144**
Archaeopteryx 155–156, 187
Archaeorhynchus 174
Archescytinidae **113**, 119, **120**
Archipsylla sinica **144**
Archizelmiridae 291
Archizygoptera 93
archosauromorphs 75, **88**, 89–92, **90**, **91**, 96
archostematan beetles 82
Arguniella 176, 177
Argyrarachne solitus 112
arthropods
    in amber 289, 290, 291, 292–296, **309**, **313–327**
    East Kirkton Lagerstätte 56–59, **58**, 62
    large size 57
    Rhynie and Windyfield cherts 3, 4, 5, 16, 17, 24–27, **25**, **26**, 30
    see also crustaceans; insects
articulatans 143
artiodactyls **262**, 280–281
arums 267
Ascalaphidae **320**
ascomycetes 18, **18**
Asiatoceratodus sharovi 84, **86**, 96
Asiatosuchus 275
Asilidae **327**
Asteroxylon 2, 3, 4, 18, **18**, 20, **21**, 28
atmospheric oxygen levels 57
Atractosteus 273
Atreipus 121
Auchenorrhyncha 82, **83**, 120
auditory ossicles 191
Aumelasia 280

343

Aurornis xui 135, 148, **149**
Austromerope brasiliensis 296
Austromerope poultoni 296
Axelrodichthys araripensis **235**, 236, 239

Bacharboilus lii 151
Bacillafaex **15**, 27
backswimmers *see* Notonectidae
Baiera 79
Baikalophyllum 181
Balanerpeton woodi 44, **45**, 52, **53**
Baltic amber 287, 289, 292, 294
baphetids 47–49, **48**, 52
Barbosania 242
Baryphracta 275
bat flies 267
Bathgate Hills Volcanic Formation 42
bats 267, 278, 281–283, **283**
bees 267, **271**
beetles
    in amber 295, **321–323**
    Madygen Lagerstätte 77, 82, **83**
    Messel Pit Fossil Site 267, **268**, **269**
    Solite Quarry 112, **113**, 118–119, **118**, **119**, 120
    Yanliao Biota 143
Beiyanerpeton jianpingensis 137, 145, **146**, 155
Belostomatidae 113–114, **113**, **115**, 120, 125
bennettitaleans **142**, 143, 151
benthic foraminifera 232
Bergisuchus 275
Berothidae **320**
Bilignea 60
biominerals, preservation of original 228, **228**
biostratigraphic ages
    Burmese amber 290–291
    Jehol Biota 176
    Madygen Lagerstätte 75
    Rhynie and Windyfield cherts 11
    Yanliao Biota 137, 138–139
bioturbation 92–93, **92**
birds
    in amber 292, **329**
    evolution of flight 187
    evolutionary origins 155–156
    feathers 263, **263**, 277–278, **277**, **278**, 292, **328**, **329**
    Jehol Biota 171, 173, 174, 176, 177, 180, 187–191, **187–190**
    Messel Pit Fossil Site 263, **263**, 276–278, **277**, **278**
bivalve molluscs 59, 78, 80–81, **80**, 135, 139, 145, **145**, 176, 177
Blattaria 143, **144**, 267
Blattida 143, 151
Blattodea 77, 82, **83**, **116**, 294–295
Blephariceridae **326**
Boidae 273
book lungs 4, 17, 24, 57
booklice 84, 143, 293
Boreidae 154
Boreomerope antiquaa 296
bothremyidid turtles 236
Bothryopneustes araripensis 224, 238
Boverisuchus 275
Brachycera 112–113, 191–193, **192**
Brachyphyllum 110
branchiopods **25**, 27, 75, 78, **80**, 81–82

Brannerion 229
Brasilemys josai 236
Brasileodactylus 242
Brazil *see* Santana Formation
Brejo Santos Formation 220
Brigantobunum listoni **58**, 59
British Association for the Advancement of Science 2
bryophytes **76**, 78–79, 291, **306**
bryozoans, freshwater 81, 93
Buprestidae 267
Burlagonegops eskovi **310**
Burmaculex antiquus **326**
Burmapogon bruckschi **327**
Burmazelmira asiatica **326**
Burmesarchaea 296
Burmesarchaea grimaldii **310**
Burmese amber 287–297
    age 290–291
    biodiversity 291–292, **306–329**
    biogeography 296
    botanical origins 288–289
    deposition 289
    evolutionary significance 292–296
    historical and geological context 290–291, **305**
    location 290, **304**
    molecular composition 288
    preservation of inclusions 289–290
Burmomerope eureka 296, **324**
Burmorthezia insolita **316**
burrows 77, 92–93, **92**
Buthidae **310**
butterflies 267
Buxolestes 280

caddisflies 84, 145, 267, **325**
Calamopleurus 234
Caloblattinidae 82
calopterygids 151
Calymperites burmensis **306**
Canadian amber 291
Caprimulgiformes 277
Carabidae 267
Carboniferous *see* East Kirkton Lagerstätte
Carbonita 56
carettochelyid turtles 275–276, **276**
Carinametria burmensis **317**
Cariri Formation 220
Caririemys violetae 236
Caririsuchus camposi 236–237
carnivorans 279
Carusia 182
Castorocauda lutrasimilis 135, 149, **150**, 152
Castracollis **25**, 27
caudates *see* salamanders
Caudipteryx 174, 186
Cearachelys placidoi 236
Cearadactylus 241
centipedes **25**, 27
Cerambycidae 267
Ceropria messelense **269**
chameleons 292
Changchengopterus pani 137, 148
Chaoboridae 267

Chaoyangopterus 182
Charentes amber 291
chelid turtles 236
Chelotriton robustus 273
Chengia laxispicata 181, 194
Chernetidae **310**
Chimeromyiidae 291
China *see* Jehol Biota; Yanliao Biota
Chinle Formation 109
Chirononomidae **308**
Chlorobiaceae 239
chondrichthyans 56, 84, **86**, 87, 96
choristoderans 169, 173, 176, 177, 179–180, **179**, 182
Chresmodidae 143, 152
chroniosuchians 87, **88**, 89
Chrysomelidae 267, **268**
Chunerpeton tianyiensis 135–136, **136**, 137, 145, **146**, 152, 155
cicadas 82, **83**, 267
Cladocyclus gardneri 232, 234, **235**, 239
Cladophlebis 79
Coahomasuchus 106
coalified compressions 59–60, **60**, 61
Coccoidea 295–296, **316**
Cockerell, T. D. A. 290–291
cockroaches 77, 82, **83**, **116**, 143, **144**, 151, 267
coelacanths 84–87, **86**, 220, **235**, 236, 239
coelurosaurian theropods
    Jehol Biota 171, 173, 174, 176, 180
    Santana Formation 237, **237**, 238
    Yanliao Biota 135, **136**, 148, **149**, 152, 153, 155–156, 157
Coleoptera
    in amber 295, **321**–**323**
    Madygen Lagerstätte 77, 82, **83**
    Messel Pit Fossil Site 267, **268**, **269**
    Solite Quarry 112, **113**, 118–119, **118**, **119**, 120
    Yanliao Biota 143
Coleorrhyncha 82
collembolans **25**, 27
Coloborhychus 241
Coloborhynchus 241, 242
Coloborhynchus spielbergi 242
colour preservation
    Jehol Biota 171, 187, **187**
    Messel Pit Fossil Site 267, **268**, **269**
compsognathid dinosaurs 237, **237**
conchostracans **80**, 81, 135, 136, 139, 145, **145**, 152, 176, 177
cone and cone scales 111
Confuciusornis 187, **187**
confuciusornithid birds 176, 177, 180
conifers 79, 110, **142**, 143, 181, 265, 291, **306**, **307**
Coniopteris 135, 139, **142**
Cope, Edward Drinker 105
copepods 232, **232**
coprolites
    Madygen Lagerstätte **92**, 93
    Messel Pit Fossil Site 267, 275
    Rhynie and Windyfield cherts **15**, 16, **26**, 27, 30
Corixidae 112, **113**, 114–117, **115**
Cow Branch Formation **106**, 107, 109
crabs 232
Cran, William 3, 16
Crato Formation **218**, 219, 220, 223, 224, 232, 238, 242, 291
creodont mammals 273, **275**

Cretaceous *see* Burmese amber; Jehol Biota; Santana Formation
Cretoscelis burmitica 293
Cretostylops engeli 294, **325**
crocodylians 273, 275
crocodyliformes 236–237
crocodylomorphs 106
crop contents, birds 174, 189, **189**
Crosaphidae 113
crustaceans
    in amber 291, 292, **309**
    East Kirkton Lagerstätte 56
    Madygen Lagerstätte 75, **80**, 81–82
    Rhynie and Windyfield cherts 3, 4, **25**, **26**, 27
    Santana Formation 232, **232**
    Yanliao Biota 135, 136, 139, 145, **145**
cryptobranchids 135–136, **136**, 137, 145, **146**, 152, 155, 177
Cryptocercus 294
Cryptodira 179, 181, 182
cryptodire turtles 236
Cteniogenidae 182
ctenochamatoid pterosaurs 241
Ctenochasmatidae 182
Culicidae 267
culicomorphans 112–113
cupedid beetles 82
Curculionidae **323**
Curvirimula 59
cyanobacteria 17–18, 42
Cycadolepis **142**
cycadophytes **76**, 79, 181
cycads **142**, 143
Cyclurus 273
cynodonts 87, **88**, 89, 106, 135, 137
Cypridea 56, 177
Cyrtoctenus 56–57
Czekanowskia 135, **142**, 143

Dabeigou Formation 138, 171
Dadianzi Formation 138, 171
Dalinghosaurus 180, 182
damselflies 151, **313**
Dan River–Danville basin **106**, 106, 107
Danaeopsis 79
Daohugou Biota 131, 133–135, **133**, 137
Daohugoucossus shii **144**
Darwinius masillae 281
Darwinopterus linglongtaensis 148, **148**
Darwinopterus modularis 135, 148
Darwinopterus robustodens 148, **148**, 152
Darwinula 135, 139, 145, 176
David, Père 169
dawn redwood 289
decapods **80**, 82, 229, 232, 239
deinonychosaurians 155–156
Delitzschala bitterfeldensis 62
Dendrerpeton 52
Dendrorhynchus mutoudengensis 147
Dentilepisosteus laevis 239
depositional environments
    East Kirkton Lagerstätte 42–44
    Madygen Lagerstätte **70**, 70, 71–75, **73**, 77
    Messel Pit Fossil Site 259–262
    Rhynie and Windyfield cherts 9–11, **10**, 30

Solite Quarry 124–125
Dermaptera 84, 143, 267, **315**
Devonian *see* Rhynie and Windyfield cherts
Diaplexa tigjanensis 81
diatreme formation 258, **259**
Dichelesthiidae 239
dichobunids **262**, 280–281
dicksoniaceans 143
Dictyophyllum 110, **111**
dicynodonts 106
Didelphidae 279
Dijungarica 177
Dilong 186
dinosaur trackways 109, 121
dinosaurs
    colour 187
    Jehol Biota 171, 173, 174, 176, 177, 180, 185–187, **185**, **186**, **187**, 191, **192**
    Santana Formation 229, 237–238, **237**
    Yanliao Biota 135, **136**, 148–149, **149**, 152, 153, 155–156, 157–158
Diplichnites 8, **9**, 27
Diplocynodon 273, 275
Diplodoselache 56
diplopods 27
Dipsocoromorpha **120**
Diptera
    in amber 291, 293, **308**, **326**–327
    Jehol Biota 191–193, **192**
    Madygen Lagerstätte 75, 84
    Messel Pit Fossil Site 267
    Solite Quarry 112–113, **113**, **114**, 125
    Yanliao Biota 143
Djungarica camarata 177
DNA, in amber 290
Dobruskina, Inna A. 66
docodonts 149, **150**, 156
Dominican amber 287, 289, 292, 294, 295
Dracochelys 182
dragonflies 84, 93, 151, 267
drepanosaurs 105
dromaeosaurid dinosaurs 186, **186**
Dromicosuchus grallator 106
Dryden Flags Formation 7, **7**, 8–9, 11
Dunsopterus 56–57

ear, mammalian 191
earwig flies 296
earwigs 84, 143, 267, **315**
East Kirkton Lagerstätte 39–62
    depositional environment 42–44
    ecosystem reconstruction 61
    fish 54–56, **56**
    fossil preparation 61–62
    geology 41–44, **41**, **43**
    hydrothermal system 41, 42–44, 61
    invertebrates 56–59, **58**, 62
    plants 59–61, **60**
    taphonomy and preservation 41, 61–62
    tetrapods 41, 44–54, **45–49**, **51**, **53**, **55**, 61
East Kirkton Limestone 41–44, **41**, **43**
Ebullitiocaris 4, **25**, **26**, 27
echinoids 224, 238

Echinostachys 79
ecosystem reconstructions
    East Kirkton Lagerstätte 61
    Madygen Lagerstätte 93–95, **94**, **95**
    Rhynie and Windyfield cherts 29, **29**, 30
    Santana Formation 238–239
    Yanliao Biota 149–153
ectoparasites 151, 154, **192**, 193
Edenia villisperma 111, **111**
edentulous pterosaurs 182, **183**, 242
Edwards, D. S. 16
Edwards, Dianne 4
Edyndella 79
eels 273
egg capsules, shark 77, **86**, 87, 96
eggs 173, 184, **184**, 189, **190**
Eifel region, Germany 259, **260**
elasmobranchs 56, 70, **70**, **86**, 87, 93, 96, 220, 229, 232, 233, **234**, 238
Elcanidae 120, **121**
Eldeceeon rolfei 47, **47**, 52, 54
Emballonuridae 283
Embioidea **314**
embolomeres 44–45, 47, 54
embryos 173, 182, **184**, 189, **190**, 281
Emmons, Ebenezer 105
Empidoidea **326**
enantiornithine birds 180, 187, **188**, 189, **190**
Eoarthropleura 27
Eocene *see* Messel Pit Fossil Site
Eoconfuciusornis 177, 187
Eoherpeton 44–45
Eoparacypris 176
Eopelobates 273
Eophalangium 4, **26**, 27
Eophyllium messelensis 267, **270**
Eoplectreurys gertschi 145
Eosestheria 169, 170, **170**, 171, 177
Eosestheria–Ephemeropsis–Lycoptera (EEL) assemblage 169, 170, **170**, 171
Eosinopteryx brevipenna 135, 148, **149**
Ephedra 181, 194
Ephedraceae 194
Ephedrites **194**
ephedroids 181, 194, **194**
Ephemeropsis 169, 170, **170**, 171
Ephemeropsis trisetalis 177
Ephemeroptera 84, 143, **144**, 152, 267, **313**
Epidendrosaurus ningchengensis 135, 148, 152, 153, 155
Epidexipteryx hui 135, **136**, 148, **149**, 152, 153, 155
equoid perissodactyls 280, **281**
Eretmophyllum 110
erinaceomorphs **264**, 279–280, **279**
Eristophyton 60
Eryops 52
Ethiopian amber 294
Eucritta melanolimnetes 47–49, **48**, 52, 54
eucryptodiran turtles 179, 181, 182
eudicots 181, 193, **193**
Euesthesia 135, 136, 139, 145, **145**, 152
eumelanosomes 187
Euraxemys essweini 236
Eurodexis 280

Eurohippus messelensis 280
Europolemur 281
Eurotamandua joresi 279
Eurynotus 55–56, **56**
eurypterids 39, 56–57, **58**, 62
eusociality, in insects 294–295
eutherian mammals 131, 135, 149, **150**, 153, 156, 180
euthycarcinoids 3, **25**, **26**, 27
eutriconodonts 191, **191**
evenkiids 84, **85**
evolutionary significance
    Burmese amber 292–296
    Jehol Biota 181–194
    Madygen Lagerstätte 96
    Messel Pit Fossil Site 283–284
    Rhynie and Windyfield cherts 30
    Santana Formation 239–245, **240**, **241**, **243**
    Yanliao Biota 153–158
Exu Formation 220, **221**, 223–224, **225**, **226**

Falsiformicidae 291
Fayolia sharovi **86**, 87
feathered dinosaurs
    colour 187
    Jehol Biota 171, 173, 176, 177, 180, 185–187, **185**, **186**, **187**, 191, **192**
    Yanliao Biota 135, **136**, 148–149, **149**, 152, 153, 155–156, 157–158
feathers
    birds 263, **263**, 277–278, **277**, **278**, 292, **328**, **329**
    dinosaurs and pterosaurs **136**, 140, 153, 156, 157–158
Fenghuangopterus lii 147, 153
Ferganiscus osteolepis 84, **85**, 87
Ferganoconcha 139, 145, **145**
Ferganodendron sauktangensis 79
ferns 59, 79, 93, 110–111, **111**, **142**, 143, 180, 291, **306**
filamentous feathers, dinosaurs and pterosaurs **136**, 140, 157–158
fish
    East Kirkton Lagerstätte 54–56, **56**
    Jehol Biota 169, 174, 176–178, **178**, 181
    Madygen Lagerstätte 70, **70**, 77, 84–87, **85**, **86**, 96
    Messel Pit Fossil Site 273
    Santana Formation 216, **217**, 220, 226–227, **226**, **227**, 228, 229, **230**, 231, **231**, 232, 233–236, **233**, **234**, **235**, 238, 239
    Solite Quarry 108–109, **110**, 121
    Yanliao Biota 133, 135, 139, 145, **147**
fleas 151, 154, **192**, 193
flies, true *see* Diptera
flight
    evolution of in birds 244–245
    pterosaurs 244–245
Florissant Formation 141
flower bugs 151
flowering plants *see* angiosperms
foetal remains 173, 182, **184**, 189, **190**, 281
footprints 109, 121, 122
foraminifera 232, 297
Formicium giganteum 267
Forschungsinstitut Senckenberg 257
fossil preparation 16–17, 61–62, 109, 262–263, **262**
fossorial mammaliaforms 152, 156

Fourier Transform Infrared Spectroscopy (FTIR) 288
francolite 229–230, 231
Fraxinopsis 111
frogs 178, **179**, 181, 273, 292, **328**
Fruitafossor 156
Frullania cretacea **306**
fungi 18–19, **18**, 291, **306**
fusain 59, 60–61
Fuzi dadao **144**

Gackstroemia 296
Gackstroemia cretacea 296
Galliformes 277
Gallodactylidae 182
Galloromatidae 291
game birds 277
gametophytes 4, 21–24, **22–23**, 30, 78
Gardner, George 216–217
gars 239
Gastornis 277
gastrointestinal contents
    amphibians 140, 152
    birds 174, 189, **189**, 277
    crocodylians 275
    dinosaurs 174, 186
    mammals 174, 191, **192**, 279, 283, **283**
    snakes 273, **274**, 275
gastroliths 174, 186, 275
gastropods
    in amber 292, **308**, **309**
    Jehol Biota 176, 177
    Madygen Lagerstätte **80**, 81
    Santana Formation 224, **225**, 228, 232, 238–239
    Yanliao Biota 145
geckos 292
Gegepterus 182
Geikie, Archibald 2
Geinitziidae 151
Geiseltaliellus 273, **274**
Gelastocoridae **316**
geoemydid turtles 276
Geological Survey of India 290
Germany *see* Messel Pit Fossil Site
Gerontoformica **318**, 319
Gerridae 267
geysers **8**, 9, **10**, 11, 17
Ginkgo **142**, 181
ginkgoaleans 79, 93, 110, 139, **142**, 143, 151–152, 181
Ginkgoites 79
gliding lizards 180, 182
gliding mammaliaforms 135, 153, 156
gliding tetrapods 89–91, **90**, 105, 109, 122–124, **123**
Glomites rhyniensis 18, 19
Glosselytrodea 84
Glossophyllum 79
gnetaleans 181, 194, **194**
Gorgetosuchus 106
Grabau, Amadeus William 169
Graciliblatta bella 151
Grallator 121
Grammolingia boi **144**
grasshoppers 82, 84, 267
Green Sulphur bacteria 239

gregarious behaviour, dinosaurs 174
ground beetles 267, **269**
Grylloblattodea 82, 143, **144**, 151
Guidraco 182, **183**
gut contents *see* gastrointestinal contents
Gwyneddichnium 109
Gwynne-Vaughan, D. T. 2
gymnosperms
    in amber 291, **306**, **307**
    East Kirkton Lagerstätte 59–61, **60**
    Jehol Biota 180, 181
    Madygen Lagerstätte **76**, 79, 93
    Messel Pit Fossil Site 265
    Solite Quarry 110, 111
    Yanliao Biota **142**, 143, 151–152, 154–155

Haidomyrmex **319**
Haifanggou Formation 132–133, 135, 137, 138–139, 140, 141, 143, 151
Haldanodon 156
Hallensia matthesi 280
hangingflies 151–152, 154–155
haplogyne spiders 155
haramiyidan mammals 135, 137, 149, **150**, 152, 156
Harbinia 232
harvestman spiders 4, **25**, **26**, 27, **58**, 59, 145, **145**
Hass, Hagen 4, 17
Hassianycteris 283
Haubold, Hartmut 66
Hebei Province, China *see* Jehol Biota; Yanliao Biota
Helminthoidichnites tenuis **92**, 93
helminthomorphs 27, 59
Hemiptera
    in amber 293–294, 295, **316–317**
    Madygen Lagerstätte 75, 82, **83**
    Messel Pit Fossil Site 267
    Solite Quarry 112, 113–117, **113**, **115**, 119, 120, **120**, 125
    Yanliao Biota 143, **144**, 151
Hessisches Landesmuseum Darmstadt 257
Heterocrania **25**, **26**, 27
Heterocrania rhyniensis 3
heterodontosaurid dinosaurs 135, 148–149, 152
Heteroptera 75, 143, 267, **316–317**
hexapods 27, **312**
Hibbertopterus scouleri 39, 56–57, **58**
Hitchcock, Edward 105
Holcoptera 112
Homoptera 143, 267
Hongshanornis 174, 189
hoopoes 277
Horneophyton 3, 20, **20**, **21**, 27–28
horses 278, 280, **281**
horsetails 143
hot-spring systems *see* hydrothermal systems
Houcheng Formation 138
Hourqia 232
Huajiying Formation 138, 171, 176
hummingbirds 277
Huperzia 20
hybodontid sharks 56, 70, **70**, **86**, 87, 93, 220, 229, 232, 233, 238
Hydrometridae **317**
Hydrophilidae 267

hydrothermal systems
    East Kirkton 41, 42–44, 61
    Rhynie 3, 6, 8, **8**, 9–11, **10**, 30
Hymenoptera 75, 84, 143, **144**, 267, **271**, 291, 292, 294, 295, **318–319**
hynobiids 145, **146**, 155, 181
Hyphalosaurus 179–180, **179**, 182
Hypuronector 105
Hyrachyus 280

Iamanja 238
Iansan beurleni 229, 232, 233, **234**, 238, 239
Icarosaurus 105, 124
ichthyodectid fish 232, 239
Iemanja palma 232
Iguania 273
iguanodontid dinosaurs 186
Ikechosaurus 179
Incisivosaurus 186
Inner Mongolia, China *see* Jehol Biota; Yanliao Biota
insectivorans **264**, 279
insects
    in amber 289, 290, 291, 292–296, **313–327**
    East Kirkton Lagerstätte 62
    eusociality 294–295
    feeding traces 93, 151
    insect–plant interactions 151–152, 154–155, 295–296
    Jehol Biota 171, 177, 191–193, **192**
    Madygen Lagerstätte 75, 77, 82–84, **83**, 96
    Messel Pit Fossil Site 267, **268–272**
    mimesis 151–152, 154–155
    phytophagy 295–296
    Rhynie and Windyfield cherts 27
    Solite Quarry 107, 108, 109, 112–120, **113–121**, 124–125
    Yanliao Biota 132, 133, 143–145, **144**, 151–152, 153, 154–155
Ipubi Formation 220–222, **221**, **222**, 223, 224, 225
Irritator challengeri 237
Isoetites madygensis **76**, 79, 93
Isoetites sixteliae **76**, 79, 93
isopods 291, 292, **308**
Isoptera 267, 294–295, **315**
Istiodactylidae 148, 182
Itasuchus 237

Jeanrogerium sornayi 82
Jehol Biota 131, 132, 133, 134, 140–141, 169–194
    amphibians 177, 178, **179**, 181
    biostratigraphic ages 176
    birds 171, 173, 174, 176, 177, 180, 187–191, **187–190**
    choristoderans 169, 173, 176, 177, 179–180, **179**, 182
    dinosaurs 171, 173, 174, 176, 177, 180, 185–187, **185**, **186**, **187**, 191, **192**
    evolutionary radiations 176–177, 180, 184–185, 187
    evolutionary significance 181–194
    fish 169, 174, 176–178, **178**, 181
    geological context 171–176, **172**, **173**, **174**
    insects 171, 177, 191–193, **192**
    lizards 169, 173, **175**, 177, 180, **180**, 182
    mammals and mammaliaforms 171, 177, 180, 191, **191**, **192**
    plants 180–181, 193–194, **193**, **194**
    pterosaurs 171, 173, 177, 180, 182–185, **183**, **184**
    taphonomy and preservation 171–173

turtles 169, 177, 178–179, 181–182
Jeholopterus ningchengensis **136**, 147, **148**, 152, 157
Jeholornis 155, 174, 187, 189
jeholornithiforms 180
Jeholotriton paradoxus 145, **146**, 152, 155
jewel beetles 267
Jianchangnathus robustus 148, **148**
Jianchangopterus zhaoianus 148, **148**
Jinzhousaurus 186
Jiufotang Formation 171, **173**, 176
Jiulongshan Formation 132–133, 135, 137, 138–139, 140, 141
Jordanian amber 294
Juglandaceae 265, **266**
Juncitarsus 278
Jurachresmoda gaskelli 152
Juramaia sinensis 135, 149, **150**, 156
Juramantophasma sinica **144**
Jurassic *see* Yanliao Biota
Jurassonurus amoenus **144**
Jurinida 84

Kabatarina pattersoni 232, **232**
katydids 153, 154
kazacharthrans 75, 78, **80**, 81–82
Keratestheria 176
kerogen 259
Kidston, Robert 2–3, **3**
Kirktonecta milnerae 50, **51**
Kokartus 155
Kopidodon macrognathus 281, **282**
Krispiromyces discoides 18, **18**
Kronocharon prendinii **311**
Kuehneosaurus 124
Kuehneosuchus 124
Kunpengopterus sinensis 148
Kurgan-Tash Mountains **67**, 68
Kyrgyzsaurus 76
Kyrgyzsaurus bukhanchenkoi 87, **90**, 91
Kyrgyzstan *see* Madygen Lagerstätte

Laccotriton 178, 181
Lacertidae 273, 292
lacewings 82, **117**, 151, 154, **192**, 267
laminites 42, **43**
lampreys 177, **178**, 181
Lancifaex **26**, 27
Lang, William H. 2–3, **3**
Lanqi Formation 135, 136, 137, 138–139, 140, 141, 143, 151
Las Hoyas Formation 177
Latimeria 236, 239
latimeroids 236
leaf beetles 267, **268**
leaf insects 267, **270**
leaf mimesis 151–152, 154–155
Lebanese amber 291
Leehermania prorova 119, **119**
Lepacyclotes zeilleri **76**, 79
Lepidocaris 3, **25**, 27
Lepidolaenaceae 296
Lepidoptera 267, **272**, **283**, 295, **325**
Lepidopteris 79
Lepidotes 220, 238
lepidotids 234

lepisosteids 273
lepospondyls 50, 54
Leptictidium nasutum 280, **280**
leptictids 280, **280**
Lesmesodon edingeri 273, **275**
Leverhulmia 27
Levoberezhye *see* Madygen Lagerstätte
lianas 267
Liaobatrachus 178, **179**, 181
Liaochelys 179, 181, 182
Liaoconodon **191**
Liaoning Province, China *see* Jehol Biota; Yanliao Biota
Liaoningopterus 182
Liaosteus hongi 133, 145
Liaotherium gracile 133, 149
Liaoxitriton 178, **179**, 181
Liaoxitriton daohugouensis 145, **146**, 155
lichens 19, **19**, 143
Limmocypridea abscondida 177
Limnoscelis 52
Limoniidae 113
Linguatormyia teletacta **327**
Lioxylon liaoningense 143
Liposcelididae 293
Little Cliff Shale 42
liverworts **76**, 78–79, 291, 296, **306**
lizards
    in amber 292, **328**
    gliding 180, 182
    Jehol Biota 169, 173, **175**, 177, 180, **180**, 182
    Messel Pit Fossil Site 273–275, **274**
    Yanliao Biota 132–133, **136**, 145, **147**, 152
Lockatong Formation 109
Lonchidion ferganensis **70**, **86**, 87, 93
lonchodectid pterosaurs 241, 242
Longirostravis 189
Longisquama 77
Longisquama insignis 87, 89, **90**, 91
Lophiodon 280
Luanpingella 176
Lucanidae 267
lungfish 84, **86**, 93, 96, 220
lycopsids 59, 61, **76**, 78, 79, 93, 143, 180
Lycoptera 169, 170, **170**, 171, 176–177, 181
Lycoptera davidi 169
Lyell, Charles 105
Lyginorachis 60
Lymnaea websteri 176
Lyon, Geoffrey 4, **4**, 5
Lyonophyton **22–23**

maar lakes 258–259, **259**, **260**, **261**
Mackay, Alexander 2
Mackie, William 1, **1**, 2, 3
Mackiella 19
Macrocranion tupaiodon **264**, 279
Macrotermes falciger 294
Madygen Lagerstätte 65–97
    biostratigraphic ages 75
    crustaceans 75, **80**, 81–82
    depositional environment **70**, **70**, 71–75, **73**, 77
    ecosystem reconstruction 93–95, **94**, **95**
    evolutionary significance 96

fish 70, **70**, 77, 84–87, **85**, **86**, 96
history of exploration 66–68
insects 75, 77, 82–84, **83**, 96
lithostratigraphy 71–75, **72**
location **67**, 68
lower aquatic invertebrates 79–81, **80**
plants 75, **76**, 78–79, 93
radiometric ages 75
regional geology 68–71, **69**
spores and pollen 78
taphonomy and preservation 75–78
tetrapods 75, 77–78, 87–92, **88**, **90**, **91**, 96
trace fossils 77, 92–93, **92**, 96
Madygenerpeton pustulatum 87, **88**, 89
Madygenia 79
Madygenopteris 79
Madysaurus sharovi 87, 89
Maimetshidae 291
malacostracan decapods **80**, 82
mammals and mammaliaforms
    Jehol Biota 171, 177, 180, 191, **191**, 192
    Messel Pit Fossil Site 273, **275**, 278–283, **279–283**
    Yanliao Biota 133, 135, 137, 149, **150**, 152, 153, 156
Manchurochelys 169, 178–179, 181
Manchurodon simplicidens 149
maniraptoran theropods 135, **136**, 148, **149**, 152, 153, 155–156, 157, 237
Manlayamia dabeigouensis 177
Mantophasmatidae **144**
Marmorerpeton 155
Marsh, Othniel Charles 105
marsupials 279
Martinmuellera tuberculata **266**
Martius, Carl Friedrich Philipp von 216, **217**
Masillabune 280
Masillosteus 273
mass-mortality events 173, 174, **175**
mating pairs
    insects 153
    turtles 275–276, **276**
Matthew, William Diller 105
Mawsonia 220, 236, 239
Mawsonia brasiliensis 236
Mawsonia gigas 236
mayflies 84, 143, **144**, 152, 267, 313
mealybugs 295–296
Mecistotrachelos 109, 122–124, **123**, 125
Meckel's cartilage 191
Mecoptera 82, 117–118, **117**, 143, 151–152, 154–155, **192**, 193, 267, 292–293, 296, **324**
Megaconus mammaliformis 135, 149, **150**, 152, 156
Megaloptera 143
Megalyridae **318**
Megaperleidus lissolepis 84
Mei long 174
melanosomes 187, 263, 277, **328**
Mengyinaia 177
Menispermaceae **266**
Merope tuber 296
Meropeidae 296, **324**
Mesenteriophyllum 79
Mesenteriophyllum kotschnevii **76**, 78, 79
Mesobunus martensi 145, **145**

Mesomyzon 177, **178**, 181
Messel Pit Fossil Site 257–284
    age 258–259
    amphibians 273
    birds 263, **263**, 276–278, **277**, **278**
    depositional environment 259–262
    evolutionary significance 283–284
    fish 273
    fossil preparation 262–263, **262**
    geological context 258–259, **259**, **260**
    history of exploration 257–258
    insects 267, **268–272**
    location 257, 258, **258**
    mammals 273, **275**, 278–283, **279–283**
    plants 265–267, **265**, **266**
    reptiles 273–276, **274**, **275**, **276**
    taphonomy and preservation 263, **263**, **264**, 277–278
'Messel rail' 278, **278**
Messelobunodon schaeferi **262**, 280–281
Messelophis 273
Messelornis cristata 278, **278**
Metarchilimonia krzemiskoroum **114**
Metasequoia 289, 291, **306**, 307
Metatheria 180
micro-stromatolites **15**, 17
microbes **15**, 17–18
microbial mats 13, 14, 15, **15**, **260**, 261
microconchids 79–80, **80**
Microraptor 174, 186, **186**, 187
microsaurs 50, **51**
middle ear, mammalian 191
millipedes 27, 59, **312**, **328**
Milton Flags Member 9
mimesis 151–152, 154–155
Minikh, Maxim G. 66
Miomoptera 82
Mirischia asymmetrica 237, **237**
Missao Velha Formation 220, **221**
mites 3, 16, **25**, 291
molluscs 59, 78, 80–81, **80**, 135, 139, 145, **145**, 176, 177, 232, 238
Monjurosuchidae 182
Monjurosuchus 169, 173, 176, 180, 182
Mordellidae **321**
Mormolucoides 112
Mortoniceras 291
mosses 78, 180, 291, **306**
moths 267, **272**, **283**, **325**
Mount Toby Formation 112
Multituberculata 180
Multramificans ovalis 152
Muscites 78
Muscites brickiae 78
mutualistic relationships 151–152, 154–155
Myanmar *see* Burmese amber
Myanmariscus deboiseae 292
mycorrhizae 13, 19, 21
myriapods **25**, 27, 59, 291, **312**

Nakamuranai 177
Nannochoristidae 154
National Museum of Scotland 39, 40
Natural History Museum, London 290, 291

Naucoridae **316**
Necrosaurus 275
Nei Mongol Autonomous Region, China *see* Jehol Biota; Yanliao Biota
Nematocera 112
nematodes 4, 13, 24, **24**, 292, **308**
Nematoplexus 4, 19
Neocalamites 79, 135, 139
Neocalamostachys 79
Neoceratodus 220
Neochelys 276
Neoproscinetes 238
Neoproscinetes penalvai 232, **233**
neopterygians 84
Nephila jurassica 145, 151, 155
Nephilidae 155
Nepomorpha 112, 113–117, **113**, **115**, 120, 125, 267
nesting behaviour, dinosaurs 174
Nestoria 176
Neuroptera 82, **117**, 143, **144**, 151, 154, 267
Nevrorthidae **320**
Newark Supergroup 105–106
  see also Solite Quarry
Nilssonia 142
North America *see* Solite Quarry
North China Craton 138
Notelops 229
Notelops brama 231
Notharctus 281
Nothia 3, 18, 20, 28
Notonectidae 112, **113**, 114–117, **115**, 267
notostracans 81, 82
notosuchians 236
Nuclear Magnetic Resonance (NMR) 288
Nycteribiidae 267

Obaichthys decoratus 239
Ochteridae **316**
Odonata 84, 93, 143, **144**, 151, 267, **313**
oil shale *see* Messel Pit Fossil Site
Old Red Sandstone (ORS) 6, **7**
Onychophora 292
Ophiderpeton kirktonense 50, **51**
Ophisauriscus 273
opilionids 4, **25**, **26**, 27, **58**, 59
Orcadian Basin 6
Ordosemys 178–179, 181
Ornatocephalus 273
ornithischian dinosaurs 135, 148–149, 152, 169, 174, 176, 185, 186
ornithocheirid pterosaurs 240, 241, 242
Ornithocheiridae 182
ornithomimosaurians 186
ornithurine birds 174, 180, 187, 189
Ortheziidae **316**
Orthoptera 75, 82, 84, 120, **121**, 143, **144**, 153, 154, 267, 295, **314**
Osborn, Henry Fairfield 105
Oshia ferganica 84, **85**
osmundaceans 143
Osteoglossomorpha 177, 181
ostracods 56, 78, **80**, 81, 135, 139, 145, 176–177, 228, 229, **229**, 232

ovaries and oviducts, bird 191
oviraptorid dinosaurs 186
oviraptorosaurians 186, 237–238

pachymeridiids 153
Pagiophyllum 110
Palaeobatrachidae 273
Palaeoblastocladia milleri 18, **18**
Palaeocharinus 4, 16, 17, 24, **26**
Palaeochiropteryx tupaiodon 283, **283**
palaeoclimatic reconstructions 93, 157–158, 176
palaeoenvironments *see* depositional environments
Palaeonema 4, 24, **24**
palaeoniscids 84, **85**
palaeonisciforms 135, 139
Palaeonitella 3, 18, **18**, 19, **19**
Palaeonitella crani 3
Palaeontinidae 144
Palaeontological Institute of the Russian Academy of Sciences 66
Palaeoperca 273
Palaeopython fischeri 273, **275**
Palaeotididae 277
Palaeoxyris alterna **86**, 87
Paleopyrenomycites devonicus 18
Palerasnitsynus 325
palms **265**, 267
palynomorphs *see* pollen; spores
Pangerpeton sinensis 135–136, 145, **146**, 155
pangolins 279
Pannaulika triassica 110–111, **111**
pantolestids 280
Paraelops cearensis **235**
Paraglauconia 224, **225**, 239
Paraneoptera 293–294
paraparchitids 56
Parapolycentropus 292–293, **324**
parasitic copepods 232, **232**
parasitic insects 151, 154, **192**, 193, 267
paravian theropods 135, **136**, 148, **149**, 152, 153, 155–156, 157
parental care, dinosaurs 174
paroxyclaenids 281, **282**
Parvodus 220
Patarchaea muralis 145
Pattersoncypris 232
Pattersoncypris micropapillosa **229**
Pederpes 52
Pedopenna daohugouensis 135, 148, 153
Peipiaosteus 177
Peipiaosteus fengningensis 176
Pelecinidae **318**
pelican spiders 296
pelomedusid turtles 236, **236**
Peltaspermum 79
pennaceous feathers, dinosaurs 153, 156
Peradectes 279
perciforms 273
Peregrinpachymeridium comitcola 153
perissodactyls 280, **281**
Perleidus 84
Permiana 81
Perochelys 179
Perochelys lamadongensis 182

phaeomelanosomes 187, **187**
Phasmatodea 82, 267, **270**
Phasmatoptera 143
Philydrosaurus 182
Phoenicopsis speciosa 135
pholadid bivalve borings, in amber 291
Pholiderpeton 54
Pholidocercus hassiacus 279–280, **279**
phreatomagmatic explosions 175, 258, **259**
Phthiraptera 293
Phylloblattidae 82
Phytophaga 295
phytophagy 295–296
phytosaurs 105, 109, 121
Placosauriops 273
Planohybodus 220
plants
    in amber 291, 296, **306**
    East Kirkton Lagerstätte 59–61, **60**
    insect feeding traces 93, 151
    insect–plant interactions 151–152, 154–155, 295–296
    Jehol Biota 180–181, 193–194, **193, 194**
    leaf mimesis 151–152, 154–155
    Madygen Lagerstätte 75, **76**, 78–79, 93
    Messel Pit Fossil Site 265–267, **265, 266**
    Rhynie and Windyfield cherts 2–3, 11–16, **12–15**, 20–24, **20–23**, 27–29
    Solite Quarry 110–111, **111**
    Yanliao Biota 135, 138, 139, 141–143, **142**, 153, 154–155
Platanus 111
Platypezidae 327
Plecoptera 84, 143, 267, **314**
pleurodire turtles 236, 276
Pleuromeiopsis kryshtofovichii 79
Podozamites 79
pollen 78
polygonalis-emsiensis Spore Assemblage Biozone (PE SAB) 11
polygonalis-wetteldorfensis Oppel Zone (PoW OZ) 11
polyodontids 177, **178**, 181
polyphagan beetles 82
Postopsyllidium 294
Postopsyllidium rebeccae **316**
Postosuchus alisonae 106
potoos 277
Pravoberezhye *see* Madygen Lagerstätte
preservation
    of inclusions in amber 289–290
    see also taphonomy and preservation
Price, Llewellyn Ivor 219
primates 281
Probaicalia vitimensis 177
Problematospermum 154
proboscides, long 292–293
Procramptonomyiidae 113, **114**
procynosuchians 89
Propalaeotherium hassiacum 280
Propalaeotherium parvulum **281**
Propalaeotherium voigti 280
prophalangopsids 151
Prosechamyia 112
Prosechamyiidae 113
Proterogyrinus 54
protists 291

Protocaris crani 3
Protonemestrius 192
Protopsephurus 177, **178**, 181
Protopsyllidiidae 293–294
Protopteryx 171, 187, **188**
Protopteryx fengningensis 177
protorosaurs 105, 122, 124, 125
Protorthoptera 82
Protostegidae 239
Prototaxites **18**, 19
Protoxenos 294
Psephenidae 267
Pseudococcidae 295–296
Pseudoctenis lanei **76**, 79
Pseudopolycentropodes virginicus 117–118, **117**
pseudopolycentropodids 151, 292–293
Pseudopulex wangi **144**
pseudoscorpions 310
Pseudotribos robustus 149, **150**
Psittacosaurus 169, 174, 176, 186, 191, **192**
Psocodea 293
Psocoptera 84, 143, 293
Psychodidae 113, **114**, 326
Psychopsidae **320**
Pteranodontidae 182
pteridophytes 265
pteridosperms 59–61, **60**, **76**, 79, 93
pterodactyloids 155, **184**
Pterofiltrus 182
Pterophyllum 110
Pterophyllum firmifolium 79
Pterophyllum pinnatifidum 79
Pterorhynchus wellnhoferi 147
pterosaur trackways 244
pterosaurs
    Jehol Biota 171, 173, 177, 180, 182–185, **183, 184**
    locomotion of 244–245
    Santana Formation **228**, 229, **229**, 240–245, **240, 241, 243**
    Yanliao Biota 135, **136**, 137, 139, 147–148, **148**, 152, 153, 155, 157–158
Ptilodactylidae **321**
Ptilozamites 79
ptycholepids 135, 139, 145, **147**
Pulmonoscorpius kirktonensis 57, **58**
Pumilanthocoris 151
pycnodonts 232, **233**, 234, 238
pyritization 141
Pyrolyis-Gas Chromatography and Mass Spectroscopy (Py-GC/MS) 288

Qinglongopterus guoi 137, 147, **148**
Quarry Hill Sandstone 7–8, **9**
Quinqueloculina 232

radiometric ages
    Burmese amber 291
    Madygen Lagerstätte 75
    Messel Pit Fossil Site 259
    Rhynie and Windyfield cherts 11
    Yanliao Biota 137, 138, 139–140
Raphidioptera 143, **320**
Raptorex 186
Raritan amber 291

ratites 276, 277
rauisuchians 106
Redfield, William C. 105
Reesidella robusta 177
refugium hypothesis 175
Regalerpeton 178
Regalerpeton weichangensis 177
regurgitates **113**, 273, **275**
Rehezamites 181
Remy, Winfried 4, **4**
Repenomamus 174, 191, **192**
reptiles
    Jehol Biota 169, 171, 173, 174, **175**, 176, 177, 178–180, **179**, **180**, 181–187, **183–186**
    Messel Pit Fossil Site 273–276, **274**, **275**, **276**
    Newark Supergroup 105–106
    Santana Formation 236–238, **236**, **237**, 239, 240–245, **240**, **241**, **243**
    Solite Quarry 108, 109, **110**, 121–124, **122**, **123**, 125
    Yanliao Biota 132–133, 135, **136**, 137, 139, 145, 147–149, **147**, **148**, **149**, 152, 153, 155–156, 157–158
    see also tetrapods
Rhacolepis **217**, **227**, 229, 232
rhamphorhynchoid pterosaurs 137, 139, 147, **148**, 180, 244
Rhenanoperca 273
rhinobatoid rays 229, 232, 233, **234**, 239
rhinocerotoids 280
rhizodonts 56, **56**
Rhynchertia 19
Rhynchosauroides 109
Rhynia 2–3, **4**, **14**, **15**, 20–24, **20**, 27–28
Rhynie and Windyfield cherts 1–30
    algae 19, **19**
    arthropods 3, 4, 5, 16, 17, 24–27, **25**, **26**, 30
    basin structure 6, **7**, **8**
    biostratigraphic ages 11
    depositional environment 9–11, **10**, 30
    ecosystem reconstruction 29, **29**, 30
    evolutionary significance 30
    fossil preparation 16–17
    fungi 18–19, **18**
    history of exploration 2–6
    hydrothermal system 3, 6, 8, **8**, 9–11, **10**, 30
    lichens 19, **19**
    microbes **15**, 17–18
    nematodes 4, 24, **24**
    plants 2–3, 11–16, **12–15**, 20–24, **20–23**, 27–29
    radiometric ages 11
    stratigraphy 7–9, **7**, **9**
    taphonomy and preservation 11–16, **12–15**
Rhyniella 27
Rhyniemonstrum 27
Rhyniognatha 27
Ricciopsis ferganica **76**, 78–79
rift systems 106–107, **106**
Rio Bateiras Formation 220, **221**
rodents 279
Romualdo Member *see* Santana Formation
root traces 92, **92**
Rosamygale 112
Russian Academy of Sciences 66
Rutiodon 105, 121

Saccogulus 27
Sagenopteris 79
Sakurasaurus 182
salamanders 135–136, **136**, 137, 145, **146**, 152, 153, 155, 177, 178, **179**, 181, 273
salamandroids 137, 145, **146**, 155
Saltatoria 267
Santana Formation 177, 215–245
    crocodyliformes 236–237
    dinosaurs 229, 237–238, **237**
    ecosystem reconstruction 238–239
    evolutionary significance 239–245, **240**, **241**, **243**
    fish 216, **217**, 220, 226–227, **226**, **227**, 228, 229, **230**, 231, **231**, 232, 233–236, **233**, **234**, **235**, 238, 239
    history of exploration 215–219, **217**, **218**
    invertebrates 224, **225**, 228, 229, **229**, 232, **232**, 238–239
    location 215, **216**
    outcrops 223–224
    palaeosalinity 239
    pterosaurs **228**, 229, **229**, 240–245, **240**, **241**, **243**
    sedimentology **222**, 224–227, **225**, **226**
    stratigraphy **216**, 219–223, **220**, **221**, **222**
    taphonomy and preservation 227–231, **228–231**, 242–244, **243**
    turtles 236, **236**, 239
    water depth 238–239
Santanachelys gaffneyi 236, 239
Santanaraptor placidus 237
Sapeornis 155, 174, 187, 189, **189**
sapeornithiforms 180
sarcopterygians 56, **56**, 84–87, **86**, 93, 96
Sauk Tanga *see* Madygen Lagerstätte
saurichthyids 84, **85**
Saurichthys orientalis **70**, 84, **85**
Saurophthirus exquisitus **192**
scale insects 295–296, **316**
Scaniacypselus szarskii 277
scansoriopterygid dinosaurs 153, 155
scaphognathid pterosaurs 139, 147–148, **148**
Scarabaeidea 267
schizophorid beetles 82
Schmeissneria 138, **142**, 143, 154
Sciaroidea **114**
Scincoidea 273
Scincomorpha 273
scleroglossans 145, **147**, 182
scorpionflies 82, 117–118, **117**, 151–152, 154–155, **192**, 193, 267, 292–293, 296, **324**
scorpions 57, **58**, 62, **310**
Scotland *see* East Kirkton Lagerstätte; Rhynie and Windyfield cherts
Scytophyllum 79
sedimentary environments *see* depositional environments
Sedovia fecunda 81
seedferns *see* pteridosperms
seeds, Solite Quarry 111, **111**
Serphitidae 291
Shaanxiconcha 139
Shaanxiconcha cliovata 135, 145
shark egg capsules 77, **86**, 87, 96
sharks 56, 70, **70**, **86**, 87, 93, 96, 220, 229, 232, 233, 238
Sharov, Alexander G. 66
Sharovipteryx 77

Sharovipteryx mirabilis 87, 89–91, **90**
Shaximiao Fauna 156–157
Shenzhousaurus 186
Shinisaurus 182
shipworm borings, in amber 291
Shishugou Fauna 140, 156–157
Shokawa 182
Sibireconcha 145
silicification processes, Rhynie and Windyfield cherts 11–16, **14, 15**
Silvanerpeton miripedes 44–47, **46**, 52, 54
Sinamia 178
Sinaranea metaxyostraca 145
Sinemydidae 181, 182
Sinemys 181, 182
Sinerpeton 181
Sinocarpus 181, 193, **193**
Sinopolycentropus rasnitsyni 151
Sinopterus 182, **183**
Sinosauropteryx 173, 174, 185, **185**, 186, 187, **187**
Sinosepididontus chifengensis 144
siphlonurids 152
Siphonaptera 143, **144**, 151, 154, **192**, 193
Siphonospermum 181
siphonostomatans 239
Sixtel, Tatiana A. 77
Sixtelia asiatica 84, **85**, 87
skin preservation 76, 77, **90**, 96, 140
slime moulds 292, **306**
Small, Horatio **218**, 219
Smithson, Tim 39–40
snakes 273, **274**, **275**
sociality, in insects 294–295
soft-tissue preservation
    Jehol Biota 171
    Madygen Lagerstätte 76, 77, 78, **90**, 96
    Santana Formation 229–231, **229**, **230**, **231**, 244
    Yanliao Biota 140, 141
Solaranthus daohugouensis 135, 143, 154
soldier flies 293
Solite Quarry 105–126
    age 109
    depositional environment 124–125
    fish 108–109, **110**, 121
    fossil preparation 109
    geological context 106–109, **106**, **108**
    history of exploration 107–109, **108**
    insects 107, 108, 109, 112–120, **113–121**, 124–125
    location **106**, 107
    plants 110–111, **111**
    spiders 112
    taphonomy and preservation 109, **110**
    tetrapods 108, 109, **110**, 121–124, **122**, **123**, 125
    trackways 109, 121, 122
Solnhofen assemblage, Germany 147, 177
Sophogramma **192**
South Kyrgyz Geological Expedition 66
Spanish amber 291, 293
Spathiopterygidae 291
Spathulopteris 60, **60**
sperm cells 13, 16, 21, **22–23**, 30
Sphaerium 177
Sphecomyrma freyi 292
Sphenobaeira 110
Sphenobaiera 79
sphenopsids 59, 79, 110
Sphenopteridium 60, **60**
spiders 112, 151, 155, 291, 296, **310**, **311**
    see also harvestman spiders
spinosaurid dinosaurs 237
spirorbiform polychaetes 79–80
Spix, Johann Baptist von 215–216, **217**
spores
    germinating 4, 13, 21, **22**, 30
    Madygen Lagerstätte 78
    Rhynie and Windyfield cherts 4, 13, 21, **22**, 28, 30
sporophyte–gametophyte life history 21–24, **22–23**, 30
springtails **25**, 27
Squamata *see* lizards; snakes
Stanwoodia 60
staphylinid beetles 119, **119**
Staphylinidae **321**
Sternorrhyncha 82, 119, **120**, 293–294, **316**
stick insects 82, 267
Stigmaphronidae 291
stink glands 59
stomach contents *see* gastrointestinal contents
stomach stones *see* gastroliths
stoneflies 84, 267, **314**
Strashila daohugouensis **144**
stratiomyids 293
Stratiomyomorpha 293
Strepsiptera 294, **325**
stridulatory structures 153, 154
stromatolitic limestones 42, 43–44, **43**
Struthioniformes 276, 277
Suchonella anybensis 81
swifts 277, **277**
Swinhoe, R. C. J. 290
Symmetrodonta 180

Tachypteron 283
tadpole shrimps 81, 82
tadpoles 273
Tanytrachelos 105
Tanytrachelos ahynis 96, 108, 109, **110**, 113, 122, **122**, 124, 125
Tapejara 242
Tapejara wellnhoferi **243**, 244
tapejarid pterosaurs 182, 241–242
taphonomy and preservation
    East Kirkton Lagerstätte 41, 61–62
    Jehol Biota 171–173
    Madygen Lagerstätte 75–78
    Messel Pit Fossil Site 263, **263**, **264**, 277–278
    Rhynie and Windyfield cherts 11–16, **12–15**
    Santana Formation 227–231, **228–231**, 242–244, **243**
    Solite Quarry 109, **110**
    Yanliao Biota 140–141
Tarbosaurus 186
Tashkent State University 66
Teiidae 292
teleosts 177, 181, 232
temnospondyls 44, **45**, 52–54
Tenebrionidae 267, **269**
termites 267, 294–295, **315**
terrestrial locomotion, pterosaurs 244

Testudines *see* turtles
Tethepomyiidae 291
Tetraedron minimum 259–260
tetrapods
    East Kirkton Lagerstätte 41, 44–54, **45–49**, **51**, **53**, **55**, 61
    Madygen Lagerstätte 75, 77–78, 87–92, **88**, **90**, **91**, 96
    Newark Supergroup 105–106
    Solite Quarry 108, 109, **110**, 121–124, **122**, **123**, 125
    see also amphibians; reptiles
Tetrigidae **121**
Thalassinoides paradoxicus 93
Thalassodromeus 242
Thalassodromeus sethi 242
thalassodromid pterosaurs 241–242
Thallites insolites 78
thallophytes 78
Thaumaturus 273
theropods
    Jehol Biota 171, 173, 174, 176, 180, 185–187, **185**, **186**, **187**
    Santana Formation 229, 237–238, **237**
    Yanliao Biota 135, **136**, 148, **149**, 152, 153, 155–156, 157
thin sections 16–17
thrips **116**, 117, 293, 295
Thysanoptera **116**, 117, 293, 295
Tianyulong confuciusi 135, 148–149, 152, 157
Tiaojishan Formation 135, 136, 137, 138–139, 140, 141, 143
Tien Shan **67**, 68–70
Tillibrachty Sandstone Formation 8, 11
Timiriasevia catenularia 135, 145
Tingidae **317**
Tipulomorpha **114**
Titanoptera 82, **83**, 84
titanosaurs 186
Tokhta-Boz Mountains **67**, 68
Tomistomidae 275
trace fossils
    Madygen Lagerstätte 77, 92–93, **92**, 96
    Rhynie and Windyfield cherts 8, **9**, 27
    Solite Quarry 109, 121, 122
trackways
    arthropod 8, **9**, 27
    dinosaur 109, 121
    protorosaur 122
    pterosaur 244
transitional taxa 293–294
Triassaraneus 112
Triassic *see* Madygen Lagerstätte; Solite Quarry
Triassopsychoda olseni 114
Triassothrips virginicus **116**, 117, 120
Triassurus sixtelae 87–88, **88**
Tribodus limae 229, 232, 233, 238
Trichopherophyton 4, 20, 21
Trichoptera 84, 145, 267, **325**
triconodonts 133, 149, 180
Tridactylidae **314**
trigonotarbids 3, 24, **25**, **26**
trionychid turtles 182, 276
Trionychidae 179
Tristychius 56
troodontid dinosaurs 174, 186
Tropeoganthus 242
Tuchengzi Formation 135, 138
Tupuxuara 242

Tupuxuara longicristatus 242
Turgai Sea 175
Turkestan Ocean 68
Turkestan Range **67**, 68
turtles
    Jehol Biota 169, 177, 178–179, 181–182
    Messel Pit Fossil Site 273, 275–276, **276**
    Santana Formation 236, **236**, 239
tyrannosauroid dinosaurs 176, 186

ungulates **262**, 278, 280–281, **281**
Unwindia 241, 242
Upupiformes 277
Uralophyllum 79
Urochishche Dzhaylyaucho *see* Madygen Lagerstätte
Urochishche Madygen *see* Madygen Lagerstätte

Van Houten cyclicity 107
velvet worms 292
Ventarura 5, 15, **15**, 16, 17, 20, 21
Vespidae 267, **318**
Vetuprostomis consimilis **321**
Vileginia tuberculata 81
Vinctifer 220, 234
Vinctifer comptoni **235**
Virgaichnus 93
Virginia *see* Solite Quarry
Virginiptera certa 114
Vitaceae **266**
Vitimopsyche **192**
Vittaephyllum **76**, 79
viviparity in reptiles 173, 182
Volaticotherium antiquum 135, **150**, 153, 156
volcanism
    East Kirkton Lagerstätte 41, 42, 44, 60–61
    and fossil preservation 140–141
    Jehol Biota 171, 173, 174–175
    Messel Pit Fossil Site 258, 259, **259**
    Yanliao Biota 138, 140–141, 157

walking, pterosaurs 244
walnuts 265, **266**
wasps 84, 267, **318**
water bugs 112, 113–117, **113**, **115**, 120, 125, 267
Weinfelder Maar, Germany **260**
Weltrichia daohugouensis 143
West Uzbek Geological Expedition 66
Westlothiana lizziae 49–50, **49**, 54
whatcheeriids 52
Windyfield chert *see* Rhynie and Windyfield cherts
Winfrenatia 19, **19**
Wood, Stan 39–40, **40**
woodlice 292, **308**
wukongopterid pterosaurs 135, 139, 148, **148**
Wukongopterus lii 135, 148, **148**

xenacanthid sharks 56, **86**, 87
Xenosaurus 182
Xenoxylon 176
Xenozorotypus 294
Xianglong 180, 182
Xiaolongwan, China 259, **261**
Xiaotingia zhengi 135, 148, **149**, 153, 155

Xingxueanthus **142**, 143, 154
xylomyids 293

Yabeiella 111
Yabeinosaurus 169, 173, 180, **180**, 182
Yabeinosaurus tenuis 133, 145
'*Yabeinosaurus*' *youngi* 132–133, 145, **147**
Yalea rectimedia **114**
Yanjiestheria 177
Yanliao Biota 131–158, 171
    amphibians 135–136, **136**, 137, 145, **146**, 152, 153, 155
    arachnids 145, **145**, 151, 155
    biostratigraphic ages 137, 138–139
    crustaceans and bivalves 135, 136, 139, 145, **145**
    dinosaurs 135, **136**, 148–149, **149**, 152, 153, 155–156, 157–158
    ecosystem reconstruction 149–153
    evolutionary significance 153–158
    fish 133, 135, 139, 145, **147**
    geological context **134**, 138–141
    insects 132, 133, 143–145, **144**, 151–152, 153, 154–155
    lizards 132–133, **136**, 145, **147**, 152
    location 131, **132**, 133–137, **133**
    mammaliaforms 133, 135, 137, 149, **150**, 152, 153, 156
    plants 135, 138, 139, 141–143, **142**, 153, 154–155
    pterosaurs 135, **136**, 137, 139, 147–148, **148**, 152, 153, 155, 157–158
    radiometric ages 137, 138, 139–140
    stratigraphy **133**, 138–139
    taphonomy and preservation 140–141
Yanliaocorixa chinensis 152
Yanornis 174, 189
Yanosteus longidorsalis 176
Yimaia 139
Yimaia capituliformis **142**, 143
yinotherians 149, **150**
Yixian Formation 134, 138, 149, 171, **172**, 173–174, **174**, 176, 177, 180–181, 182, 291
Yixianornis 187
Yuchoulepis 139
Yutyrannus 176, 186

Zamites 110, 135
Zhangjiakou Formation 138
Zhangsolvidae 291, 293
Zoraptera 294, **314**
Zoropelecinus zigrasi **318**
Zorotypus 294
Zorotypus nascimbenei **314**
Zygaenidae 267, **272**
Zygokaratawia reni 151
Zygoptera 151, **313**